METHODS IN MOLECULAR BIOLOGY

For further volumes:
http://www.springer.com/series/7651

For over 35 years, biological scientists have come to rely on the research protocols and methodologies in the critically acclaimed *Methods in Molecular Biology* series. The series was the first to introduce the step-by-step protocols approach that has become the standard in all biomedical protocol publishing. Each protocol is provided in readily-reproducible step-by-step fashion, opening with an introductory overview, a list of the materials and reagents needed to complete the experiment, and followed by a detailed procedure that is supported with a helpful notes section offering tips and tricks of the trade as well as troubleshooting advice. These hallmark features were introduced by series editor Dr. John Walker and constitute the key ingredient in each and every volume of the *Methods in Molecular Biology* series. Tested and trusted, comprehensive and reliable, all protocols from the series are indexed in PubMed.

Protein Folding

Methods and Protocols

Edited by

Victor Muñoz

Department of Bioengineering and Center for Cellular and Biomolecular Machines, University of California Merced, Merced, CA, USA

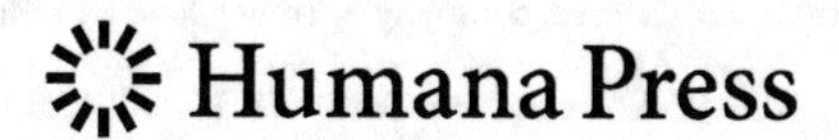

Editor
Victor Muñoz
Department of Bioengineering and
Center for Cellular and Biomolecular
Machines
University of California Merced
Merced, CA, USA

ISSN 1064-3745 ISSN 1940-6029 (electronic)
Methods in Molecular Biology
ISBN 978-1-0716-1718-2 ISBN 978-1-0716-1716-8 (eBook)
https://doi.org/10.1007/978-1-0716-1716-8

This Humana imprint is published by the registered company Springer Science+Business Media, LLC part of Springer Nature.
The registered company address is: 1 New York Plaza, New York, NY 10004, U.S.A.

Preface

The investigation of the mechanisms by which proteins spontaneously fold into their biologically functional three-dimensional structures still constitutes one of the major open questions in modern molecular biology and biochemistry. There are multiple reasons in support of this assessment. Proteins are true nanomachines in charge of most of the key cellular processes. Thus, understanding the physical-chemical principles that govern protein self-assembly is key for our ability to interpret in molecular terms the quickly increasing wealth of information provided by genomic initiatives. As we advance in our understanding of protein behavior, it is becoming increasingly clear that the classical structure-function paradigm needs to be expanded to include folding as a powerful mechanism to control and exert biological function as well as protein homeostasis. Moreover, harnessing the capacity to engineer protein structure and function a la carte remains a major drive for research in this area, as it would have tremendous impact in the development of molecular biotechnology and biomedical applications.

During the last 20 years of intense research in protein folding large strides have been made. The theoretical principles governing folding reactions are nowadays firmly established by borrowing concepts from statistical physics. On the practical side, there have been multiple technological, methodological, and analytical developments, which have expanded our horizon by providing exciting opportunities to investigate folding mechanisms in unprecedented depth. Many such advances have come from the experimental front, adding novel capacities to the panoply of existing protein folding experimental methods. We have now techniques that reach ultrafast temporal, atomic-level structural, or single-molecule resolutions. In addition, the discovery of fast and downhill folding has sparked the implementation of various analytical methods that exploit the limited folding cooperativity of fast folding proteins to extricate key mechanistic information previously thought to be inaccessible via conventional folding experiments. Similarly impressive advances have come from the computational side, in which the combination of developments in hardware, software, and analytical methods has considerably strengthened atomistic simulation methods. Atomistic simulations can now efficiently sample folding landscapes under different environmental conditions and reach the timescales required to resolve multiple folding and unfolding events in individual trajectories. Consequently, we have finally reached the exciting and sought-after point at which simulations and experiments are directly comparable and can work synergistically by combining the wealth of mechanistic detail offered by atomistic simulations with the reality checks provided by experiments. All these advances have also had significant impact in our ability to predict the behavior of proteins from their amino acid sequence and/or native structure, hence bringing us closer than ever to cracking the second half of the genetic code.

In this volume we aim at providing a comprehensive coverage of both the experimental and computational methods that are being used nowadays to probe protein folding reactions and mechanisms. The chapters are written to provide potential users with practical guidelines on how to perform and analyze the experiments, how to use existing software for running molecular simulations and/or making predictions, and how to interpret the results from such simulations.

The book is divided into five parts, each covering a specific group of methodologies used in the modern investigation of protein folding reactions. Part I presents methods based on protein engineering and protein chemistry. Part II focuses on experimental approaches to investigate the thermodynamics and kinetics of protein folding transitions using advanced analytical techniques such as differential scanning calorimetry, ultrafast kinetics methods, infrared spectroscopy, mass spectrometry, and nuclear magnetic resonance. Part III introduces the new experimental methods to probe protein folding at the single-molecule level using fluorescence or force spectroscopy. Part IV is dedicated to the analysis and interpretation of computer simulations, ranging from coarse grained simulations of folding and folding coupled to binding to the most detailed, all atom molecular dynamics simulations in explicit solvent. Finally, Part V centers on procedures and tools for the prediction of protein folding properties, including empirical methods to predict rates and effects of mutations, the energetic frustration of folding landscapes, and the *ab initio* prediction of folding from the sequence.

Merced, CA, USA ***Victor Muñoz***

Contents

Contributors

ISMAIL A. AHMED • *Department of Biochemistry and Molecular Biophysics, University of Pennsylvania, Philadelphia, PA, USA*
ALVARO ALONSO-CABALLERO • *CIC nanoGUNE, San Sebastián, Spain*
MICHAEL C. BAXA • *Department of Biochemistry and Molecular Biology, University of Chicago, Chicago, IL, USA*
ROBERT B. BEST • *Laboratory of Chemical Physics, National Institute of Diabetes and Digestive and Kidney Diseases, National Institutes of Health, Bethesda, MD, USA*
LUIS ALBERTO CAMPOS • *Centro Nacional de Biotecnología (CNB-CSIC), Madrid, Spain*
MICHELE CERMINARA • *Imdea Nanociencia, Madrid, Spain; Physics of Biological Function, Institut Pasteur, Paris, France*
XIAKUN CHU • *IMDEA Nanosciences, Madrid, Spain*
HOI SUNG CHUNG • *Laboratory of Chemical Physics, National Institute of Diabetes and Digestive and Kidney Diseases, National Institutes of Health, Bethesda, MD, USA*
VINÍCIUS G. CONTESSOTO • *Rice University, CTBP/BRC, Houston, TX, USA*
CEZARY CZAPLEWSKI • *Faculty of Chemistry, University of Gdańsk, Gdańsk, Poland*
NICOLA D'AMELIO • *Unité Génie Enzymatique et Cellulaire, Université Picardie Jules Verne, Amiens, France*
EVA DE ALBA • *Department of Bioengineering, School of Engineering, University of California, Merced, Merced, CA, USA*
VINÍCIUS M. DE OLIVEIRA • *São Paulo State University, IBILCE/UNESP, São José do Rio Preto, SP, Brazil; Brazilian Biosciences National Laboratory, LNBio/CNPEM, Campinas, SP, Brazil*
DAVID DE SANCHO • *Polimero eta Material Aurreratuak: Fisika, Kimika eta Teknologia, Kimika Fakultatea, Euskal Herriko Unibertsitatea UPV/EHU & Donostia International Physics Center (DIPC) PK 1072, Donostia-San Sebastián, Spain*
DIEGO U. FERREIRO • *Protein Physiology Lab, Facultad de Ciencias Exactas y Naturales-Universidad de Buenos Aires. IQUIBICEN/CONICET. Intendente Güiraldes 2160 – Ciudad Universitaria – C1428EGA, Buenos Aires, Argentina*
FENG GAI • *Department of Chemistry, University of Pennsylvania, Philadelphia, PA, USA*
ANGEL E. GARCIA • *Center for NonLinear Studies, Los Alamos National Laboratory, Los Alamos, NM, USA*
EWA GOŁAŚ • *Institute of Physics, Polish Academy of Sciences, Warsaw, Poland*
IRINA V. GOPICH • *Laboratory of Chemical Physics, National Institute of Diabetes and Digestive and Kidney Diseases, National Institutes of Health, Bethesda, MD, USA*
A. BRENDA GUZOVSKY • *Protein Physiology Lab, Facultad de Ciencias Exactas y Naturales-Universidad de Buenos Aires. IQUIBICEN/CONICET. Intendente Güiraldes 2160 – Ciudad Universitaria – C1428EGA, Buenos Aires, Argentina*
ANDREA HOLLA • *Department of Biochemistry, University of Zurich, Zurich, Switzerland*
BEATRIZ IBARRA-MOLERO • *Facultad de Ciencias, Departamento de Quimica Fisica, Unidad de Excelencia de Química Aplicada a Biomedicina y Medioambiente (UEQ), University of Granada, Granada, Spain*
AGNIESZKA S. KARCZYŃSKA • *Faculty of Chemistry, University of Gdańsk, Gdańsk, Poland*
PAWEŁ KRUPA • *Institute of Physics, Polish Academy of Sciences, Warsaw, Poland*

GINKA S. KUBELKA • *Department of Chemistry, University of Wyoming, Laramie, WY, USA*
JAN KUBELKA • *Department of Chemistry, University of Wyoming, Laramie, WY, USA*
LISA J. LAPIDUS • *Department of Physics and Astronomy, Michigan State University, East Lansing, MI, USA*
VITOR B. P. LEITE • *São Paulo State University, IBILCE/UNESP, São José do Rio Preto, SP, Brazil*
AGNIESZKA G. LIPSKA • *Faculty of Chemistry, University of Gdańsk, Gdańsk, Poland*
ADAM LIWO • *Faculty of Chemistry, University of Gdańsk, Gdańsk, Poland*
EMILIA A. LUBECKA • *Faculty of Electronics, Telecommunications and Informatics, Gdańsk University of Technology, Gdańsk, Poland*
MARIUSZ MAKOWSKI • *Faculty of Chemistry, University of Gdańsk, Gdańsk, Poland*
POOJA MALHOTRA • *National Centre for Biological Sciences, Tata Institute of Fundamental Research, Bengaluru, Karnataka, India*
NIVIN MOTHI • *Chemistry and Chemical Biology Graduate Program, University of California at Merced, Merced, CA, USA; NSF CREST Center for Cellular and Biomolecular Machines (CCBM), University of California at Merced, Merced, CA, USA*
MAGDALENA A. MOZOLEWSKA • *Institute of Computer Science, Polish Academy of Sciences, Warsaw, Poland*
DEBOPREETI MUKHERJEE • *Department of Chemistry, University of Pennsylvania, Philadelphia, PA, USA*
VICTOR MUÑOZ • *Department of Bioengineering and Center for Cellular and Biomolecular Machines, University of California, Merced, Merced, CA, USA*
ATHI N. NAGANATHAN • *Department of Biotechnology, Bhupat & Jyoti Mehta School of Biosciences, Indian Institute of Technology Madras, Chennai, India*
SUHANI NAGPAL • *Department of Bioengineering, School of Engineering, University of California, Merced, CA, USA*
STANISŁAW OŁDZIEJ • *Laboratory of Biopolymer Structure, Intercollegiate Faculty of Biotechnology, University of Gdańsk and Medical University of Gdańsk, Gdańsk, Poland*
RAUL PEREZ-JIMENEZ • *CIC nanoGUNE, San Sebastián, Spain; IKERBASQUE, Basque Foundation for Science, Bilbao, Spain*
JOSE M. SANCHEZ-RUIZ • *Facultad de Ciencias, Departamento de Quimica Fisica, Unidad de Excelencia de Química Aplicada a Biomedicina y Medioambiente (UEQ), University of Granada, Granada, Spain*
NICHOLAS P. SCHAFER • *Department of Chemistry, Rice University, Houston, TX, USA; Department of Physics, Rice University, Houston, TX, USA; Department of Biosciences, Rice University, Houston, TX, USA; Center for Theoretical Biological Physics, Rice University, Houston, TX, USA*
JÖRG SCHÖNFELDER • *CIC nanoGUNE, San Sebastián, Spain; IMDEA Nanosciences, Madrid, Spain*
BENJAMIN SCHULER • *Department of Biochemistry, University of Zurich, Zurich, Switzerland; Department of Physics, University of Zurich, Zurich, Switzerland*
ADAM K. SIERADZAN • *Faculty of Chemistry, University of Gdańsk, Gdańsk, Poland*
TOBIN R. SOSNICK • *Department of Biochemistry and Molecular Biology, University of Chicago, Chicago, IL, USA*
JAYANT B. UDGAONKAR • *National Centre for Biological Sciences, Tata Institute of Fundamental Research, Bengaluru, Karnataka, India*
SIVANANDAM VEERAMUTHU NATARAJAN • *IMDEA Nanociencia, Madrid, Spain*

ZIFAN WANG • *Department of Bioengineering, University of California at Merced, Merced, CA, USA; NSF CREST Center for Cellular and Biomolecular Machines (CCBM), University of California at Merced, Merced, CA, USA*

TOMASZ WIRECKI • *Laboratory of Bioinformatics and Protein Engineering, International Institute of Molecular and Cell Biology in Warsaw, Warsaw, Poland*

DAVID WITALKA • *Department of Physics and Astronomy, Michigan State University, East Lansing, MI, USA*

PETER G. WOLYNES • *Department of Chemistry, Rice University, Houston, TX, USA; Department of Physics, Rice University, Houston, TX, USA; Department of Biosciences, Rice University, Houston, TX, USA; Center for Theoretical Biological Physics, Rice University, Houston, TX, USA*

XIANG YE • *Department of Biochemistry and Biophysics, University of Pennsylvania, Philadelphia, PA, USA*

FRANZISKA ZOSEL • *Department of Biochemistry, University of Zurich, Zurich, Switzerland; Novo Nordisk A/S, Måløv, Denmark*

Part I

Protein Engineering and Protein Chemistry Methods

Chapter 1

Mutational Analysis of Protein Folding Transition States: Phi Values

Luis Alberto Campos

Abstract

The analysis of protein folding reactions by monitoring the kinetic effects of specifically designed single-point mutations, the so-termed phi-value analysis, has been a favorite technique to experimentally probe the mechanisms of protein folding. The idea behind phi-value analysis is that the effects that mutations have on the folding and unfolding rate constants report on the energetic/structural features of the folding transition state ensemble (TSE), which is the highest point in the free energy surface connecting the native and unfolded states, and thus the rate limiting step that ultimately defines the folding mechanism. For single-domain, two-state folding proteins, the general procedure to perform the phi-value analysis of protein folding is relatively simple to implement in the lab. Once the mutations have been produced and purified, the researcher needs to follow a few specific guidelines to perform the experiments and to analyze the data so produced. In this chapter, a step-by-step description of how to measure and interpret the effects induced by site-directed mutations on the folding and unfolding rate constants of a protein of interest is provided. Some possible solutions to the most typical problems that arise when performing phi-value analysis in the lab are also provided.

Key words Phi-value, Protein folding, Transition state, Energy barrier, Stability, Kinetics, Folding mechanism, Stopped-flow kinetics

1 Introduction

Protein folding is the process by which a protein spontaneously forms its native, functional state starting as an unstructured polypeptide chain (e.g., as it is emerges from the ribosome during translation). The native state is typically a well-defined and unique 3D structure, and thus, the process involves a significant loss of configurational entropy. Protein folding reactions can be complex, involving the formation of one or more intermediate states (*see* **Note 1**), but many single-domain proteins fold via a simple two-state process. This chapter focuses on how to apply phi-value (ϕ) analysis to two-state folding proteins. For two-state folding, whether one conceptualizes the process using a classical enzyme kinetics description [1] or statistical energy landscape theory [2],

Victor Muñoz (ed.), *Protein Folding: Methods and Protocols*, Methods in Molecular Biology, vol. 2376,
https://doi.org/10.1007/978-1-0716-1716-8_1,

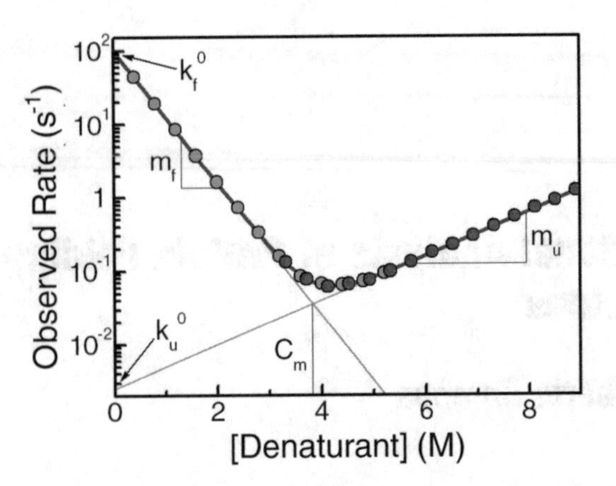

Fig. 1 Chevron plot representing the log(k_{obs}) versus denaturant concentration. Data from unfolding (red dots) and refolding (orange dots) kinetic experiments are represented, together with the parameters derived from fits to Eq. 8

the protein populates only two states, either the fully unfolded ensemble or the native 3D structure, at all conditions (although the conditions change their populations). These two states are separated by a high free energy barrier for which its maximum value, or barrier top, defines the folding transition state ensemble (TSE, or ‡). The difference in free energy between the TSE and the ground state (whether the unfolded or the native) determines the folding and unfolding rates, respectively (Fig. 1).

To decipher the molecular mechanisms of two-state folding, it is thus important to characterize structurally and energetically the three relevant kinetic species (U, N, TSE). Characterizing the unfolded and native states is relatively straightforward because they correspond to the two minima in the folding free energy surface, so one can always find conditions in which either one is the predominant species in equilibrium (Fig. 1) [3]. Therefore, one can investigate the structural properties of the ground states using spectroscopic methods such as circular dichroism, fluorescence, or even at atomic resolution using nuclear magnetic resonance. Probing the properties of the species with intermediate degrees of structure, which are the ones that define the mechanism, is not so straightforward. For the special case in which the protein folds downhill (crosses a very small free energy barrier) [4], when the protein is near the denaturation midpoint, these intermediate species have free energy values comparable to those of the ground states and thus their population becomes measurable [5]. However, for two-state folding proteins, the population of partially folded intermediates is insignificant and one can only infer the properties of the folding TSE indirectly from the relative changes in the folding and unfolding rate constants that are induced by structurally specific free energy perturbations.

A powerful tool to perturb the folding landscape of a protein in structurally specific ways is the introduction of site-directed mutations in which one amino-acid in the protein is replaced by another one, thus altering the particular set of interactions in which this residue is engaged in the native structure (e.g. an isoleucine in a core position substituted by valine eliminates a methyl group, which will result in a cavity in the native structure with the concomitant loss of van der Waals contacts and thus a net destabilization). ϕ value analysis interrogates what fraction of the total change in free energy induced by mutation is recovered on the folding direction (TSE-U, or folding rate constant) and on the unfolding direction (TSE-N, or unfolding rate constant). The ϕ value is defined as the fraction of the perturbation free energy recovered on the folding rate constant; it takes values from 0 to 1 and reports on the energetics of the mutated site at the TSE relative to the unfolded state. Similarly, the fraction of the perturbation free energy recovered on the unfolding rate constant $(1 - \phi)$ reports on the energetics of the mutated site at the TSE relative to the native state. In principle, by performing such analysis on a vast number of well-designed mutations, it becomes possible to map phi-values as a function of the location in the native structure, and even to investigate individualized interactions by comparing multiple mutations performed in the same location (e.g. substitution of a leucine to valine, alanine and glycine). A major strength of the approach is its simplicity, since mutations can be easily designed using the native 3D structure (the structural coordinates deposited in the protein databank) and introduced in the protein via straightforward site-directed mutagenesis procedures. Direct comparison of the folding-unfolding kinetics measured as a function of chemical denaturant—the so-called chevron plot [6, 7]—of the mutant and wild-type proteins yields the ϕ value for the mutation. This is, however, only strictly true when the mutants do not change the slopes of the chevron plot. Otherwise, the ϕ value is not a constant, but changes as a function of the concentration of denaturant (warning: this is a common observation) [8].

The relative simplicity, and the lack of alternative methods have made the ϕ value analysis an extremely popular tool for investigating protein folding. Accordingly, there are many proteins that have already been studied using this methodology (*see* **Note 1**). Here I introduce the general assumptions behind ϕ value analysis, describe simple protocols to properly perform the experiments and interpret the results, and also discuss some of its shortcomings.

2 Materials

A complete ϕ value analysis implies performing a series of equilibrium and kinetic experiments on a sufficiently large amount of protein variants, each one of them carrying a specifically designed single point mutation (*see* **Note 2**). The typical procedure consists on monitoring the folding status of the protein using a given spectroscopic method (usually circular dichroism or fluorescence) at different concentrations of a chemical denaturant (typically urea or GdmCl). The approach may also include a thermodynamic characterization, i.e., experiments performed in steady-state or equilibrium conditions, to independently estimate the populations of folded and unfolded state as a function of the thermodynamic variable (denaturant concentration). Then the researcher shall measure the relaxation kinetics of the protein's spectroscopic signal in response to rapid (millisecond) changes in chemical denaturant induced by dilution experiments, thereby producing the so-called chevron plot. The most common and direct methodology to measure a protein folding chevron plot is the stopped-flow kinetic technique, and thus the focus of this chapter is on proteins whose folding and unfolding rates are in the time scales accessible to stopped-flow measurements (from 1 ms to about 100 s, *see* **Note 3**). Chevron plots have also been measured on a few microsecond folding proteins using ultrafast kinetic methods (*see* **Note 4**, and also Chapter 6 in this volume). The procedures for stopped-flow kinetic experiments include preparing a set of unfolding and refolding buffers that should be prepared with ultrapure water and high purity reagents.

1. Concentrated buffer: 100 mM of the preferred buffer (*see* **Note 5**). Adjust the pH to the selected value [it will depend on the particular protein (*see* **Note 6**)]. A widely used buffer is phosphate buffer at pH 7–7.5.
2. Denaturant-free buffer: dilute five times the concentrated buffer (200 ml of concentrated in 1 l) to a final concentration of 20 mM. Adjust the pH after dilution.
3. High denaturant concentration buffer: 8 M GdmCl or 10 M urea in 20 mM of the previously selected buffer (*see* **Note 7**). Dissolve 764.24 g of guanidine hydrochloride (MW 95.53) or 600.6 g of urea (MW 60.06) in a final volume of 1 l, adding first 200 ml of concentrated buffer. Its dissolution is endothermic and produces a large volume increase, so warming the concentrated solution before dilution and carefully adjusting the volume are useful precautions to take. Confirm the concentration by measuring the index of refraction of each solution (an online calculator of GdmCl and urea concentration from indices of refraction can be found at sosnick.uchicago.edu/gdmcl.html).
4. Protein stock solutions: prepare a concentrated solution of the protein of interest in the denaturant-free buffer and in the high denaturant concentration buffer (8 M GdmCl or 10 M urea) (*see* **Note 8**). As a reference, the concentration range 10–100 μM can be used (*see* **Note 9**).

5. Prepare buffers with intermediate denaturant concentrations by mixing the previously prepared solutions (*see* Table 1 for equilibrium experiments or Table 2 for kinetic experiments with guanidine hydrochloride). Confirm denaturant concentrations by measuring the index of refraction.

Table 1
Volumes (in μl) of solutions of 8 M guanidine hydrochloride buffer, 0 M guanidine hydrochloride buffer and protein stock without denaturant required to prepare a set of final denaturant concentrations ($[GdmCl]_{final}$) to perform an equilibrium chemical denaturation experiment using guanidine hydrochloride

$[GdmCl]_{final}$	8 M GdmCl buffer (μl)	0 M GdmCl buffer (μl)	Protein stock (no denaturant) (μl)
0.0	0	900	100
0.3	38	862	100
0.6	75	825	100
0.9	113	787	100
1.2	150	750	100
1.5	188	712	100
1.8	225	675	100
2.1	263	637	100
2.4	300	600	100
2.7	338	562	100
3.0	375	525	100
3.3	413	487	100
3.6	450	450	100
3.9	488	412	100
4.2	525	375	100
4.5	563	337	100
4.8	600	300	100
5.1	638	262	100
5.4	675	225	100
5.7	713	187	100
6.0	750	150	100
6.3	788	112	100
6.6	825	75	100
6.9	863	37	100
7.2	900	0	100

Table 2
Volumes (in ml) of solutions of 8 M guanidine hydrochloride buffer and 0 M guanidine hydrochloride buffer solutions required to prepare a set of final denaturant concentrations ($[GdmCl]_{final}$) to perform a series of stopped-flow kinetic (un)folding experiments

$[GdmCl]_{final}$	8 M GdmCl buffer (ml)	0 M GdmCl buffer (ml)
0.0	0.0	50.0
0.3	1.9	48.1
0.6	3.8	46.2
0.9	5.6	44.4
1.2	7.5	42.5
1.5	9.4	40.6
1.8	11.3	38.7
2.1	13.1	36.9
2.4	15.0	35.0
2.7	16.9	33.1
3.0	18.8	31.2
3.3	20.6	29.4
3.6	22.5	27.5
3.9	24.4	25.6
4.2	26.3	23.7
4.5	28.1	21.9
4.8	30.0	20.0
5.1	31.9	18.1
5.4	33.8	16.2
5.7	35.6	14.4
6.0	37.5	12.5
6.3	39.4	10.6
6.6	41.3	8.7
6.9	43.1	6.9
7.2	45.0	5.0
7.5	46.9	3.1
7.8	48.8	1.2

3 Methods and Analysis

3.1 Thermodynamic and Kinetic Conventions for ϕ Value Analysis

The idea of inferring mechanistic information from the changes in kinetic behavior of site-directed mutations on a given protein was independently proposed by the groups of David Goldenberg [9], Robert Matthews [6], and Alan R. Fersht [1, 10]. However, the general guidelines on how to efficiently implement the procedure and the first full implementations of ϕ value analysis for protein folding come from the Fersht lab, about three decades ago [1]. The method consists of measuring the kinetic effects that the perturbation in free energy caused by a given site-specific perturbation produces on the folding and unfolding rate constants of the protein under study. From these measurements on mutant and wild-type proteins, it is possible to infer the fraction of the perturbation free energy induced by mutation that is already present on the folding TSE.

Practically, the procedure requires the researcher to measure the folding relaxation rate at different concentrations of a chemical denaturant (urea or GdmCl), most of the times using a stopped-flow apparatus (*see* **Note 10**). For a two-state folding protein, the folding relaxation rate as a function of the concentration of chemical denaturant (whether measured in denaturant dilution—refolding—or denaturant adding—unfolding—kinetic experiments) is simply the sum of the values for the two microscopic rate constants for folding and unfolding at each denaturant concentration. The observation of a V-shaped curve for the logarithm of the relaxation rate versus denaturant concentration (i.e., the chevron plot) is taken as indication of the existence of linear free energy relationships (LFER) between the two thermodynamic states (unfolded and native) and the TSE (Fig. 1). In this plot of the logarithm of the relaxation rate versus denaturant concentration, the left limb corresponds to conditions at which the folding rate constant dominates. The increase in denaturant progressively slows down the folding rate, resulting on a negative slope. The plot then levels off with a tipping point that corresponds to the denaturation midpoint (when both rates are equal). At higher denaturant concentrations, the plot becomes linear again with a positive slope (the right limb of the chevron) that corresponds to the regime in which the unfolding rate constant dominates, becoming faster with denaturant. The linear extrapolation of the two chevron limbs to the ordinate axis renders the value for the rate constants in the absence of denaturant (k_f^0 and k_u^0). Once the two denaturant-free rate constants have been obtained for the mutant and the wild-type, their ratios are easily transformed onto differences in free energy using the expressions:

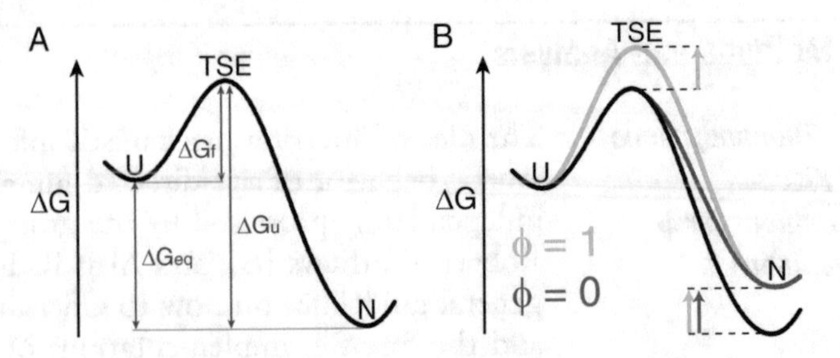

Fig. 2 Free energy diagram for the folding/unfolding transition, where the unfolded (U), transition state ensemble (TSE) and the native (N) states are shown. (**a**) General diagram with the free energy changes for the transition U↔TSE (ΔG_f) and the transition N↔TSE (ΔG_u), as well as the thermodynamic protein stability (ΔG_{eq}). (**b**) Exemplary free energy diagrams showing the two extreme scenarios of how the perturbation in free energy can change the levels of the native state and the TSE: (black) wild-type protein; (red) a mutant with ϕ value equal to 0; (green) a mutant with ϕ value equal to 1

$$\Delta\Delta G_f|_{\text{mutation}} = \Delta G_{f,\text{mut}} - \Delta G_{f,\text{wt}} = \text{RT}\, ln\left(\frac{k^0_{f,\text{mut}}}{k^0_{f,\text{wt}}}\right) \tag{1}$$

for the folding rate constants and

$$\Delta\Delta G_u|_{\text{mutation}} = \Delta G_{u,\text{mut}} - \Delta G_{u,\text{wt}} = \text{RT}\, ln\left(\frac{k^0_{u,\text{mut}}}{k^0_{u,\text{wt}}}\right) \tag{2}$$

for the unfolding rate constants. These expressions provide the difference in perturbation free energy caused by the mutation between the unfolded state and TSE ($\Delta\Delta G_f|_{\text{mutation}}$) and the difference in perturbation free energy between the native state and TSE with inverted sign ($-\Delta\Delta G_u|_{\text{mutation}}$) (*see* **Note 11** and Fig. 2a). The total perturbation free energy induced by the mutation is simply

$$\begin{aligned}\Delta\Delta G_{eq}\big|_{\text{mutation}} &= \Delta G_{eq,\text{mut}} - \Delta G_{eq,\text{wt}} \\ &= \Delta\Delta G_f|_{\text{mutation}} - \Delta\Delta G_u|_{\text{mutation}}\end{aligned} \tag{3}$$

The total perturbation free energy induced by mutation can also be obtained from the equilibrium denaturation experiments on the mutant and wild-type proteins. Agreement between the values obtained with both methods is taken as confirmation for the two-state behavior of the protein. At this point, the ϕ value is calculated as

$$\phi = \frac{\Delta\Delta G_f}{\Delta\Delta G_{eq}}\bigg|_{\text{mutation}} \tag{4}$$

and the $(1 - \phi)$ value as

$$(1 - \phi) = \frac{\Delta\Delta G_u}{\Delta\Delta G_{eq}}\bigg|_{\text{mutation}} \tag{5}$$

consistently with the principle of microscopic reversibility, assuming that there are not important contributions from non-native interactions to the free energy of folding the protein.

3.2 Interpretation of the ϕ Value

In principle, the ϕ value should range between 0 and 1 (i.e., a fractional value of the perturbation free energy). A value of 0 implies that the perturbation does not affect the TSE more than it does the unfolded state, i.e., the native interactions affected by the mutation are not present yet in the TSE. On the contrary, a value of 1 indicates that the native state and the TSE are equally affected by the perturbation or that the interactions are fully made in the TSE (Fig. 2b).

The interpretation of fractional ϕ values is less straightforward, as they can arise from multiple scenarios. For example, the fractional value may arise from a scenario in which a mutation alters many interactions, some of which are fully made, whereas others are fully broken. It is also possible that all the interactions affected by the mutation are simultaneously weakened in a similar degree (homogeneous TSE). The latter is more likely to happen for localized, relatively minor structural perturbations, such as the structural reorganization around the cavity arising by truncation of a methyl group. It is also important to realize that the reaction also includes changes in conformational entropy that compensate those in enthalpy from the interactions. The TSE also has an intermediate degree of conformational entropy between the native and unfolded states [11], and as such, it may be structurally heterogeneous with an ensemble in which different conformations are affected by the perturbation in different ways.

Finally, sometimes the experiment renders ϕ values that are either below 0 or above 1. These cases are termed "non-canonical" ϕ values. One obvious source for these type of results is experimental error, more significant with smaller perturbations (i.e. a good rule of thumb is for the perturbation to be >4–5 kJ/mol to ensure reliable experimental determination of ϕ values) [12]. However, it is also in occasion possible to obtain non-canonical ϕ values from reliable measurements (large perturbation and/or little experimental error). Such cases are usually interpreted by assuming that the TSE is stabilized by transient non-native interactions (interactions that are not present in the native state), which are also perturbed by the mutation, but are only detectable in the kinetic experiments [6, 13–15].

3.3 Design of Mutations

One of the critical steps for an effective ϕ value analysis is the mutation design strategy (which positions to mutate and what substitutions to make). The mutational strategy is key to facilitate the interpretation of the ϕ values in structural/mechanistic terms. In this section, I provide some basic guidelines of how to select the protein locations and the type of amino acid substitution.

The first consideration is to seek mutations that are likely to affect the fewest possible number of native interactions and hence that are as structurally specific as possible. The more interactions are perturbed by the mutation, the more averaged out is the measured ϕ value, and thus the more likely it is to be close to the universal value of about 0.3 [8]. This scenario is typically found in mutations that create cavities in the native hydrophobic core by truncating aliphatic or aromatic side chains.

Moreover, one should avoid mutating positions in the protein that are likely to produce a large disruption of the native structure and too strong a perturbation. These positions can be identified via simple structural analysis of the native 3D structure from the coordinates deposited in the protein databank (ϕ value analysis should only be attempted on proteins for which there is an atomic resolution structure available) using widely available molecular viewing programs (e.g., Swiss-PdbViewer [16] or Pymol [17]). Another useful tool to single out positions with key structural roles is to use a multiple sequence alignment of orthologs of the protein of interest to rule out those that are highly conserved. The intrinsic native stability of the protein is another important consideration. A highly stable protein will be able to accommodate much more drastic mutations than a protein that is marginally stable in its wild-type form.

Generally, the set of ϕ values should be comprehensive, involving a sufficiently extensive set of mutations that are distributed over the whole protein structure so that each structural element is probed by several mutations. For two-state single-domain proteins (50–100 amino acids), a minimum of about 30 mutations should be performed (Fig. 3). In addition to the obvious considerations of improving structural sampling, the more mutations are performed the more robust is the interpretation of ϕ values: self-consistency checks (nearby mutations should produce similar results). For instance, the most accurate way to measure a ϕ value for a given protein position is to perform multiple single point mutations in that location and obtain the ϕ value from the slope of the linear correlation between $\Delta\Delta G_{\mathrm{f}}|_{\mathrm{mutation}}$ and $\Delta\Delta G_{\mathrm{eq}}|_{\mathrm{mutation}}$ [18].

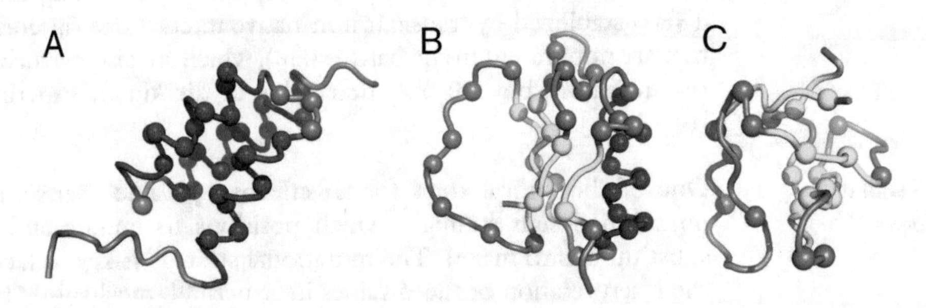

Fig. 3 Examples of possible distribution of mutations to be used for ϕ-value analysis on the structure of the target protein: (**a**) the B domain of protein A, an α-helical protein, (**b**) chymotrypsin inhibitor 2, an alpha/beta protein, and (**c**) SH3 domain from α-spectrin, a β-sheet protein. Secondary structure elements are shown in red (alpha helices) and yellow (beta strands) and the mutation locations are represented as spheres

Once all the mutation locations have been selected, the next step is to choose the amino acid replacement for each position. The most common option is to select mutations that truncate the side chain without changing the chemistry (e.g., isoleucine to valine) or to just perform simple alanine/glycine scanning (any amino acid changed to alanine or glycine). This is a most conservative methodology that avoids large distortions of the protein structure. However, its drawback is that it typically produces small perturbations, sometimes nearly negligible, that seriously impact the ability to determine the experimental ϕ values with useful accuracy. An analysis of all of the mutations from previous studies (more than 1000 in over 30 proteins) provides a useful guide to decide which type of substitutions should be attempted in any given position (Table 3).

Table 3
Mutations most frequently used for ϕ value analysis. For each mutation, the type of mutation and the frequency of use in a set of about 1000 mutations collected from bibliography are given

Mutation	Type	%
Val to Ala	Hydrophobic truncation	12.5
Leu to Ala	Hydrophobic truncation	10.8
Ala to Gly	Hydrophobic truncation	9.3
Ile to Val	Hydrophobic truncation	7.2
Ile to Ala	Hydrophobic truncation	6.9
Thr to Ala	Polar truncation	3.7
Phe to Ala	Aromatic truncation	3.3
Tyr to Ala	Aromatic truncation	2.7
Phe to Leu	Aromatic truncation	2.4
Asp to Ala	Charge deletion	2.3
Lys to Ala	Charge deletion	2.3
Gly to Ala	Methyl insertion	2.3
Asn to Ala	Polar truncation	2.0
Glu to Ala	Charge deletion	1.9
Pro to Ala	Hydrophobic truncation	1.9
Ser to Ala	Polar truncation	1.7
Val to Thr	Polar replacement	1.7
Thr to Ser	Hydrophobic truncation	1.3
Leu to Val	Hydrophobic truncation	1.1
Met to Ala	Hydrophobic truncation	1.0
Gln to Ala	Polar truncation	1.0

The table illustrates that substitutions to alanine are by far the most frequent.

Once the set of mutations has been designed, they are produced by standard site directed mutagenesis procedures (the QuickChange® kit from Stratagene is commonly used) on the cloned gene encoding for the subject protein. Finally, all protein mutants and the wild type need to be expressed and purified using the protocols adequate for the protein of interest (*see* **Note 12**).

3.4 Tests to Confirm the Conservation of the Native Structure

Once the proteins have been purified, it is important to quickly determine whether the mutation has introduced significant changes in the native 3D structure. The best option to obtain the structure of a protein is the use of X-ray crystallography or NMR techniques. However, these methods are time consuming and is often unrealistic to perform them in the large collection of mutants required to complete the ϕ value analysis. An alternative is to use simple spectroscopic methods to check for certain diagnostic signatures of a large structural perturbation. In this regard, far-UV circular dichroism is most useful because it provides a quick and easy test of the conservation of secondary structure from just the superimposition of the wild-type and mutant spectra (the far-UV circular dichroism spectrum is a molecular fingerprint for the presence of structural changes in the protein) (*see* **Note 13**). Typically, CD spectra from 190 to 250 nm are collected and compared in order to discard possible mutants with large distortions of the protein structure.

3.5 Equilibrium Denaturation Experiments

3.5.1 Experimental Setup

The estimation of the total perturbation free energy caused by mutation ($\Delta\Delta G_{eq}|_{mutation}$) is needed to determine the ϕ value (Eq. 4). Experimentally, one can do this by comparing the stability of the wild type and the mutant in equilibrium denaturation experiments (denaturant titrations). The first step is the selection of the spectroscopic technique able to follow the folding/unfolding transition. Among them, fluorescence, far-UV circular dichroism, and near-UV circular dichroism are commonly used (*see* **Note 14**). A good technique to use should provide a clear spectroscopic signature between the folded and the unfolded to be used to fit the data and obtain the thermodynamic stability of the proteins (wild type and mutants). In this chapter, I will illustrate the protocols using intrinsic tryptophan fluorescence as spectroscopic technique.

3.5.2 General Protocol

Repeat the following protocol for each mutant in the ϕ value collection:

1. Prepare a full set of samples, for the wild type and the mutant, with different concentrations of Guanidine hydrochloride (Table 1).

2. Measure the fluorescence spectrum for all the samples (*see* **Note 15**).
3. Calculate the fluorescence area under the spectrum and represent graphically the full set of data.
4. For a two-state folding transition, the data can be fitted to the equation:

$$S = \frac{(S_U + dS_U[den]) + (S_N + dS_N[den]) \cdot e^{(\Delta G_{eq} - m_o[den])/RT}}{1 + e^{(\Delta G_{eq} - m_o[den])/RT}} \quad (6)$$

where S_u and S_N are the baseline signals for the unfolded and native states without denaturant, and dS_U and dS_N are the linear slopes of the changes in the unfolded and native state signals with denaturant, m_0 is the dependence of the native stability on denaturant concentration, R is the gas constant, T is the temperature in Kelvin and ΔG_{eq} is the change in free energy upon unfolding (folding stability) of the protein.

5. For each mutant, the value of $\Delta\Delta G_{eq}|_{mutation}$ is just the difference in ΔG_{eq} between the mutant and the wild type (Eq. 3).

3.6 Kinetic Chevron Experiments

3.6.1 Experimental Setup

The other needed parameter is the difference in perturbation free energy caused by the mutation between the unfolded state and TSE for folding ($\Delta\Delta G_f|_{mutation}$). Experimentally, this value is estimated analyzing the effect of the denaturant on the folding/unfolding rates of the protein (chevron plot) (Fig. 1), both for the wild type and for the mutant. The most common method to obtain kinetic rates is the stopped-flow technique (*see* **Note 16**), where a spectroscopic signal (commonly fluorescence, *see* **Note 17**) is collected over time immediately after a change in the concentration of denaturant has been triggered in a fast mixing (dilution) experiment (Fig. 4). The instrument available in our lab is a SX20-LED stopped-flow apparatus from Applied Photophysics (www.photophysics.com), but any other rapid mixing device can be used for this purpose.

3.6.2 General Protocol

The equilibrium denaturation experiment (e.g., its denaturation midpoint) can be used to select the range of denaturant concentrations required to measure both the unfolding and the refolding limbs of the chevron plot (Fig. 1) (*see* **Note 18**). With such range in mind, follow the next steps:

1. Prepare a full set of denaturant concentration samples (Table 2) by mixing the denaturant-free buffer and the high denaturant concentration buffer from Subheading 2. These solutions, together with the protein stock solutions described in Subheading 2, will be mixed to obtain all the kinetic data necessary

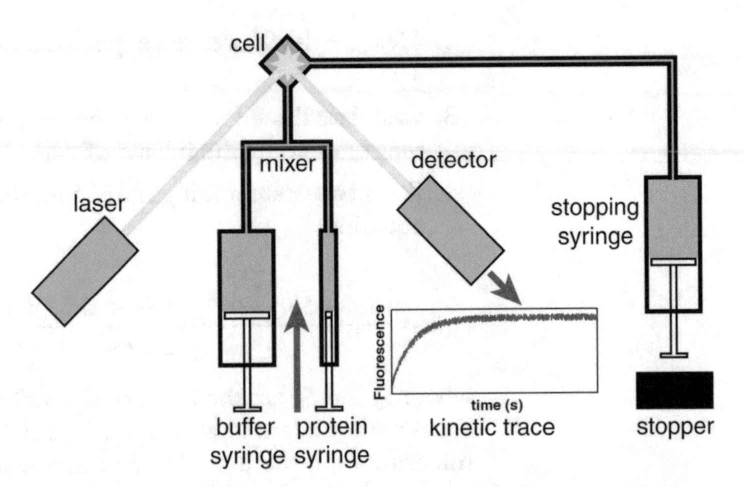

Fig. 4 Schematic representation of a typical stopped-flow instrument. The protein and buffer solutions are mixed before entering the measurement cell, where a laser excites the protein fluorophores and then the fluorescence is collected by a photodetector as a function of time to produce a kinetic trace. The flow from the main mixing syringes is stopped when the piston of the stopping syringe contacts the stopper, at which point the trigger starts the measurement

to calculate the relaxation rate as a function of denaturant (*see* **Note 19**).

2. Perform unfolding experiments by mixing the high Guanidine hydrochloride concentration solutions with the protein prepared in denaturant-free buffer. These experiments will provide the unfolding (left) limb of the chevron plot. Collect at least ten individual stopped-flow traces to obtain an average decay. It is usually convenient to start with the highest denaturant concentration available and finish when you reach the first concentration at which there is no apparent decay in the stopped-flow traces (*see* **Note 20**).
3. Perform refolding experiments by mixing the low guanidine hydrochloride concentration solutions with the protein prepared in a buffer with very high denaturant concentration. These experiments will provide the folding (right) limb of the chevron plot. It is usually convenient to start this series of experiments from the highest final denaturant concentration that still produces a measurable signal change in the stopped-flow experiment (*see* **Note 20**).
4. Assuming a single exponential process (*see* **Note 21**), fit each averaged kinetic trace to the equation:

$$F = F_{\infty} + Ae^{-k_{\mathrm{obs}}t} \quad (7)$$

where F_∞ is the fluorescence value at infinite time, A is the amplitude of the signal change, and k_{obs} is the observed rate (relaxation rate), which is the sum of the folding and the unfolding rates.

5. Represent graphically the logarithm of k_{obs} versus the concentration of denaturant (chevron plot) (Fig. 1). Fit the chevron dataset to the following equation:

$$\log k_{obs}([\mathrm{den}]) = \log\left(k_f^0 10^{m_f[\mathrm{den}]} + k_u^0 10^{m_u[\mathrm{den}]}\right) \quad (8)$$

where k_f^0 and k_u^0 correspond to the folding and unfolding rates at 0 M denaturant, and m_f and m_u are the slopes of the folding and unfolding rates with denaturant concentration (*see* **Note 22**).

Once this protocol has been completed for the wild type and all the mutants, calculate the difference in perturbation free energy caused by the mutation between the unfolded state and TSE ($\Delta\Delta G_f|_{mutation}$) using Eq. 1. Calculate the difference between the native state and the TSE ($\Delta\Delta G_u|_{mutation}$) with Eq. 2.

3.7 Applications of the ϕ Value Analysis

3.7.1 Folding Transition State Characterization

The ϕ value analysis was first applied to study the TSE of barnase, a protein with a folding intermediate [1]. Since then, the method has been widely used for a large list of single-domain, two-state folding proteins. The catalog comprises proteins of many sizes (37–169), native folds (alpha, alpha/beta, and beta proteins), and kinetic properties (rates ranging from 0.2 to 2×10^5 s^{-1} for folding and from 1×10^{-5} to 600 s^{-1} for unfolding) (*see* Table 4) [19–51].

Here I have focused on the application of the methodology to two-state folding proteins with rates accessible to stopped-flow kinetic experiments (<1000 s^{-1}). More complex folding scenarios have also been studied, including proteins with the accumulation of equilibrium or kinetic intermediates (and obtaining information for one of the transition states) [10, 52], [53, 54], and sub-millisecond folding proteins in which faster kinetic techniques have been used (continuous flow [47, 51], joule-heating temperature jump [46, 47, 49, 51], or NMR techniques [20, 27]).

The ϕ value analyses performed so far suggest that there is high variability in the structural properties of folding TSEs and folding mechanisms [55, 56]. These experiments have been interpreted with phenomenological mechanisms such as the nucleation-condensation folding model (where there are a few specific regions with very high ϕ values, whereas the remainder manifest an averaged fractional value (near 0.3) that is interpreted to reflect a diffuse overall compactness [19]) and the framework folding mechanism in which the high values are for local interactions and the lower values for tertiary interactions [49].

Table 4
Two-state single-domain proteins that have been subjected to ϕ value analysis for their folding transition states ensembles, sorted by year of publication. The table includes name, pdb code, secondary structure composition, size (number of amino acids), number of mutations used in the ϕ value analysis and the folding and unfolding rate constants of the wild type protein at zero denaturant

Year	Protein	pdb	Fold	Size	Mutations	k^0_{fold} (s^{-1})	k^0_{unf} (s^{-1})
1995	Chymotrypsin inhibitor 2	3Cl2	α/β	65	65	$5.6 \cdot 10^1$	$1.2 \cdot 10^{-4}$
1997	N-term 1c repressor	1LRP	α	80	7	$9.6 \cdot 10^4$	$3.5 \cdot 10^1$
1998	ADA2h	106X	α/β	81	18	$7.6 \cdot 10^2$	$4.8 \cdot 10^{-1}$
1999	Acylphosphatase	2ACY	α/β	98	26	$2.4 \cdot 10^{-1}$	$6.5 \cdot 10^{-5}$
1999	FKBP12	1FKB	α/β	107	34	$4.2 \cdot 10^{-1}$	$2.1 \cdot 10^{-4}$
1999	Acyl-coA binding domain	2ABD	α	86	30	$2.2 \cdot 10^2$	$1.1 \cdot 10^{-4}$
1999	α-spectrin SH3	1SHG	β	62	14	$7.7 \cdot 10^1$	$1.1 \cdot 10^{-2}$
1999	Src SH3	1SRM	β	64	54	$4.7 \cdot 10^1$	$3.4 \cdot 10^{-1}$
1999	U1A	1URN	α/β	102	13	$2.5 \cdot 10^2$	$1.0 \cdot 10^{-5}$
2000	Villin 14T	2VIL	α/β	126	25	$3.0 \cdot 10^3$	$1.4 \cdot 10^{-2}$
2000	Fibronectin type Ill domain	1TEN	β	92	45	2.9	$4.6 \cdot 10^{-4}$
2000	Sso7d	1SSO	β	62	20	$1.0 \cdot 10^3$	$3.9 \cdot 10^{-2}$
2000	Protein L	1HZ6	α/β	62	68	$6.1 \cdot 10^1$	$2.0 \cdot 10^{-2}$
2000	Protein G	1PGB	α/β	56	31	$2.4 \cdot 10^2$	$3.3 \cdot 10^{-1}$
2000	Cold shock protein A	1MJC	β	69	6	$2.6 \cdot 10^2$	1.9
2002	Rd apocytochrome b562	1YYJ	α	106	39	$1.5 \cdot 10^2$	$2.0 \cdot 10^{-2}$
2002	Fyn SH3	1SHF	β	67	34	$3.0 \cdot 10^1$	$5.1 \cdot 10^{-1}$
2002	Protein S6	1RIS	α/β	101	28	$3.3 \cdot 10^2$	$4.6 \cdot 10^{-4}$
2003	Colicin E9 immunity protein	1IMQ	α	86	25	$1.2 \cdot 10^3$	$1.2 \cdot 10^{-2}$
2003	c-Myb DNA binding domain	1IDY	α	54	18	$6.2 \cdot 10^3$	5.3
2004	Cold shock protein B	1CSP	β	67	21	$1.1 \cdot 10^3$	1.6
2004	Ribosomal protein L23	1N88	α/β	96	17	$2.0 \cdot 10^1$	$7.4 \cdot 10^{-5}$
2005	N-term ribosomal protein L9	1DIV	α/β	56	24	$8.7 \cdot 10^2$	$9.0 \cdot 10^{-1}$
2005	Ubiquitin	1UBQ	α/β	76	27	$3.5 \cdot 10^2$	$1.6 \cdot 10^{-3}$
2005	Apoazurin	1AZU	β	126	8	$1.6 \cdot 10^2$	$6.7 \cdot 10^{-3}$
2005	Yeast acyl-coA binding domain	1ST7	α	86	18	$5.0 \cdot 10^3$	$1.6 \cdot 10^{-3}$
2006	Apoflavodoxin	1FTG	α/β	169	34	$6.9 \cdot 10^1$	$7.0 \cdot 10^{-2}$
2006	FBP28WW	1E0L	β	37	45	$2.3 \cdot 10^4$	$3.4 \cdot 10^2$
2006	E3BD	1W4E	α	45	22	$2.7 \cdot 10^4$	$2.0 \cdot 10^1$

(continued)

Table 4 (continued)

Year	Protein	pdb	Fold	Size	Mutations	k^0_{fold} (s^{-1})	k^0_{unf} (s^{-1})
2007	C-term ribosomal protein L9	1DIV	α/β	92	24	$2.6 \cdot 10^1$	$3.9 \cdot 10^{-4}$
2007	B-domain protein A	1SS1	α	60	85	$9.7 \cdot 10^4$	$2.5 \cdot 10^1$
2007	C-Raf-1 Ras binding domain	1RFA	α/β	78	47	$2.3 \cdot 10^3$	$4.9 \cdot 10^{-2}$
2008	POB	1W4J	α	51	22	$2.1 \cdot 10^5$	$5.7 \cdot 10^2$

3.7.2 Experimental Errors, Energetic Perturbations, and Average ϕ Values

The ϕ value analysis of folding has been widely used. However, there are some important issues about the analysis of data and the conclusions that need to be considered. First, one should not forget that the accuracy of the calculated ϕ values is proportional to the size of the perturbation produced by the mutation. An analysis of the intrinsic experimental error based on the direct comparison between data on the same protein/mutants produced by different laboratories indicated that mutations with perturbations below 7 kJ/mol are not reliable for ϕ value analysis. The main source of error was found to be in the extrapolation of the rates to zero denaturant concentration [57]. Some empirical ways to limit such errors (and extend the applicability of the analysis to perturbations >5 kJ/mol) are to reference the ϕ values to a given denaturant concentration near the midpoint (provided that the slopes of the chevron do not change), or use average m_f and m_u vales for all the mutants.

The inherent problems in the estimation of ϕ value errors were confirmed applying global statistical analysis methods to large collections of mutations from multiple proteins [8, 12]. For more than 800 mutations, it was observed that the whole set of experimental ϕ values could be fitted to the equation:

$$\Delta\Delta G_u = 0.76 \cdot \Delta\Delta G_{eq} \tag{9}$$

with a correlation coefficient higher than 0.9 and a standard deviation of 1.8 kJ/mol, not far from the estimated experimental precision between labs of ~1.3 kJ/mol [57]. This correlation produces an average ϕ value of 0.24, supporting the idea of a uniform structurally weak transition state [8]. Importantly, this study found that the ϕ values decrease with protein stability (and hence denaturant concentration) and proposed that the best strategy is to report ϕ values at the denaturation midpoint, where the conditions of iso-stability and the absence of extrapolations makes the comparison among mutants most straightforward. At the midpoint the average ϕ value for the entire dataset was 0.36 ± 0.11. ϕ values close to 1, and values below 0 or above 1, have been typically interpreted as key indicators of the TSE structure and folding

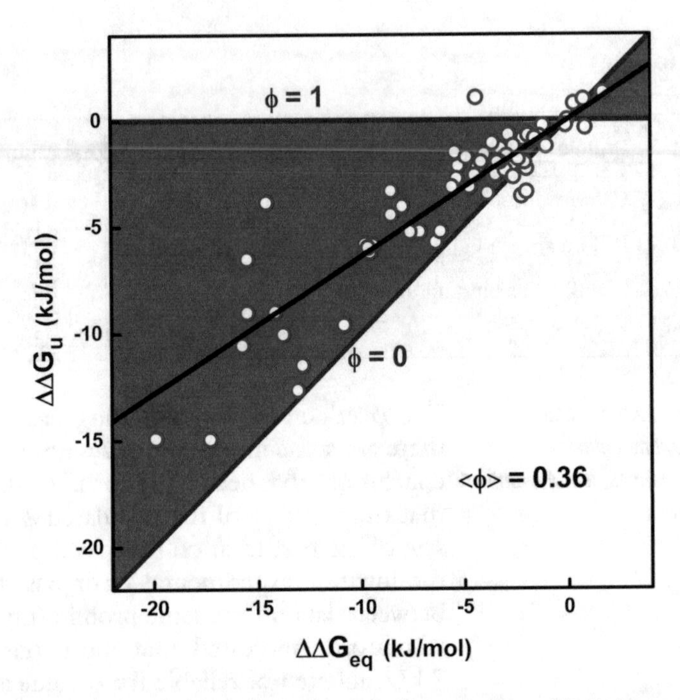

Fig. 5 $\Delta\Delta G_u$ versus $\Delta\Delta G_{eq}$ values for a set of 65 mutants on CI2 that have been used for ϕ value analysis (blue dots) [19]. The lines represent $\phi = 1$ and $\phi = 0$ (dark red), and the $\phi = 0.36$ (black) that corresponds to the average ϕ value of more than 800 mutants on 24 different proteins [8]

mechanism. However, the global statistical analysis reveals that many of the most extreme deviations from the average ϕ value are for mutations with very small perturbations (unreliable) or more experimental uncertainty (noisy datasets). These observations support the idea that folding TSE represents broad ensembles with diffuse native structure in which local interactions are more consolidated and long-range interactions largely absent (Fig. 5).

3.7.3 Leffler Analysis: Multiple Mutations at the Same Position

The previous discussion (Subheading 3.7.2) highlights some of the perils of interpreting experimental ϕ values in mechanistic terms, as well as the importance of carefully establishing the experimental error in their determination.

An approach that minimizes many of these issues is to obtain a position-specific average ϕ value from data arising from multiple mutations performed at the same protein position. A Leffler analysis consisting in plotting the $\Delta\Delta G_f$ versus $\Delta\Delta G_{eq}$ experimental data for the series of mutations in each position permits to accurately determine the average ϕ value for that position from the fitted slope of the plot (Fig. 6) (*see* **Note 23**). Practically, this means that many more mutations are needed because probing each structural position in the protein requires several amino acid substitutions, greatly adding to the workload: mutagenesis, expression, purification, and

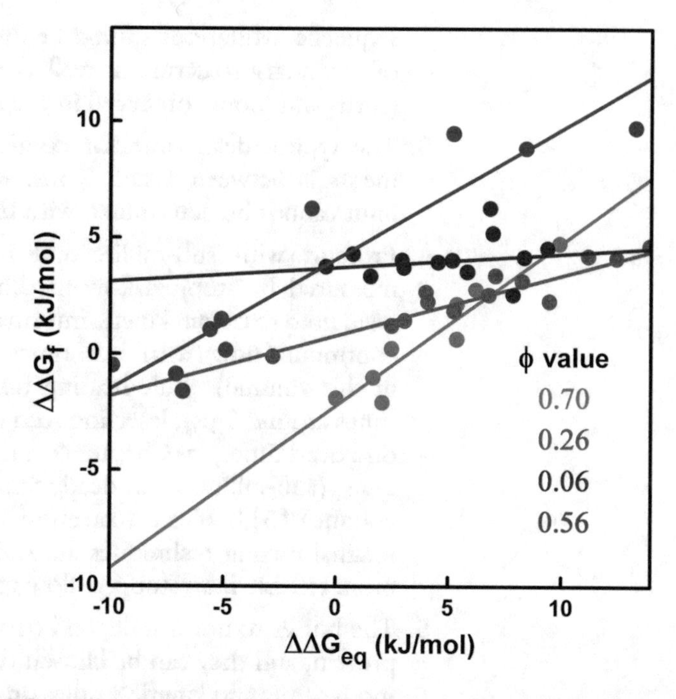

Fig. 6 Examples of Leffler plot ($\Delta\Delta G_f$ versus $\Delta\Delta G_{eq}$) for multiple mutations on given protein locations: T22 (orange), E24 (red) and S41 (dark red) positions on SH3 domain and R48 (blue) on CI2. The ϕ values indicated in the figure are the slopes from the linear fits of the data for each position. The data was collected from previous studies [58–60]

thermodynamic and kinetic characterization. This becomes too onerous for a comprehensive structural analysis, and thus, it is best applied on the second round of experiments on just those positions that appear to have the most interesting results in one mutation per site preliminary analysis (e.g., locations with more extreme individual ϕ values).

4 Notes

1. Most of the proteins used for ϕ value analysis exhibit two-state folding kinetics without any visible accumulation of intermediates (given in Table 4). However, other proteins that exhibit the accumulation of a folding intermediate have been studied as well (in fact, barnase, the first candidate to apply the ϕ value methodology exhibits an intermediate [1]). Other proteins that have been studied with just a handful of mutations are not included in Table 4.
2. For a comprehensive scanning of the 3D structure of a 50–150 amino acid protein, the minimal number of mutants that should be studied is about 30, distributed throughout the

sequence. Mutations should be inserted in all of the elements of secondary structure as well as in the connecting segments (turns and loops) observed in the 3D structure.

3. The typical dead time for commercial stopped-flow instruments is between 1 and 2 ms. Relaxation times below this limit cannot be determined with this technique.
4. Proteins with sub-millisecond relaxation times cannot be measured by stopped-flow methods. Other techniques have been used to obtain kinetic information in such cases, including continuous flow (with dead times around 50 μs, *see* Chapter 7 in this volume), joule-heating temperature jump (with dead times around 1 μs), laser-induced temperature jump (nanosecond dead times, *see* Chapter 6 in this volume) or NMR techniques (sub-millisecond dead time, *see* Chapter 10 in this volume) [61]. If the relaxation time is longer than 1 min, manual mixing techniques are sufficiently fast and potentially more reliable than stopped-flow techniques.
5. The buffer to use will depend on the properties of the subject protein, and they can be chosen on the basis of previous thermodynamic and kinetic studies on the wild type. Here I do not mention salt effects specifically, but some proteins need relatively higher ionic strengths to be stably folded. Phosphate is a favorite choice as neutral pH buffer, which can be used for circular dichroism measurements without running into background absorbance issues.
6. Every protein has a pH range over which it is stably folded. The pH of each protein should be selected accordingly.
7. The selection of denaturant agent to use for these experiments can be guided by the intrinsic stability of the wild-type protein. Of the two most commonly used denaturants, urea and guanidine hydrochloride, the latter is significantly stronger (about twofold stronger on average). Hence, proteins that are intrinsically very stable may need the addition of very high concentrations of urea to become fully unfolded. In such cases, it is advisable to use the stronger denaturant guanidine hydrochloride. For less stable proteins, urea is the preferable denaturant as it does not change the ionic strength of the solution (simpler denaturation mechanism). Once a denaturant has been chosen, the whole set of experiments should be carried out with the same denaturant to allow for direct comparison (in this chapter, all the protocols have been described with the use of guanidine hydrochloride, but they are easily adapted to urea).
8. The denaturant concentration required for the protein stock solution to be use in refolding experiments can be reduced for proteins that are not very stable (e.g., if the denaturation midpoint of the protein is 4 M GdmCl, a concentration of

6 M GdmCl is enough for the protein stock solution in high denaturant concentration).

9. The protein concentration to use in the stock solutions will vary depending on the sensitivity of the spectroscopic technique used to monitor (un)folding (i.e., fluorescence versus circular dichroism) and the magnitude of the signal change undergone by the test protein between the folded and unfolded states. The larger the signal change, the lower the signal to noise of the decays (and fewer traces needed to obtain the average decay) and the more accurate the parameters obtained from the fitting of the data. In setting up the whole experimental protocol, it is also useful to perform preliminary experiments at exemplary minimal signal conditions (around the denaturation midpoint) and evaluate the amount of signal and the quality of the fits. The stock solutions described here are 10 times concentrated (both for kinetic and equilibrium experiments) to match the 1:10 dilution experiment of a pneumatic stopped-flow apparatus (e.g., Applied Photophysics). Stopped-flow systems equipped with stepped motors (e.g., Biologic) allow more flexibility in the volumes, and thus the concentrations. For the experiments described here, a final concentration in the 1–10 μM is usually adequate for fluorescence experiments provided that the protein has at least one tryptophan residue.

10. The relaxation rates in refolding conditions are measured by mixing a solution of protein incubated in high denaturant concentration buffer (unfolded) and buffers with concentrations of denaturant that are below the denaturation midpoint for the protein. Using a pneumatic stopped-flow instrument, a typical dilution ratio is 1:10 ratio. The relaxation rates in unfolding conditions are measured by mixing the protein incubated in a buffer without denaturant added (folded) with buffers in which the denaturant concentration is higher than the denaturation midpoint (different buffers to render different datapoints using a constant 1:10 dilution ratio).

11. In this chapter, the changes in free energy are all defined using the folded/native state as the reference. Therefore, the sign convention implies that a positive change in free energy ($\Delta G > 0$) means conditions at which the protein is stable in its native state. For mutations, a positive change in free energy ($\Delta\Delta G > 0$) corresponds to an increased stabilization by the mutation.

12. The mutations proposed for the ϕ value analysis should be conservative in the sense that they should not make the protein intrinsically unstable. Ideally, the most destabilizing mutations should still have an ~5 kJ/mol repository of native stability.

Problems in the expression and purification of the mutants using the standard protocol for the wild-type protein of interest could be an indication that the mutation has produced a drastic destabilization, and thus may not be suitable for the analysis (it can be preemptively discarded before going through the efforts to purify it and characterize its folding properties).

13. Every type of secondary structure (e.g., α helices and β strands) produces a signature far-UV circular dichroism spectrum. Hence, each protein will exhibit a characteristic spectrum arising from the linear combination of the spectra from its secondary structure elements. An estimate of the secondary structure composition of the protein can be obtained from the analysis of the far-UV circular dichroism spectrum measured between 190 and 250 nm.

14. Intrinsic fluorescence in proteins arises from its aromatic residues (tryptophan (Trp), tyrosine (Tyr), and phenylalanine (Phe)). However, only Trp and Tyr have emission quantum yields that are high enough to be used as signal to track the folding status of the protein. Typical conditions used to measure an intrinsic fluorescence spectrum of a protein are to excite at 280 nm (excitation of both Trp and Tyr) and collect fluorescence in the range between 290 and 400 nm. An alternative is to excite at 295 nm, which excites only Trp, to obtain more specific/local information about the protein tertiary structure (single-domain proteins have few Trp residues 0–3, and many have only 1). Folding usually increases the fluorescence emission of Trp and Tyr and shifts the maximum to the blue due to the desolvation of the aromatic residues concomitant to the formation of tertiary structure. However, in some proteins, the tertiary structure may place residues near the fluorophore (Trp or Tyr) that quench their fluorescence emission. Therefore, the fluorescence signal associated to the unfolded state could in principle be higher or lower than for the folded state. This must be determined empirically by measuring the fluorescence spectra of the protein of interest in the extreme buffers (fully folded and fully unfolded).

    Another technique to monitor folding spectroscopically is circular dichroism (CD), which is based on the difference in absorption that occurs in molecules with chirality when excited with left-handed or right-handed circularly polarized light. In proteins, the peptide backbone is composed of chiral centers (each peptide bond in a protein is a chiral center) that show characteristic far-UV signals in the 190–250 nm range and monitor the local conformation of the protein (secondary structure). In addition, aromatic residues (Trp and Tyr) become chirally oriented when their side chain mobility is restrained by the rest of the protein structure, showing a characteristic CD spectrum in the near-UV region (250–300 nm) as well as contributions to the far-UV region (adding to the far-UV CD spectrum corresponding to the secondary structure). Thus, secondary structure changes can be followed with far-UV CD and tertiary structure variations in the vicinity of aromatic residues with near-UV CD.

Alternative spectroscopic techniques can be used if there is enough signal change upon unfolding, and there is no interference from the denaturant. Infrared absorption in the amide I band region is difficult to use for these experiments as both urea and guanidine hydrochloride show strong absorption in this part of the vibrational spectrum. Another option is to use extrinsic fluorophores and perform Förster Resonance Energy Transfer (FRET) measurements, in which case the protein has to be chemically modified to incorporate a pair of fluorophores suitable for FRET (donor and acceptor). In addition to the added complexity in the protocols to prepare the samples, the bulkiness and hydrophobicity of extrinsic fluorophores makes them relatively likely to further destabilize the folded state or introduce other folding artifacts.

15. Collect the spectrum from 290 to 400 nm, exciting at 280 nm. Optimize the fluorescence parameters (excitation and collection bandwidth, acquisition time, and photomultiplier voltage) to increase the signal to noise ratio. The instrument available in our lab is a PTI QuantaMaster 400 from Horiba, but any other fluorometer can be used for this purpose.
16. For proteins with relaxation rates higher than 1 ms.
17. Although fluorescence is the most common spectroscopic signal to follow the folding/unfolding transition of proteins by the stopped-flow technique, stopped-flow instruments can be attached to a circular dichroism spectrophotometer to incorporate CD measurements of folding-unfolding kinetics. The caveat here is that the addition of urea or Guanidine hydrochloride limits the far-UV range to wavelengths >220 nm, due to intrinsic absorbance of the denaturants below that value.
18. The denaturation midpoint is the concentration of denaturant (it is denaturant specific) at which the folded and unfolded states are equally populated (50% to 50%). The unfolding kinetic experiments can be performed starting at the highest denaturant concentration (most unfolding, largest signal change) and end at a concentration slightly below the midpoint (a minimal conversion to the unfolded state is necessary for bulk detection). The refolding kinetic experiments can be performed starting at 0 M denaturant final concentration (maximum refolding and signal change) to a concentration slightly above the midpoint. The advantage of that scheme is to ensure

that there is a region of overlap between both experimental sets (i.e., around the midpoint) so that consistency can be simply assessed.

19. In general, the best way to perform the (un)folding stopped-flow experiments is by mixing 1 part of the protein solution with 10 parts of the corresponding buffer. This 1:10 dilution is best suited for the typical syringes used in pneumatic stopped-flows and is easily implemented in other types of instruments as well. The 1:10 dilution implies that the minimal denaturant concentration accessible to the refolding experiments is one-eleventh of the denaturant concentration used for the protein stock solution in high denaturant concentration. Even lower denaturant concentrations can be accessed by performing pH jump experiments (if the protein can be fully unfolded by pH). Such experiments are performed preparing the protein at a pH where it is unfolded, and then mixing it with a buffer with a pH adjusted so that the final pH is equivalent to the pH for all the other experiments and at which the protein is stably folded. This experiment has to be previously calibrated on a pH meter and in a steady-state instrument to determine that the mixing produces the desired pH.
20. More information can be obtained by analyzing the total amplitude of the signal decay. But, in order to compare data from different denaturant concentrations, it is essential to maintain the photomultiplier voltage at a fixed value throughout the experiments. The appropriate voltage should be fixed when the protein signal is maximal. If the maximal signal occurs at the highest denaturant concentration, the chevron plot experiment has to be acquired from the highest to the lowest concentration, starting with the unfolding limb. If it occurs at the lowest denaturant concentration, the experiment has to be performed from the lowest to the highest concentration, starting with the refolding limb.
21. In some instances, the curves measured in refolding experiments cannot be well fitted to a single exponential equation, indicating that there are additional processes taking place. This could be an indication of the transient accumulation of a folding intermediate. However, it is also frequent for two-state folding proteins that have proline residues to exhibit multiple minor kinetic phases in refolding experiments due to the heterogeneity in the unfolded state caused by proline isomerization. Proline isomerization takes place in fractions of a second (cis to trans) to tens of seconds (trans to cis), and thus is typically slower than refolding reactions. Therefore, even two-state folding proteins may exhibit several, slower kinetic phases that represent the populations of unfolded

protein molecules in which the prolines have a non-native conformation. For prolines that are in trans in the native state, these extra phases usually have small amplitudes (the population of a proline in cis in the unfolded state is typically <10%). However, for prolines that are natively in cis, the phase with the largest amplitude will be very slow (tens of seconds to minutes) and limited by the trans to cis isomerization of the proline rather than by folding. Because these isomerization processes are characteristically slow, the contribution of proline isomerization phases to refolding stopped-flow kinetic traces longer than 1 s can be approximated as a linear contribution [62]. In such case, the researcher should substitute the equation to fit the trace from Eq. 7 to the following one:

$$F = F_{\infty} + m_{\text{pro}}t + Ae^{-k_{\text{obs}}t}$$

22. In some cases, chevron plots show nonlinear slopes, also named rollovers, that are most evident at the extreme values of denaturant. This phenomenon is more commonly observed in the refolding limb, but it has also been observed for some proteins in the unfolding limb. In such case, the chevron plot cannot be fit with the standard equations. There are several possible sources for this behavior, including the accumulation of on-pathway intermediates [63], relaxation rates that are too close to the effective dead time of the instrument, the shift of the transition state along the reaction coordinate induced by denaturant (Hammond or anti-Hammond effects) [27], the presence of pre-aggregates material in the stock protein solution, an ultrafast folding process in which the protein does not cross an effective free energy barrier (downhill folding) [64] and the effect of ionic strength on the kinetics (Debye-Hückel effect) [65]. Because these rollover effects are more common in refolding conditions, at which the partially folded protein is more likely to transiently aggregate, some groups have proposed to discard the nonlinear segment of the chevron refolding limb (lowest concentrations) to fit the data and estimate the kinetic parameters expected in the absence of such effects. However, a word of caution is required here because such practice can lead to serious overestimates of the folding rate and/or underestimates of the unfolding rate in native conditions. These effects have to be taken into consideration to calculate phi-values as well.
23. The design of the sets of mutations that can be used to probe each single protein position using a Lefler analysis is not necessarily limited by the mutational rules presented in this chapter. This is because in this type of analysis, an average value is obtained from the slope of all the data such that specific mutation effects are effectively not considered. Therefore, the

researcher can in principle use any of the other 19 amino acids and then correlate the perturbations obtained for the whole mutant set (Fig. 6). In this case, the best option is to aim to produce both stabilizing and destabilizing mutations so the range of changes in stability is as large as possible.

References

1. Matouschek A, Kellis JT Jr, Serrano L, Fersht AR (1989) Mapping the transition state and pathway of protein folding by protein engineering. Nature 340(6229):122–126
2. Onuchic JN, Wolynes PG (2004) Theory of protein folding. Curr Opin Struct Biol 14 (1):70–75
3. Campos LA, Sadqi M, Liu J, Wang X, English DS, Muñoz V (2013) Gradual disordering of the native state on a slow two-state folding protein monitored by single-molecule fluorescence spectroscopy and NMR. J Phys Chem B 117(42):13120–13131
4. Bryngelson JD, Onuchic JN, Socci ND, Wolynes PG (1995) Funnels, pathways, and the energy landscape of protein folding: a synthesis. Proteins 21(3):167–195
5. Muñoz V, Campos LA, Sadqi M (2016) Limited cooperativity in protein folding. Curr Opin Struct Biol 36:58–66
6. Matthews CR (1987) Effect of point mutations on the folding of globular proteins. Methods Enzymol 154:498–511
7. Jackson SE, Fersht AR (1991) Folding of chymotrypsin inhibitor 2. 1. Evidence for a two-state transition. Biochemistry 30 (43):10428–10435
8. Naganathan AN, Muñoz V (2010) Insights into protein folding mechanisms from large scale analysis of mutational effects. Proc Natl Acad Sci U S A 107(19):8611–8616
9. Goldenberg DP, Frieden RW, Haack JA, Morrison TB (1989) Mutational analysis of a protein-folding pathway. Nature 338 (6211):127–132
10. Serrano L, Matouschek A, Fersht AR (1992) The folding of an enzyme. III. Structure of the transition state for unfolding of barnase analysed by a protein engineering procedure. J Mol Biol 224(3):805–818
11. Akmal A, Muñoz V (2004) The nature of the free energy barriers to two-state folding. Proteins 57(1):142–152
12. Sanchez IE, Kiefhaber T (2003) Origin of unusual phi-values in protein folding: evidence against specific nucleation sites. J Mol Biol 334 (5):1077–1085
13. Li L, Mirny LA, Shakhnovich EI (2000) Kinetics, thermodynamics and evolution of non-native interactions in a protein folding nucleus. Nat Struct Biol 7(4):336–342
14. Viguera AR, Vega C, Serrano L (2002) Unspecific hydrophobic stabilization of folding transition states. Proc Natl Acad Sci U S A 99 (8):5349–5354
15. Ventura S, Vega MC, Lacroix E, Angrand I, Spagnolo L, Serrano L (2002) Conformational strain in the hydrophobic core and its implications for protein folding and design. Nat Struct Biol 9(6):485–493
16. Guex N, Peitsch MC (1997) SWISS-MODEL and the Swiss-PdbViewer: an environment for comparative protein modeling. Electrophoresis 18(15):2714–2723
17. DeLano WL (2002) Pymol: an open-source molecular graphics tool. CCP4 Newslet Protein Crystallograph 40:82–92
18. Raleigh DP, Plaxco KW (2005) The protein folding transition state: what are Phi-values really telling us? Protein Pept Lett 12 (2):117–122
19. Itzhaki LS, Otzen DE, Fersht AR (1995) The structure of the transition state for folding of chymotrypsin inhibitor 2 analysed by protein engineering methods: evidence for a nucleation-condensation mechanism for protein folding. J Mol Biol 254(2):260–288
20. Burton RE, Huang GS, Daugherty MA, Calderone TL, Oas TG (1997) The energy landscape of a fast-folding protein mapped by Ala-->Gly substitutions. Nat Struct Biol 4 (4):305–310
21. Villegas V, Martinez JC, Aviles FX, Serrano L (1998) Structure of the transition state in the folding process of human procarboxypeptidase A2 activation domain. J Mol Biol 283 (5):1027–1036
22. Chiti F, Taddei N, White PM, Bucciantini M, Magherini F, Stefani M, Dobson CM (1999) Mutational analysis of acylphosphatase suggests the importance of topology and contact order in protein folding. Nat Struct Biol 6 (11):1005–1009

23. Fulton KF, Main ER, Daggett V, Jackson SE (1999) Mapping the interactions present in the transition state for unfolding/folding of FKBP12. J Mol Biol 291(2):445–461
24. Kragelund BB, Osmark P, Neergaard TB, Schiodt J, Kristiansen K, Knudsen J, Poulsen FM (1999) The formation of a native-like structure containing eight conserved hydrophobic residues is rate limiting in two-state protein folding of ACBP. Nat Struct Biol 6 (6):594–601
25. Martinez JC, Serrano L (1999) The folding transition state between SH3 domains is conformationally restricted and evolutionarily conserved. Nat Struct Biol 6(11):1010–1016
26. Riddle DS, Grantcharova VP, Santiago JV, Alm E, Ruczinski I, Baker D (1999) Experiment and theory highlight role of native state topology in SH3 folding. Nat Struct Biol 6 (11):1016–1024
27. Ternstrom T, Mayor U, Akke M, Oliveberg M (1999) From snapshot to movie: phi analysis of protein folding transition states taken one step further. Proc Natl Acad Sci U S A 96 (26):14854–14859
28. Choe SE, Li L, Matsudaira PT, Wagner G, Shakhnovich EI (2000) Differential stabilization of two hydrophobic cores in the transition state of the villin 14T folding reaction. J Mol Biol 304(1):99–115
29. Hamill SJ, Steward A, Clarke J (2000) The folding of an immunoglobulin-like Greek key protein is defined by a common-core nucleus and regions constrained by topology. J Mol Biol 297(1):165–178
30. Guerois R, Serrano L (2000) The SH3-fold family: experimental evidence and prediction of variations in the folding pathways. J Mol Biol 304(5):967–982
31. Kim DE, Fisher C, Baker D (2000) A breakdown of symmetry in the folding transition state of protein L. J Mol Biol 298(5):971–984
32. McCallister EL, Alm E, Baker D (2000) Critical role of beta-hairpin formation in protein G folding. Nat Struct Biol 7(8):669–673
33. Rodriguez HM, Vu DM, Gregoret LM (2000) Role of a solvent-exposed aromatic cluster in the folding of Escherichia coli CspA. Protein Sci 9(10):1993–2000
34. Chu R, Pei W, Takei J, Bai Y (2002) Relationship between the native-state hydrogen exchange and folding pathways of a four-helix bundle protein. Biochemistry 41 (25):7998–8003
35. Northey JG, Di Nardo AA, Davidson AR (2002) Hydrophobic core packing in the SH3 domain folding transition state. Nat Struct Biol 9(2):126–130
36. Otzen DE, Oliveberg M (2002) Conformational plasticity in folding of the split beta-alpha-beta protein S6: evidence for burst-phase disruption of the native state. J Mol Biol 317(4):613–627
37. Friel CT, Capaldi AP, Radford SE (2003) Structural analysis of the rate-limiting transition states in the folding of Im7 and Im9: similarities and differences in the folding of homologous proteins. J Mol Biol 326 (1):293–305
38. Gianni S, Guydosh NR, Khan F, Caldas TD, Mayor U, White GW, DeMarco ML, Daggett V, Fersht AR (2003) Unifying features in protein-folding mechanisms. Proc Natl Acad Sci U S A 100(23):13286–13291
39. Garcia-Mira MM, Boehringer D, Schmid FX (2004) The folding transition state of the cold shock protein is strongly polarized. J Mol Biol 339(3):555–569
40. Hedberg L, Oliveberg M (2004) Scattered Hammond plots reveal second level of site-specific information in protein folding: phi' (beta++). Proc Natl Acad Sci U S A 101 (20):7606–7611
41. Anil B, Sato S, Cho JH, Raleigh DP (2005) Fine structure analysis of a protein folding transition state; distinguishing between hydrophobic stabilization and specific packing. J Mol Biol 354(3):693–705
42. Went HM, Jackson SE (2005) Ubiquitin folds through a highly polarized transition state. Protein Eng Des Sel 18(5):229–237
43. Wilson CJ, Wittung-Stafshede P (2005) Role of structural determinants in folding of the sandwich-like protein Pseudomonas aeruginosa azurin. Proc Natl Acad Sci U S A 102 (11):3984–3987
44. Teilum K, Thormann T, Caterer NR, Poulsen HI, Jensen PH, Knudsen J, Kragelund BB, Poulsen FM (2005) Different secondary structure elements as scaffolds for protein folding transition states of two homologous four-helix bundles. Proteins 59(1):80–90
45. Bueno M, Ayuso-Tejedor S, Sancho J (2006) Do proteins with similar folds have similar transition state structures? A diffuse transition state of the 169 residue apoflavodoxin. J Mol Biol 359(3):813–824
46. Petrovich M, Jonsson AL, Ferguson N, Daggett V, Fersht AR (2006) Phi-analysis at the experimental limits: mechanism of beta-hairpin formation. J Mol Biol 360(4):865–881
47. Ferguson N, Sharpe TD, Johnson CM, Fersht AR (2006) The transition state for folding of a

peripheral subunit-binding domain contains robust and ionic-strength dependent characteristics. J Mol Biol 356(5):1237–1247

48. Li Y, Gupta R, Cho JH, Raleigh DP (2007) Mutational analysis of the folding transition state of the C-terminal domain of ribosomal protein L9: a protein with an unusual beta-sheet topology. Biochemistry 46 (4):1013–1021
49. Sato S, Fersht AR (2007) Searching for multiple folding pathways of a nearly symmetrical protein: temperature dependent phi-value analysis of the B domain of protein A. J Mol Biol 372(1):254–267
50. Campbell-Valois FX, Michnick SW (2007) The transition state of the ras binding domain of Raf is structurally polarized based on Phi-values but is energetically diffuse. J Mol Biol 365 (5):1559–1577
51. Sharpe TD, Ferguson N, Johnson CM, Fersht AR (2008) Conservation of transition state structure in fast folding peripheral subunit-binding domains. J Mol Biol 383(1):224–237
52. Campos LA, Bueno M, Lopez-Llano J, Jimenez MA, Sancho J (2004) Structure of stable protein folding intermediates by equilibrium phi-analysis: the apoflavodoxin thermal intermediate. J Mol Biol 344(1):239–255
53. Krantz BA, Sosnick TR (2001) Engineered metal binding sites map the heterogeneous folding landscape of a coiled coil. Nat Struct Biol 8(12):1042–1047
54. Sosnick TR, Krantz BA, Dothager RS, Baxa M (2006) Characterizing the protein folding transition state using psi analysis. Chem Rev 106(5):1862–1876
55. Fersht AR, Sato S (2004) Phi-value analysis and the nature of protein-folding transition states. Proc Natl Acad Sci U S A 101(21):7976–7981
56. Nolting B, Andert K (2000) Mechanism of protein folding. Proteins 41(3):288–298
57. de los Rios MA, Muralidhara BK, Wildes D, Sosnick TR, Marqusee S, Wittung-Stafshede P, Plaxco KW, Ruczinski I (2006) On the precision of experimentally determined protein folding rates and phi-values. Protein Sci 15 (3):553–563
58. Mok YK, Elisseeva EL, Davidson AR, Forman-Kay JD (2001) Dramatic stabilization of an SH3 domain by a single substitution: roles of the folded and unfolded states. J Mol Biol 307 (3):913–928
59. Northey JG, Maxwell KL, Davidson AR (2002) Protein folding kinetics beyond the phi value: using multiple amino acid substitutions to investigate the structure of the SH3 domain folding transition state. J Mol Biol 320 (2):389–402
60. Lawrence C, Kuge J, Ahmad K, Plaxco KW (2010) Investigation of an anomalously accelerating substitution in the folding of a prototypical two-state protein. J Mol Biol 403 (3):446–458
61. Muñoz V, Cerminara M (2016) When fast is better: protein folding fundamentals and mechanisms from ultrafast approaches. Biochem J 473(17):2545–2559
62. Jackson SE, Fersht AR (1991) Folding of chymotrypsin inhibitor 2. 2. Influence of proline isomerization on the folding kinetics and thermodynamic characterization of the transition state of folding. Biochemistry 30 (43):10436–10443
63. Sanchez IE, Kiefhaber T (2003) Evidence for sequential barriers and obligatory intermediates in apparent two-state protein folding. J Mol Biol 325(2):367–376
64. Kaya H, Chan HS (2003) Origins of chevron rollovers in non-two-state protein folding kinetics. Phys Rev Lett 90(25 Pt 1):258104
65. de Los Rios MA, Plaxco KW (2005) Apparent Debye-Huckel electrostatic effects in the folding of a simple, single domain protein. Biochemistry 44(4):1243–1250

Chapter 2

Engineered Metal-Binding Sites to Probe Protein Folding Transition States: Psi Analysis

Michael C. Baxa and Tobin R. Sosnick

Abstract

The formation of the transition state ensemble (TSE) represents the rate-limiting step in protein folding. The TSE is the least populated state on the pathway, and its characterization remains a challenge. Properties of the TSE can be inferred from the effects on folding and unfolding rates for various perturbations. A difficulty remains on how to translate these kinetic effects to structural properties of the TSE. Several factors can obscure the translation of point mutations in the frequently used method, "mutational Phi analysis." We take a complementary approach in "Psi analysis," employing rationally inserted metal binding sites designed to probe pairwise contacts in the TSE. These contacts can be confidently identified and used to construct structural models of the TSE. The method has been applied to multiple proteins and consistently produces a considerably more structured and native-like TSE than Phi analysis. This difference has significant implications to our understanding of protein folding mechanisms. Here we describe the application of the method and discuss how it can be used to study other conformational transitions such as binding.

Key words Heterogeneity, Contact order

1 Introduction

The characterization and prediction of protein folding pathways remains a challenge despite numerous advances in experimental and computational methodologies. Many questions persist including the order of events, the existence of preferred or highly heterogeneous pathways, and the general properties of species along the folding pathway [1]. Many single domain proteins fold in a kinetically two-state manner (U↔N) without measurably populating intermediate states [2, 3]. For these proteins, the transition state ensemble (TSE) remains the only entity that is readily accessible experimentally.

The traditional method of probing the transition state, ϕ analysis, utilizes site-directed mutagenesis to introduce point mutants and measures the effect on the folding and unfolding rates [4]. Despite the ready application of this procedure, translating

Victor Muñoz (ed.), *Protein Folding: Methods and Protocols*, Methods in Molecular Biology, vol. 2376, https://doi.org/10.1007/978-1-0716-1716-8_2, © Springer Science+Business Media, LLC, part of Springer Nature 2022

the kinetic response to transition state structure is non-trivial [5–12]. Often the magnitude of the effects is low to moderate ($\phi \leq 0.3$), leading to ambiguity in the interpretation [5, 7, 11, 13–19]. The low values could be due to structural relaxation around the side chain, which minimizes the energetic effect of the perturbation on the folding rate, k_f, or the lack of interactions involving the mutated side chain itself even when the residue backbone is ordered and hydrogen bonded [5–12].

We have developed ψ analysis in part to avoid many of the complicating issues with ϕ, so as to more accurately characterize TSEs [16, 20]. In ψ analysis, one introduces individual bi-Histidine (biHis) metal binding sites on the surface of the protein. The addition of metal ions stabilizes conformations where the biHis site is binding competent, which produces changes in k_f and stability, ΔG_{eq}. Since the stability is perturbed in an isosteric and isochemical manner, the resulting series of ion concentration-dependent data can be justifiably combined and used to identify whether a specific pair-wise interaction is present in the TSE. The ion-induced changes in the folding rate and stability are used to calculate ψ. In analogy to ϕ, ψ is the instantaneous change in the activation free energy for folding, $\Delta\Delta G_f$, relative to the change in stability $\Delta\Delta G_{eq}$, i.e.,

$$\phi = \left.\frac{\Delta\Delta G_f}{\Delta\Delta G_{eq}}\right|_{\text{mutation}};\psi = \left.\frac{\partial\Delta\Delta G_f}{\partial\Delta\Delta G_{eq}}\right|_{\Delta\Delta G_{eq}}. \tag{1}$$

Potential issues related to the addition of metal ions are alleviated by evaluating ψ in the limit of zero perturbation,

$$\psi_0 = \left.\frac{\partial\Delta\Delta G_f}{\partial\Delta\Delta G_{eq}}\right|_{\Delta\Delta G_{eq}=0}. \tag{2}$$

As defined, ψ_0 reflects the intrinsic degree of contact formation in the TSE in the absence of metal ions. This ability to extrapolate to zero ion concentration addresses a potential misconception that metal binding induces structure in the TSE and, therefore, biases the outcome (a valid objection when one introduces *covalent* cross-links [21, 22]). Rather than biasing the outcome, ψ analysis facilitates the characterization of a specific interaction in the TSE by stabilizing conformations having the two histidines positioned to bind a metal ion, a reaction that typically is in fast equilibrium relative to the folding reaction [23]. In contrast, ϕ values are determined using mutations that can affect multiple interactions (e.g., native- and non-native contacts, altered Ramachandran maps), and therefore are not always translatable to specific structure formation.

Generally, ψ_0 values of zero or one indicate that the biHis site has the same ion binding affinity in the TSE as it has in the unfolded or native site, respectively. These two limits are interpreted as the biHis site being absent or native-like in the TSE, respectively. A fractional ψ_0 value indicates that the biHis site either is native-like in a subpopulation of the TSE or contains non-native binding affinity (e.g., a distorted site with less favorable binding geometry or a flexible site that must be restricted prior to ion binding), or some combination thereof [5, 11, 16, 24]. The detailed treatment of ψ in these situations is presented below. In spite of these issues, a specific pair of residues is being probed in ψ analysis, and hence, a kinetic response with metal indicates that the pair is interacting in some manner. Accordingly, the method is well-suited for identifying contacts that help define the topology and structure of the TSE.

We have applied ψ to seven globular proteins, including acyl phosphatase [25–27], ubiquitin [5, 17, 28], the B domain of Protein A (BdpA) [18, 29], Protein L [15], Protein G and NuG2b [30–34], and λ repressor [35, 36] (Fig. 1), and observed a consistent theme of an extensive TSE structure (Fig. 1). Specifically, the seven TSEs share a common and high degree of native topology, defined according to relative contact order: $RCO^{TSE} \approx 0.7 \cdot RCO^{N}$. This finding rationalizes the well-known correlation between k_f and RCO [37], which provides additional support for the use of ψ in identifying folding principles. Also, the TSE's high level of native topology implies that only a limited amount of heterogeneity is possible at the rate limiting step in folding for these proteins.

In contrast, the TSE deduced from ϕ analysis often barely defines a protein's fold, and the ensuing RCO level of the TSE is variable for different proteins (Fig. 1). We suspect underreporting also occurs with other proteins, particularly those characterized as having a polarized TSE, such as cold shock protein [38] or src SH3 [39]. Besides energy minimization through structural relaxation in the TSE, an additional situation where ϕ analysis can underreport structure occurs when a residue's side chain is buried in the native state but not in the TSE due to a portion of the protein being unfolded. This situation applies to residues on the hydrophobic face of the four β strands in Protein G & L, as the helix is absent in their TSEs [14]. As a result, a substitution on a strand can yield a smaller energy signature in the TSE than in the native state. Consequently, a small ϕ is observed even for residues participating in the sheet, leading to erroneous inferences about the degree of sheet structure in the TSE of β sheet containing proteins. Also, ϕ can underreport the structural content if the TSE involves non-native features [6, 40–42].

The implementation of ψ analysis using biHis sites comes with some issues. Fractional ψ_0 values raise the same interpretational issues as fractional ϕ values, including the possibility that they arise from either TS heterogeneity or partial structure formation [5], as noted above. Nevertheless (and significantly), the conclusion that the ψ-determined TSE has near-native topology emerges

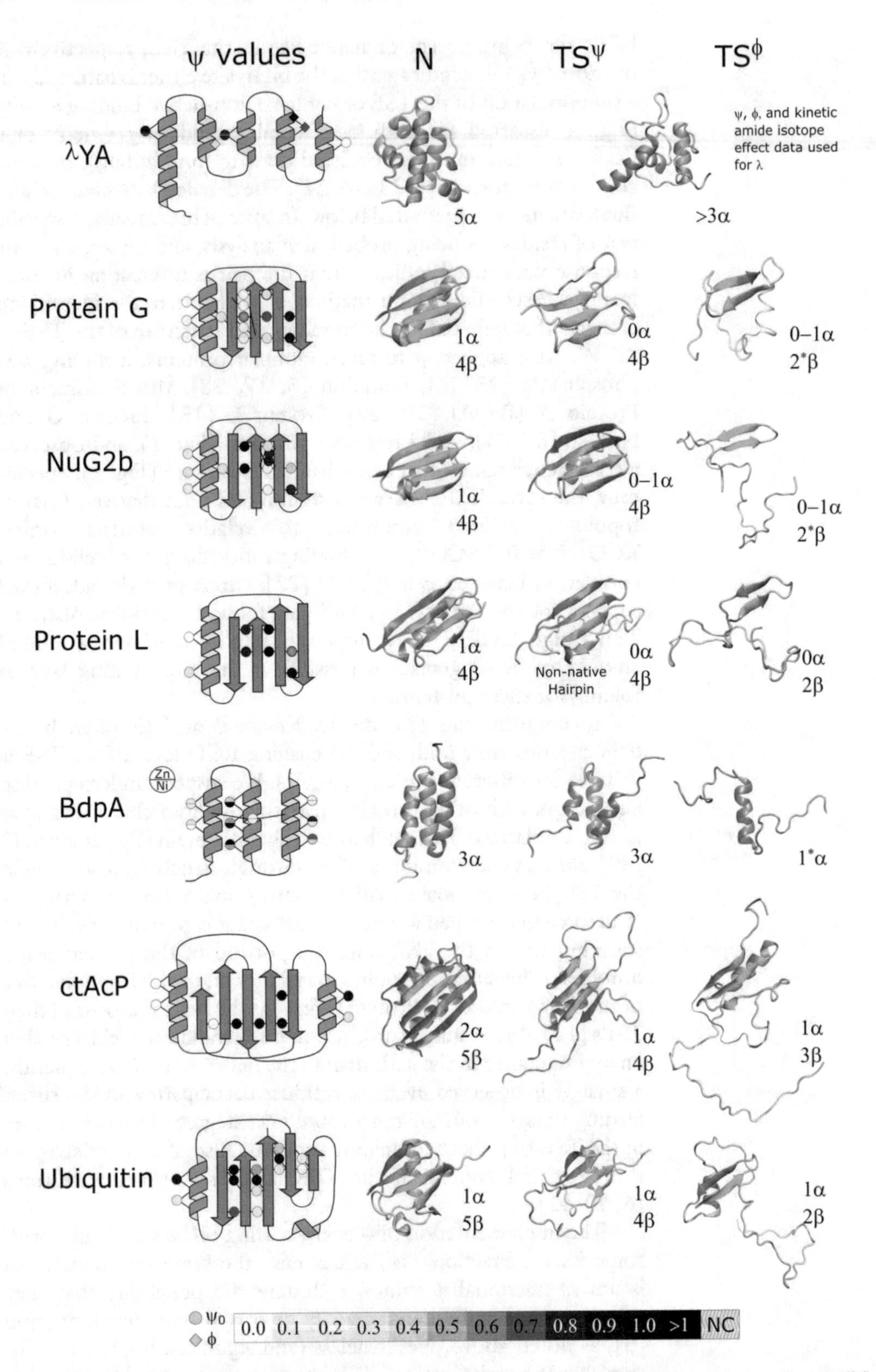

Fig. 1 TSEs of several small proteins characterized by ψ analysis. The TSEs of λ-repressor, Protein G/L/NuG2b, BdpA, Ubiquitin, and ctAcP have been characterized from ψ values and exhibit significant more

for α/β proteins even when only the sites with near-unity ψ_0 values are considered. Furthermore, for the seven globular proteins studied, each has at least one, and often more, ψ_0 values of unity. This finding implies that at least one site is formed in the entire TSE.

Another factor to consider is the change in stability for the substitution, whether it be a biHis site or a point mutant. While having a highly destabilizing mutation may permit an accurate determination of ϕ (e.g., $\Delta\Delta G_{\text{mutation}}$ ~ 2 kcal/mol [7, 43]), this level of perturbation could distort the structural content of the TSE. The introduction of a model biHis site should ideally produce no change in stability, but this is not always the case. Regardless, the addition of cations usually stabilizes the protein by at least 1 kcal/mol, which mitigates the potential destabilization resulting from the introduction of the biHis site. Also, the ability to measure a plethora of rates across multiple cation concentrations permits the accurate determination of ψ even when $\Delta\Delta G_{\text{bind}} < 1$ kcal/mol.

Although biHis sites are limited to surface positions, one can combine ψ with ϕ measured at core positions so as to produce a more complete description of the TSE [5, 17, 18]. Also, the use of kinetic amide isotope effects can probe the helical hydrogen bond content in the TSE to further characterize the TSE [18, 44, 45]. Overall, the experimentalist should use whatever tools are available to generate a complete picture as possible of the ephemeral TSE.

2 Materials

The application of ψ analysis requires measuring kinetic and equilibrium folding behavior in both the presence and the absence of divalent metal. Since ψ requires using biHis sites that bind metal, the histidines should be deprotonated using a buffer pH above their pK_a to maximize metal ion binding (intrinsic pK_a is ~6.5) (*see* **Note 1**).

One can employ nickel, zinc, or cobalt ions. In our experience, zinc and nickel are the most stabilizing and well behaved. They also have different electron orbital geometries, which allows for evaluating heterogeneity versus affinity differences in the TSE (Subheading 3.5.2).

1. *Zinc chloride* ($ZnCl_2$) readily precipitates at high concentration and neutral pH, so buffers should be prepared with care. For

Fig. 1 (continued) structure than what has been determined from mutational ϕ analysis. Model TS structures except for λ_{YA} were generated using the Upside simulation package [67, 68]. The TSE model for λ_{YA} (and BdpA to a lesser extent) utilized a combination of ψ, ϕ, and kinetic amide isotope effect data [18, 35]. NC indicates ψ_0 values for which $\Delta\Delta G_{eq}\rightarrow 0$, i.e., $\psi_0\rightarrow\infty$

example, when making a zinc-containing buffer, add the buffer stock solution, dilute to just under the final volume, and add the zinc stock solution. The addition of acidic metal solution often lowers the pH, so it should be rechecked immediately before use.

2. *Nickel chloride* ($NiCl_2$) solutions can be prepared at 0.25 M in 25 mM HCl. We have found that nickel can be readily combined with buffers as this metal is not as susceptible to precipitation problems as the other metal ions.
3. *Cobalt chloride* ($CoCl_2$) solutions appear burgundy red, and stocks can be prepared at 1 M in water. Using buffers with cobalt at guanidine hydrochloride (GdmCl) concentration of ~5.5 M can cause metal precipitation and erratic kinetic results. Excessive amount of chloride ions, as present in high-concentration GdmCl buffers, will cause cobalt solutions to appear blue, which may affect results.

Other divalent ions that we have not necessarily evaluated include copper and magnesium.

Buffer solutions: All solutions are prepared using ultrapure water and high purity reagents. With the exception of the High GdmCl stock (#1), all buffers should be filtered (0.2 μm) to remove contaminants and stored at 4 °C to prolong shelf life.

1. High (8.5 M) GdmCl stock: unbuffered (*see* **Note 2**). To make a 500 mL stock of unbuffered GdmCl, mix 406 g of guanidine hydrochloride (GdmCl, MW 95.5 g/mol) with 193.36 mL of water (*see* **Notes 3** and **4**). Verify the concentration by measuring the indices of refraction of water (which is not necessarily 1.33) and the GdmCl stock [46] (a calculator for measuring [GdmCl] and [urea] from indices of refraction can be found at sosnick.uchicago.edu/gdmcl.html).
2. Denaturant-free buffer: 50 mM HEPES (*see* **Note 5**), 100 mM NaCl (*see* **Note 6**), pH 7.5 *at the temperature of study*. Dissolve 5.9575 g of HEPES free acid (MW = 238.3 g/mol) and 2.922 g NaCl (MW = 58.44 g/mol) in 450 mL water (*see* **Note 7**). Titrate with NaOH to appropriate pH to account for temperature difference between preparation and final experimental conditions (*see* **Note 8**). Adjust volume to 500 mL using water.
3. High GdmCl buffer: 50 mM HEPES, 100 mM NaCl, 8 M GdmCl, pH 7.5 at the experimental temperature (*see* **Note 9**). Mix 188.25 mL High GdmCl stock with 6.75 mL water. Dissolve 2.383 g of HEPES free acid (MW = 238.3 g/mol) and 1.1688 g NaCl (MW = 58.44 g/mol). Adjust pH to match the denaturant-free buffer. Adjust volume to 200 mL using water.

4. Prepare buffers with intermediate GdmCl concentrations by counter-mixing the 0 M and the high molar GdmCl buffers. Confirm pH of counter-mixed buffers. Verify all GdmCl concentrations by measuring each index of refraction.
5. Zn^{2+} stock solution: 250 mM Zn^{2+}, 25 mM HCl. Slowly add 41 μL of 12.1 N HCl to 5 mL water and adjust volume to 19 mL with water. Dissolve 0.6815 g of $ZnCl_2$ (MW = 136.29 g/mol) in the solution and adjust volume to 20 mL with water. The metal solutions can be stored at room temperature but should be remade every 1–2 months.
6. Ni^{2+} stock solution: 250 mM Ni^{2+}, 25 mM HCl. Slowly add 41 μL of 12.1 N HCl to 5 mL water and adjust volume to 19 mL with water. Dissolve 0.648 g of $NiCl_2$ (MW = 129.6 g/mol) in the solution, which will turn green once the Ni is dissolved. Adjust volume to 20 mL using water.
7. Metal standard buffer: 1 mM Zn^{2+}/Ni^{2+}. Add 0.4 mL of metal stock solution to 100 mL buffer to make 1 mM metal solution. Double-check pH after adding the metal solution.
8. Cross-linking buffer #1: 200 mM triethanolamine, pH 5.5. Dissolve 2.9838 g of triethanolamine (MW = 149.19 g/mol) in water and make up to 90 mL. Adjust pH to 5.5 and then make solution up to final volume of 100 mL.
9. Cross-linking buffer #2: 250 mM sodium borate, pH 8.3 chilled on ice. Dissolve 9.535 g sodium borate decahydrate ($Na_2B_4O_7 + 10H_2O$; MW = 381.4 g/mol) in water and adjust to 90 mL. Titrate pH to 8.3 and then make solution up to a final volume of 100 mL using water.
10. Cross-linking agent: 100 mM 1,3-Dichloroacetone (DCA) in dimethylformamide (DMF). Dissolve 0.1270 g (MW = 126.97 g/mol) in DMF and adjust volume to 10 mL.
11. Quenching buffer: 2-Mercaptoethanol.

3 Methods and Analysis

3.1 Derivation of ψ

3.1.1 Models of ψ

We present a general scheme for the psi value according to the scheme laid out in Figs. 2 and 3. The effect of metal binding on the equilibrium stability can be described as a linked equilibrium expression [47], i.e.,

$$\Delta\Delta G_{eq}\left(\left[Me^{2+}\right]\right) = RT\ln\frac{1+\left[Me^{2+}\right]/K_N}{1+\left[Me^{2+}\right]/K_U} \quad (3)$$

where K_N and K_U are the metal dissociation constants of the native and unfolded states, respectively (*see* **Note 10**). The folding scheme in Fig. 3 assumes that the TSE can be decomposed into two main

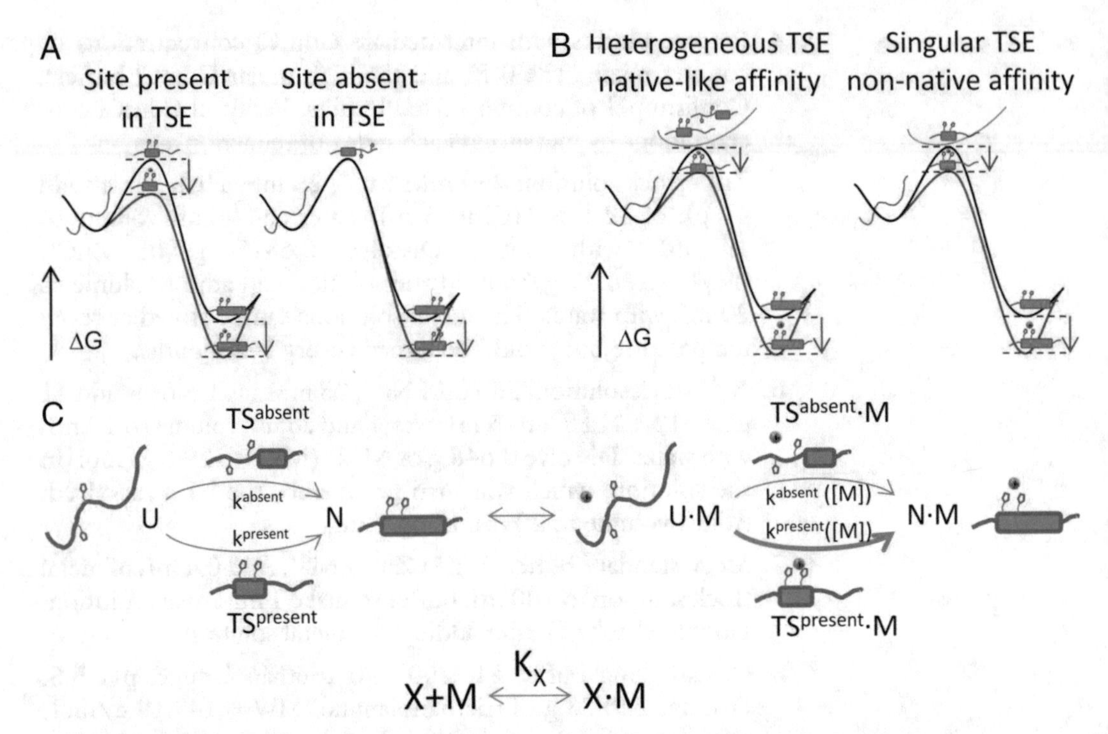

Fig. 2 Thermodynamic states considered in ψ analysis. (**a**) A metal binding biHis site can bind metal at every point along a free energy reaction surface, even in the TSE. In the surface on the left, the biHis site is present in the TSE, so both the TSE and N state are stabilized relative to *U*. In the case on the right, the biHis site is not present, and therefore, only the Native state is stabilized. (**b**) There are two extreme limits in interpreting a fractional ψ_0 value, where the TSE is stabilized to a different extent than the native state. Either ψ_0 reflects the population fraction of the TSE having the site formed with native-like binding affinity (left) or the site is binding metal in a non-native like manner (right). (**c**) The general reaction scheme in ψ analysis assumes a partitioned TSE according to the biHis site being present or absent. All states are allowed to have their respective binding affinities, as in Eq. 4, which affects the relative flux through $TS^{present}$ and TS^{absent}

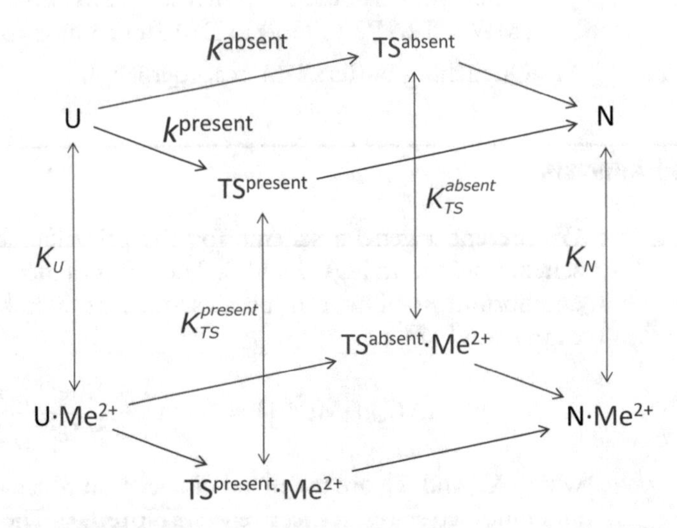

Fig. 3 General reaction scheme considered in ψ analysis. The TSE is bifurcated according to a site's ability to bind metal ($TS^{present}$) or not bind metal (TS^{absent}). The *U*, $TS^{present}$, TS^{absent}, and *N* states have metal binding affinity K_U, $K_{TS}^{present}$, K_{TS}^{absent}, K_N, respectively

classes, a binding competent $TS^{present}$ and non-competent TS^{absent}, which have metal dissociations constants $K_{TS}^{present}$ and K_{TS}^{absent}, respectively. As per Eyring Reaction Rate Theory [48], the overall reaction rate is taken to be proportional to the relative populations of the TS and U ensembles, $k_f \propto [TS]/[U]$. The net folding rate is the sum of the rates going down each of the two routes, $k_f = k^{present} + k^{absent}$. In terms of the TS and U ensembles,

$$\frac{[TS_{total}]}{[U_{total}]} = \frac{[TS_{total}^{present}]}{[U_{total}]} + \frac{[TS_{total}^{absent}]}{[U_{total}]}$$
$$= \frac{[TS^{present}](1 + [Me^{2+}]/K_{TS}^{present})}{[U](1 + [Me^{2+}]/K_U)}$$
$$+ \frac{[TS^{absent}](1 + [Me^{2+}]/K_{TS}^{absent})}{[U](1 + [Me^{2+}]/K_U)}$$

$$k_f = k_0^{present} \frac{1 + [Me^{2+}]/K_{TS}^{present}}{1 + [Me^{2+}]/K_U} + k_0^{absent} \frac{1 + [Me^{2+}]/K_{TS}^{absent}}{1 + [Me^{2+}]/K_U} \quad (4)$$

where $k_0^{present} \propto [TS^{present}]/[U]$ and $k_0^{absent} \propto [TS^{absent}]/[U]$ are the rates through each TS class prior to the addition of metal.

Given the folding rate in the absence of metal, $k_f^0 = k_0^{present} + k_0^{absent}$, and solving Eq. 3 for $[Me^{2+}]$, we can change Eq. 4 to relate $\Delta\Delta G_f$ to $\Delta\Delta G_{eq}$, i.e.,

$$\Delta\Delta G_f = RT\, ln\left(\psi_0 e^{\Delta\Delta G_{eq}/RT} + 1 - \psi_0\right) \quad (5)$$

where ψ_0 is a single parameter that relates $\Delta\Delta G_f$ to $\Delta\Delta G_{eq}$, such that

$$\psi_0 = \frac{K_N}{K_N - K_U}\left(1 - \frac{K_U}{K_{TS}^{absent}}\right) - \frac{K_N K_U}{K_N - K_U}$$
$$\times \left(\frac{1}{K_{TS}^{present}} - \frac{1}{K_{TS}^{absent}}\right) \frac{k_0^{present}}{k_0^{present} + k_0^{absent}}. \quad (6)$$

The curvature between $\Delta\Delta G_f$ and $\Delta\Delta G_{eq}$ is described more generally by ψ (Fig. 4, **Note 11**), i.e.,

$$\psi = \frac{\partial \Delta\Delta G_f}{\partial \Delta\Delta G_{eq}} = \frac{\psi_0 e^{\Delta\Delta G_{eq}/RT}}{\psi_0 e^{\Delta\Delta G_{eq}/RT} + 1 - \psi_0}$$
$$= \frac{\psi_0}{\psi_0 + (1 - \psi_0) e^{-\Delta\Delta G_{eq}/RT}}. \quad (7)$$

ψ_0 is also the curvature of $\Delta\Delta G_f$ in the limit $\Delta\Delta G_{eq} \rightarrow 0$, i.e., $\psi(\Delta\Delta G_{eq} \rightarrow 0) = \psi_0$ (Eq. 2).

It is important to stress that according to Eq. 6, ψ_0 does not depend on $[Me^{2+}]$ and is evaluated in the limit of *no perturbation*.

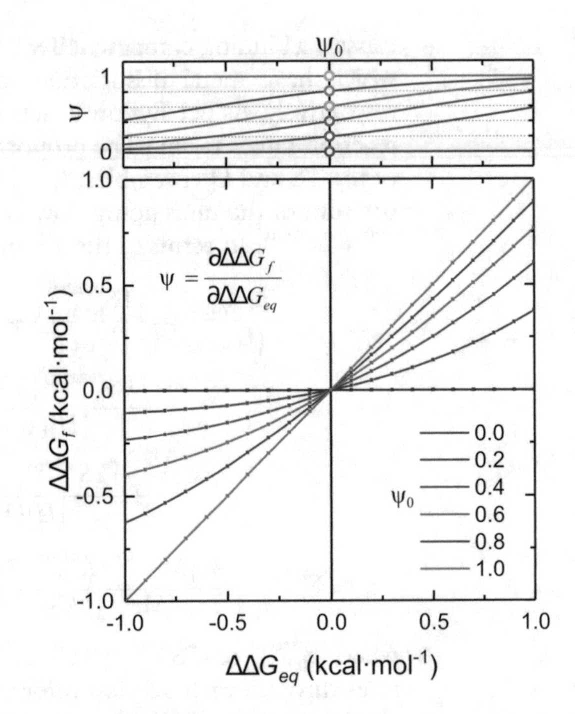

Fig. 4 Relationship of ψ and ψ_0 to Leffler Plot. The change in activation energy $\Delta\Delta G_f$ is nonlinearly related to the change in stability $\Delta\Delta G_{eq}$ through the parameter ψ_0, the instantaneous slope at $\Delta\Delta G_{eq} = 0$. The slope ψ is plotted above the Leffler plot of $\Delta\Delta G_f$ vs $\Delta\Delta G_{eq}$ showing that with sufficiently large values $\Delta\Delta G_{eq}$, $\psi \rightarrow 1$

Rather, ψ_0 is a function of the extent to which the biHis site is formed in the TSE, $\frac{k_0^{\text{present}}}{k_0^{\text{present}} + k_0^{\text{absent}}}$, and the metal dissociation constants of the different states on the folding pathway. In the ideal heterogeneous limit, where $K_{\text{TS}}^{\text{present}} = K_{\text{N}}$ and $K_{\text{TS}}^{\text{absent}} = K_U$,

$$\psi_0 = \frac{k_0^{\text{present}}}{k_0^{\text{present}} + k_0^{\text{absent}}} \tag{8}$$

i.e., ψ_0 reflects the fraction of the TSE with the biHis site formed. Conversely, if there is no heterogeneity in the TSE, i.e., $K_{\text{TS}}^{\text{absent}} \rightarrow \infty$ and $k_0^{\text{absent}} \rightarrow 0$, then

$$\psi_0 = \frac{K_{\text{N}}(K_{\text{TS}} - K_{\text{U}})}{K_{\text{TS}}(K_{\text{N}} - K_{\text{U}})}. \tag{9}$$

One can also consider a situation where the biHis site in TS^{present} has non-native binding affinity ($K_{\text{TS}}^{\text{present}} \neq K_{\text{N}}$), while the site in TS^{absent} has unfolded-like affinity ($K_{\text{TS}}^{\text{absent}} = K_{\text{U}}$). Now, the initial slope is the degree of heterogeneity multiplied by an additional factor representing the differential binding affinity between TS^{present} and N, i.e.,

$$\psi_0 = \frac{K_N}{K_{TS}^{present}} \frac{K_{TS}^{present} - K_U}{K_N - K_U} \frac{k_0^{present}}{k_0^{present} + k_0^{absent}}. \quad (10)$$

The curvature can be due to TS heterogeneity, non-native binding affinity in a singular TS, or a combination thereof [5, 16, 24] (D. Goldenberg, private communication; *see* also Fersht [49] for a comparison of the ψ and ϕ analysis methods using an alternative kinetic model which focuses on unfolded state population shifts while omitting any consideration of TS binding).

3.1.2 Interpretation of ψ_0 Values

When $\psi_0 \rightarrow 1$ in the first model, $k_0^{absent} \rightarrow 0$, implying that the biHis site is present in the entire TSE. In the second model, a unity psi implies that $K_{TS} = K_N$, so that the biHis site is formed in a native-like manner. In the final model, a ψ_0 of 1 implies that either the site is formed 100% with native-like affinity, or (perhaps less likely) the biHis site is formed in a fraction of the TSE, but with a tighter affinity than the native state.

In the limit that $\psi_0 \rightarrow 0$ in the first model, $k_0^{present} \rightarrow 0$, i.e., the biHis site is absent from the TSE. In the purely homogenous model, a zero ψ_0 value implies an unfolded-like binding affinity in the TSE, which implies that the biHis site is unfolded-like in the TSE. In the final model, either $k_0^{present} \rightarrow 0$ (the site is not formed in the TSE) or $K_{TS}^{present} \rightarrow K_U$ (the site is binding metal in the TSE with the same affinity as in the denatured state).

While the interpretation of ψ_0 values of 0 or 1 are independent of the models described above, there is ambiguity in the interpretation of fractional ψ values. In the purely heterogeneous model, a fractional psi value reflects the population fraction of the TSE with the biHis site formed. However, a fractional ψ_0 in a homogenous TSE implies that the biHis site binds the metal with a non-native-like affinity, and a precise structural interpretation becomes difficult.

3.2 Designing biHis Sites

1. Construct the pseudo-wild-type (psWT) sequence by removing any endogenous His residues from the WT sequence, i.e., His→Asn. Bi-histidine mutations typically are manually identified using a molecular viewing program, preferably one that allows for mutations (e.g., Swiss-PdbViewer, http://www.expasy.org/spdbv/ [50]). The sites are located along the surface of the protein either along *i*,*i* + 4 helices, across β-strands, or replacing side chain-side chain tertiary contacts such as salt bridges (*see* Fig. 5, **Notes 12** and **13**).
2. Generate mutants using either gene fragment insertions (e.g., gBlocks® from IDT) or site-directed mutagenesis (e.g., QuikChange® from Stratagene). While it is possible to introduce an *i*,*i* + 4 biHis site in a single Quikchange® step with a longer fragment, sites with larger sequence separations will require two separate Quikchange® steps. In these instances, we have

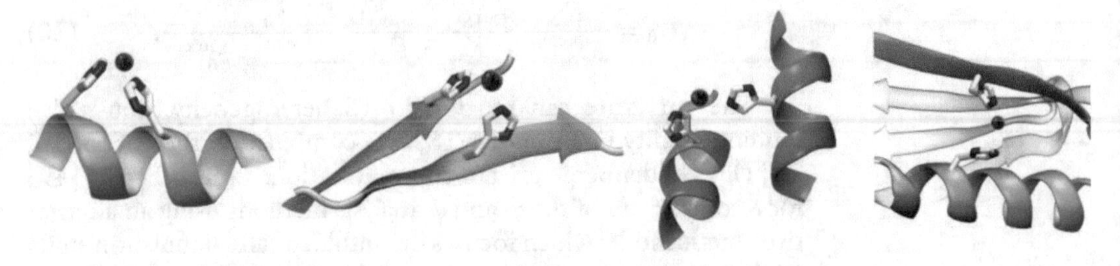

Fig. 5 Example biHis sites. In helices, the histidines should be introduced at solvent exposed *i*, *i* + *4* positions (i.e., H-X-X-X-H, *see* Subheading 3.2). Metal sites can be introduced across β hairpins or either parallel or anti-parallel β strands although site selection does not appear to be very stringent. For example, a histidine on one strand of a β sheet can form a binding-competent biHis site with residues on either adjacent strand in ubiquitin. However, β sheets often are quite twisted, and care must be taken to use two positions where the side chains are not angled away from each other. We have also successfully engineered biHis sites in sites outside of traditional secondary structural units, e.g., replacing an endogenous salt bridge [18] or side-chain–sidechain docking between strands and helices [14]. These experiences suggest that preexisting well-defined side chain interactions may also be good candidates for biHis metal sites (*see* **Notes 12** and **13**)

found that gene fragments offer a viable alternative strategy as the time to complete two QuikChange steps (PCR, mini-prep, sequence) is not too different from ordering a single gene fragment and inserting into a plasmid vector using Gibson assembly® [51].

Proteins are expressed and purified using standard methods.

3.3 Equilibrium Denaturation

3.3.1 General Protocol

For each variant do the following:

1. Make a protein stock solution so as to control for variations in protein concentrations in the individual denaturation titrations (*see* **Note 14**).
2. Measure native circular dichroism (CD) spectrum in the 0 M buffer with no metal ion (*see* **Notes 15** and **16**).
3. (Optional) Measure the CD spectrum under denaturing conditions (*see* **Note 17**).
4. Measure denaturation titration in the absence of metal (*see* **Notes 18** and **19**).
5. Assuming that the equilibrium reaction is two-state, U ↔ N, fit the data with the following:

$$S([den]) = \frac{S_{\mathrm{U}} + S_{\mathrm{N}} e^{-(\Delta G + m_0[den])/\mathrm{RT}}}{1 + e^{-(\Delta G + m_0[den])/\mathrm{RT}}} \tag{11}$$

where S_{U} and S_{N} are the baseline signals of U and N and may be linear functions of denaturant. Confirm that the m_0 value for each biHis variant is consistent with the psWT value (*see* **Note 20**).

A sample script for fitting the chevron using the nonlinear algorithm in Origin (OriginLab) (*see* **Note 21**).

Fixed Parameters

R	// gas constant, i.e. 0.001987 kcal/mol K
T	// Temperature (°C)

Variable Parameters

DG	// stability of protein
m0	// denaturant dependence
aN, aU	// linear slopes of native and unfolded baselines, respectively
bN, bU	// linear intercepts of the native and unfolded baselines, respectively

Script

```
rt=R*(273.15+T); // define fixed parameters
sN = aN*x + bN;
sU = aU*x + bU;
dG = DG - m0*x;
myexp = exp(dG/rt);
y = (sU+sN*myexp)/(1+myexp);
```

6. Measure the native CD spectrum in the 0 M buffer with near-saturating metal concentrations 1 mM [Me^{2+}].
7. Measure denaturation titration in the presence of metal (*see* **Notes 22** and **23**). In practice, one only needs to measure a denaturation profile in near-saturating metal concentrations, e.g., 1 mM.

3.3.2 Application of Equilibrium Data

Ideally, equilibrium denaturant titrations should be conducted at several metal concentrations, including near-saturating concentrations (e.g., 1 mM). However, in practice one minimally needs a denaturation measurement in the absence of metal and in the presence of ~1 mM metal. These data provide a number of useful quantities:

1. Quantifying the maximal amount of metal ion-induced stabilization. The change in free energy of ion binding under saturating conditions identifies the experimental limits of metal-induced stabilization and the sensitivity to minor folding pathways. Values of $\Delta\Delta G_{eq}$ typically range from 0.5 to 3 kcal/mol. We have found that cobalt ions stabilize α-helices to a greater extent while zinc and nickel ions tend to prefer β-sheet sites, although this correlation is inconsistent, and multiple metals should be tested at the outset. The use of the most

stabilizing ion increases the accuracy in which fractional ψ values can be determined.

2. The value of $\Delta\Delta G_{eq}$ is to be compared to that obtained from chevron analysis to confirm that metal binding is in equilibrium during the kinetic measurements, a requirement for implementation of the method (*see* Subheading 3.4.3). Also, the m_0 values should match the kinetic data as well.

3.4 Kinetic Chevron Analysis

Our lab uses a four-syringe SFM-4000 stopped-flow apparatus, which allows for easy changes in metal or denaturant concentration (www.bio-logic.net). Nevertheless, other rapid mixing equipment can be used.

3.4.1 General Experimental Protocol

1. Equilibrium denaturation measurements should be performed first, so that one knows the stability and m_0-values, which can be used in the design of the stopped-flow experiments, specifically the denaturant concentrations used for generating the folding and unfolding arms of the chevron.
2. A concentrated protein stock will be used in syringe S4 (assuming a 4 syringe SFM), so that the protein concentration can be constant across all denaturant concentrations.
3. Measure unfolding arm first by configuring syringe buffers similar to Fig. 6 (*see* **Notes 24** and **25**).
4. Measure the folding arm.

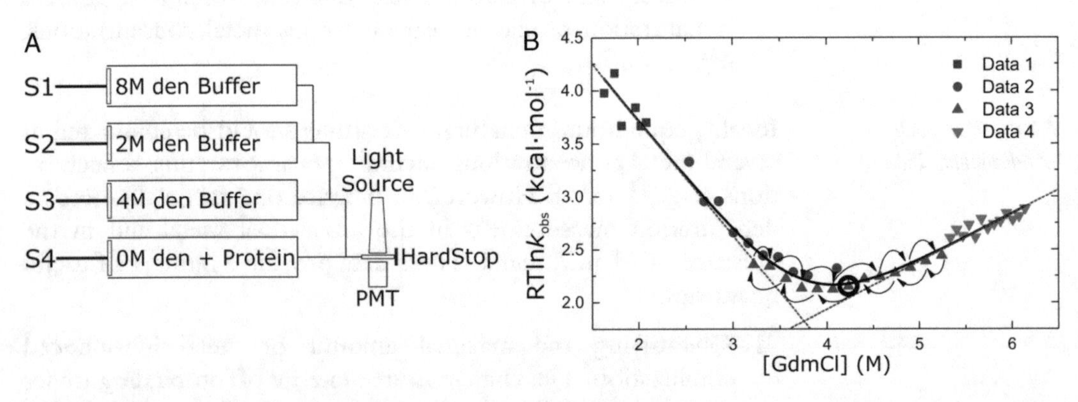

Fig. 6 General scheme for collecting chevron data. (**a**) Sample arrangement of buffer concentrations in a typical stopped-flow experiment for measuring unfolding arm of chevron. Syringe S1 can be counter-mixed with syringe S2 or S3 to titrate the final unfolding condition. (**b**) Sample folding chevron of a BdpA variant in 50 mM HEPES, 100 mM NaCl, pH 7.5 at 10 °C. The circled blue triangle data point in "Data 3" represents the condition where the two counter-mixing syringes are equal in volume. This tends to be the best condition to begin collecting data to minimize buffer loss to troubleshooting any issues with data collection. The arrows indicate a variable order of collecting points, which helps identify any systematic errors (e.g., temperature changes during the measurement): identify the vertex and then move away from the vertex, but skipping denaturant points to test for self-consistency in the chevron

5. Fit the chevron according to the following equations [52]:

$$\begin{aligned} y([den]) &= \mathrm{RT}\ln k_{obs}([den]) = \mathrm{RT}\ln\left(k_f([den]) + k_u([den])\right) \\ &= RTln\left(e^{(\Delta G_{\mathrm{f}} - m_{\mathrm{f}}\cdot[den])/\mathrm{RT}} + e^{(\Delta G_{\mathrm{u}} + m_{\mathrm{u}}\cdot[den])/\mathrm{RT}}\right) \\ &= \mathrm{RT}\ln\left(e^{(\Delta G_{\mathrm{f}} - m_{\mathrm{f}}\cdot[den])/\mathrm{RT}} + e^{(\Delta G_{\mathrm{f}} - \Delta G_{\mathrm{eq}} + m_{\mathrm{u}}\cdot[den])/\mathrm{RT}}\right) \end{aligned} \tag{12}$$

where $\Delta G_{\mathrm{f}} = \mathrm{RT}\ln k_{\mathrm{f}}^{\mathrm{H_2O}}$, ΔG_{eq} is the stability, m_{f} and m_{u} are the denaturant dependences of the folding and unfolding rates, respectively, and R is the gas constant (*see* **Note 26**).

A sample script for fitting the chevron using the nonlinear algorithm in Origin (OriginLab).

Fixed Parameters

R	// gas constant, i.e. 0.001987 kcal/mol·K
T	// Temperature (°C)

Variable Parameters

DGeq	// stability of protein
DGf	// folding free energy, RT ln kf(0 M)
mf, mu	// folding and unfolding arm slopes, respectively

Script

```
rt=R*(273.15+T); // define fixed parameters
dgf = DGf - mf*x;
dgu = DGf - DGeq + mu*x;
kf=exp(dgf/rt);
ku=exp(dgu/rt);
y=rt*ln(kf+ku);
```

6. Verify that the no metal chevron parameters are consistent with the equilibrium denaturation, i.e., ΔG_{eq}(kinetic) = ΔG_{eq}(equilibrium) and $m_{\mathrm{f}} + m_{\mathrm{u}} = m_0$.
7. Repeat chevron measurement using the 1 mM metal buffers in all syringes (*see* **Notes 27** and **28**).

Additional chevrons should be measured under near-saturating metal conditions to determine if metal binding grossly alters the folding pathway of the biHis protein (Fig. 7a). These kinetic studies can be used to calculate the maximal stabilization imparted by the binding of several different metal ions. Chevron data are acquired at a single high metal concentration for each cation, and the $\Delta\Delta G_{\mathrm{eq}}$ is determined from the change in k_{f} and k_{u}. These values should match those obtained in the equilibrium measurements to confirm that cation binding is in fast equilibrium (*see* Subheading 3.4.3).

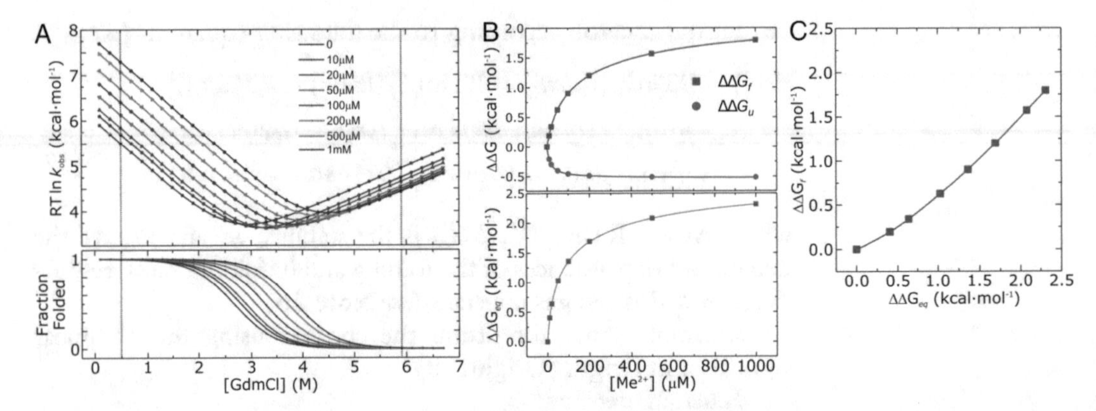

Fig. 7 Idealized chevrons illustrating metal dependence of folding. (**a**) Idealized chevrons for a hypothetical protein with a biHis site having $\psi_0 = 0.4$, and native and unfolded state binding affinities of 10 mM and 1 mM, respectively. The corresponding equilibrium fraction plots are shown beneath to accentuate the need to choose x_f and x_u such that strongly folding and unfolding conditions, respectively, are maintained throughout all metal concentrations. (**b**) (Upper) Metal-dependent changes in folding and unfolding rates are plotted assuming $x_f = 0.5$ M and $x_u = 5.9$ M GdmCl, respectively. Fits are shown assuming a singular TSE with a non-native binding affinity according to Eq. 4 (*see* **Note 36**). (Lower) The net change in stability $\Delta\Delta G_{eq}(Me^{2+})$ is fit according to Eq. 3. (**c**) From the data points in **b**, a Leffler plot may be constructed, and a ψ_0 value can be calculated using Eq. 5

The refolding and unfolding denaturant m_f and m_u values should be largely unchanged compared to the wild-type, signifying no major change in surface area burial during the course of the reaction for the pathways with or without cations [53]. Just as in mutational ϕ analysis, data from systems with changing m values should be interpreted cautiously as the biHis substitution may have altered the conformation of the denatured, native, or transition states.

3.4.2 Applications and Interpretation

Additional denaturant chevrons can be obtained, each at a fixed metal ion concentration traversing the range of interest (e.g., 0.0, 0.1, 0.2, 0.4, 1 mM $[Me^{2+}]$) (Fig. 7a and Table 1, **Note 27**). However, if the change in stability of the kinetic studies is in accord with the equilibrium studies, then one can calculate ψ_0 simply from the two chevrons. If the high-metal chevron shifts only the folding arm up (folding is faster) while leaving the unfolding arm unchanged, then metal binding stabilizes both the TS and the native state equally, and ψ_0 is unity. Conversely, if the presence of metal only shifts the unfolding rate down (unfolding is slower), then metal only stabilizes the native state, implying $\psi_0 \sim 0$. When both arms shift, the ψ_0 value is fractional.

From Eq. 5, we see that the $\Delta\Delta G_f(Me^{2+}) = \Delta\Delta G_f(\psi_0, \Delta\Delta G_{eq}(Me^{2+}))$ so the ψ_0 value can actually be obtained from simultaneously fitting the no metal and high metal chevrons, provided that the $\Delta\Delta G_{eq}$ derived from the chevrons matches the equilibrium studies. One can incorporate Eq. 5 into Eq. 12 to yield:

Table 1
Sample shot protocol for a denaturant chevron at high metal ion concentration[a]

Metal (μM)	GdmCl, (M)	Buffer Syringe 1 (μL)	4 M GdmCl Syringe 2 (μL)	Protein 4 M GdmCl 6 mM Me^{2+} Syringe 3 (μL)	Total volume (μL)
1000	0.83	250	0	50	300
1000	1.50	200	50	50	300
1000	2.17	150	100	50	300
1000	2.83	100	150	50	300
1000	3.50	50	200	50	300
1000	4.17	0	250	50	300

[a]The metal concentration is designed to stay constant while the denaturant changes. For a three-syringe protocol, Syringe 1 contains buffer, Syringe 2 contains 4 M GdmCl with buffer, and Syringe 3 contains Protein, 5 M GdmCl, 6 mM $MeCl_2$, and buffer. Of course, this table may be applied to the case where all syringes have the same $[Me^{2+}]$ as well

$$y([den], Me^{2+}) = RT\ln\left(e^{(\Delta G_f + \Delta\Delta G_f(Me^{2+}) - m_f\cdot([den]-x_f))/RT} + e^{(\Delta G_f - \Delta G_{eq} + \Delta\Delta G_f(Me^{2+}) - \Delta\Delta G_{eq}(Me^{2+}) + m_u\cdot([den]-x_u))/RT}\right) \quad (13)$$

where x_f and x_u are denaturant concentrations corresponding to strongly folding and unfolding conditions, respectively. The choice in x_f and x_u should be in the linear regions of the folding and unfolding arms of the chevron (*see* **Note 29**, Fig. 7).

In the case of the no metal chevron, $\Delta\Delta G_f$ (Me^{2+}) & $\Delta\Delta G_{eq}(Me^{2+}) = 0$, while in the metal chevron, $\Delta\Delta G_{eq}$ becomes a fitting parameter, and the $\Delta\Delta G_f$ is fit according to Eq. 5. Global fitting of both data sets allows for a direct fit of ψ_0 while minimizing error propagation.

Sample script for fitting two chevrons simultaneously with parameter sharing implemented as Origin nonlinear fitting function:

Shared Fixed Parameters

R	// gas constant, i.e., 0.001987 kcal/mol·K
T	// Temperature (°C)
xf, xu	// denaturant concentrations for folding and unfolding conditions, respectively.

Local Fixed Parameters

switch	// controls whether to include $\Delta\Delta G_f$ and $\Delta\Delta G_{eq}$.
	// For no metal data, switch = 0
	// For the metal chevron, switch = 1

Variable Parameters

psi	// ψ_0 value (not shared; for no metal data, fix psi to 0)
DDGeq	// change in stability between the chevrons (not shared, fix to 0 for no metal data)
DGf, DGu	// no metal chevron values at xf and xu, respectively. (shared)

Script

```
rt=R*(273.15+T); // define fixed parameters
ddgf=rt*ln(psi*exp(DDGeq/rt)+1-psi); // ΔΔGf (ΔΔGeq)
ddgu=ddgf-DDGeq; // ΔΔGu = ΔΔGf - ΔΔGeq
dgf=DGf+switch*ddgf-mf*(x-xf); // switch controls including
the metal chevron.
dgu=DGu+switch*ddgu+mu*(x-xu);
kf=exp(dgf/rt);
ku=exp(dgu/rt);
y=rt*ln(kf+ku);
```

3.4.3 Testing for Fast Ion Binding Equilibrium

If folding and unfolding processes obey single exponential kinetics, and the metal-enhanced stability derived from their changes matches the value obtained from equilibrium measurements, it is highly likely that the assumption that metal binding is in fast equilibrium is valid [23]. However, if the folding rates are fast compared to metal binding on- and off-rates, binding may not be in fast equilibrium during the course of the folding reaction. As a result, the assumption that binding stabilizes the TSE according to Eq. 4 may not hold, and the folding and binding processes will be convoluted. For very fast folding rates, metal ions may even no longer stabilize the TS as the biHis site is kinetically inaccessible for ion binding. For ubiquitin, the on-rate binding constant and binding affinity for a site located along an α-helix was ~10^7 M^{-1} s^{-1} and 3 μM, respectively [23]. When binding is no longer in fast equilibrium, the ψ-analysis formalism will no longer be valid (*see* **Notes 30** and **31**).

When metal release rates are slower than unfolding rates, multiple populations will be observed in unfolding experiments; some molecules will unfold having metal bound while other molecules will unfold as if there is no metal ion present in solution. Nevertheless, a slower unfolding rate for the pre-bound population indicates that the biHis site is present and bound in a subpopulation of the TSE. This information is useful in its own right.

3.5 Metal-Dependent Folding Kinetics: Leffler Plot

Once the denaturant chevrons have been acquired, a ψ_0 value can be qualitatively evaluated by comparing the chevron measured without metal to one at saturating metal concentration. More detailed metal-dependent folding measurements can also be conducted to determine a more precise value of ψ_0. This is accomplished by fixing the denaturant concentration and conducting numerous folding and unfolding measurements at multiple metal concentrations (Fig. 7b). The folding and unfolding measurements are conducted at a relevant single denaturant concentration under strongly folding and unfolding conditions, respectively (Fig. 7a) (*see* **Note 29**).

This strategy—varying metal concentration at a fixed denaturant value—effectively is a vertical slice through a multitude of denaturant chevrons in the presence of differing metal ion concentrations. The advantage of the second method is that many more points are obtained for the Leffler plot, which is used to calculate the ψ_0 value (*see* Fig. 7c). The advantage of the first method using denaturant chevrons is that one can monitor changes in the *m* values as a function of metal ion concentration.

The metal dependence of folding and unfolding rates can be collected with a stopped-flow apparatus, similar to the chevron measurements, using a three-syringe mixing protocol where one syringe contains a protein/denaturant mixture. The two other syringes have identical denaturant concentration, but one syringe contains metal ions (Table 2). By varying the relative delivery volumes of these otherwise identical buffers, a large amount of finely spaced data can be obtained over a range of ion concentrations using a single set of buffer solutions (*see* **Note 32**).

Table 2
Sample shot protocol for measuring metal-dependent folding rates[a]

Metal (μM)	GdmCl (M)	0 M GdmCl Syringe 1 (μL)	0 M GdmCl 1 mM Me^{2+} Syringe 2 (μL)	Protein 5 M GdmCl Syringe 3 (μL)	Total volume (μL)
0	0.83	250	0	50	300
166.7	0.83	200	50	50	300
333.3	0.83	150	100	50	300
500	0.83	100	150	50	300
666.7	0.83	50	200	50	300
833.3	0.83	0	250	50	300

[a]The shot protocol is designed to vary metal concentration while keeping denaturant concentration constant. Syringe 1 contains buffer, Syringe 2 contains buffer with 1 mM $MeCl_2$, and Syringe 3 contains Protein in 5 M GdmCl

General Protocol

1. Configure syringes as described in Table 2 for monitoring folding (*see* **Notes 33** and **34**).
2. Similar to the chevron protocol, collect kinetic data across a range of mixing ratios of the metal and no metal buffers (*see* **Note 35**).
3. Mix equal volumes of the no metal and metal buffers to make a new metal buffer with lower [Me^{2+}] (*see* **Note 36**). Load Syringe 2 with the new lower [Me^{2+}] buffer.
4. Repeat **steps 2** and **3** until the rate converges on the no metal kinetic rate.
5. Load Syringe 2 with the no metal buffer and collect the zero metal folding point.
6. Repeat the protocol for titrating metal in strongly unfolding conditions. The data should be taken at the same metal concentrations as the folding data to allow calculation of $\Delta\Delta G_{eq}$ directly from the difference of $\Delta\Delta G_f$ and $\Delta\Delta G_u$ data points according to

$$\Delta\Delta G_{eq}([Me^{2+}]) = \Delta\Delta G_f - \Delta\Delta G_u = RT\ln(k_f([Me^{2+}])/k_f(0)) - RT\ln(k_u([Me^{2+}])/k_u(0)) \quad (14)$$

where R is the gas constant and T is the absolute temperature, $k_f([Me^{2+}])$ and $k_u([Me^{2+}])$ are the relaxation rates in the presence of the same concentration of cation, respectively. The folding and unfolding rates at zero metal should also be well established as this rate serves as the reference point in the Leffler plot [54].

3.5.1 Analysis of Leffler Plot

Metal Stabilization

The degree of stabilization due to metal binding, $\Delta\Delta G_{eq}$, depends upon the difference in metal dissociation constants between the biHis sites in the native state, K_N, and in the unfolded state, K_U (*see* Fig. 7b). The increase in protein stability upon the addition of metal is fit to the linked equilibrium expression in Eq. 1 [47].

A sample script for fitting $\Delta\Delta G_{eq}([Me^{2+}])$ data (*see* **Note 37**).

Fixed Parameters

R	// gas constant, i.e. 0.001987 kcal/mol·K
T	// Temperature (°C)

Variable Parameters

KN	// Native binding affinity
KU	// Unfolded state binding affinity

Script

```
rt=R*(273.15+T); // define fixed parameters
ddgf = 1+x/KN;
ddgu = 1+x/KU;
y=rt*ln(ddgf/ddgu);
```

Obtaining ψ Values from the Leffler Plot

When multiple chevrons are obtained, each at a different metal ion concentration, one Leffler data point is obtained for each chevron. When folding and unfolding data are obtained at fixed denaturant concentrations at varying metal ion concentration, one Leffler data point is obtained for each metal concentration.

A sample script for fitting the ($\Delta\Delta G_{eq}$, $\Delta\Delta G_f$) data.

Fixed Parameters

R	// gas constant, i.e. 0.001987 kcal/mol·K
T	// Temperature (°C)

Variable Parameters

psi	// ψ_0 value

Script

```
rt=R*(273.15+T); // define fixed parameters
y=rt*ln(psi*exp(x/rt)+1-psi);
```

As previously mentioned, there are two scenarios where the slope in the Leffler plot is linear. In the first scenario, the slope is zero across all metal concentrations ($\psi = \psi_0 = 0$). This behavior occurs when metal ion binding does not increase the population of the biHis site in the TS ensemble relative to *U*. In this case, the entire TS ensemble lacks the binding site, or more rigorously, the site has the same binding affinity as the unfolded state. At the other limit, the slope is one ($\psi = \psi_0 = 1$), indicating that the entire ensemble has the binding site formed with native-like affinity. Otherwise, the Leffler plot will be curved as metal continuously increases the population configurations having the biHis site formed in the TS ensemble. In such cases, the curvature can be due to TS heterogeneity, non-native binding affinity in a singular TS, or a combination thereof.

3.5.2 Delineation Between the Heterogeneous and Homogeneous TSE Scenarios

The question of whether fractional ψ_0 values reflect TS heterogeneity or non-native binding affinity remains unresolved and may be site dependent. This interpretive ambiguity plagues ϕ analysis as well; however, in the case of ψ, the fact that a specific contact pair is being probed provides some advantage.

Method 1: Measure ψ values with different metal ions.

We believe that discrimination between the two extreme models described in Subheading 3.1 is possible through the use of at least two different metal ions, which have different coordination geometries (e.g., Zn^{2+} and Ni^{2+} have tetrahedral and octahedral electron orbitals geometries, respectively). The two ions are likely to manifest the same ψ_0 value only in the case of TS heterogeneity [25], because the same fractional binding affinity is unlikely to be realized with both ions. However, ψ_0 values should depend on the type of metal ion if the site is distorted [14, 17, 18] (*see* **Note 38**).

Method 2: Destabilize TS and remeasure the ψ value.

Another test for TS heterogeneity in a region that may be present in the TSE involves altering the stability of this region, e.g., via mutation, and remeasuring the ψ region of the TSE. If the ψ_0 value responds accordingly, as we observed in the dimeric GCN4 coiled coil [20], the heterogeneity model is the most parsimonious interpretation. For the coiled coil, the introduction of the destabilizing glycine (A24G) shifted the pathway flux away from this region so that most nucleation events occurred near the biHis site, which was located at the other end of the protein. As expected, the ψ_0 value increased to 0.5, indicating that half of the nucleation events occurred with the biHis site formed [20] (*see* **Note 39**).

On the other hand, a homogenous TSE in BdpA was deduced using a similar strategy [18]. A L45A mutation destabilized a possible mini-core by 2 kcal/mol, and the ψ_0 value on the other side of the protein was remeasured. Despite being topologically designed for alternative TSEs, the ψ_0 value at the alternate site did not change, suggesting a relatively homogenous TSE with non-native binding affinity in the TSE (Fig. 8). The general strategy of introducing destabilizing mutations or loop insertions followed by ϕ or ψ analysis was applied to src [55], α spectrin SH3 [56], and ubiquitin [21]. The four globular proteins were found to have a robust or "mechanic nucleus."

Method 3: Link-Psi—stabilize one site using a covalent cross-link.

An additional strategy for testing heterogeneity versus homogeneity is to measure the effect of cross-linking at one site in the protein on the ψ_0 value of another site. Rather than measuring the effect of a destabilization, the Link-Psi method measures the response of a site to stabilization in another region of the protein using cross-linking (Fig. 9) [22]. Through the utilization of this method, the TSE of Ubiquitin was found to be singular [21].

Choose site for cross-linking by inserting biCys mutations instead of biHis sites, as done in Subheading 3.2, **step 1**.

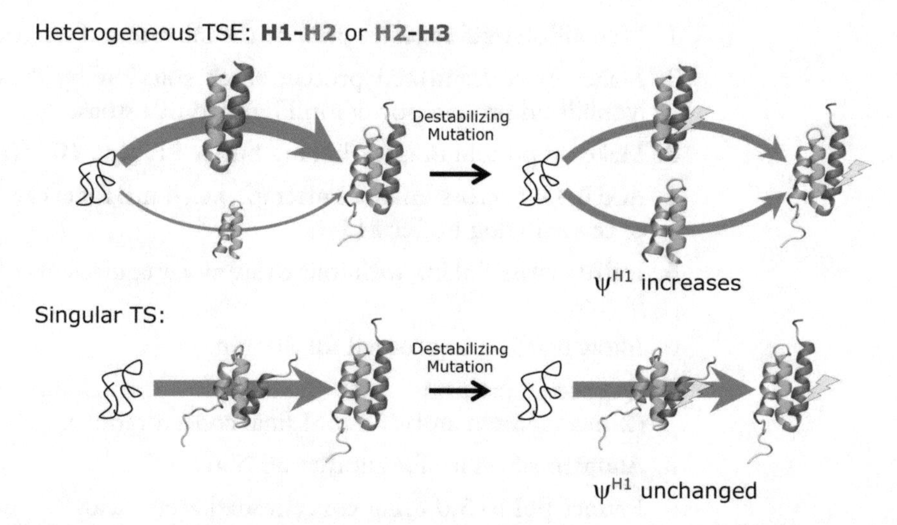

Fig. 8 Evaluating heterogeneity versus homogeneity via destabilizing mutations. A fractional ψ_0 value in the helix H1 (blue) in BdpA may be interpreted as two mini-TS existed centering around H1-H2 or H2-H3, of which H2-H3 is the dominant route (upper). In this limit, a destabilizing mutation to the H2-H3 interface would lead to a significant increase in the flux through the H1-H2 TS path (and cause an increase in the ψ_0 value at the C-terminus of H1). However, if the TSE was composed of partially formed helices docked in a native-like topology (lower), then the destabilizing mutation to H2-H3 would not change the ψ_0 value in H1.

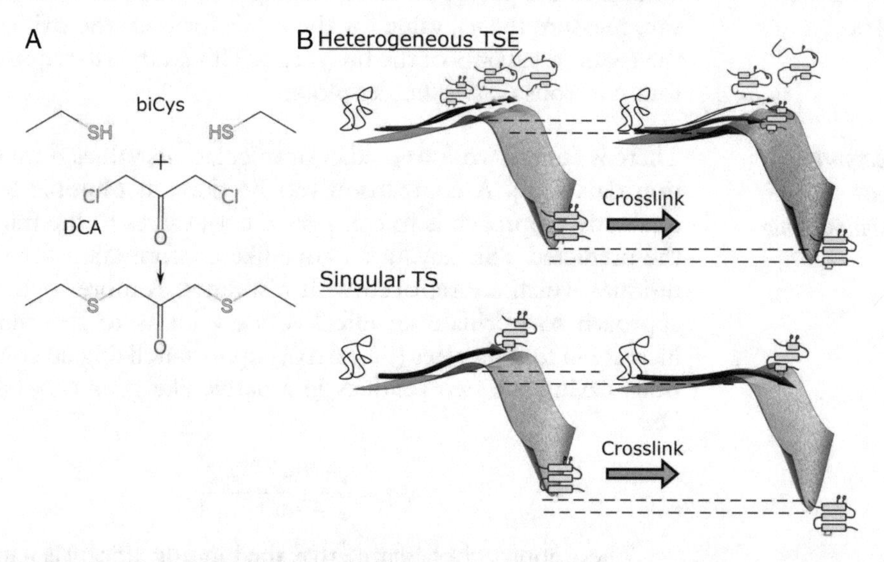

Fig. 9 Link-ψ strategy for characterizing TS heterogeneity. (**a**) One region of the protein is stabilized by a covalent DCA cross-link and a bi-Cysteine residue pair. (**b**) Illustrative TSE scenarios that Link-ψ is able to probe. In the upper case, the TSE is heterogeneous, and only a fraction of the TSE is stabilized by crosslinking. This could result in a new ψ_0 value at a biHis site located in another region of the protein. In the lower scenario, a single TSE exists, and cross-linking stabilizes the entire TSE, thus leaving ψ_0 unchanged

Cross-link biCys Sites

This reaction is a modified version of the protocol described elsewhere [57]. Perform the reaction on ice to eliminate occurrence of side reactions.

1. Pre-chill all buffers on ice prior to the initiation of the reaction.
2. Make up concentrated protein stock solution by dissolving lyophilized protein powder in High GdmCl stock.
3. Make up protein in cross-linking buffer #1, e.g., 10 mL.
4. Add 0.4× of cross-linking buffer #2, i.e., 4 mL for every 10 mL of cross-linking buffer #1.
5. Initiate cross-linking with four to six molar equivalents of DCA buffer.
6. Allow reaction to proceed for 30 min.
7. Quench reaction with 50 μL quenching buffer (2-mercaptoethanol) (50 mM final concentration).
8. Allow to sit on ice for another 30 min.
9. Reduce pH to 5.0 using concentrated acetic acid.
10. Isolate cross-linked species using reverse-phase chromatography.
11. Confirm successful cross-linking using mass spectrometry. The mass should change by 54 Da.

Measure the Effects of Cross-link

Similar to the protocol for measuring a ψ_0 value for a single biHis site, measure the ψ_0 value for the biHis for both the oxidized and the reduced variants of the biCys site. This study will require either four chevrons or two Leffler plots.

3.6 Comparing ψ_0 Values to Simulation Data

There is limited work to predict or calculate ψ_0 values from simulation data [17]. A comparison can be done at multiple levels. A qualitative approach is to compare the ψ_0 values to the fraction of the predicted TSE having a native-like conformation for the two residues which are replaced with histidines. A more sophisticated approach to calculate an effective ψ_0 value is to determine the increase in folding rates (i.e., $\Delta\Delta G_f$) upon stabilizing all conformations having the two residues in a native-like geometry by $\Delta\Delta G$, i.e.,

$$\psi_0 = \frac{e^{\Delta\Delta G_f/RT} - 1}{e^{\Delta\Delta G_{eq}/RT} - 1}. \quad (15)$$

These approaches assume that the binding affinity is native-like for a native-like conformation and do not accurately account for highly populated but distorted sites having weaker binding affinity. In this situation, the ψ_0 values can be considered to be a lower bound for the population having the site in a native-like conformation. This level of interpretation can be sufficient to make detailed comparisons with simulations, e.g., testing for the presence of helices and strands, and their contacts [58].

3.7 Applications of ψ Beyond Protein Folding

This methods paper has described the use of ψ in the context of protein folding. The general principles are not limited to denaturant dependent kinetics. Equations 3–5 specifically detail how the relationship between an activation energy and the change in stability can be probed through measuring metal dependent kinetics and equilibrium. ψ has been used to characterize the transition states for both mechanical unfolding [59] and protein-protein docking, where a precise orientation rather than side chain interactions, is the critical property of a successful binding encounter [60].

4 Notes

1. Phosphate must not be used as a buffer. The divalent metal ions associate very strongly with phosphate and immediately precipitate.
2. The solubility of GdmCl is 8.5 M at 25 °C and will decrease with decreasing temperature.
3. The dissolution of GdmCl is endothermic such that making a buffer at high concentration will be greatly aided by using a heated stirring plate. Also, GdmCl displaces volume when it dissolves, e.g., 100 mL of water and 100 g of GdmCl will produce ~175 mL of 6 M GdmCl. Care should be taken in determining the molarity of the solution (rather than molality).
4. Urea can be used as a denaturant, e.g., if the protein is relatively unstable or if one wishes to minimize ionic effects.
5. Tris, i.e., tris(hydroxymethyl)aminomethane, is more amenable than HEPES for measuring CD spectra due to lower absorbance in the far-UV. HEPES strongly absorbs light below 230 nm. Tris is slightly better, but absorbance becomes problematic at 220 nm. The disadvantage with Tris is that the temperature dependence of pH is stronger ($\Delta pH/\Delta T = -0.028/°C$).
6. To minimize the effects of varying ionic strength at low denaturant concentrations (which can alter kinetic rates), add salts to the GdmCl-containing solution, e.g., 100–300 mM NaCl, to increase the net ionic strength.
7. Although one can make several buffers, it is preferable to use as few buffers as possible to span the range of denaturant. In this manual, we will be following the scheme laid out in Fig. 5a, wherein one prepares 50+ mL of four buffers. One buffer should be at 0 M denaturant, while another should be a relatively high amount denaturant, e.g., 7.5 M GdmCl. Proteins more readily dissolve in high denaturant conditions (e.g., lyophilized ubiquitin does not dissolve well in low denaturant conditions). Note, however, that the actual concentration of

the higher GdmCl-containing buffer can be much smaller for less stable proteins.

8. For example, if the HEPES buffer is prepared at 25 °C but is to be used at 5 °C, then adjust the pH to 7.22 at 25 °C ($\Delta pH/\Delta T = -0.014/°C$ for HEPES).
9. Depending on the stability of the protein, a concentration lower than 8 M GdmCl can be used.
10. Throughout this manuscript, we use the sign convention that $\Delta G > 0$ refers to the condition where the protein is stable, and $\Delta\Delta G > 0$ refers to increased stability. Furthermore, the sign convention is set so that both m_f and m_u are positive.
11. This formulation of ψ differs from the original derivation presented [11, 20], where curvature was associated only with the heterogeneous model and is more in line with [16]. Readers should note that according to Eq. 5, $\Delta\Delta G_{eq} \gg 1 \Rightarrow \psi \rightarrow 1$, i.e., the Leffler plot will always eventually have a unity slope regardless of any curvature near $\Delta\Delta G_{eq} = 0$.
12. Generally, biHis sites require that the imidazole nitrogen (N_ε) of the histidines be located within 3–5 Å of each other to be binding competent, which can be accomplished using residues with a C_α-C_α distance less than 13 Å [61]. However, the placement of histidine residues has a substantial effect on the binding affinity and the degree of energetic stabilization for each type of ion [61]. We have successfully employed sites located across a turn ($i,i + 4$) of a helix, between two β strands, and at salt bridges.

    These properties generally are not applicable to proteins with buried sites, such as zinc finger proteins, where metal ions are required for the cooperative folding of the entire protein [62]. Buried sites typically have four side-chain ligands arranged in a precise geometry. The introduction of such sites often requires a substantial amount of protein design [63–65], which is not required for the surface biHis sites.
13. Regardless of the biHis design, the protein variant must be able to fold in the absence of metal.
14. Often it is advantageous to make a high concentration protein stock in strongly denaturing conditions as the protein stock is less likely to aggregate.
15. Measure reference spectra of the buffers as well.
16. If your protein has a tryptophan, when taking the native CD spectrum, measure the spectrum and absorption (high tension, HT) up to 320 nm. The HT can be converted to absorbance, and then the protein concentration can be directly calculated.
17. This step will allow for determining the wavelength that would be best for tracking the denaturation. This is important since

the high concentrations of denaturant will lead to high tension issues that will affect the reliability of the CD measurement.

18. If possible, one can simultaneously monitor tryptophan fluorescence in these experiments as well by using a secondary photomultiplier tube. If it is not possible to alternate between a CD wavelength (e.g., $\lambda \sim 222$ nm) and canonical Trp excitation for fluorescence ($\lambda \sim 280$ nm), then one can simultaneously monitor CD and excite the Trp near 226 nm (though the fluorescence signal will be significantly reduced).
19. The advantage of starting unfolded in high denaturant and adding denaturant-free buffer is that then one does not have to worry about denaturant in the titrator lines. However, this will impact the equilibration time. On the other hand, one can start in strongly folding conditions, which also allows one to measure the native CD spectrum, and then easily equilibrate as the denaturant changes.
20. Change in denaturant m_0 value. This parameter, which reflects the amount of surface burial in the folding transition, should be largely unchanged upon the addition of metal. A decrease in the m_0 value may indicate the formation of residual structure in the denatured state or a perturbation of the native structure. Furthermore, the m_0 value should not change significantly between the WT and the biHis variant. In most cases, surface histidine substitutions have been found to have little apparent effect on the structure or folding behavior of proteins, compared to the more common use of core residue substitutions in other folding studies [5]. Preferably, $|\Delta m_0| = |m_0^{WT} - m_0^{variant}| \sim \sigma$, where σ is the fitting error of m_0 (typically, ~0.1 kcal/mol·M).
21. A good practice is to first fit the data assuming flat baselines (aN, aU = 0) and then use the converged values of ΔG and m_0 as the initial values in a new fit with all parameters allowed to vary.
22. To determine the change in denaturation midpoint upon the addition of metal ions, one measures the CD signal at the denaturant concentration where 20% of the molecules are folded (i.e., $K_{eq} = [N]/[U] = 1/4$). Then in a 2 mL solution, inject 8 μL of the metal stock solution into the cuvette and measure the new CD value. One can use the no metal denaturation parameters to estimate $\Delta\Delta G_{eq} = RT\ln K_{eq}'/K_{eq}$.
23. It is entirely possible to insert a biHis site only to discover that adding metal will destabilize the protein structure because the native backbone arrangement is not amenable to binding metal. However, this does not change the application of ψ or the interpretation. In this case, you will actually project the Leffler analysis into another quadrant of ($\Delta\Delta G_{eq}$, $\Delta\Delta G_f$) space (Fig. 3).

24. It is not necessary to measure the unfolding arm first. In our experience, the unfolding arm tends to be the easier of the two chevron arms to characterize first, especially when measuring the folding chevron for the first time.
25. It is the best practice to first measure a folding rate where the two counter-mixing syringes (e.g., S1 and S2) are of equal volume, then to move toward the vertex of the chevron. Also, skip denaturant points in order to control for systematic errors, e.g., protein aggregation in the reservoir, the equipment not being fully equilibrated, etc. (Fig. 6).
26. A double-site $\phi^{\text{WT-biHis}}$ value can be calculated from a comparison of the chevrons,

$$\phi^{\text{WT-biHis}} = \frac{\Delta G_{\text{f}}^{\text{biHis}} - \Delta G_{\text{f}}^{\text{WT}}}{\Delta G_{\text{eq}}^{\text{biHis}} - \Delta G_{\text{eq}}^{\text{WT}}} \tag{16}$$

    However, the translation of this ϕ-value to structure in the TS may be difficult, as the perturbation is due to the introduction of the biHis site which reflects both the interactions between the two mutated positions and between them and other parts of the protein.
27. Several different cations can be readily tested using a configuration where metal ion is placed only in one syringe, taking into account the dilution factor of the final mix (*see* Table 1). The testing of different metal ions using this protocol requires only changing the contents of the single metal-containing syringe. However, metal concentrations in this syringe will be higher than if the ion was placed in all syringes, and metal precipitation may become an issue. The addition of metal can stabilize a protein to the point where higher than convenient levels of denaturant are required to unfold the protein (e.g., either due to denaturant solubility limits or dilution factors on the stopped-flow). Configuring all syringes to contain buffer with Me^{2+} will require preparing more buffer and washing syringes more often. Also, maintaining constant Me^{2+} in all syringes allows the user flexibility to vary dilution ratios to achieve a desired denaturant concentration without sacrificing the final $[Me^{2+}]$.
28. The m_f and m_u values reflect the amount of surface burial from the denatured state to the transition state and from the transition state to the native state, respectively. A change in the values may reflect a shift in the transition state in either Hammond or anti-Hammond behavior. Similar to **Note 20**, changes in m values should be small, e.g., below 0.1 kcal/mol·M. Even if there are (minor) changes in the m values, one still should ensure that $\beta_T = m_f/(m_f + m_u)$ is relatively unchanged, as this represents the fraction of surface area buried in the TSE.

29. The choice in xf and xu must be in the linear regions of the folding and unfolding arms, respectively, so that one is specifically measuring $\Delta\Delta G_f$ and $\Delta\Delta G_u$, respectively.
30. However, ψ can still be applied to the regions of the chevron where fast equilibrium does hold to provide characterization of the TSE.
31. An additional test to confirm fast equilibrium is that the metal-induced stabilization determined from the equilibrium studies should match that from kinetic measurements (Eq. 1). If metal binding is not in fast equilibrium, different metals can be tested as their binding properties may be more suitable. Also, folding rates can be manipulated by working at lower temperatures or higher denaturant concentrations.
32. When choosing the final denaturant concentration for the unfolding measurements, bear in mind that metal stabilizes the protein and shifts the chevron to the right. On rare occasion, the metal can actually destabilize the site as well [14]. As a result, unfolding conditions in the absence of metal ions could become folding conditions with the addition of metal (Fig. 7a, [GdmCl] ~ 4 M). For this reason, it is advisable to measure a chevron under saturating metal conditions first so that appropriate folding and unfolding conditions can be chosen at the outset.
33. A standard two syringe, fixed volume ratio stopped-flow apparatus can be used as well, although the solutions must be changed for each metal ion condition.
34. Whether one measures folding or unfolding data first is up to the experimentalist. One strategy is to choose the condition that showed the greatest effect due to metal, e.g., compare $\Delta\Delta G_f$ and $\Delta\Delta G_u$ in the 1 mM metal chevron.
35. Analogous to **Note 25**, it is best practice to begin with a near equal ratio of the counter-mixing syringes, e.g. the 100/150/50 configuration as shown in Table 2. Furthermore, when collecting data points, skip metal concentrations to control for systematic errors, e.g., protein or metal aggregation, the equipment not being fully equilibrated, etc.
36. Stopped-flow equipment often have limited dilution ratios due to inaccuracies arising from low-volume delivery. In our experimental setup, one set of buffers provides data over approximately one decade in metal ion concentration. To cover a wider range, only the metal-containing buffer needs be changed. This change readily can be accomplished using dilutions of the metal buffer using the no-metal buffer, e.g., lower ranges are obtained by fivefold serial dilutions of the metal-containing buffer, e.g., 5 mM, 1 mM, 200 μM, and 40 μM [Me^{2+}].

37. In the limit that the TSE is singular, then from Eq. 4, we can write $\Delta\Delta G_f\left(\left[Me^{2+}\right]\right) = RT\ln\frac{1+\left[Me^{2+}\right]/K_{TS}}{1+\left[Me^{2+}\right]/K_U}$, and this script can be used to fit the $\Delta\Delta G_f\left(\left[Me^{2+}\right]\right)$ data.
38. Sites having fractional ψ_0 values that are topologically located near sites with unity ψ values are more likely to reflect distortion. Sites along helices and possibly at the edges of β-hairpins are less topologically constrained and therefore capable of exhibiting heterogeneity [17]. We believe that distorted sites will especially be prevalent in β-sheets since the β-sheets are often twisted in native structures.
39. A quantitative comparison indicated that the change in the degree of pathway heterogeneity recapitulated the destabilizing effect of the glycine substitution. The A24G mutation increased the amount of flux going through the N-terminal biHis site. The ratio of the heterogeneity in these two molecules reflected the loss in stability for this mutation,

$$\begin{aligned}\Delta\Delta G_{eq}^{Ala\rightarrow Gly} &= RT\ln\left(k_{Ala}^{present}/k_{Ala}^{absent}\right) - RT\ln\left(k_{Gly}^{present}/k_{Gly}^{absent}\right)\\ &= 2.5\ \text{kcal/mol}.\end{aligned}$$

This shift was consistent with the decrease in stability for the mutation backgrounds (1.7–2.4 ± 0.1 kca/mol) [66]. Hence, in this case, ψ analysis successfully quantified the level of TS heterogeneity. Invoking a homogeneous model with non-native binding affinity in the TS would require that the A24G mutation cause the biHis the site to acquire native-like binding affinity. This is an unlikely scenario given the distance between the substitution and the biHis site. Potentially, binding sites introduced into well-defined helices will have native-like binding affinities in the TS. In which case, fractional ψ values will be due to TS heterogeneity.

References

1. Englander SW, Mayne L (2014) The nature of protein folding pathways. Proc Natl Acad Sci U S A 111(45):15873–15880. https://doi.org/10.1073/pnas.1411798111
2. Jackson SE, Fersht AR (1991) Folding of chymotrypsin inhibitor 2. 1. Evidence for a two-state transition. Biochemistry 30 (43):10428–10435
3. Krantz BA, Mayne L, Rumbley J, Englander SW, Sosnick TR (2002) Fast and slow intermediate accumulation and the initial barrier mechanism in protein folding. J Mol Biol 324 (2):359–371
4. Jackson SE, elMasry N, Fersht AR (1993) Structure of the hydrophobic core in the transition state for folding of chymotrypsin inhibitor 2: a critical test of the protein engineering method of analysis. Biochemistry 32 (42):11270–11278
5. Sosnick TR, Dothager RS, Krantz BA (2004) Differences in the folding transition state of ubiquitin indicated by phi and psi analyses. Proc Natl Acad Sci U S A 101 (50):17377–17382
6. Feng H, Vu ND, Zhou Z, Bai Y (2004) Structural examination of Phi-value analysis in protein folding. Biochemistry 43 (45):14325–14331
7. Sanchez IE, Kiefhaber T (2003) Origin of unusual phi-values in protein folding: evidence

against specific nucleation sites. J Mol Biol 334 (5):1077–1085
8. Bulaj G, Goldenberg DP (2001) Phi-values for BPTI folding intermediates and implications for transition state analysis. Nat Struct Biol 8 (4):326–330
9. Ozkan SB, Bahar I, Dill KA (2001) Transition states and the meaning of Phi-values in protein folding kinetics. Nat Struct Biol 8(9):765–769
10. Fersht AR, Sato S (2004) Phi-value analysis and the nature of protein-folding transition states. Proc Natl Acad Sci U S A 101(21):7976–7981
11. Krantz BA, Dothager RS, Sosnick TR (2004) Discerning the structure and energy of multiple transition states in protein folding using psi-analysis. J Mol Biol 337(2):463–475
12. Raleigh DP, Plaxco KW (2005) The protein folding transition state: what are phi-values really telling us? Protein Pept Lett 12 (2):117–122
13. Naganathan AN, Muñoz V (2010) Insights into protein folding mechanisms from large scale analysis of mutational effects. Proc Natl Acad Sci U S A 107(19):8611–8616. https://doi.org/10.1073/pnas.1000988107
14. Baxa MC, Yu W, Adhikari AN, Ge L, Xia Z, Zhou R, Freed KF, Sosnick TR (2015) Even with nonnative interactions, the updated folding transition states of the homologs Proteins G & L are extensive and similar. Proc Natl Acad Sci U S A 112(27):8302–8307. https://doi.org/10.1073/pnas.1503613112
15. Yoo TY, Adhikari A, Xia Z, Huynh T, Freed KF, Zhou R, Sosnick TR (2012) The folding transition state of protein L is extensive with nonnative interactions (and not small and polarized). J Mol Biol 420(3):220–234. https://doi.org/10.1016/j.jmb.2012.04.013
16. Sosnick TR, Krantz BA, Dothager RS, Baxa M (2006) Characterizing the protein folding transition state using psi analysis. Chem Rev 106(5):1862–1876
17. Baxa MC, Freed KF, Sosnick TR (2009) Psi-constrained simulations of protein folding transition states: implications for calculating Phi values. J Mol Biol 386(4):920–928
18. Baxa M, Freed KF, Sosnick TR (2008) Quantifying the structural requirements of the folding transition state of protein A and other systems. J Mol Biol 381:1362–1381
19. Pandit AD, Krantz BA, Dothager RS, Sosnick TR (2007) Characterizing protein folding transition states using Psi-analysis. Methods Mol Biol 350:83–104
20. Krantz BA, Sosnick TR (2001) Engineered metal binding sites map the heterogeneous folding landscape of a coiled coil. Nat Struct Biol 8(12):1042–1047
21. Shandiz AT, Baxa MC, Sosnick TR (2012) A "Link-Psi" strategy using crosslinking indicates that the folding transition state of ubiquitin is not very malleable. Prot Sci 21(6):819–827. https://doi.org/10.1002/pro.2065
22. Shandiz AT, Capraro BR, Sosnick TR (2007) Intramolecular cross-linking evaluated as a structural probe of the protein folding transition state. Biochemistry 46(48):13711–13719
23. Bosco G, Baxa M, Sosnick T (2009) Metal binding kinetics of bi-Histidine sites used in Psi-analysis: evidence for high energy protein folding intermediates. Biochemistry 48 (13):2950–2959. https://doi.org/10.1021/bi802072u
24. Krantz BA, Dothager RS, Sosnick TR (2005) Erratum to Discerning the structure and energy of multiple transition states in protein folding using psi-analysis. J Mol Biol 347 (5):1103
25. Pandit AD, Jha A, Freed KF, Sosnick TR (2006) Small proteins fold through transition states with native-like topologies. J Mol Biol 361(4):755–770
26. Taddei N, Chiti F, Fiaschi T, Bucciantini M, Capanni C, Stefani M, Serrano L, Dobson CM, Ramponi G (2000) Stabilisation of alpha-helices by site-directed mutagenesis reveals the importance of secondary structure in the transition state for acylphosphatase folding. J Mol Biol 300(3):633–647
27. Chiti F, Taddei N, White PM, Bucciantini M, Magherini F, Stefani M, Dobson CM (1999) Mutational analysis of acylphosphatase suggests the importance of topology and contact order in protein folding. Nat Struct Biol 6 (11):1005–1009
28. Went HM, Jackson SE (2005) Ubiquitin folds through a highly polarized transition state. Protein Eng Des Sel 18(5):229–237
29. Sato S, Religa TL, Daggett V, Fersht AR (2004) Testing protein-folding simulations by experiment: B domain of protein A. Proc Natl Acad Sci U S A 101(18):6952–6956
30. Gu H, Kim D, Baker D (1997) Contrasting roles for symmetrically disposed beta-turns in the folding of a small protein. J Mol Biol 274 (4):588–596

31. Kim DE, Fisher C, Baker D (2000) A breakdown of symmetry in the folding transition state of protein L. J Mol Biol 298(5):971–984
32. McCallister EL, Alm E, Baker D (2000) Critical role of beta-hairpin formation in protein G folding. Nat Struct Biol 7(8):669–673
33. Nauli S, Kuhlman B, Baker D (2001) Computer-based redesign of a protein folding pathway. Nat Struct Biol 8(7):602–605
34. Kuhlman B, O'Neill JW, Kim DE, Zhang KY, Baker D (2002) Accurate computer-based design of a new backbone conformation in the second turn of protein L. J Mol Biol 315 (3):471–477. https://doi.org/10.1006/jmbi.2001.5229
35. Yu W, Baxa MC, Gagnon I, Freed KF, Sosnick TR (2016) Cooperative folding near the downhill limit determined with amino acid resolution by hydrogen exchange. Proc Natl Acad Sci U S A. https://doi.org/10.1073/pnas.1522500113
36. Burton RE, Huang GS, Daugherty MA, Calderone TL, Oas TG (1997) The energy landscape of a fast-folding protein mapped by Ala-->Gly substitutions. Nat Struct Biol 4 (4):305–310
37. Plaxco KW, Simons KT, Baker D (1998) Contact order, transition state placement and the refolding rates of single domain proteins. J Mol Biol 277(4):985–994
38. Garcia-Mira MM, Boehringer D, Schmid FX (2004) The folding transition state of the cold shock protein is strongly polarized. J Mol Biol 339(3):555–569
39. Grantcharova VP, Riddle DS, Santiago JV, Baker D (1998) Important role of hydrogen bonds in the structurally polarized transition state for folding of the src SH3 domain. Nat Struct Biol 5(8):714–720
40. Neudecker P, Zarrine-Afsar A, Choy WY, Muhandiram DR, Davidson AR, Kay LE (2006) Identification of a collapsed intermediate with non-native long-range interactions on the folding pathway of a pair of Fyn SH3 domain mutants by NMR relaxation dispersion spectroscopy. J Mol Biol 363:958–976
41. Zarrine-Afsar A, Dahesh S, Davidson AR (2012) A residue in helical conformation in the native state adopts a beta-strand conformation in the folding transition state despite its high and canonical Phi-value. Proteins. https://doi.org/10.1002/prot.24030
42. Di Nardo AA, Korzhnev DM, Stogios PJ, Zarrine-Afsar A, Kay LE, Davidson AR (2004) Dramatic acceleration of protein folding by stabilization of a nonnative backbone conformation. Proc Natl Acad Sci U S A 101 (21):7954–7959. https://doi.org/10.1073/pnas.0400550101
43. Ruczinski I, Sosnick TR, Plaxco KW (2006) Methods for the accurate estimation of confidence intervals on protein folding phi-values. Protein Sci 15(10):2257–2264
44. Krantz BA, Moran LB, Kentsis A, Sosnick TR (2000) D/H amide kinetic isotope effects reveal when hydrogen bonds form during protein folding. Nat Struct Biol 7(1):62–71
45. Krantz BA, Srivastava AK, Nauli S, Baker D, Sauer RT, Sosnick TR (2002) Understanding protein hydrogen bond formation with kinetic H/D amide isotope effects. Nat Struct Biol 9 (6):458–463
46. Scholtz JM, Grimsley GR, Pace CN (2009) Chapter 23 Solvent Denaturation of Proteins and Interpretations of the m Value. In: Johnson ML, Ackers GK, Holt JM (ed) Biothermodynamics, Part B. Methods in Enzymology, vol 466. Academic Press, pp 549–565. https://doi.org/10.1016/S0076-6879(09)66023-7
47. Sharp KA, Englander SW (1994) How much is a stabilizing bond worth? Trends Biochem Sci 19(12):526–529
48. Eyring H (1935) The activated complex in chemical reactions. J Chem Phys 3:107–115
49. Fersht AR (2004) ϕ value versus ψ analysis. Proc Natl Acad Sci U S A 101 (50):17327–17328
50. Guex N, Peitsch MC (1997) SWISS-MODEL and the Swiss-PdbViewer: an environment for comparative protein modeling. Electrophoresis 18(15):2714–2723
51. Gibson DG, Young L, Chuang RY, Venter JC, Hutchison CA 3rd, Smith HO (2009) Enzymatic assembly of DNA molecules up to several hundred kilobases. Nat Methods 6 (5):343–345. https://doi.org/10.1038/nmeth.1318
52. Matthews CR (1987) Effects of point mutations on the folding of globular proteins. Methods Enzymol 154:498–511
53. Myers JK, Pace CN, Scholtz JM (1995) Denaturant m values and heat capacity changes: relation to changes in accessible surface areas of protein unfolding. Protein Sci 4 (10):2138–2148
54. Leffler JE (1953) Parameters for the description of transition states. Science 107:340–341
55. Martinez JC, Pisabarro MT, Serrano L (1998) Obligatory steps in protein folding and the conformational diversity of the transition state. Nat Struct Biol 5(8):721–729
56. Viguera AR, Serrano L (1997) Loop length, intramolecular diffusion and protein folding. Nat Struct Biol 4(11):939–946

57. Yin L, Krantz B, Russell NS, Deshpande S, Wilkinson KD (2000) Nonhydrolyzable diubiquitin analogues are inhibitors of ubiquitin conjugation and deconjugation. Biochemistry 39(32):10001–10010
58. Reddy G, Thirumalai D (2015) Dissecting ubiquitin folding using the self-organized polymer model. J Phys Chem B 119 (34):11358–11370. https://doi.org/10.1021/acs.jpcb.5b03471
59. Shen T, Cao Y, Zhuang S, Li H (2012) Engineered bi-histidine metal chelation sites map the structure of the mechanical unfolding transition state of an elastomeric protein domain GB1. Biophys J 103(4):807–816. https://doi.org/10.1016/j.bpj.2012.07.019
60. Horn JR, Sosnick TR, Kossiakoff AA (2009) Principal determinants leading to transition state formation of a protein-protein complex, orientation trumps side-chain interactions. Proc Natl Acad Sci U S A 106(8):2559–2564. https://doi.org/10.1073/pnas.0809800106
61. Higaki JN, Fletterick RJ, Craik CS (1992) Engineered metalloregulation in enzymes. TIBS 17(3):100–104
62. Kim CA, Berg JM (1993) Thermodynamic beta-sheet propensities measured using a zinc-finger host peptide. Nature 362 (6417):267–270
63. Benson DE, Wisz MS, Hellinga HW (1998) The development of new biotechnologies using metalloprotein design. Curr Opin Biotechnol 9(4):370–376
64. Dwyer MA, Looger LL, Hellinga HW (2003) Computational design of a Zn2+ receptor that controls bacterial gene expression. Proc Natl Acad Sci U S A 100(20):11255–11260
65. Regan L (1995) Protein design: novel metal-binding sites. Trends Biochem Sci 20 (7):280–285
66. Moran LB, Schneider JP, Kentsis A, Reddy GA, Sosnick TR (1999) Transition state heterogeneity in GCN4 coiled coil folding studied by using multisite mutations and crosslinking. Proc Natl Acad Sci U S A 96(19):10699–10704
67. Jumper JM, Faruk NF, Freed KF, Sosnick TR (2018) Trajectory-based training enables protein simulations with accurate folding and Boltzmann ensembles in cpu-hours. PLOS Computational Biology 14(12):e1006578. https://doi.org/10.1371/journal.pcbi.1006578
68. Jumper JM, Faruk NF, Freed KF, Sosnick TR (2018) Accurate calculation of side chain packing and free energy with applications to protein molecular dynamics. PLOS Computational Biology 14(12):e1006342. https://doi.org/10.1371/journal.pcbi.1006342

Chapter 3

Site-Specific Interrogation of Protein Structure and Stability

Debopreeti Mukherjee, Ismail A. Ahmed, and Feng Gai

Abstract

To execute their function or activity, proteins need to possess variability in local electrostatic environment, solvent accessibility, structure, and stability. However, assessing any protein property in a site-specific manner is not easy since native spectroscopic signals often lack the needed specificity. One strategy that overcomes this limitation is to use unnatural amino acids that exhibit distinct spectroscopic features. In this chapter, we describe several such unnatural amino acids (UAAs) and their respective applications in site-specific interrogation of protein structure and stability using standard biophysical methods, including circular dichroism (CD), infrared (IR), and fluorescence spectroscopies.

Key words Unnatural amino acid, Site-specific spectroscopic probe, Protein stability, Protein structure, Infrared spectroscopy, Circular dichroism spectroscopy, Fluorescence spectroscopy, Unnatural amino acid incorporation

1 Introduction

The structural integrity of a folded protein is maintained by many weak interactions, and as a result, it may be subject to local and global conformational fluctuations (or folding and unfolding) [1], and its thermodynamic stability may vary across regions or domains [2]. In addition, a protein may be intrinsically disordered, sampling a dynamic conformational ensemble without forming well-defined and long-lasting tertiary contacts [3]. Furthermore, in order to carry out its function, a protein may need to transiently vary its conformation, either locally or globally, in response to a functional stimulus (e.g., a binding event or change in an environmental property) [4]. Therefore, to provide a complete description of the conformational energy landscape of the protein system in question, from either a protein folding or a functional point of view, we need experimental methods that can characterize protein structure, environment, dynamics, and stability in a site-specific manner. While

Debopreeti Mukherjee and Ismail A. Ahmed contributed equally with all other contributors.

Victor Muñoz (ed.), *Protein Folding: Methods and Protocols*, Methods in Molecular Biology, vol. 2376, https://doi.org/10.1007/978-1-0716-1716-8_3, © Springer Science+Business Media, LLC, part of Springer Nature 2022

NMR spectroscopy is perhaps the method of choice in this regard, its required expertise, cost, and sample requirement often limit its widespread use in the field. Herein, we describe methods that are based on commonly used biophysical techniques, including CD, IR, and fluorescence spectroscopies, for assessing various protein structural and thermodynamic properties with site specificity.

The most commonly used method to determine protein secondary structure and stability is CD spectroscopy. Depending on the secondary structure content, the far-UV CD spectrum (~190–260 nm) of a protein varies, which provides the basis for using CD spectroscopy to determine protein folding-unfolding thermodynamics and structural changes [5]. For example, the CD spectrum of a helical protein is characterized by two negative bands at 209 and 222 nm, respectively, with roughly equal amplitude, while its unfolded counterpart exhibits a CD spectrum that is dominated by a negative band at ~200 nm. Therefore, by monitoring a helical protein's CD signal at 222 nm as a function of either temperature or denaturant concentration, one can obtain a CD unfolding curve that can be further analyzed to determine its folding/unfolding stability [6]. Similarly, the amide I band, which arises from protein's amide backbone units, is also widely used as an IR reporter of protein structure and stability [7]. However, the structural sensitivity of these intrinsic protein spectroscopic signals originates from the dependence of the underlying electronic or vibrational couplings across multiple sites on backbone configurations, and therefore, they cannot be used to reveal environmental or structural details at a specific amino acid site. One intrinsic spectroscopic signal that does offer this ability is tryptophan fluorescence [8]. Nevertheless, the less desirable photophysical properties of tryptophan often limit its practical utility [9–12] in protein biophysical studies.

To increase the information content obtainable via CD, IR, or fluorescence spectroscopy, one strategy is to introduce one or multiple external labels into the protein system of interest that possess unique spectroscopic properties that are not only distinguishable from the rest of the protein but also dependent on the local environment. However, besides the isotope-editing strategy used in IR spectroscopy [13], incorporation of a foreign moiety into the protein sequence in question could induce significant and undesirable perturbations in the native structure and function. Therefore, the past 15 years have seen an increased interest in the development of UAA-based spectroscopic probes and their applications to study various biochemical and biophysical questions in a site-specific manner. This is because UAAs can be easily incorporated into proteins using either chemical or genetic methods, and perhaps more importantly, when replacing a native amino acid of similar size and structure, the potential structural perturbation by an UAA can be minimized. Below, we describe the currently available UAA-based spectroscopic probes (Table 1), with a focus on their utility to reveal site-specific environmental, structural, or stability information via commonly used spectroscopic techniques, including IR, CD, and fluorescence spectroscopies.

2 Materials

The key materials are summarized in Table 1.

3 Methods

3.1 Site-Specific IR Probes

IR spectroscopy (both linear and nonlinear) is an exceedingly useful tool to study the structure and dynamics of biological molecules. This is because the IR spectrum of a molecule is a distinct manifestation of its chemical identity and structure, as well as its environment. In addition, since molecular vibrations occur on the femtosecond timescale, IR spectroscopy has no temporal resolution limit in the study of biological processes. However, the IR spectrum of a native protein is typically dominated by spectral features that arise from overlapping vibrational transitions, due to degeneracy and/or vibrational couplings, hence making many of the naturally occurring IR bands of proteins unsuitable or unusable to reveal any site-specific information. Therefore, to increase the information content obtainable via IR spectroscopy, a straightforward approach is to label the protein in question with an appropriate external IR probe, most commonly in the form of an UAA [14]. This approach, in conjunction with different vibrational spectroscopic techniques, has been used to study a wide variety of protein biochemical and biophysical properties in a site-specific manner, including electrostatic field, hydration status and dynamics, hydrogen-bonding status and dynamics, conformation and conformational stability, intermolecular interactions, ligand binding, and amyloid formation. Currently, there are a large number of UAAs that have been shown to be useful as site-specific IR probes of proteins [14–43], and many of them have an aromatic sidechain (Table 1) [32, 33, 36–38, 41–44]. In addition, several methods are available to incorporate these UAAs into peptides and proteins, including (A) solid-phase peptide synthesis, (B) posttranslational modification, for instance by reacting a suitable reagent with the thiol group of a cysteine residue [34, 45, 46], (C) native chemical ligation, and (D) genetic incorporation through amber codon suppression [47, 48] or through auxotrophs. In Table 2, we summarize the relevant IR property of each UAA, as well as the corresponding method of incorporation.

Table 1
Commonly used UAA-based IR, CD, and fluorescence probes

UAA	Structure (name)	Application
1	*p*-Cyanophenylalanine	IR, fluorescence, and CD probe
2	*p*-Tolylthiocyanate	IR probe
3	*p*-Tolyl selenocyanate	IR probe
4	Cyanocysteine	IR probe
5	5-Cyanotryptophan	IR, fluorescence, and CD probe
6	Phenylcyanate	IR probe
7	Azidohomoalanine	IR probe
8	*p*-Azidophenylalanine	IR probe

(continued)

Table 1 (continued)

UAA	Structure (name)	Application
9	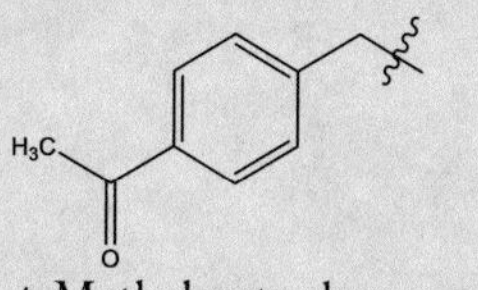 H_3C, O *p*-Methylacetophenone	IR probe
10	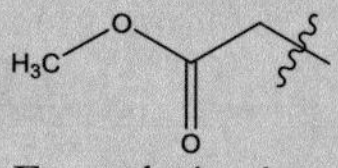 H_3C, O, O Ester derivative of aspartic and glutamic acid	IR probe
11	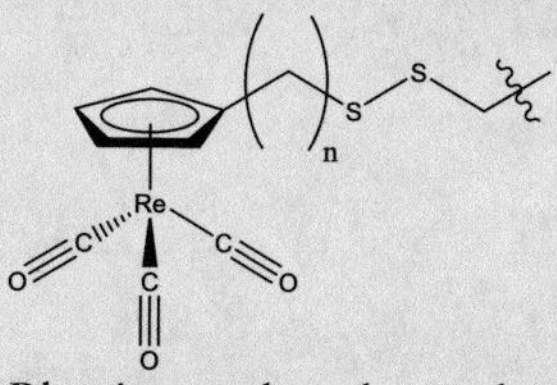 n, S, S, Re, C, O, C, O, C, O Rhenium carbonyl complex	IR probe
12	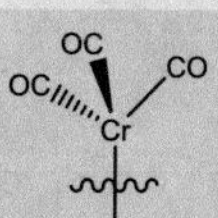 OC, CO, OC, Cr Chromium tricarbonyl complex	IR probe
13	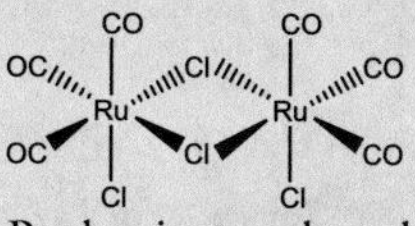 CO, CO, OC, Cl, CO, Ru, Ru, OC, Cl, CO, Cl, Cl Ruthenium carbonyl complex	IR probe
14	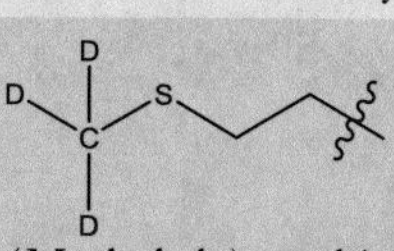 D, D, C, S, D (Methyl-d_3) methionine	IR probe
15	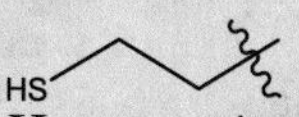 HS Homocysteine	IR probe
16	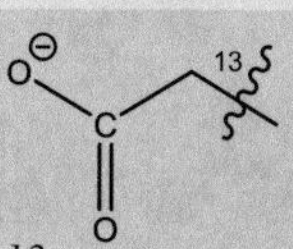 $O^{\ominus}$, 13, C, O ^{13}C-labeled aspartic acid	IR probe
17	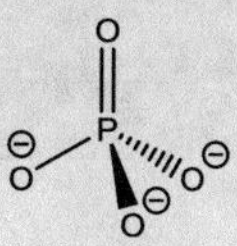 O, P, $O^{\ominus}$, $O^{\ominus}$, $O^{\ominus}$ Phosphate	IR probe

(continued)

Table 1
(continued)

UAA	Structure (name)	Application
18	C-F Carbon-fluorine	IR probe
19	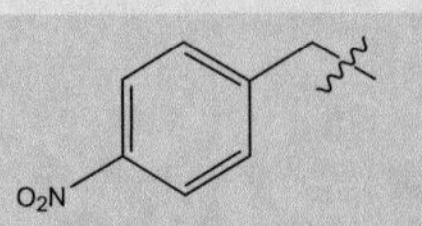 O_2N 4-Nitrophenylalanine	IR probe
20	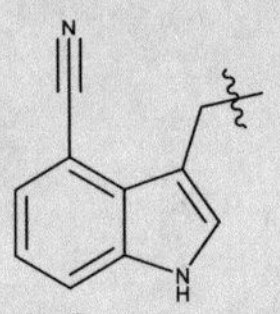 N H 4-Cyanotryptophan	Fluorescence and IR probe
21	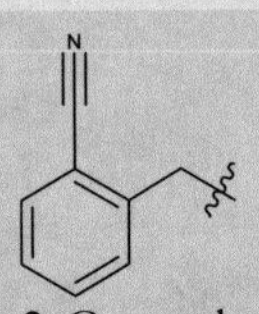 N 2-Cyanophenylalanine	Fluorescence probe
22	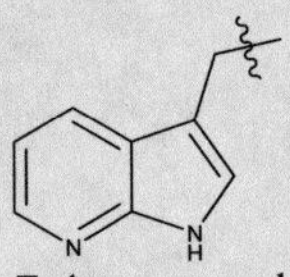 N H 7-Azatryptophan	Fluorescence probe
23	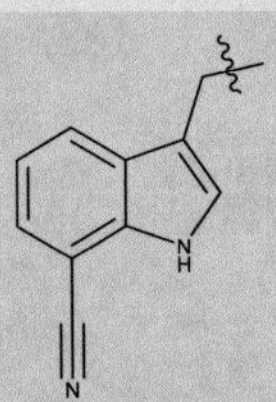 H N 7-Cyanotryptophan	Fluorescence probe
24	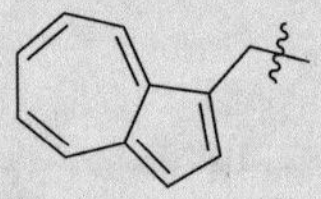 β-(1-Azulenyl)-L-alanine	Fluorescence probe
25	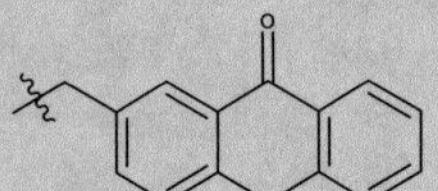 O H Acridon-2-ylalanine	Fluorescence probe

Table 2
Structures and key spectroscopic properties of the commonly used UAA-based IR probes, such as peak frequencies (ν) and bandwidths (w), measured in aqueous (ν_A, w_A) and organic (ν_0, w_0) solutions, as well as molar extinction coefficient (ε) and $|\Delta\nu| = |\nu_A - \nu_0|$

UAA	Structure (name)	Oscillator	ν_A, w_A (cm^{-1}) (solvent)	ν_0, w_0 (cm^{-1}) (solvent)	$\|\Delta\nu\|$ (cm^{-1})	ε (M^{-1} cm^{-1}) (solvent)	Incorporation method	References
1	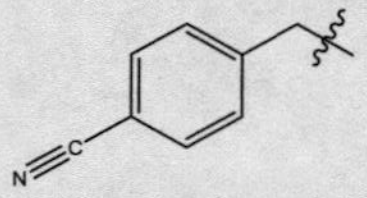 *p*-Cyanophenylalanine	–CN	2237, 10 (H_2O)	2228, 5 (THF)	9	~220 (H_2O)	A, B, C, D	[36, 37, 42]
2	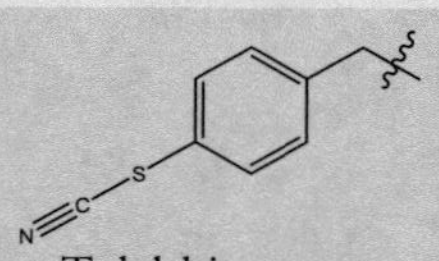 *p*-Tolylthiocyanate	–SCN	2152, 8 (THF/H_2O)	2155, 6 (THF)	3	~50–120 (THF)	A, B, C	[16]
3	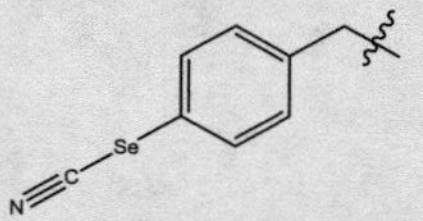 *p*-Tolyl selenocyanate	–SeCN	2150, 8 (THF/H_2O)	2152, 6 (THF)	2	~50–120 (THF)	A, B	[16]
4	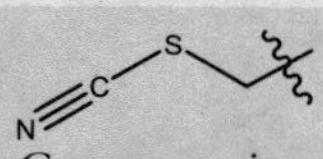 Cyanocysteine	–SCN	2159, 8 (Gly/H_2O)[a]	–	–	~120 (Gly/H_2O)[a]	C	[34, 35]
5	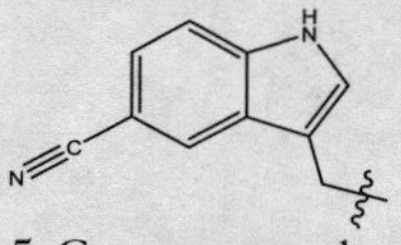 5-Cyanotryptophan	–CN	2222, 18 (THF/H_2O)	2224, 8 (THF)	2	~160 (Gly/H_2O)[a]	A, B, D	[33, 38]
6	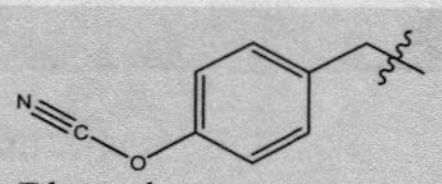 Phenylcyanate	–OCN	2291, 15 (H_2O)	2281, 11 (THF)	10	~800 (THF)	A, B	[32]

(continued)

Table 2 (continued)

UAA	Structure (name)	Oscillator	ν_A, w_A (cm^{-1}) (solvent)	ν_0, w_0 (cm^{-1}) (solvent)	$\|\Delta\nu\|$ (cm^{-1})	ε (M^{-1} cm^{-1}) (solvent)	Incorporation method	References
7	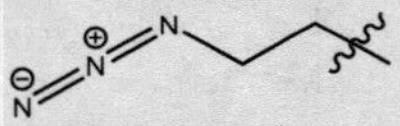 Azidohomoalanine	$-N_3$	2094, 15 (H_2O)	–	–	350–400 (D_2O)	A, B, D	[17, 30, 31]
8	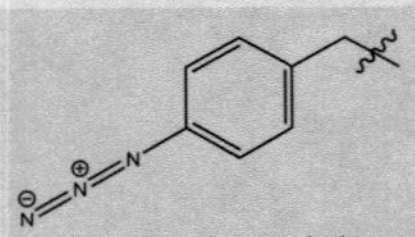 *p*-Azidophenylalanine	$-N_3$	2111, 33 (H_2O)	2098, 28 (DMSO)	13	~610 (H_2O)	A, B, D	[29, 42, 43]
9	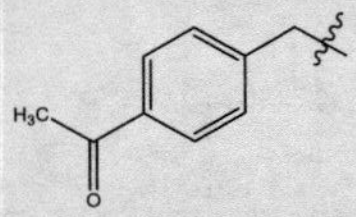 *p*-Methylacetophenone	–CO	1668, 35 (H_2O)	1690, 8 (THF)	22	~1800 (2-MeTHF)[b]	A, B, D	[28, 44]
10	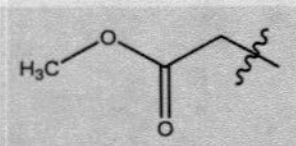 Ester derivative of aspartic and glutamic acid	–CO	~1710, ~40 (D_2O)	1735, 8 (DMSO)	25	~290 (H_2O)	A, B	[27, 45]
11	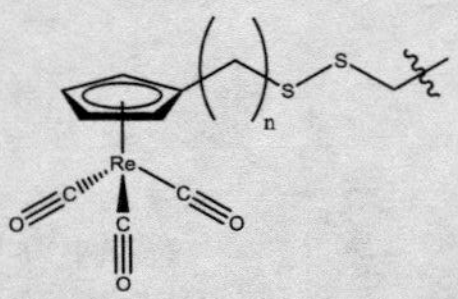 Rhenium carbonyl complex	–CO	–	2008, 10 (DMSO)	–	~4100 (H_2O)	B	[26]
12	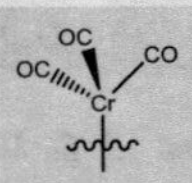 Chromium tricarbonyl complex	–CO	1910, 25 1980, 10 (MeOH)	1895, 30 1965, 10 (DMSO)	15	–	B	[18]

13	Ruthenium carbonyl complex	–CO	2004, 20 2080, 10 (D_2O)	–	–	–	B	[24, 25]
14	(Methyl-d_3) methionine	–CD	1955, 30 (D_2O)	–	–	5–10 (H_2O)	B	[23]
15	Homocysteine	–SH	2550, – (H_2O)	–	–	5–150 (H_2O)	A, B, D	[22]
16	^{13}C-labeled aspartic acid	$–^{13}COO^-$	1540, 10 (D_2O)	–	–	–	A	[15]
17	Phosphate	–PO	1235, 25 (H_2O)	–	–	~500 (H_2O)	A, B, D	[21]
18	C-F Carbon-fluorine	–CF	1215, 8 (H_2O)	–	–	~700 (H_2O)	A, B	[20]
19	4-Nitrophenylalanine	$–NO_2$	1351.1, 9.3 (H_2O)	1347.3, 7.6 (THF)	3.8	–	D	[41]
20	4-Cyanotryptophan	–CN	2219, 12 (H_2O)	2215, 9 (THF)	4	~160	A	[38]

[a]Gly: glycerol

[b]2-MeTHF: 2-methyltetrahydrofuran

3.1.1 Site-Specific CD Probes

When two chromophores (A and B), such as two aromatic residues, are in close proximity in a protein, a bisignate CD couplet can arise due to exciton coupling. The strength of the resultant exciton CD couplet (i.e., $\Delta\varepsilon$) depends on the distance between A and B (r_{AB}) and the magnitudes (μ_A, μ_B) and relative orientation (θ) of their transition dipole moments: [49].

$$\Delta\varepsilon \propto \pm \frac{\mu_A^2 \mu_B^2}{r_{AB}^2} \theta(a, b, c) \quad (1)$$

where a and b represent the angles between r_{AB} and the transition dipole vectors of A and B, respectively, whereas c represents the angle between the two transition dipole moments.

As indicated (Fig. 1), if the two transition dipole moments are oriented in such a way that a clockwise (counterclockwise) rotation is required to bring the one in the front onto the one in the back, the longer-wavelength component of the corresponding CD couplet would have a positive (negative) molar ellipticity value. Therefore, it is possible to use two chromophores to confer site-specificity in protein CD spectroscopy [49]. For example, the π-π* transition (at ~229 nm) of tryptophan has been exploited in this regard and used to probe tryptophan-tryptophan interactions in proteins [50], as well as to determine the stability of designed β-hairpins [51], via CD spectroscopy. However, the exciton CD bands arising from naturally occurring aromatic side chains are typically weak and also overlap with the CD spectrum of the protein backbone, thereby making them less useful in practice to acquire site-specific information in the study of protein structure and stability.

As shown in Eq. 1, in order to produce a CD couplet that is strong and distinguishable from other protein CD signals, one needs to use chromophores that give rise to a (relatively) narrow absorption band with a large molar extinction coefficient and also a

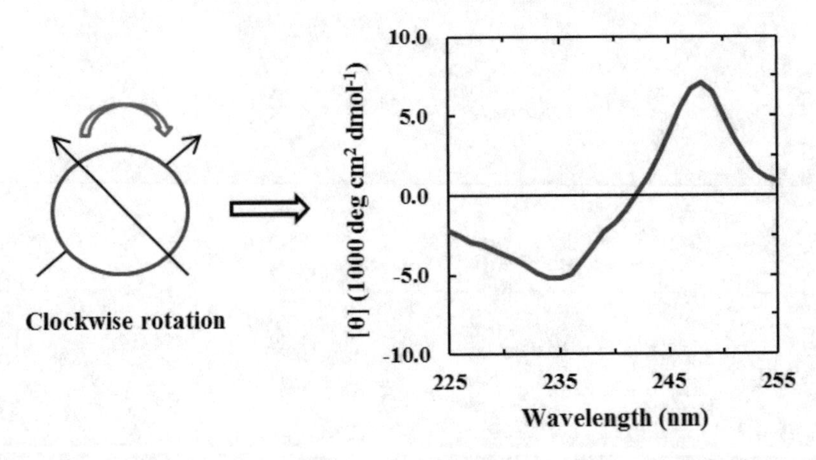

Fig. 1 A CD couplet arising from exciton coupling between two electronic transition dipole moments in a peptide or protein environment

peak wavelength greater than 229 nm. It turns out that, among the site-specific IR probes discussed above, two UAAs, **1** and **5** (Table 2) meet these requirements and can be used as site-specific CD probes [52]. In comparison, probe **5** (i.e., 5-cyanotryptophan) is particularly useful in this regard since the far-UV absorption spectrum of its chromophore (i.e., 5-cyanoindole) is not only red-shifted from that of indole (i.e., the chromophore of tryptophan) but also has a much larger extinction coefficient (Fig. 2). Hence, a pair of 5-cyanotryptophan residues in close proximity can produce a distinct CD couplet that does not significantly overlap with any intrinsic protein CD signals (Fig. 3). Therefore, two 5-cyanotryptophan residues constitute a unique, site-specific CD reporter of protein structure and stability. When placed at strategically selected positions in a protein, their exciton CD signal can be used to probe the stability of either an individual structural element, such as an α-helix, or a specific tertiary contact within the protein following the standard protein denaturation protocols [53]. In addition, when incorporated into different proteins, their CD signature can be used to monitor protein-protein interactions or complex formation. Because of the wide availability of CD spectrometers, we expect that this method will be extensively used to probe protein conformations and conformational changes.

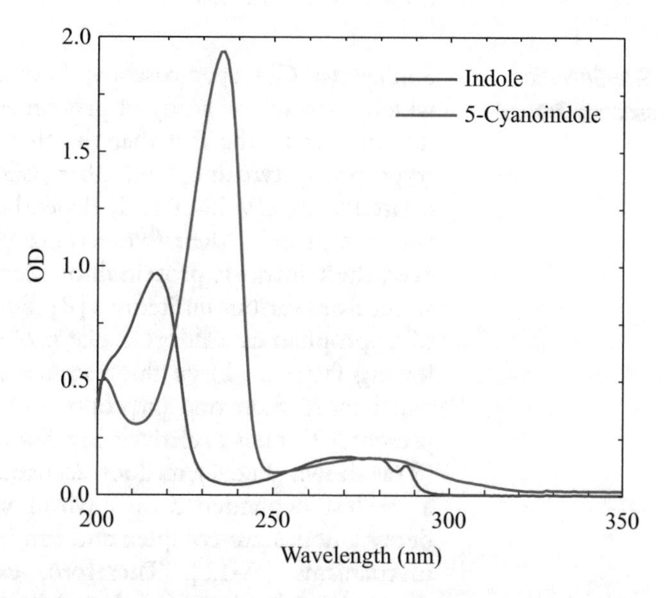

Fig. 2 Absorption spectra of 5-cyanoindole and indole in methanol. Reprinted with permission from Mukherjee D, Gai F (2016) Exciton circular dichroism couplet arising from nitrile-derivatized aromatic residues as a structural probe of proteins. Anal. Biochem. 507:74–78. Copyright (2016) Elsevier

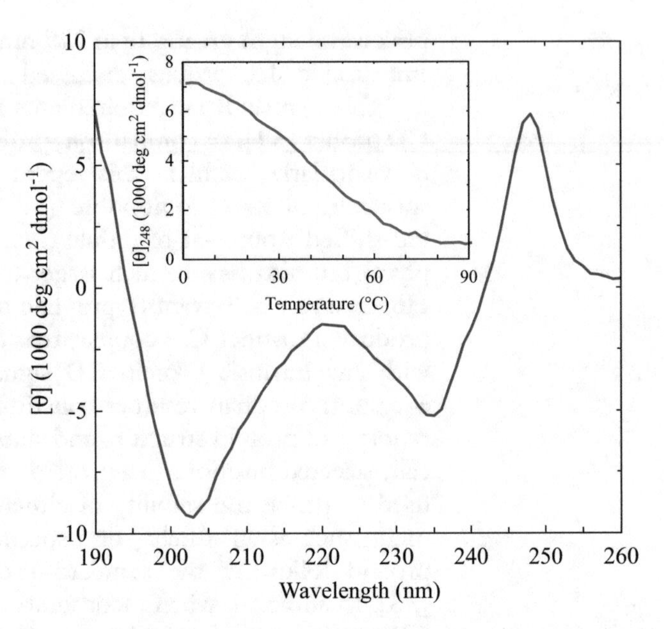

Fig. 3 CD spectrum of a tryptophan zipper peptide that contains two closely spaced 5-cyanotryptophan residues. The T-melt curve of the peptide monitored at 248 nm is shown in the inset. Reprinted with permission from Mukherjee D, Gai F (2016) Exciton circular dichroism couplet arising from nitrile-derivatized aromatic residues as a structural probe of proteins. Anal. Biochem. 507:74–78. Copyright (2016) Elsevier

3.2 Site-Specific Fluorescence Probes

Similar to CD spectroscopy, fluorescence spectroscopy is also widely used in the study of protein structure and stability. This is due in part to the fact that the three aromatic amino acids (i.e., tryptophan, tyrosine, and phenylalanine) are fluorescent when excited with UV light and, depending on their specific location within a protein, their fluorescence properties can change. However, these intrinsic protein fluorescence probes, while very useful, suffer from various limitations [8]. For example, the practical utility of tryptophan as a fluorescence probe is often limited by the following factors: (1) its fluorescence cannot be selectively excited when more than one tryptophan or other aromatic residues are present, (2) it has a relatively low fluorescence quantum yield (QY), (3) as shown (Fig. 4), its fluorescence spectrum and QY only exhibit a modest dependence on environment, and (4) its exited-state decay kinetics are complex and can be affected by many different mechanisms [9–11]. Therefore, extensive efforts have been devoted to develop UAA-based fluorophores that possess much improved photophysical properties and hence offer enhanced utility over that of naturally occurring fluorescent amino acids. As shown (Table 3), these extrinsic fluorophores are simple derivatives of tryptophan, phenylalanine, or tyrosine [54, 55], hence having the smallest size in comparison to those commonly used bulky

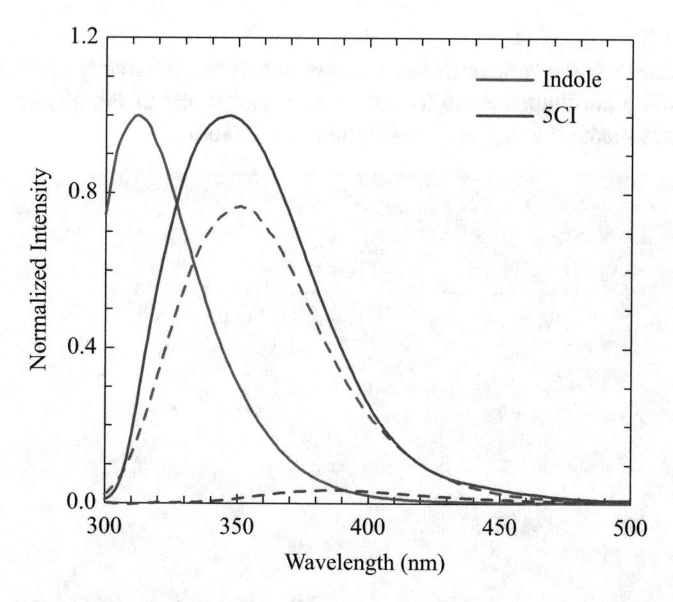

Fig. 4 Normalized fluorescence spectra of indole (sidechain of tryptophan) and 5-cyanoindole (5CI, sidechain of 5-cyanotryptophan) obtained in 1,4-dioxane (solid lines) and H_2O (dashed lines), as indicated. Reprinted with permission from Markiewicz BN, Mukherjee D, Troxler T, Gai F (2016) Utility of 5-cyanotryptophan fluorescence as a sensitive probe of protein hydration. J. Phys. Chem. B 120:936–944. Copyright (2016) American Chemical Society

organic dyes or fluorescent proteins [8]. This characteristic can be quite advantageous in applications where any potential structural perturbation should be minimized. Moreover, it is worth noting that among these fluorescent UAAs 4-cyanotryptophan exhibits distinct photophysical properties, making it a useful blue fluorescent amino acid for biological spectroscopy and microscopy [56]. In addition, a recent study [40] suggests that it is possible to further red-shift the emission wavelength of tryptophan-based fluorophores by using different substitution groups at the fourth position of the indole ring.

In addition to being used as stand-alone fluorophores, these UAAs can also form novel fluorescence donor–acceptor pairs with other natural and unnatural amino acids. These donor–acceptor pairs (Table 4), where the fluorescence intensity of the donor is quenched by the acceptor through the mechanism of either fluorescence resonance energy transfer (FRET) or electron transfer (ET), are exceedingly useful in providing site-specific structural information as the quenching efficiency is distance dependent [8]. However, in comparison to commonly used FRET pairs composed of fluorescent dyes or proteins, these UAA-based FRET pairs have a smaller Förster distances (R_0) and hence are more amenable for site-specific interrogation of local interactions. In addition, a detailed data analysis procedure can be found in Ref. 69 on how to use denaturant concentration-dependent FRET signals to determine protein folding stability.

Table 3
Structures and key spectroscopic properties of the commonly used UAA-based fluorescence probes, including the fluorescence QY and peak wavelengths of the absorption (λ_{abs}) and emission (λ_{em}) spectra measured in water and an organic solvent

UAA	Structure (name)	λ_{abs} (nm) (solvent)	λ_{em} (nm), QY (solvent)	λ_{em} (nm), QY (solvent)	Incorporation method	References
1	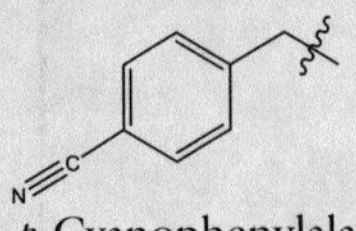 *p*-Cyanophenylalanine	240	295, 0.11 (H_2O)	295, 0.026 (Acetonitrile)	A, B, C, D	[47, 48, 54]
5	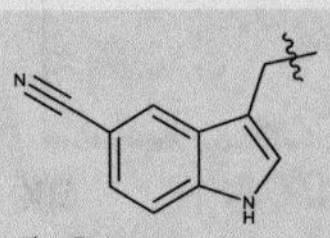 5-Cyanotryptophan	280	391, 0.01 (H_2O)	361, 0.11 (1,4-Dioxane)	A, B, D	[55]
20	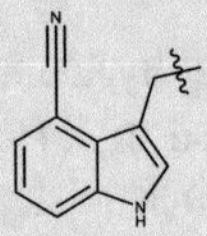 4-Cyanotryptophan	325	405, 0.84 (H_2O)	405, 0.62 (THF)	A, B	[38, 56–58]
21	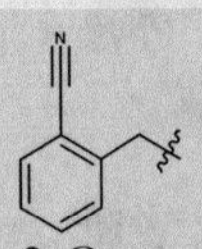 2-Cyanophenylalanine	280	305, 0.14 (H_2O)	305, – (Acetonitrile)	A, B	[59]
22	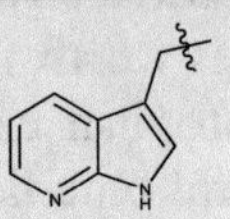 7-Azatryptophan	295	400, 0.01 (H_2O)	362, 0.25 (Acetonitrile)	A, B, D	[60–63]
23	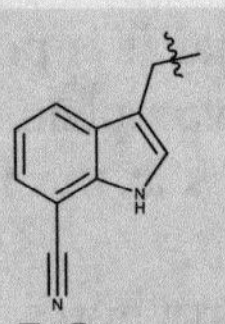 7-Cyanotryptophan	308	400, 0.05 (H_2O)	375, 0.54 (THF)	A, B, D	[57, 64]
24	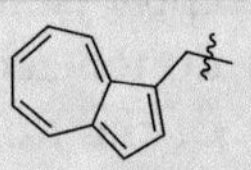 β-(1-Azulenyl)-L-alanine	342	381, 0.07 (H_2O)	384, 0.06 (Ether)	A, B, D	[65, 66]
25	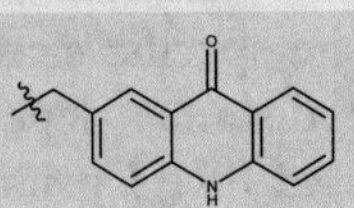 Acridon-2-ylalanine	407	430, 0.95 (H_2O)	425, 0.29 (THF)	A, B, C, D	[67, 68]

Table 4
Structures and key spectroscopic properties of the commonly used UAA-based fluorescence donor–acceptor pairs, including the Förster distance (R_0) and peak wavelengths of the absorption (λ_{abs}) and emission (λ_{em}) spectra

Donor	Acceptor	Mechanism	R_0 (Å)	Donor $\lambda_{abs}/\lambda_{em}$ (nm)	Acceptor $\lambda_{abs}/\lambda_{em}$ (nm)	References
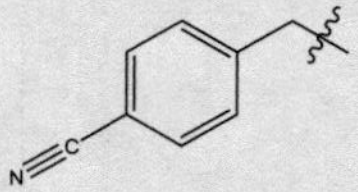 *p*-Cyanophenylalanine	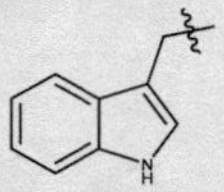 Tryptophan	FRET	16.0	240/295	280/350	[69]
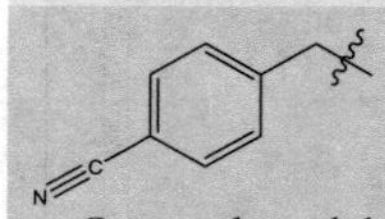 *p*-Cyanophenylalanine	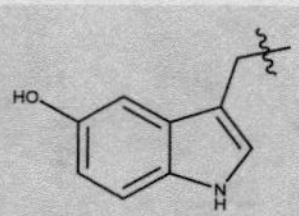 5-Hydroxytryptophan	FRET	18.5	240/295	315/340	[70]
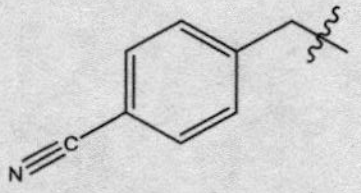 *p*-Cyanophenylalanine	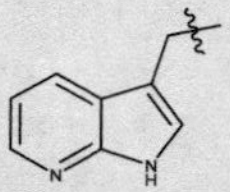 7-Azatryptophan	FRET	18.5	240/295	295/400	[70]
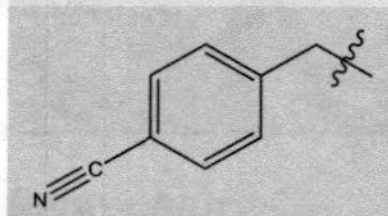 *p*-Cyanophenylalanine	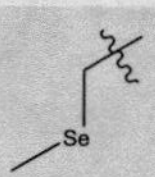 Selenomethionine	ET	–	240/295	219/–	[71–73]
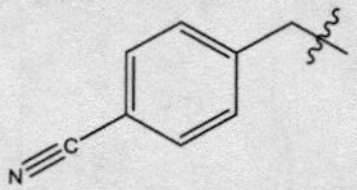 *p*-Cyanophenylalanine	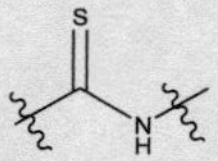 Thioamide	FRET	20	240/295	266/–	[74]
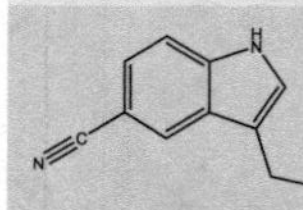 6-Cyanotryptophan	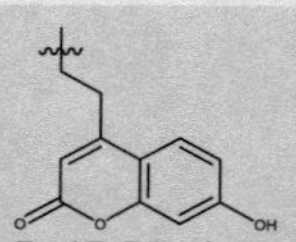 L-(7-Hydroxycoumarin-4-yl) ethylglycine	FRET	28.3	290/370	345/450	[64]

(continued)

Table 4 (continued)

Donor	Acceptor	Mechanism	R_0 (Å)	Donor $\lambda_{abs}/\lambda_{em}$ (nm)	Acceptor $\lambda_{abs}/\lambda_{em}$ (nm)	References
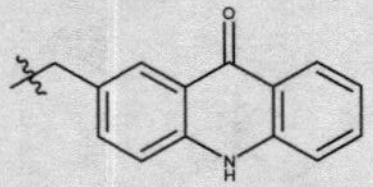 Acridon-2-ylalanine	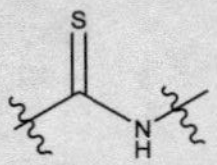 Thioamide	ET	–	407/430	266/–	[67, 68]
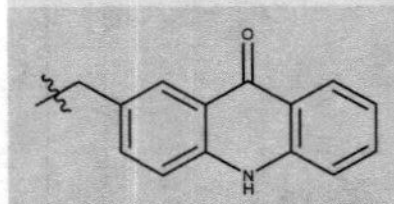 Acridon-2-ylalanine	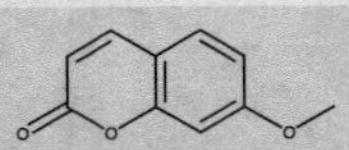 Methoxycoumarin	FRET	25	407/430	340/390	[67, 68]

4 Notes

Although a vast majority of UAAs have been shown to be useful as site-specific IR probes of proteins; [14–46] however, as indicated (Table 2), these probes differ in size, vibrational property, sensitivity to environment, and the ease of being incorporated into proteins. Therefore, the following factors need to be considered when choosing an UAA for a specific application.

1. Size and structure: Except UAA **11**, **13**, and **15**, incorporation of any UAA will unavoidably perturb the structure and/or other property of the protein system in question. Therefore, to minimize any potential perturbations, the chosen UAA should be similar, in both structure and size, to the targeted native residue for mutation or substitution. In this regard, UAA **1**, **2**, **3**, **6**, **8**, and **9** can be used to replace phenylalanine and tyrosine, **5** can be used to replace tryptophan, **10** and **16** can be used to replace aspartic acid and glutamic acid, and **14** can be used to replace methionine.
2. Sensitivity to environment: For a vibrational mode to be useful as a site-specific spectroscopic reporter, one of its properties, including peak frequency (ν), bandwidth (w), molar absorptivity (ε), and lifetime (τ), should exhibit a dependence on the environment. Except **15**, whose oscillator strength has been shown to increase by a factor of ~30 upon moving from a hydrated to a dehydrated environment [22], the utility of all other listed IR probes stems from the dependence of their ν and/or w on the environment. In practice, this dependence is often evaluated by the difference of frequencies ($|\Delta\nu|$) measured in H_2O/D_2O and an organic solvent, such as tetrahydrofuran (THF) or dimethyl sulfoxide (DMSO). If the width of the IR band in question is sufficiently narrow, a larger $|\Delta\nu|$ would mean a higher sensitivity and hence a larger dynamic range. Some of the probes in Table 2 have been examined in those solvents. For example, upon changing the solvent from H_2O to THF, the CN stretching frequency of *p*-cyanophenylalanine is shifted from 2237 to 2228 cm^{-1}, thus resulting in a $|\Delta\nu|$ of 9.0 cm^{-1} (Fig. 5) [37]. Based on the $|\Delta\nu|$ values, the environmental sensitivities of these probes can be ranked in the following order: $\mathbf{10 > 9 > 12 > 8 > 6 > 1 > 20 \approx 21 > 2 > 3 \approx 5}$.
3. Frequency range and solvent: Water is an essential solvent for biological molecules. However, water absorbs ubiquitously in the mid-IR region, hence producing a solvent background in all IR measurements. Due to the isotopic effect, the vibrational spectrum of D_2O is shifted from that of H_2O; therefore, to reduce this solvent background, H_2O should be used for

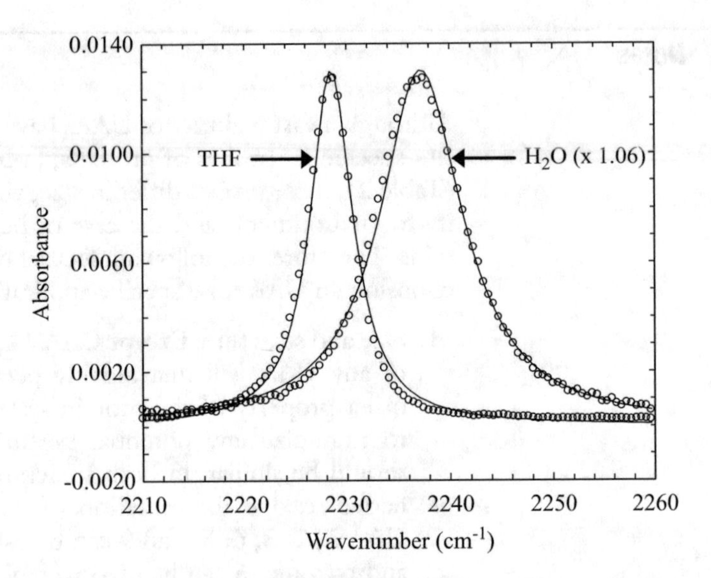

Fig. 5 FTIR spectra of *p*-cyanophenylalanine in water and THF. Reprinted with permission from Getahun Z, Huang CY, Wang T, De León B, DeGrado WF, Gai F (2003) Using nitrile-derivatized amino acids as infrared probes of local environment. J. Am. Chem. Soc. 125:405–411. Copyright (2003) American Chemical Society

probes **1–8, 15**, **20**, and **21**, whereas for probes **9–13**, one should consider using D_2O as the solvent. In addition, for studies involving lipid membranes, probes **9** and **10** are less applicable due to strong spectral overlapping between their IR bands and those of the lipid headgroups.

4. Sample concentration: Due to constraints arising from protein availability, solubility, and/or aggregation, sample concentration is a main factor that determines which probe is better suited for a specific application. When structural perturbation is not a major concern, using probes that have a high extinction coefficient, including UAA **6–11**, **17**, and **18**, would help lower the amount of protein sample used in a single experiment. While azide-based probes (i.e., **7** and **8**) enjoy advantages in this regard, it could be difficult to quantitatively interpret their signals due to different effects.
5. Application: Depending on the nature of the question being studied, one probe can be better suited than others. While all of them, when placed at the right location in a protein, can be used to follow a well-defined structural transition, such as protein folding and unfolding, due to the dependence of their frequencies on the environment, only probes **1**, **4**, **5**, **8**, **10**, **14**, **15**, and **18** can be used to provide quantitative information about the local electric field because the corresponding frequency-field maps or Stark tuning rates have been

determined. Whereas for studies that employ nonlinear IR techniques to investigate protein hydration dynamics, probes **1**, **5**, and **11–13** are better choices because of their larger transition dipole moments. On the other hand, probes **3** and **6** are more useful in applications where a long vibrational lifetime is desirable or required.

Similar to those highlighted in the IR section above, below we discuss factors concerning selection of fluorescent UAA probes.

1. Size and structure: All of the fluorescent UAAs listed in Table 3 are simple derivatives of tryptophan, tyrosine, or phenylalanine. Therefore, ideally they should replace the respective native residues. If that is not possible, the chosen UAA should be placed at a location where it is unlikely to induce a large structural change or impede the native function.
2. Brightness: In order for a UAA to be useful as a fluorescent probe, it must have a certain brightness (B) which is the product of the molar extinction coefficient (ε) at the excitation wavelength (λ_{ex}) and the corresponding fluorescence QY. For all the UAAs listed, their ε values are in the range of ~1000 M^{-1} cm^{-1} to 8000 M^{-1} cm^{-1}, and their maximum fluorescence QYs vary (Table 3). For applications where a low protein concentration is required, a probe that has a larger B value is more desirable (if it also meets other requirements).
3. Sensitivity to environment: The applicability of a fluorescent probe in studying protein structure and stability stems from the environmental dependence of its fluorescence properties, including fluorescence QY, peak emission wavelength (λ_{em}), and lifetime (τ_F). In other words, for an ideal fluorophore, one or all of those properties should undergo a drastic change, hence providing a large signal contrast, upon going from a hydrated to a dehydrated environment. In this regard, UAA **1**, **5**, **21**, and **22** are the most useful environment-sensitive fluorescence reporters. For example, as indicated (Fig. 4), the total fluorescence intensity of 5-cynaoindole, the sidechain of UAA **5** is increased by a factor of ~23 upon going from water to 1,4-dioxane, much larger than that observed for indole (the sidechain of tryptophan). Besides applications to study protein conformational changes, another widely pursued utility of fluorescence spectroscopy is to determine protein distribution or localization in the cellular environment. For this purpose, UAA **20** and **25** are the most applicable.
4. Excitation wavelength: When multiple fluorescent amino acids are present in the protein system in question, the ability to selectively excite the UAA fluorophore is crucial for attaining site-specific information. Among those listed UAAs, **1**, **5**, and

20–25 can all be specifically excited, even in the presence of other canonical aromatic amino acids.

5. Incorporation method: As discussed above, several methods are available to incorporate UAAs into peptides and proteins, and for those fluorescent UAAs, the methods of incorporation are given in Table 3.

Acknowledgments

We gratefully acknowledge financial support from the National Institutes of Health (GM-065978 and P41-GM104605). I.A.A. is supported by a NIH T32 Interdisciplinary Cardiovascular Training Grant (T32-HL007954).

References

1. Kim PS, Baldwin RL (1990) Intermediates in the folding reactions of small proteins. Annu Rev Biochem 59:631–660
2. Garcia-Mira MM, Sadqi M, Fischer N, Sanchez-Ruiz JM, Muñoz V (2002) Experimental identification of downhill protein folding. Science 298:2191–2195
3. Tompa P (2012) Intrinsically disordered proteins: a 10-year recap. Trends Biochem Sci 37:509–516
4. Fersht A (1999) Structure and mechanism in protein science: a guide to enzyme catalysis and protein folding. W.H. Freeman
5. Berova N, Nakanishi K, Woody R (2000) Circular dichroism : principles and applications. Wiley-VCH
6. Greenfield NJ (1999) Applications of circular dichroism in protein and peptide analysis. Trends Anal Chem 18:236–244
7. Yang H, Yang S, Kong J, Dong A, Yu S (2015) Obtaining information about protein secondary structures in aqueous solution using Fourier transform IR spectroscopy. Nat Protoc 10:382–396
8. Lakowicz JR (2006) Principles of fluorescence spectroscopy, 3rd edn. Springer
9. Adams PD, Chen Y, Ma K, Zagorski MG, Sönnichsen FD, McLaughlin ML, Barkley MD (2002) Intramolecular quenching of tryptophan fluorescence by the peptide bond in cyclic hexapeptides. J Am Chem Soc 124:9278–9286
10. Chen Y, Liu B, Yu HT, Barkley MD (1996) The peptide bond quenches indole fluorescence. J Am Chem Soc 118:9271–9278
11. Gudgin E, Lopez-Delgado R, Ware WR (1981) The tryptophan fluorescence lifetime puzzle. A study of decay times in aqueous solution as a function of pH and buffer composition. Can J Chem 59:1037–1044
12. Petrich JW, Chang MC, McDonald DB, Fleming GR (1983) On the origin of nonexponential fluorescence decay in tryptophan and its derivatives. J Am Chem Soc 105:3824–3832
13. Decatur SM (2006) Elucidation of residue-level structure and dynamics of polypeptides via isotope-edited infrared spectroscopy. Acc Chem Res 39:169–175
14. Ma J, Pazos IM, Zhang W, Culik RM, Gai F (2015) Site-specific infrared probes of proteins. Annu Rev Phys Chem 66:357–377
15. Abaskharon RM, Brown SP, Zhang W, Chen J, Smith AB III, Gai F (2017) Isotope-labeled aspartate sidechain as a non-perturbing infrared probe: application to investigate the dynamics of a carboxylate buried inside a protein. Chem Phys Lett 683:193–198
16. Levin DE, Schmitz AJ, Hines SM, Hines KJ, Tucker MJ, Brewer SE, Fenlon EE (2016) Synthesis and evaluation of the sensitivity and vibrational lifetimes of thiocyanate and selenocyanate infrared reporters. RSC Adv 6:36231–36237
17. Taskent-Sezgin H, Chung J, Banerjee PS, Nagarajan S, Dyer RB, Carrico I, Raleigh DP

(2010) Azidohomoalanine: a conformationally sensitive IR probe of protein folding, protein structure, and electrostatics. Angew Chem Int Ed 49:7473–7475
18. Osborne DG, Dunbar JA, Lapping JG, White AM, Kubarych KJ (2013) Site-specific measurements of lipid membrane interfacial water dynamics with multidimensional infrared spectroscopy. J Phys Chem B 117:15407–15414
19. Nie B, Stutzman J, Xie A (2005) A vibrational spectral maker for probing the hydrogen-bonding status of protonated Asp and Glu residues. Biophys J 88:2833–2847
20. Suydam IT, Boxer SG (2003) Vibrational stark effects calibrate the sensitivity of vibrational probes for electric fields in proteins. Biochemistry 42:12050–12055
21. Levinson NM, Bolte EE, Miller CS, Corcelli SA, Boxer SG (2011) Phosphate vibrations probe local electric fields and hydration in biomolecules. J Am Chem Soc 133:13236–13239
22. Kozi M, Garrett-Roe S, Hamm P (2008) 2D-IR spectroscopy of the sulfhydryl band of cysteines in the hydrophobic core of proteins. J Phys Chem B 112:7645–7650
23. Zimmermann JO, Thielges MC, Yu W, Dawson PE, Romesberg FE (2011) Carbon-deuterium bonds as site-specific and nonperturbative probes for time-resolved studies of protein dynamics and folding. J Phys Chem Lett 2:412–416
24. King JT, Arthur EJ, Brooks CL, Kubarych KJ (2012) Site-specific hydration dynamics of globular proteins and the role of constrained water in solvent exchange with amphiphilic cosolvents. J Phys Chem B 116:5604–5611
25. King JT, Kubarych KJ (2012) Site-specific coupling of hydration water and protein flexibility studied in solution with ultrafast 2D-IR spectroscopy. J Am Chem Soc 134:18705–18712
26. Woys AM, Mukherjee SS, Skoff DR, Moran SD, Zanni MT (2013) A strongly absorbing class of non-natural labels for probing protein electrostatics and solvation with FTIR and 2D IR spectroscopies. J Phys Chem B 117:5009–5018
27. Pazos IM, Ghosh A, Tucker MJ, Gai F (2014) Ester carbonyl vibration as a sensitive probe of protein local electric field. Angew Chem Int Ed 126:6194–6198
28. Fried SD, Bagchi S, Boxer SG (2013) Measuring electrostatic fields in both hydrogen-bonding and non- hydrogen-bonding environments using carbonyl vibrational probes. J Am Chem Soc 135:11181–11192
29. Bazewicz CG, Liskov MT, Hines KJ, Brewer SH (2013) Sensitive, site-specific, and stable vibrational probe of local protein environments: 4-Azidomethyl-L-Phenylalanine. J Phys Chem B 117:8987–8993
30. Bloem R, Koziol K, Waldauer SA, Buchli B, Walser R, Samatanga B, Jelesarov I, Hamm P (2012) Ligand binding studied by 2D IR spectroscopy using the azidohomoalanine label. J Phys Chem B 116:13705–13712
31. Choi JH, Raleigh D, Cho M (2011) Azidohomoalanine is a useful infrared probe for monitoring local electrostatistics and side-chain solvation in proteins. J Phys Chem Lett 2:2158–2162
32. Tucker MJ, Kim YS, Hochstrasser RM (2009) 2D IR photon echo study of the anharmonic coupling in the OCN region of phenyl cyanate. Chem Phys Lett 470:80–84
33. Waegele MM, Tucker MJ, Gai F (2009) 5-Cyanotryptophan as an infrared probe of local hydration status of proteins. Chem Phys Lett 478:249–253
34. Fafarman AT, Webb LJ, Chuang JI, Boxer SG (2006) Site-specific conversion of cysteine thiols into thiocyanate creates an IR probe for electric fields in proteins. J Am Chem Soc 128:13356–13357
35. Alfieri KN, Vienneau AR, Londergan CH (2011) Using infrared spectroscopy of cyanylated cysteine to map the membrane binding structure and orientation of the hybrid antimicrobial peptide CM15. Biochemistry 50:11097–11108
36. Gao Y, Zou Y, Ma Y, Wang D, Sun Y, Ma G (2016) Infrared probe technique reveals a millipede-like structure for Aβ (8–28) amyloid fibril. Langmuir 32:937–946
37. Getahun Z, Huang C, Wang T, De León B, DeGrado WF, Gai F (2003) Using nitrile-derivatized amino acids as infrared probes of local environment. J Am Chem Soc 125:405–411
38. van Wilderen LJ, Brunst H, Gustmann H, Wachtveitl J, Broos J, Bredenbeck J (2018) Cyano-tryptophans as dual infrared and fluorescence spectroscopic labels to assess structural dynamics in proteins. Phys Chem Chem Phys 20:19906–19915
39. Jia B, Sun Y, Yang L, Yu Y, Fan H, Ma G (2018) A structural model of the hierarchical assembly of an amyloid nanosheet by an infrared probe technique. Phys Chem Chem Phys 20(43):27261–27271
40. Huang XY, You M, Ran GL, Fan HR, Zhang WK (2018) Ester-derivatized indoles as fluorescent and infrared probes for hydration environments. Chin J Chem Phys 31:477–484

41. Smith EE, Linderman BY, Luskin AC, Brewer SH (2011) Probing local environments with the infrared probe: L-4-nitrophenylalanine. J Phys Chem B 115:2380–2385
42. Deiters A, Cropp TA, Mukherji M, Chin JW, Anderson JC, Schultz PG (2003) Adding amino acids with novel reactivity to the genetic code of *Saccharomyces Cerevisiae*. J Am Chem Soc 125:11782–11783
43. Deiters A, Cropp TA, Summerer D, Mukherji M, Schultz PG (2004) Site-specific PEGylation of proteins containing unnatural amino acids. Bioorg Med Chem Lett 14:5743–5745
44. Wang L, Zhang Z, Brock A, Schultz PG (2003) Addition of the keto functional group to the genetic code of Escherichia coli. Proc Natl Acad Sci 100:56–61
45. Ahmed IA, Gai F (2017) Simple method to introduce an ester infrared probe into proteins. Protein Sci 26:375–381
46. Jo H, Culik RM, Korendovych IV, DeGrado WF, Gai F (2010) Selective incorporation of nitrile-based infrared probes into proteins via cysteine alkylation. Biochemistry 49:10354–10356
47. Miyake-Stoner SJ, Miller AM, Hammill JT, Peeler JC, Hess KR, Mehl RA, Brewer SH (2009) Probing protein folding using site-specifically encoded unnatural amino acids as FRET donors with tryptophan. Biochemistry 48:5953–5962
48. Schultz KC, Supekova L, Ryu Y, Xie J, Perera R, Schultz PG (2006) A genetically encoded infrared probe. J Am Chem Soc 128:13984–13985
49. Berova N, Di Bari L, Pescitelli G (2007) Application of electronic circular dichroism in configurational and conformational analysis of organic compounds. Chem Soc Rev 36:914–931
50. Gasymov OK, Abduragimov AR, Glasgow BJ (2015) Double tryptophan exciton probe to gauge proximal side chains in proteins: augmentation at low temperature. J Phys Chem B 119:3962–3968
51. Cochran AG, Skelton NJ, Starovasnik MA (2001) Tryptophan zippers: stable, monomeric beta-hairpins. Proc Natl Acad Sci U S A 98:5578–5583
52. Mukherjee D, Gai F (2016) Exciton circular dichroism couplet arising from nitrile-derivatized aromatic residues as a structural probe of proteins. Anal Biochem 507:74–78
53. Neurath H, Greensteix JP, Putnam FW, Erickson JO (1944) The chemistry of protein denaturation. Chem Rev 34:157–265
54. Tucker MJ, Oyola R, Gai F (2006) A novel fluorescent probe for protein binding and folding studies:p-cyano-phenylalanine. Biopolymers 83:571–576
55. Markiewicz BN, Mukherjee D, Troxler T, Gai F (2016) Utility of 5-cyanotryptophan fluorescence as a sensitive probe of protein hydration. J Phys Chem B 120:936–944
56. Hilaire MR, Ahmed IA, Lin CW, Jo H, DeGrado WF, Gai F (2017) Blue fluorescent amino acid for biological spectroscopy and microscopy. Proc Natl Acad Sci U S A 114:6005–6009
57. Mukherjee D, Ortiz Rodriguez LI, Hilaire MR, Troxler T, Gai F (2018) 7-cyanoindole fluorescence as a local hydration reporter: application to probe the microheterogeneity of nine water-organic binary mixtures. Phys Chem Chem Phys 20:2527–2535
58. Hilaire MR, Mukherjee D, Troxler T, Gai F (2017) Solvent dependence of cyanoindole fluorescence lifetime. Chem Phys Lett 685:133–138
59. Martin JP, Fetto NR, Tucker MJ (2016) Comparison of biological chromophores: photophysical properties of cyanophenylalanine derivatives. Phys Chem Chem Phys 2016 (18):20750–20757
60. Soumillion P, Jespers L, Vervoort J, Fastrez J (1995) Biosynthetic incorporation of 7-azatryptophan into the phage lambda lysozyme: estimation of tryptophan accessibility, effect on enzymatic activity and protein stability. Protein Eng Des Sel 8:451–456
61. Schlesinger S (1968) The effect of amino acid analogues on alkaline phosphatase. Formation in Escherichia coli K-12. II. Replacement of tryptophan by azatryptophan and by tryptazan. J Biol Chem 243:3877–3883
62. Ross JB, Senear DF, Waxman E, Kombo BB, Rusinova E, Huang YT, Laws WR, Hasselbacher CA (1992) Spectral enhancement of proteins: biological incorporation and fluorescence characterization of 5-hydroxytryptophan in bacteriophage lambda cI repressor. Proc Natl Acad Sci U S A 89:12023–12027
63. Guharay J, Sengupta PK (1996) Characterization of the fluorescence emission properties of 7-azatryptophan in reverse micellar environments. Biochem Biophys Res Commun 219:388–392
64. Talukder P, Chen S, Roy B, Yakovchuk P, Spiering MM, Alam MP, Madathil MM, Bhattacharya C, Benkovic SJ, Hecht SM (2015) Cyanotryptophans as novel fluorescent probes for studying protein conformational changes and DNA-protein interaction. Biochemistry 54:7457–7469

65. Moroz YS, Binder W, Nygren P, Caputo GA, Korendovych IV (2013) Painting proteins blue: β-(1-azulenyl)-l-alanine as a probe for studying protein–protein interactions. Chem Commun 49:490–492
66. Shao J, Korendovych IV, Broos J (2015) Biosynthetic incorporation of the azulene moiety in proteins with high efficiency. Amino Acids 47:213–216
67. Szymańska A, Wegner K, Łankiewicz L (2003) Synthesis of N-[(tert-Butoxy)carbonyl]-3-(9,10-dihydro-9-oxoacridin-2-yl)-L-alanine, a new fluorescent amino acid derivative. Helv Chim Acta 86:3326–3331
68. Speight LC, Muthusamy AK, Goldberg JM, Warner JB, Wissner RF, Willi TS, Woodman BF, Mehl RA, Petersson EJ (2013) Efficient synthesis and in vivo incorporation of acridon-2-ylalanine, a fluorescent amino acid for lifetime and förster resonance energy transfer/luminescence resonance energy transfer studies. J Am Chem Soc 135:18806–18814
69. Glasscock JM, Zhu Y, Chowdhury P, Tang J, Gai F (2008) Using an amino acid fluorescence resonance energy transfer pair to probe protein unfolding: application to the villin headpiece subdomain and the LysM domain. Biochemistry 47:11070–11076
70. Rogers JMG, Lippert LG, Gai F (2010) Non-natural amino acid fluorophores for one- and two-step fluorescence resonance energy transfer applications. Anal Biochem 399:182–189
71. Peran I, Watson MD, Bilsel O, Raleigh DP (2016) Selenomethionine, p-cyanophenylalanine pairs provide a convenient, sensitive, non-perturbing fluorescent probe of local helical structure. Chem Commun 52:2055–2058
72. Watson MD, Peran I, Raleigh DP (2016) A non-perturbing probe of coiled coil formation based on electron transfer mediated fluorescence quenching. Biochemistry 55:3685–3691
73. Mintzer MR, Troxler T, Gai F (2015) *p*-cyanophenylalanine and selenomethionine constitute a useful fluorophore–quencher pair for short distance measurements: application to polyproline peptides. Phys Chem Chem Phys 17:7881–7887
74. Wissner RF, Batjargal S, Fadzen CM, Petersson EJ (2013) Labeling proteins with fluorophore/thioamide förster resonant energy transfer pairs by combining unnatural amino acid mutagenesis and native chemical ligation. J Am Chem Soc 135:6529–6540

Chapter 4

Purification and Handling of the Chaperonin GroEL

Xiang Ye

Abstract

GroEL is an important model molecular chaperone. Despite being extensively studied, several critical aspects of its functionality are still in dispute due partly to difficulties in obtaining protein samples of consistent purity. Here I describe an easy-to-carry-out purification protocol that can reliably produce highly purified and fully functional GroEL protein in large quantities. The method takes advantage of the remarkable stability of the GroEL tetradecamer in 45% acetone which efficiently extracts and removes tightly bound substrate proteins that cannot be separated from GroEL by the usual chromatographic methods. The efficiency of the purification method can be assessed by the amount of residual tryptophan fluorescence associated with the purified GroEL sample. The functionality of the thus obtained GroEL sample is demonstrated by measuring its ATPase turnover both in the presence and absence of the GroEL model substrate protein α-lactalbumin.

Key words GroEL, Purification, Acetone treatment, ATP turnover, Allostery

1 Introduction

E. coli chaperonin GroEL is a ring-shaped molecular chaperone. It assists refolding of mis-folded substrate proteins (SP) that makes up ~10% of *E. coli* proteome [1]. GroEL consists of 14 identical subunits in two 7-protomer back-to-back ring configurations. Its cochaperonin GroES is a dome-shaped homoheptamer [2]. Assisted SP refolding is achieved via GroEL natural functional cycle driven by ATP binding and hydrolysis. In the cycle, SP is initially captured by the SP binding sites at the GroEL apical domain; ATP and GroES binding displaces SP from its binding site and results in SP encapsulation inside the cavity formed between the GroEL seven-protomer ring and GroES [3–6]. The cycle continues until SP reaches its native conformation no longer recognized by GroEL [7, 8].

Despite being one of the most extensively studied chaperone systems, many aspects of the functionality of GroEL are still in dispute, e.g., the predominant active folding species in the functional cycle, the role played by GroEL in assisting SP refolding, etc.

Victor Muñoz (ed.), *Protein Folding: Methods and Protocols*, Methods in Molecular Biology, vol. 2376,
https://doi.org/10.1007/978-1-0716-1716-8_4,

[6, 9, 10]. If not entirely, a large component of the dispute can be blamed to varying purity of GroEL samples used in studies conducted by different research groups [11]. After usual chromatographic separation, despite appearing clean on SDS gels (Fig. 1 Lane 8), the proteins that co-purify are predominantly denatured and tightly associated with GroEL. These strong GroEL binding SPs can seriously obscure and complicate the functional assessment: (1) they themselves are potent GroEL allosteric effectors, (2) their mere presence can compete for GroEL binding sites with SPs that are added by researchers in their studies. Therefore, the importance of eliminating contaminants in GroEL preparations cannot be overstated.

Quite a few purification protocols have been published over the years; however, they are either ineffective in removing tightly bound contaminating proteins [12] or require fully/partially denaturation of GroEL to remove SPs by chromatographic methods and subsequent renaturation to recover the functional chaperonin [13, 14]. Albeit the latter can result in highly purified GroEL samples, it is usually a tedious and difficult to perform procedure. Here, we present a purification protocol that is relatively mild but also effective in stripping off tightly associated SPs from GroEL binding sites. The efficiency of SP removal by this method can be assessed by measuring residual tryptophan fluorescence originating from the small amounts of SPs that survive such treatment (Fig. 3).

GroEL is well-known as a molecular machine utilizing energy of ATP binding and hydrolysis to drive refolding of its SPs. Not surprisingly, SPs, in turn, have a profound effect on GroEL's ATPase activity. Therefore, comparing measured ATP turnover numbers both in the presence and absence of SP is an important gauge as to the purity and functionality of purified GroEL. To measure steady-state ATP turnover by GroEL, we employ the commonly used coupled enzyme assay system consisting of pyruvate kinase and lactate dehydrogenase (Scheme 1) which, as a result of GroEL ATPase activity, converts one equivalent of ADP to ATP, and in the meanwhile, oxidizes one equivalent of NADH. Decrease of UV absorption at 340 nm is measured by using a UV-Vis spectrophotometer capable of controlling the temperature in the cuvette.

2 Materials

The pH of all buffers is adjusted to 7.5 using a pH meter unless otherwise specified.

2.1 Chromatographic Purification of GroEL

1. GroEL is overexpressed from BL21(DE3) *E. coli* strain cells transformed with a plasmid hosting GroEL sequence. The transformed cells carry ampicillin resistance. Protein

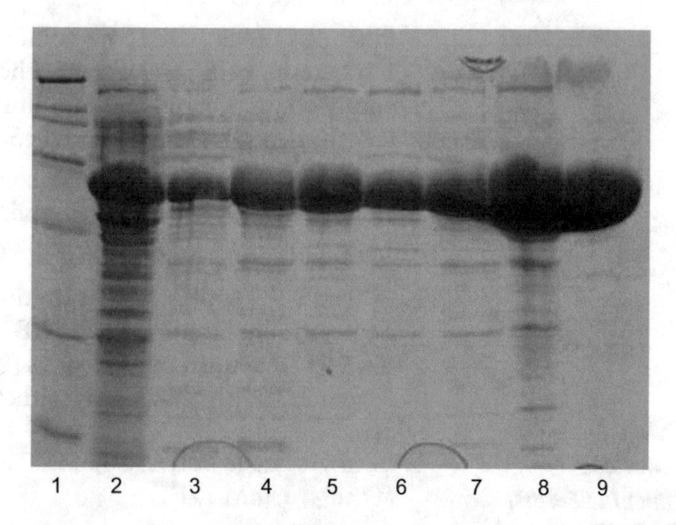

Fig. 1 12% SDS polyacrylamide gel of GroEL samples corresponding to the various purification steps. The lanes are identified as follows: molecular weight standards (lane 1), crude lysate (lane 2), DEAE fractions (lanes 3–7), pre-acetone-treated GroEL (lane 8), pure GroEL following acetone treatment (lane 9)

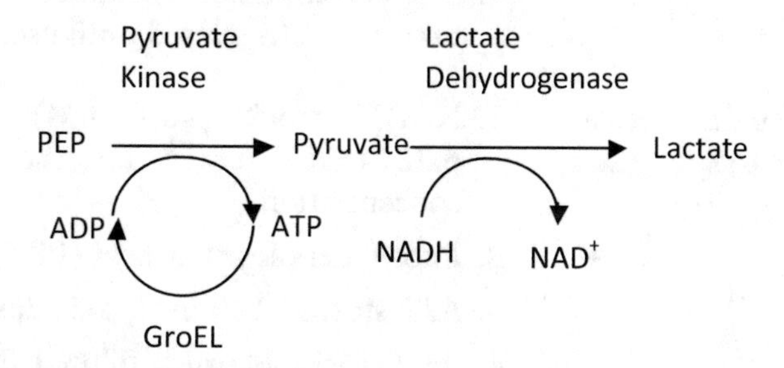

Scheme 1 GroEL ATPase activity measured by the coupled enzyme assay

overexpression is induced by addition of isopropyl β-D-1-thiogalactopyranoside (IPTG).

2. LB culture broth: 10 g tryptone, 10 g NaCl, 5 g yeast extract for each 1 l culture medium. Autoclave for 20 min at ~120 °C and store at 4 °C until use.
3. 1 M IPTG (isopropyl β-D-1-thiogalactopyranoside) stock solution.
4. Lysis buffer: 50 mM Tris–HCl, 1 mM EDTA, 1 mM DTT (dithiothreitol), and 1 tablet per 50 ml lysis buffer protease inhibitor cocktail tablets (Roche).
5. 100 mg/ml streptomycin sulfate stock solution.
6. DEAE buffer A: 50 mM Tris–HCl pH = 8.0, 10 mM $MgCl_2$, and 1 mM DTT.
7. DEAE buffer B: DEAE buffer A plus 1 M NaCl.

8. 500 ml DEAE Fast Flow column: packed with DEAE Fast Flow resin (GE Healthcare Life Science) and XK 50/30 empty column barrel (50 mm in inner diameter and 30 cm in length, also GE Healthcare Life Science).
9. Saturated ammonium sulfate solution with pH adjusted to close to neutral by using ammonia.
10. S300 buffer: 50 mM Tris–HCl, 10 mM $MgCl_2$, 1 mM DTT.
11. 300 ml S300 Sephacryl gel filtration column: packed with S300 Sephacryl gel filtration resin (GE Healthcare Life Science) and XK 26/70 empty column barrel (26 mm in inner diameter and 70 cm in length, also GE Healthcare Life Science).

2.2 Acetone Treatment to Further Purify GroEL

1. GroEL storage buffer: 10 mM Tris–acetate (OAc), 10 mM Mg $(OAc)_2$, 1 mM DTT.
2. 10 mM Tris–HCl.
3. Bovine serum albumin (BSA) stock solution in 10 mM Tris–HCl: using the extinction coefficient of 43,824 M^{-1} cm^{-1} at 280 nm to determine the exact concentration.
4. 8 M Guanidinium chloride (GnHCl) stock solution. Store frozen or refrigerated until use.

2.3 Coupled-Enzyme ATPase Assay

1. NADH stock (~30 mM): using extinction coefficient 6.22 mM^{-1} cm^{-1} for the determination of the exact concentration.
2. Phosphoenolpyruvic acid (PEP) stock (~120 mM) pH ~7.5.
3. ATP stocks ~100 mM, pH adjusted to ~7.5.
4. ATP stocks ~1 mM: diluted from 100 mM ATP stock with water, using extinction coefficient 15.4 mM^{-1} cm^{-1} at 260 nm to calculate the exact concentration.
5. 5× ATPase assay buffer: 250 mM Tris–OAC, 50 mM magnesium acetate, 500 mM potassium acetate.
6. Pyruvate kinase (PKA) from rabbit muscle preserved in 50% glycerol.
7. Lactic dehydrogenase (LDH) from bovine heart preserved in 50% glycerol.
8. Lyophilized α-lactalbumin (from bovine milk).
9. 2 mM DTT and 1 mM Tris–HCl solution.
10. 20 mM HCl solution.

3 Methods

All protein concentrations refer to concentrations of monomer.

The protocol is written for protein purification from 6 l culture medium. Scale all the volumes and masses proportionally for a different total culture volume.

3.1 Column Chromatographic Purification of GroEL

1. Streak glycerol stocks of *E. coli* BL21 cells containing GroEL plasmid onto a LB plate containing 100μg/ml ampicillin and grow it at 37 °C overnight.
2. Pick up and transfer a single colony to a starter culture containing 15 ml LB media with 100μg/ml ampicillin and grow it at 37 °C for at least 6 h with vigorous shaking. In the meanwhile, equilibrate the 500 ml DEAE column with 2500 ml DEAE buffer A.
3. Dilute the starter 100 times to inoculate 6 l of expression culture media (same with the starter culture). Grow at 37 °C until the absorption at 600 nm reaching 0.6 with 1 cm light path (usually takes 2–2.5 h). Add IPTG to a final concentration of 0.5 mM to induce protein overexpression. Continue to grow the culture for another 12–15 h at 30 °C with vigorous shaking (225 rpm).
4. Harvest cells by centrifugation at ~3000 × *g* for 30 min and resuspend them in 150 ml lysis buffer. Sonicate the cells in 50 ml per portion by using a Branson sonicator for three rounds on ice. The sonication program for each round is: 1 min sonication plus 1 min rest with power level at 5 and 50% duty cycle.
5. Remove cell debris by centrifugation (10,000 × *g* and 25 min at 4 °C). Precipitate the nucleic acid by adding 1.5 ml streptomycin sulfate stock solution into the supernatant dropwise with constant stirring.
6. Remove the precipitate by centrifugation at 32,500 × *g* for 25 min at 4 °C. Filter the supernatant with 0.45-μm syringe filter (*see* **Note 1**).
7. Load all crude lysate (150 ml) onto the DEAE column. Elute GroEL with a gradient of 0–0.5 M of NaCl (0–50% Buffer B) over 2 l at a flow rate of 6 ml/min. Fractions containing GroEL usually come out at around 35% Buffer B (*see* **Note 2**).
8. Pool all fractions containing GroEL together (total volume is typically ~180 ml). Concentrate it by adding saturated ammonium sulfate solution to a final concentration of 65% (v/v) and allow it to sit overnight at 4 °C (or at least for an hour). Spin down and collect the pellets by centrifugation at 10,000 × *g* and 4 °C for 25 min. Resuspend the proteins in 20 ml of S300 buffer. In the meanwhile, equilibrate the S300 column with 900 ml of S300 buffer at a flow rate of 3 ml/min.

9. Load the concentrated protein solution onto the S300 column equilibrated with 300 ml S300 buffer to remove residual $(NH_4)_2SO_4$ as well as low molecular weight contaminating proteins from GroEL. Pool factions containing GroEL and concentrate the protein up to 10 mg/ml with centrifugal concentrator (*see* **Note 3**).

3.2 Acetone Treatment of Chromatographic Purified GroEL Sample (See Note 4)

1. Each batch of acetone treatment is done with 30 ml of the concentrated chromatographically purified GroEL.
2. Supplement 30 ml of GroEL sample with KCl and ATP (*see* **Note 5**) up to the final concentrations of 10 mM and 1 mM, respectively, and incubate the solution on ice for ~10 min to weaken interaction between GroEL and SPs.
3. Transfer the solution into a 100-ml graduated cylinder. At room temperature, pump pure acetone into the cylinder as illustrated in Fig. 2 at a constant rate of 1 ml/min with fast and even stir (*see* **Note 6**).
4. Keep on stirring for 5 min after the final acetone concentration reaching 45% (~24.6 ml acetone are needed for 30 ml of GroEL solution). Pelleting the precipitate by centrifugation at 10,000 × *g* for 25 min at room temperature (*see* **Note 7**). Discard the supernatant and resuspend the pellet in 20 ml storage buffer.
5. Only GroEL can be re-solubilized whereas contaminant SPs cannot and remain in the form of aggregates. Remove the

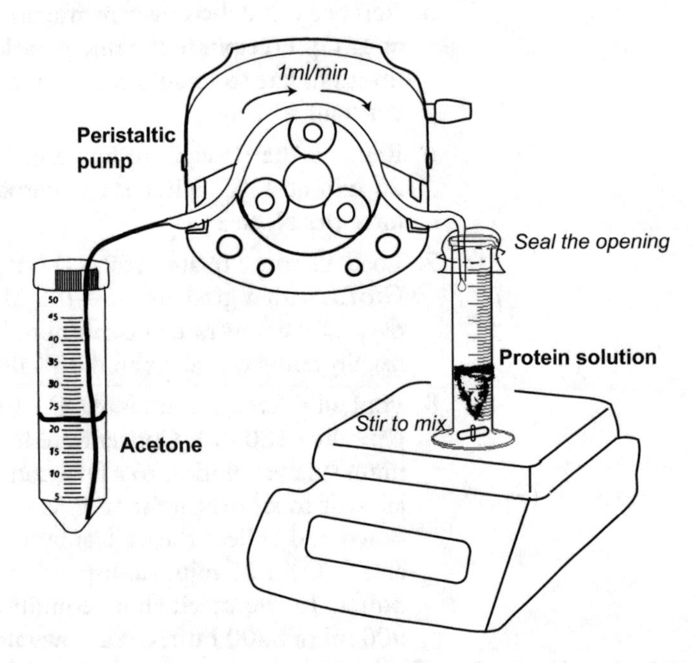

Fig. 2 Illustration of the experimental setup to perform acetone treatment of impure GroEL samples

aggregates by centrifugation again at 10,000 × *g* for 25 min at room temperature.

6. Remove the residual acetone by precipitating GroEL with 65% ammonium sulfate at 4 °C overnight (or at least for an hour). Centrifuge at 10,000 × *g* and 4 °C for 25 min to recover the GroEL pellet and resuspend it in 10 ml protein storage buffer.
7. Remove residual ammonium sulfate by gel-filtration using PD-10 column pre-equilibrated with the same protein storage buffer (*see* **Note 8**).

3.3 GroEL Purity Assessed by Tryptophan Fluorescence

GroEL does not have any tryptophan, making tryptophan fluorescence a very sensitive method of detecting and quantifying the presence of contaminating SPs.

1. Determine the concentration of GroEL in the eluant from PD-10 by UV-Vis absorption at 280 nm in the presence of 6 M GnHCl. Use an extinction coefficient of 9600 M^{-1} cm^{-1} to calculate [GroEL] (in μM) as in Eq. 1.

$$[\text{GroEL}] = \frac{V_{\text{Total}}}{V_{\text{GroEL}}} \cdot \frac{\text{Abs}_{\text{GroEL}} - \text{Abs}_{\text{bg}}}{1\ \text{cm}\cdot 9600\ (\text{cm}^{-1}\text{M}^{-1})} \times 10^6 \quad (1)$$

in which V_{GroEL} and V_{Total} are the volume of GroEL sample and the final volume in the quartz curvette for UV-Vis measurement, respectively (both are in μl), and $\text{Abs}_{\text{GroEL}}$ and Abs_{bg} are the absorption readings at 280 nm with and without GroEL.

2. Prepare a series of diluted BSA solutions using 10 mM Tris–HCl. Depending on the sensitivity of the available fluorimeter, the concentration of BSA should be chosen so that the fluorescence signal falls within the linear response range of the instrument. The following concentration range of BSA works well for a Perkin Elmer fluorimeter: 0.25μM, 0.5μM, 0.75μM, 1μM, 2μM. Mix these BSA samples with 8 M GnHCl with a ratio of 1:3 and incubate at room temperature for ~1 h before taking fluorescence measurements.
3. Excite fluorescence at 295 nm (*see* **Note 9**) and record signal from 300 nm to 400 nm. Integrate the area under the peak and plot the area against the corresponding [BSA] as the standard curve.
4. Dilute the GroEL sample to 20μM by using 10 mM Tris–HCl solution. Then mix it with 8 M GnHCl 1:3 and incubate at room temperature for ~1 h. Determine the tryptophan fluorescence and obtain the integrated peak area in the same way as described in **step 3**.
5. Use the standard curve to convert the area into the corresponding [BSA]. Multiply [BSA] by 4 (there are four

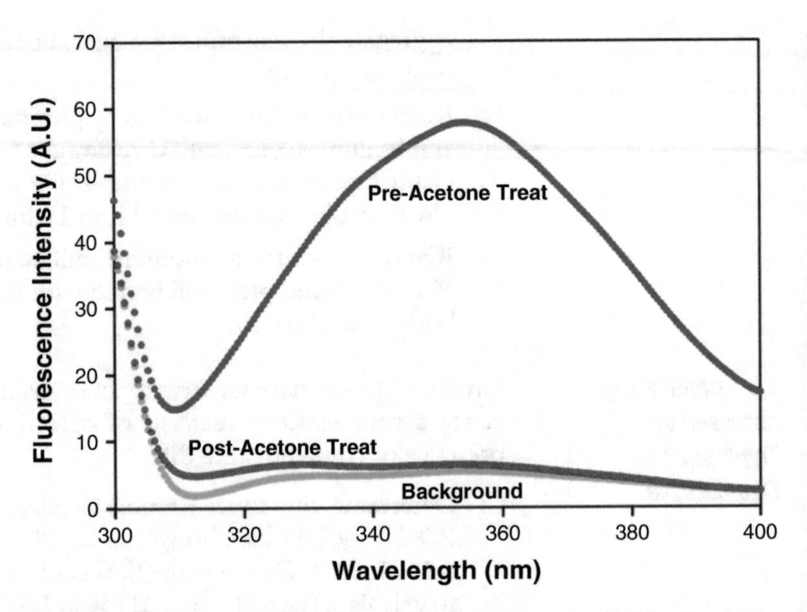

Fig. 3 Tryptophan fluorescence of pre- and post-acetone-treated GroEL samples

tryptophan residues per BSA) to get the equivalent [Trp] in the GroEL sample. The percentage of GroEL ring ($GroEL_7$) occupied by SP can be calculated by Eq. 2 assuming that the whole proteome of *E. coli* contains on average 3.3 tryptophan residues per protein (Fig. 3):

$$\%GroEL_7 occupied = \frac{[Trp]/\mu M}{3.3} \cdot \frac{7}{20\ \mu M} \times 100\% \qquad (2)$$

6. If the protein is pure enough (*see* **Note 10**), it can be concentrated >0.5 mM by centrifugal filter devices and stored in small aliquots at −80 °C (*see* **Note 11**).

3.4 GroEL ATPase Activity Measured by Coupled-Enzyme Assay

1. The assay solution is split into two parts: for each measurement, 1 ml containing all the assay components but ATP, and 0.2 ml containing only ATP at varying amounts.
2. Prepare the first part of the assay solution using the concentrated stock solution of NADH, PEP, and buffer salts to the final concentration (regarding the final 1.2 ml combined assay solution) of 50 mM Tris–OAc pH 7.5, 10 mM magnesium acetate, 100 mM potassium acetate (*see* **Note 12**), 0.5 mM PEP, 0.2 mM NADH. Also include 1μM GroEL and 5 units each of PKA and LDH in this part of the assay solution (*see* **Notes 13** and **14**).
3. Prepare the other part of the assay solution containing varying [ATP]s (such as from 2μM to 100μM as used in Fig. 4) using

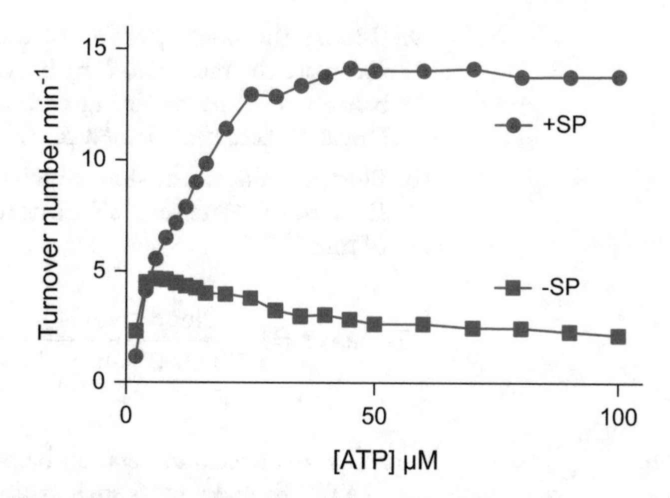

Fig. 4 Steady-state ATP hydrolysis by acetone-treated GroEL in the presence and absence of denatured α-lactalbumin. All the measurements were performed as described in the presence of 100 mM potassium and at 37 °C

the 1 mM ATP stock. Add enough MilliQ water to bring the volume to 0.2 ml per each point (*see* **Note 15**).

4. Incubate the two parts of the assay solution separately at 37 °C for at least 5 min. Transfer Part 1 solution into a plastic cuvette with 1.5 ml maximum sample volume and place it into the spectrophotometer with temperature also set at 37 °C.
5. Record the absorption at 340 nm for approximately a minute to make sure (1) the recorded trace is reasonably flat, (2) the reading is close to 1.5 which is the expected absorption of 0.24 mM NADH contained in this part of the assay solution.
6. Pause data collection. Initiate the reaction by adding the 0.2 ml ATP solution. Mix thoroughly and quickly while avoiding generating bubbles. Resume data collection for another 5 min.
7. (Optional) Prepare denatured α-lactalbumin from lyophilized powder. Dissolve ~2 mg lyophilized α-lactalbumin (from bovine milk) in 1 ml 2 mM DTT and 1 mM Tris–HCl solution, and incubate at room temperature for 15 min. Then mix it with an equal volume of 20 mM HCl and incubate on ice for an hour to denature the protein. Buffer exchange denatured α-lactalbumin into 10 mM Tris–HCl, 1 mM DTT. Determine the exact concentration of the protein using an extinction coefficient of 28,400 M^{-1} cm^{-1} at 280 nm.
8. (Optional) Pause data collection and add in denatured α-lactalbumin (*see* **Note 16**) to the final concentration of 2μM (14 times in excess of GroEL heptameric ring). Then resume data collection for another 2–3 min before stopping the measurement.

9. Choose the linear portion of each of the collected datasets. Calculate the rate of ATP hydrolysis from the change in absorbance over time by fitting each data set by linear regression. Discard traces that yielded R^2 values less than 0.99.
10. Plug the value of the slope obtained from linear regression into Eq. 3 to calculate the ATP turnover number by GroEL in units of min^{-1}.

$$\text{Turnover} = \frac{\text{Slope}\left(\text{min}^{-1}\right)}{6.22\ \text{mM}^{-1}\text{cm}^{-1}.\ 1\ \text{cm}} \cdot \frac{1\text{ml}}{10^3\ \mu\text{l}} \cdot \frac{1}{[\text{GroEL}](\mu\text{M})} \quad (3)$$

11. The turnover numbers can be plotted against corresponding [ATP] to make plots such as the one shown in Fig. 4 where the ATPase profile obtained with and without SP can be compared (*see* **Notes 17** and **18**).

4 Notes

1. It is important to move onto the next purification step as soon as possible. Avoid keeping crude lysate in refrigerator overnight to minimize the amount of time GroEL in contact with *E. coli* proteins that may be denatured and serve as potential GroEL SPs.
2. Because GroEL does not contain any tryptophan, fractions containing it usually have low UV absorption and are difficult to separate from the baseline. Therefore, it is better to identify GroEL-containing fractions by SDS-PAGE gel electrophoresis.
3. The Bradford assay can be used instead of UV-Vis absorption. At this stage, the GroEL sample contains some fraction of contaminating proteins with tryptophans. Consequently these contaminants may render the concentration calculated using the GroEL extinction coefficient inaccurate. Results obtained by Bradford assay calibrated with BSA can provide a rough idea of the total protein concentration.
4. We have found that the acetone precipitation method, modified from Voziyan and Fisher [15], is the best way of removing the remaining contaminant proteins. The principle behind this method is that incubating GroEL in the presence of low to medium concentration of acetone helps to dislodge/extract SPs from GroEL binding sites that have substantial hydrophobic exposure. The high stability of the GroEL tetradecamer prevents individual monomers from falling out off the complex and become unfolded. At 45% acetone (v/v) which is the final acetone concentration applied in the protocol, GroEL is

precipitated as tetradecamers (or heptamers), which allows readily re-solubilization in the native buffer condition. The method presented here is generally efficient and effective for GroEL wild-type, but caution and appropriate modifications may be required if it were to be applied to a GroEL mutant with diminished stability.

5. The pH of the ATP stock solution should be adjusted to a value close to 7.5 before adding to the GroEL sample.
6. Stripped off SPs are prone to formation of aggregates, so that the solution begins to appear cloudy at acetone concentrations of ~15%. As acetone concentration increases, intact GroEL oligomers are precipitated to completion at concentrations close to 45%.
7. The centrifuge tube should be made of acetone resistant material such as Teflon or glass.
8. This step is necessary given that ammonium ions can act as K^+ surrogate for GroEL [16]. It is well accepted that K^+ is an important allosteric effector in GroEL function [16–18], and therefore, its concentration should be controlled carefully (*see* also **Note 12**).
9. Exciting at 295 nm, instead of at 280 nm corresponding to the absorption peak of tryptophan, minimizes excitation of tyrosine which still has substantial absorption at 280 nm but little at 295 nm.
10. The contamination level should be as low as possible: 10% SP per GroEL ring is the upper limit of acceptance. If the sample contains a higher level of contamination, it needs to go through more rounds of acetone treatment until it reaches the desired level of purity.
11. For short-term storage, GroEL can be kept at 4 °C for up to a month without obvious loss of activity so long as enough concentration of a suitable reducing reagent is included and the solution is sterile filtered.
12. Potassium plays an important role in nucleotide binding and hydrolysis by GroEL. Consequently, the steady-state ATP hydrolysis rates are strongly affected by [K^+] [16–19]. For reproducibility purposes, [K^+] should be controlled carefully. [K^+] between 1 mM and 10 mM will not adversely affect the GroEL intrinsic ATPase activity, but they still are problematic because PKA requires potassium for catalysis, and [K^+] $<$ 10 mM can substantially inhibit its enzymatic activity [20] and thus diminish the efficiency of the coupled enzyme system.
13. To reduce variation between different measurements, such as in the amount of GroEL, we recommend preparing the assay

solution in sufficient amounts to perform all measurements at once and make aliquots.

14. Long-term storage of the combined assay solution should be avoided. Best practice is to prepare enough assay solution just for a single day's use.

15. To make sure the actual [ATP] is as close as possible to the intended value, it is a good idea to make serial dilutions in small steps (e.g., by half) from the stock instead of a single large dilution.

16. In the absence of calcium and with the four pairs of disulfide bonds in reduced form, α-lactalbumin exists in a denatured molten globular state that has substantial hydrophobic exposure and can be readily recognized and captured by GroEL [21]. α-Lactalbumin is an ideal model SP to study SP allosteric effects on GroEL. The α-lactalbumin molten globular state is stable and highly soluble at neutral pH and can be concentrated >500μM. Moreover, α-lactalbumin remains in the molten globular state despite GroEL action so long as no calcium or zinc is present and the cysteine residues remain reduced by agents such as DTT or Tris(2-carboxyethyl)phosphine hydrochloride (TCEP) [21, 22].

17. In the presence of high [K^+], hydrolysis of ATP by pure GroEL with minimal contaminating SP follows unusual characteristic non-Michaelis-Menten enzyme kinetics: ATP turnover number increases as [ATP] increases at low [ATP]s but decreases at higher [ATP]s until it levels off. Such behavior was initially proposed to be caused by negative cooperativity of ATP binding between the two GroEL rings, i.e., ATP binding to one ring decreases affinity for ATP of the other ring [17]. It was later found that other factors such as the rate of ADP release may also be responsible for the observed unusual GroEL ATPase properties [18].

18. Addition of SP eliminates the inhibitory effect observed at high [ATP] and causes the ATP turnover number to further increase to a level that is ~5-fold higher than the steady-state rate in its absence (Fig. 4). This difference demonstrates the allosteric effect of SP on GroEL and highlights the importance of using clean GroEL samples to study its function.

References

1. Balchin D, Hayer-Hartl M, Hartl FU (2016) In vivo aspects of protein folding and quality control. Science 353:aac4354
2. Xu Z, Horwich AL, Sigler PB (1997) The crystal structure of the asymmetric GroEL-GroES-(ADP)7 chaperonin complex. Nature 388: 741–750
3. Horwich AL, Farr GW, Fenton WA (2006) GroEL-GroES-mediated protein folding. Chem Rev 106:1917–1930

4. Lin Z, Rye HS (2006) GroEL-mediated protein folding: making the impossible, possible. Crit Rev Biochem Mol Biol 41:211–239
5. Hartl FU, Bracher A, Hayer-Hartl M (2011) Molecular chaperones in protein folding and proteostasis. Nature 475:324–332
6. Taguchi H (2015) Reaction cycle of chaperonin GroEL via symmetric "Football" intermediate. J Mol Biol 427:2912–2918
7. Todd MJ, Lorimer GH, Thirumalai D (1996) Chaperonin-facilitated protein folding: optimization of rate and yield by an iterative annealing mechanism. Proc Natl Acad Sci U S A 93: 4030–4035
8. Yang D, Ye X, Lorimer GH (2013) Symmetric GroEL:GroES2 complexes are the protein-folding functional form of the chaperonin nanomachine. Proc Natl Acad Sci U S A 110: E4298–E4305
9. Gruber R, Horovitz A (2016) Allosteric mechanisms in chaperonin machines. Chem Rev 116:6588–6606
10. Hayer-Hartl M, Bracher A, Hartl FU (2016) The GroEL-GroES chaperonin machine: a nano-cage for protein folding. Trends Biochem Sci 41:62–76
11. Marchenkov VV, Semisotnov GV (2009) GroEL-assisted protein folding: does it occur within the chaperonin inner cavity? Int J Mol Sci 10:2066–2083
12. Weissman JS, Hohl CM, Kovalenko O, Kashi Y, Chen S, Braig K, Saibil HR, Fenton WA, Horwich AL (1995) Mechanism of GroEL action: productive release of polypeptide from a sequestered position under GroES. Cell 83:577–587
13. Walti MA, Clore GM (2018) Disassembly/reassembly strategy for the production of highly pure GroEL, a tetradecameric supramolecular machine, suitable for quantitative NMR, EPR and mutational studies. Protein Expr Purif 142:8–15
14. Ryabova N, Marchenkov V, Kotova N, Semisotnov G (2014) Chaperonin GroEL reassembly: an effect of protein ligands and solvent composition. Biomol Ther 4:458–473
15. Voziyan PA, Fisher MT (2000) Chaperonin-assisted folding of glutamine synthetase under nonpermissive conditions: off-pathway aggregation propensity does not determine the co-chaperonin requirement. Protein Sci 9: 2405–2412
16. Todd MJ, Viitanen PV, Lorimer GH (1993) Hydrolysis of adenosine 5′-triphosphate by Escherichia coli GroEL: effects of GroES and potassium ion. Biochemistry 32:8560–8567
17. Horovitz A, Willison KR (2005) Allosteric regulation of chaperonins. Curr Opin Struct Biol 15:646–651
18. Grason JP, Gresham JS, Widjaja L, Wehri SC, Lorimer GH (2008) Setting the chaperonin timer: the effects of K+ and substrate protein on ATP hydrolysis. Proc Natl Acad Sci U S A 105:17334–17338
19. Clark AC, Karon BS, Frieden C (1999) Cooperative effects of potassium, magnesium, and magnesium-ADP on the release of Escherichia coli dihydrofolate reductase from the chaperonin GroEL. Protein Sci 8:2166–2176
20. Laughlin LT, Reed GH (1997) The monovalent cation requirement of rabbit muscle pyruvate kinase is eliminated by substitution of lysine for glutamate 117. Arch Biochem Biophys 348:262–267
21. Okazaki A, Ikura T, Nikaido K, Kuwajima K (1994) The chaperonin GroEL does not recognize apo-alpha-lactalbumin in the molten globule state. Nat Struct Biol 1:439–446
22. Murai N, Taguchi H, Yoshida M (1995) Kinetic analysis of interactions between GroEL and reduced alpha-lactalbumin. Effect of GroES and nucleotides. J Biol Chem 270: 19957–19963

Part II

Kinetic and Thermodynamic Analysis of Protein Folding

Chapter 5

Folding Free Energy Surfaces from Differential Scanning Calorimetry

Jose M. Sanchez-Ruiz and Beatriz Ibarra-Molero

Abstract

Protein folding/unfolding processes involve a large number of weak, non-covalent interactions and are more appropriately described in terms of the movement of a point representing protein conformation in a plot of internal free energy versus conformational degrees of freedom. While these energy landscapes have an astronomically large number of dimensions, it has been shown that many relevant aspects of protein folding can be understood in terms of their projections onto a few relevant coordinates. Remarkably, such low-dimensional free energy surfaces can be obtained from experimental DSC data using suitable analytical models. Here, we describe the experimental procedures to be used to obtain the high-quality DSC data that are required for free-energy surface analysis.

Key words Differential scanning calorimetry (DSC), Experimental guidelines for DSC, Protein folding, Absolute heat capacity, Free energy barriers, Fast folding

1 Introduction

Differential scanning calorimetry (DSC) for diluted solutions of biomacromolecules was developed in the 1960s [1] and has contributed enormously to our understanding of protein biophysics. Early calorimetric studies were in fact pivotal to establishing the foundations of the protein stability field, and many efforts have been devoted since then to dissect protein energetics in molecular contributions and to use this information as a basis for protein engineering [2–6].

Technological advances incorporated in commercial calorimeters have transformed DSC into a common technique in protein biophysics laboratories. Furthermore, the current availability of fully automated, easy-to-run calorimeters has expanded DSC uses to more practically oriented goals, such as the medium-throughput screening for protein thermal stability and its application to drug development in the biotechnological industry. Usually, these applications of DSC rely on "quick-and-dirty" experiments and very

Victor Muñoz (ed.), *Protein Folding: Methods and Protocols*, Methods in Molecular Biology, vol. 2376,
https://doi.org/10.1007/978-1-0716-1716-8_5,

simple data analyses aimed at determining basic parameters of protein thermal denaturation, such denaturation temperatures and enthalpy changes.

At the other end, there are some applications of DSC that require high-quality experimental data and sophisticated data analysis [7]. These include the characterization of ligand-binding effects, investigations on residual structure in the unfolded state, and most notably, the estimation of free energy surfaces in protein folding. Protein folding/unfolding processes involve the formation/breakage of a large number of weak, non-covalent interactions and are more appropriately described in terms of the movement of a point representing protein conformation in a plot of internal free energy versus conformational degrees of freedom. In principle, these energy landscapes are hyper-dimensional, representing all the relevant degrees of freedom. However, it has been shown that many relevant aspects of protein folding can be understood in terms of their projections onto just a few relevant coordinates. Such free energy surfaces of highly reduced dimensionality can be obtained from experimental DSC data using suitable models. The key feature behind this possibility is the equivalence of equilibrium DSC data and the relevant protein partition function. Free energy surfaces derived from experimental DSC data have been used to estimate the size of the marginal free energy barriers for fast and ultra-fast folders and have provided evidence supporting the existence of barrierless (i.e., downhill) folding, one of the fundamental predictions of the energy landscape theory of protein folding [8–11].

Various protein folding theoretical models amenable to deriving free energy surfaces from DSC data have been developed, and their details are discussed in several publications [11, 12]. Here, we focus in one aspect that has not been described in sufficient detail in the published literature: the experimental procedures to be used to obtain the high-quality DSC data that are required for free-energy surface analysis. In particular, we will provide experimental guidelines to the determination of high-quality absolute heat capacity values as a function of temperature and will discuss under which conditions it is possible to arrive to this kind of data, as well as the most common pitfalls and how to circumvent them.

To illustrate the experimental procedures, we will use the small α-helical protein PDD (peripheral subunit-binding domain of dihydrolipoamide acetyltransferase (E2) component of the pyruvate dehydrogenase complex), as example. PDD has been previously shown to be an ultra-fast folder, within the microsecond time scale, with a free energy surface that approaches the downhill folding (barrierless) regime [13].

2 Materials

1. PDD is a 49-residue all-helical domain from the dihydrolipoamide acetyltransferase subunit (E2) of the pyruvate dehydrogenase multienzyme complex from *Bacillus stearothermophilus* (our PDD variant is actually six residues longer than the wild-type form described as 2PDD.pdb: MDNRRVIAMPSVR KYAREKGVDIRLVQGTGKNGRVLKEDIDAFLAGGAK ; MW: 5401.3 Da). PDD protein was synthesized using standard solid phase methods and F-moc chemistry. Purity greater than 96% was achieved by HPLC.
2. NaH_2PO_4 and $Na_2HPO_4{\cdot}2H_2O$, analytical grade.
3. Milli-Q ultrapure water.
4. Slide-A-Lyzer dialysis cassettes of MWCO of 2K from Thermo-Fisher Scientific or dialysis tubing membranes of similar cutoff.
5. Syringe filters with a 0.45-μm pore size.
6. pH-meter.
7. Refrigerated benchtop centrifuge.
8. UV-Vis Spectrophotometer and UV-grade quartz cells.
9. A Microcal VP-Capillary DSC from Malvern instruments was used for the experiments detailed below.

3 Methods

3.1 Buffer Preparation

- Buffer solutions were prepared using Milli-Q ultrapure water and filtered with a 0.45-μm pore size filters.
- 20 mM sodium phosphate buffer pH 7 was prepared by mixing two solutions of identical concentration of NaH_2PO_4 and $Na_2HPO_4{\cdot}2H_2O$ so the final pH-meter reading was 7.0 (*see* **Note 1**).

3.2 Protein Preparation

- PDD stock solution (~1.5 mL at 6 mg/mL approximately) was exhaustively dialyzed against phosphate buffer, at 4 °C. Typically, three buffer changes were required of about 12 h each, as diffusion across the membrane was expected to take longer due to the small membrane cutoff (2000 Da) required for a small protein such as PDD. The buffer from the last dialysis change (i.e., the buffer that was equilibrated with the protein solution) was kept to fill the reference cell in the calorimetric experiment (*see* **Note 2**).
- Once the dialysis process was over, the protein solution was centrifuged in a benchtop centrifuge at 4 °C and 12,000 × *g*

for 15 min, in order to remove any insoluble aggregates. The supernatant was carefully withdrawn (*see* **Note 3**).

- Protein concentrations were determined from the UV-Vis spectrum collected in the 210–400 nm range (*see* **Note 4**). The value used for the molar extinction coefficient for PDD at 280 nm was 1490 $M^{-1} \cdot cm^{-1}$.

3.3 DSC Experiment (See Note 5)

Before running the actual DSC experiment, it is important to ascertain that the instrument is working under optimal conditions, so its performance achieves maximum sensitivity and repeatability. The conditions of the cells are critical to ensure high-quality data. It is advisable to perform an exhaustive rinsing/cleaning procedure after each round of experiments, following the manufacturer directions (*see* **Note 6**).

- Setting up the calorimetric parameters. The VP Cap-DSC is fully automated, but a number of instrumental parameters have to be set up. For PDD experiments: pre-scan equilibration time was 10 min, filtering period was 10 s, and the feedback mode was set to passive mode. The scans were performed from 2 °C to 120 °C, at a scan rate of 200 K/h (*see* **Note 7**). Overpressure was applied to prevent boiling above 100 °C.
- Buffer-buffer baselines: This is the actual instrumental baseline, and it is normally subtracted from the protein-buffer thermogram to correct for the fact that cells are not identical and other instrumental contributions.

  It is essential to use the buffer from the final dialysis step to fill both cells up and monitor a number of buffer-buffer baselines (typically overnight for automatic calorimeters), refreshing both solutions each run. When baseline repeatability is within manufacturer indications, the thermal history of the instrument is optimum, and we can proceed to run the protein-buffer experiment. Note that the more stable is the baseline the more accurate absolute heat capacity measurements are (*see* upper panel in Fig. 1). Therefore, do not hesitate to spend more time collecting buffer-buffer baselines until the instrument reaches the best performance: typically baseline reproducibility is on the order of a few μcal/K.
- PDD-buffer run: Following instrumental baseline determination, the protein experiment is run with fresh protein (PDD in this case) and buffer solutions. As mentioned before, we aimed to obtain an accurate profile of PDD absolute heat capacity versus temperature and, to this end, we followed the approach described by Kholodenko and Freire [14]. A number of protein thermograms at different concentrations need to be obtained, typically from 0.1 to ~3 mg/mL (*see* upper panel in Fig. 1 for representative examples). It is highly advisable to check for

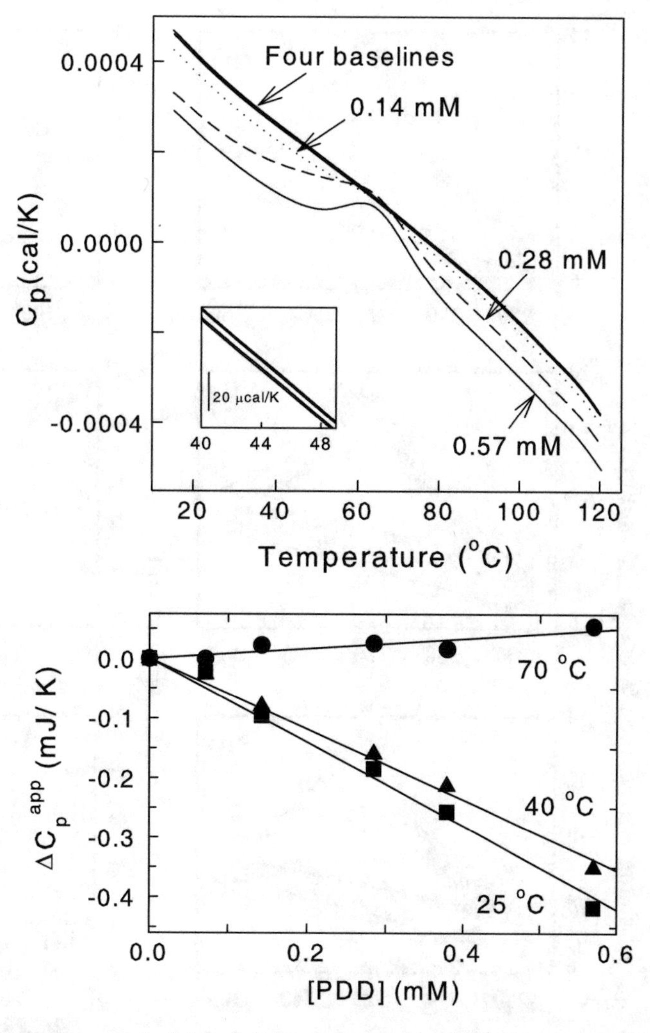

Fig. 1 (Upper panel) Original DSC thermograms for PDD at the indicated concentrations. Also, four buffer-buffer baselines obtained prior to the protein experiments are shown. Note that the reproducibility for the instrumental baseline was within the microcalories per Kelvin range (*see* inset). This high-quality data is required when aiming at absolute heat capacity data, especially in the case of small proteins that approach the downhill folding regime as they may exhibit broad calorimetric transitions that at low concentrations may not even be visually apparent (*see*, for instance, the thermogram at 0.14 mM). (Lower panel) Plots of apparent heat capacity as a function of PDD concentration at 25, 40, and 70 °C. Symbols represent experimental data and continuous lines are the best linear fits, considering zero intercept, to Eq. 1. As explained in the text, it is possible from the slopes of these profiles to estimate values for the absolute heat capacity of PDD at that particular temperature. Note that the PDD form used for the representative experiments reported in this figure is slightly different (it has six additional residues) than that employed in ref. [11] and used in Fig. 2 to illustrate data analysis

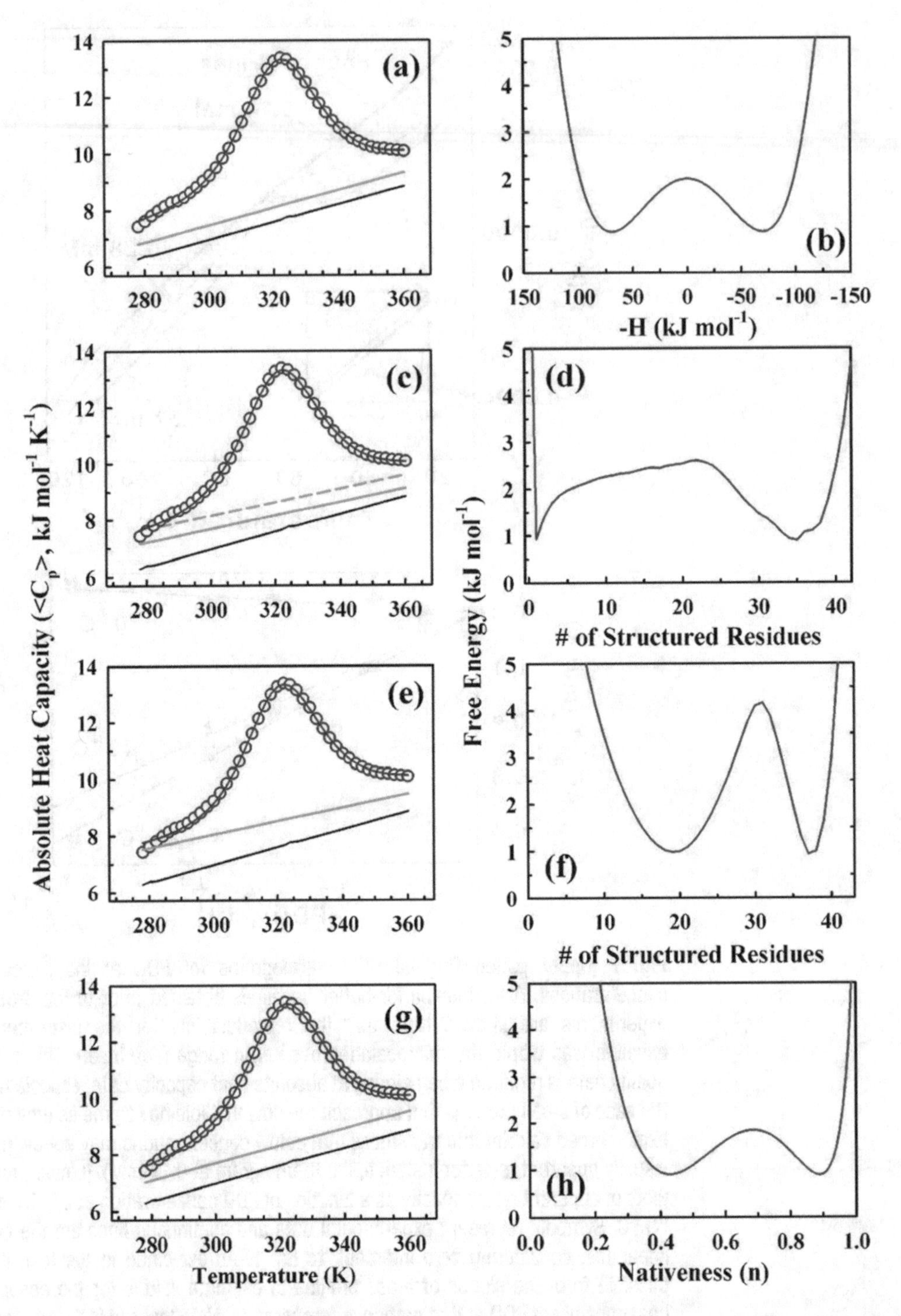

Fig. 2 Comparison of the fits of various free energy surface models to heat capacity profile of PDD. Circles represent experimental data in absolute heat capacity units, obtained according to the Kholodenko and Freire approach [14]. The models used are: the Variable-Barrier Model (panels **a**, **b**), the Muñoz-Eaton Model-Single Sequence Approximation (panels **c**, **d**), the Muñoz-Eaton Model-Exact Solution (panels **e**, **f**), and the Mean Field Free Energy Surface Model (panels **g**, **h**). In panels (**a**), (**c**), (**e**), and (**g**), the fits are shown in red lines. The straight lines are

baseline reproducibility between protein scans. For the PDD example, we run a buffer-buffer experiment after each protein experiment in order to rule out that instrumental drifts were affecting the quality of the data.

- Reheating run: Experiments carried out at the highest and lowest protein concentrations should be followed by the cooling down of the solutions inside their respective cells and a subsequent second run to confirm calorimetric reversibility within the protein concentration range under studied (*see* **Note 8**).

3.4 Absolute Heat Capacity Calculations

Using the Origin software provided by the manufacturer of the calorimeter, raw protein thermograms were corrected for the instrumental baseline and instrument time-response to obtain the apparent heat capacity change (ΔC_p^{app}in units of cal/K) of the protein solution as a function of temperature. Note that, at this point, different temperature dependence profiles of ΔC_p^{app} for a number of protein concentrations are available. The concentration dependence of ΔC_p^{app} at a particular temperature can be written as [16]:

$$\Delta C_p^{app} = C\tilde{n}V_0\tilde{n}\left(C_{p,P} - \frac{V_P}{V_W}\tilde{n}C_{p,W}\right) \quad (1)$$

where C is the protein concentration (mM), PDD in this case, V_0 represents the volume of the calorimetric cell (mL), $C_{p,P}$ and $C_{p,W}$ stand for the absolute heat capacity of protein (PDD) and buffer, respectively, and finally V_P and V_W are the molar volumes of protein (PDD) and buffer. V_P of PDD is calculated according to Makhatadze et al. [17] following a simple additive scheme. According to Eq. 7 in [17], the molar volumes of the protein constituent groups (*see* Table VI in [17] for values corresponding to the peptide unit and amino acid side chains) are added in order to obtain the total molar volume of the protein. By using the molecular weight, one can easily arrive to 0.746 mL/g for the specific volume of PDD. $C_{p,W}$ and V_W are properties of water that can be easily found in the published literature. The lower panel in Fig. 1 shows a plot of experimental ΔC_p^{app} values versus protein (PDD) concentration at different temperatures. In agreement with Eq. 1, these

Fig. 2 (continued) the heat capacity of the native state, the so-called native baseline. Native baselines derived from the fittings are shown in green, while the black straight lines are theoretical predictions calculated using the equation proposed by Freire [15]. In panels (**b**), (**d**), (**f**), and (**h**), the corresponding plots of free energy versus an order parameter are shown. These free energy surfaces are qualitatively similar, revealing two basins (corresponding to the native and unfolded states) separated by marginal barrier that is on the order of the thermal energy. (Figure reproduced from [11] with permission from the PCCP Owner Societies)

profiles are well described by straight lines with zero intercept. Furthermore, $C_{p,P}$ can be easily calculated at that temperature from the slope of the line. This calculation is then repeated for many temperatures to arrive to a profile of $C_{p,P}$ (T), as shown in Fig. 2.

A few additional comments are warranted here: (1) the lower panel in Fig. 1 displays negative values for the apparent heat capacity change within most of the temperature range. These negative values arise from the fact that the content of water in the sample cell is lower than that in the reference cell, and water has very high heat capacity. The corresponding water-displacement correction is in fact implicit in Eq. 1; (2) it is certainly possible to obtain absolute heat capacity values from a single DSC experiment. However, the approach proposed by Kholodenko and Freire [14], as described above, allows for more accurate determinations because it relies on multiple thermograms carried out at different protein concentrations so experimental errors are minimized. In addition, as the concentration dependence is evaluated over an extended range, deviations from ideal behavior (due to association/aggregation processes or poor performance of the instrument) can be readily detected; (3) the calorimetric trace for PDD denaturation exhibits the typical signatures of a protein that folds over a marginal thermodynamic barrier (*see* experimental data in Fig. 2). Thus, the unfolding peak is very broad, as expected for a marginally cooperative process, and, as a consequence, the native and unfolded baselines are not well defined.

3.5 Data Analysis: Estimation of the Thermodynamic Free Energy Barrier

Once the PDD absolute heat capacity profile as a function of temperature is available (Fig. 2), one can move on to the fitting procedure to obtain a free energy surface. A variety of protein folding models are available for this purpose, including a phenomenological model based on Landau theory of critical transitions, mean-field one-dimensional models and Ising-like models that use the native structure as template for the estimation of the energetics of partially unfolded states. These models are based on different theoretical premises and use different properties as order parameters to describe the extent of protein unfolding. Yet, when used to fit the DSC data for PDD denaturation (*see* [11] for details), they lead to qualitatively similar free energy surfaces (Fig. 2). That is, in all cases, the resulting free energy exhibits two basins, corresponding to the native and unfolded states, and a marginal free energy barrier separating them. The estimated barrier is on the order of the kJ/mol, i.e., in the scale of thermal energy. Such a marginal free energy barrier is consistent with the ultra-fast folding kinetics of PDD, and places this protein near or at the downhill folding regime.

4 Notes

1. The choice of an appropriate buffer for the calorimetric experiment is an important issue. Ideally, its ionization enthalpy change should be as low as possible, so its pK_a remains unchanged within the temperature scan, thus avoiding strong pH changes with temperature. *See* Goldberg et al. [18] for a list of thermodynamic quantities for the ionization reaction of most common buffers in water. At neutral pH, phosphate and HEPES buffers are commonly used for DSC.
2. Dialysis is key in the design of the calorimetric experiment as the protein solution must be precisely in identical conditions to the reference solution. This is because the DSC instrument works in a differential mode, that is, the actual calorimetric output signal is proportional to the difference in heat capacity between the sample and the reference cells. Hence, even slight differences in terms of buffer composition between the reference and sample cells introduce artifactual contributions to the thermogram.
3. In general, filtration is not recommended as significant amount of protein might be retained in the membrane. Thus, depending on the material the membrane is made of, unwanted protein interactions likely of hydrophobic nature may take place with the concomitant loss of sample.
4. Errors in protein concentration measurements are actually one of the most common pitfalls when aiming at a rigorous analysis of the DSC data. Note that accurate protein concentration determinations are crucial in order to derive high-quality partial heat capacity values (as described in Subheading 3.4), as it is our purpose. Also, general thermodynamic parameters of the unfolding transition, in particular the denaturation enthalpy, can be highly affected by errors in protein concentration.

   A few points to check out are: (1) Make sure the spectrophotometer signal at 280 nm is within the linear range of the instrument, otherwise spectra of different PDD dilutions might be required. (2) It is highly recommended to collect the entire UV-Visible absorbance of the protein (typically from 210 to 400 nm) rather than a single measurement at 280 nm. The spectroscopic baseline at high wavelengths should be flat and centered around zero indicating that no light scattering effects are significant. This is actually the ideal situation, although not always achieved. Thus, it is quite common that the baseline exhibits some degree of slope, which is a clear indication of significant scattering contributions. Accurate protein quantification under these conditions is not possible as the absorbance at 280 nm is significantly distorted, given that the Rayleigh scatter intensity changes with λ^{-4} and increases rapidly at low wavelengths. It is not recommended to carry out mathematical corrections for the scattering contribution but, rather, we recommend centrifuging the sample thoroughly or, as a last resort, filtrating it. (3) Finally, check for the ratio A_{280}/A_{260} as a good indicator of the absence of nucleic acids contamination (a value of about 1.6 is expected if the protein contains tryptophan in its sequence, while significantly lower values may indicate substantial contamination by nucleic acids in the protein preparation).

5. Note that for the VP Cap-DSC, although the cell volume is 130 μL, each well of the 96-well tray was loaded with approximately 400 μL of solution to allow the calorimetric cells to be rinsed and filled up without introducing air bubbles.
6. A standard rinsing/cleaning procedure can be performed by clicking the "Maintenance" tab of the "Autosampler setup" box in the VPViewer program. Under "Cleaning Controls," the syringe, the injector, and the cells can be rinsed thoroughly with degassed ddH_2O from the wash stations. If protein aggregation occurs, the "Advance" tab should be used to set up an extensive cleaning protocol including, if necessary, the use of detergent (10% Decon 90) from the solvent reservoirs. It is also possible to incubate the detergent inside the cells for a certain period of time at constant temperature. Make sure the baselines are stable afterwards.
7. What are the general guidelines to select the scan rate for the calorimetric experiment?

   It is important to always keep in mind that to analyze the DSC data on the basis of equilibrium thermodynamics, one must ensure that the collected data reflect an equilibrium denaturation process. Maintaining proper equilibrium conditions during the scan is the main requirement for obtaining absolute heat capacity values and deriving free energy surfaces from them. In order to demonstrate equilibrium, two different experimental tests must be carried out: (1) Investigate the scan rate effect on the calorimetric peak. A number of protein-buffer scans need to be obtained at different scan rates within the instrumental range, typically 15–200 K/h (note that buffer-buffer baselines must also be obtained at each scan rate). A clear effect of scan rate on transition temperature indicates that the denaturation process is under kinetic control and does not reflect the equilibrium thermodynamics. Therefore, the transition temperature should remain essentially constant, indicating no substantial scan rate effect. The observation of a substantial scan rate effect on the transition temperature indicates that the calorimetric transition is kinetically distorted probably because the time scale of the folding/

unfolding kinetics of the protein is comparable to the time scale of the calorimetric experiment or because of the influence of irreversible denaturation processes (*see* below). Under these conditions, kinetic models are required in the data analysis and the use of equilibrium thermodynamics models is typically out of question (*see* Fig. 1 in [19]).

(2) Test the calorimetric reversibility by performing a second, reheating run of the protein solution (this aspect is extensively described in **Note 8**). In summary, the lack of any scan rate effect and the demonstration of calorimetric reversibility support the validity of using thermodynamics analysis and are at the top of any calorimetrist's wish list. However, we must admit these characteristics are not found in many cases. PDD is a small fast-folding protein [13], which does not exhibit significant scan rate dependence in DSC. Therefore, the experiments described in this chapter were performed at the highest scan rate, 200 K/h. In this case, dynamic correction of the data owing to the (small) distortion introduced by the time response of the instrument was advisable (this correction can be easily performed using the software provided by Microcal). In addition, as described in **Note 8**, PDD thermal denaturation was highly reversible, thus further supporting analyses based on equilibrium thermodynamics.

8. The experimental test of calorimetric reversibility is a must in differential scanning calorimetry. As elaborated in **Note 7**, calorimetric reversibility is a necessary, but not sufficient, condition for the applicability of equilibrium thermodynamics to the analysis of the data (*see* Fig. 1 in [19] for a detailed description of possible scenarios for the DSC data analysis).

   To estimate calorimetric reversibility in PDD denaturation, we first subtracted the last instrumental baseline collected from the PDD thermogram. Then, and using the tools provided by the program Microcal Origin ("Baseline Session"), we connected the pre- and post-transition baselines with a smooth line ("progress baseline" option). Finally, we integrated the area under the peak from the baseline. Typically, the unfolding process is considered to be calorimetrically reversible if at least 80% of the area under the peak is recovered in the second run. In the case of PDD, the unfolding transition was highly reversible, a much more common observation for small proteins. Be aware that reversibility may depend on the final temperature reached in the first protein scan. Thus, the higher the final temperature, the longer the protein stays under conditions where irreversible processes (aggregation, chemical modifications, etc.) are fast [20]. Therefore, if a poor reversibility is achieved in the reheating scan, it might be helpful to set a lower final temperature, just right after the unfolding peak, and carry out the standard protocol described above. If this is the case, note that the thermal history of the instrument should be established again. For PDD, calorimetric reversibility was excellent, contrary to the common behavior observed in large, more complex proteins.

References

1. Privalov PL, Plotnikov VV (1989) Three generations of scanning microcalorimeters for liquids. Thermochim Acta 139:257–277
2. Becktel WJ, Schellman JA (1987) Protein stability curves. Biopolymers 26:1859–1877
3. Privalov PL (1990) Cold denaturation of proteins. Crit Rev Biochem Mol Biol 25:281–305
4. Makhatadze GI, Privalov PL (1995) Energetics of protein structure. Adv Protein Chem 47:307–425
5. Robertson AD, Murphy KP (1997) Protein structure and the energetics of protein stability. Chem Rev 97:1251–1268
6. Freire E (2001) The thermodynamic linkage between protein structure, stability, and function. Methods Mol Biol 168:37–68
7. Ibarra-Molero B, Sanchez-Ruiz JM (2006) Differential scanning calorimetry of proteins: an overview and some recent developments. In: Arrondo JLR, Alonso A (eds) Advanced techniques in biophysics. Springer-Verlag, Berlin, pp 27–48
8. Muñoz V, Sanchez-Ruiz JM (2004) Exploring protein-folding ensembles: a variable-barrier model for the analysis of equilibrium unfolding experiments. Proc Natl Acad Sci U S A 101:17646–17651
9. Naganathan AN, Sanchez-Ruiz JM, Muñoz V (2005) Direct measurement of barrier heights in protein folding. J Am Chem Soc 127:17970–17971
10. Godoy-Ruiz R, Henry ER, Kubelka J, Hofrichter J, Muñoz V, Sanchez-Ruiz JM, Eaton WA (2008) Estimating free-energy barrier heights for an ultrafast folding protein from calorimetric and kinetic data. J Phys Chem B 112:5938–5949
11. Naganathan AN, Perez-Jimenez R, Muñoz V, Sanchez-Ruiz JM (2011) Estimation of protein folding free energy barriers from calorimetric data by multi-model Bayesian analysis. Phys Chem Chem Phys 13:17064–17076
12. Sanchez-Ruiz JM (2011) Probing free-energy surfaces with differential scanning calorimetry. Annu Rev Phys Chem 62:231–255
13. Naganathan AN, Li P, Perez-Jimenez R, Sanchez-Ruiz JM, Muñoz V (2010) Navigating the downhill protein folding regime via structural homologues. J Am Chem Soc 132:11183–11190
14. Kholodenko V, Freire E (1999) A simple method to measure the absolute heat capacity of proteins. Anal Biochem 270:336–338
15. Gomez J, Hilser VJ, Xie D, Freire E (1995) The heat capacity of proteins. Proteins 22:404–412
16. Guzman-Casado M, Parody-Morreale A, Robic S, Marqusee S, Sanchez-Ruiz JM (2003) Energetic evidence for formation of a pH-dependent hydrophobic cluster in the denatured state of *Thermus thermophilus* ribonuclease H. J Mol Biol 329:731–743
17. Makhatadze GI, Medvedkin VN, Privalov PL (1990) Partial molar volumes of polypeptides and their constituent groups in aqueous solutions over a broad temperature range. Biopolymers 30:1001–1010
18. Goldberg RN, Kishore N, Lennen RM (2002) Thermodynamic quantities for the ionization reactions of buffers. J Phys Chem Ref Data 31:231–370
19. Ibarra-Molero B, Naganathan AN, Sanchez-Ruiz JM, Muñoz V (2016) Modern analysis of protein folding by differential scanning calorimetry. Methods Enzymol 567:281–318
20. Sanchez-Ruiz JM (2010) Protein kinetic stability. Biophys Chem 148:1–15

Chapter 6

Fast-Folding Kinetics Using Nanosecond Laser-Induced Temperature-Jump Methods

Michele Cerminara

Abstract

The development of ultrafast kinetic methods is one of the factors that allowed the research on protein folding to flourish over the last 20 years. The introduction of new optical triggering techniques enabled to experimentally investigate the protein dynamics at the nanosecond to millisecond timescale, allowing researchers to test theoretical predictions and providing experimental benchmarks for computer simulations. In this work, the details of how to perform kinetic experiments by the laser-induced temperature-jump technique, using the two most commonly used probing techniques (namely infrared absorption and fluorescence spectroscopy) are given, with a strong emphasis on the practical details.

Keywords Protein folding, Folding kinetics, Pump and probe, Temperature jump, Infrared absorption, Fluorescence, Förster resonance energy transfer

1 Introduction

Protein folding is a central problem in protein science: in order to be able to perform their function, proteins have to fold rapidly and reliably (i.e., avoiding misfolding and aggregation) to a specific three-dimensional structure that has been designed by evolution to accomplish that particular task. The theory of protein folding states that the process does not occur following a single well-defined pathway analogous to elementary chemical reaction, while it is a rather heterogeneous process in which individual polypeptide molecules can follow very different pathways. Energetically, the problem is described by a hyperdimensional energy landscape, which for well-evolved proteins has an overall funnel shape where the unfolded ensemble corresponds to the broader end (large conformational entropy) located at higher energy, whereas the folded state lies at the tip of the funnel, corresponding to the global energy minimum and a minimal number of conformational sub-states [1]. The folding of a protein is then described as the diffusive motion on such energy landscape, which on top of the

Victor Muñoz (ed.), *Protein Folding: Methods and Protocols*, Methods in Molecular Biology, vol. 2376, https://doi.org/10.1007/978-1-0716-1716-8_6,

overall funnel shape contains a rough topography arising by the transient occurrence of non-native interactions that will need to be broken in order to reach the correct folded structure. Roughness plays a crucial role in protein folding kinetics because escape from such local minima can represent the main bottleneck for the process, especially when it is not thermally activated, i.e. when the energy barrier between the folded and unfolded ensembles becomes small or even negligible. Predictions of protein folding kinetics pointed to the existence of fast processes, even complete folding when the global diffusive process goes downhill, that take place below the ms timescale of conventional stopped-flow kinetic experiments. These theoretical predictions were the main inspiration for the development of new ultrafast kinetic experiments that could test them and which have resulted in a major turning point in protein folding research [2].

Time-resolved experiments monitor the relaxation of a protein to its equilibrium condition in response to an external triggering event (perturbation). The triggering event defines the zero time, which is limited by how fast the perturbation can be applied. The first kinetic study on a fast-folding protein used a photochemical trigger to rapidly initiate the folding process, namely the photodissociation of carbon monoxide from denatured cytochrome c [3]. Other optically initiated trigger methods based on photo-induced isomerization [4] or photolyzable groups [5] have been proposed. These methods are attractive due to the very high time resolution that is limited only by the onset of the photo-induced reaction, but they require to modify the protein to introduce the reactive groups for the photo-triggering, with the possibility to introduce structural perturbation to the system under study, thus limiting their use to a small number of proteins.

A class of generalizable optical triggers are the laser-induced temperature-jump techniques. The basic concept is to use a pump and probe approach in which two beams are focused on the same spot on the sample; the pump beam induces the perturbation and the probe beam reports about its effect on the sample. As pump beam, the laser-induced temperature jump uses intense infrared laser pulses at a frequency tuned to match the energy of the vibrational modes of the solvent molecules (typically water or deuterated water), thus inducing a local increase in the temperature by vibrational excitation of the solvent, typically in less than 10 ns (when a Q-switched laser is used as pump). The subsequent relaxation of the protein to adapt to the increased temperature is then monitored by using a spectroscopic probe, most commonly in this case infrared absorption or fluorescence spectroscopy [6, 7]. These methods can be applied to virtually any protein, since the only requirement is that the perturbation of the folding–unfolding equilibrium induced by the temperature jump (which is universal since protein folding involves a significant change in enthalpy) is reflected in a change of the probe spectroscopic signal. Furthermore, the temperature-jump technique has an excellent dynamic range, permitting to track the process from the onset of the temperature jump (few ns) up to ms.

1.1 Infrared Absorption

Infrared spectroscopy is one of the classical biophysical methods used to characterize the structural features of peptides and proteins. In particular, Fourier transform infrared absorption is a popular ensemble technique used to characterize protein unfolding at equilibrium conditions. The amide bond exhibits nine vibrational modes, identified as amide A, B and I–VII bands: these absorption bands are very sensitive to the protein conformation and from their analysis it is possible to extract information about the secondary structure of the protein [8]. The most useful band for the analysis of protein secondary structure is the amide I band, which falls in the range 1600–1700 cm^{-1}, and is mainly due to stretching of the C=O bond. The actual value of the amide I resonance depends on the local structure of the amide bond; thus, it is possible to correlate the position of the peak with the secondary structure of the peptide bond, which can be forming α-helices, β-sheets, β-strands, loops, turns, or disordered regions. The IR absorption of a typical protein is usually a broad band that results from the summed contributions of all its peptide bonds, each in its particular type of secondary structure. This spectrum can be deconvolved to calculate the average secondary structure contents of the protein.

In the laser-induced temperature-jump experiment, the goal is not necessarily to measure the whole IR spectrum of the protein, but to characterize the time response of the protein to the conformational readjustment upon partial thermal unfolding. Therefore, the probe beam is a laser whose wavelength is tuned to match the maximum of the amide I band of the protein to be studied; the experiment can be repeated at different wavelengths to extract information from different structural properties (keeping in mind that the bands for various secondary structures usually overlap).

1.2 Fluorescence Spectroscopy

Fluorescence spectroscopy is a common biophysical technique [9] complementary to IR absorption. The intrinsic fluorescence of proteins is normally due to the presence of tryptophan, the amino acid with the strongest fluorescent quantum yield, whereas other amino acids are either nonfluorescent or very weekly fluorescent (tyrosine and phenylalanine). The position and intensity of the tryptophan emission peak is strongly affected by the local environment (i.e., the degree of exposure to the solvent or interactions with other residues that can act as quenchers), characteristic that can be used as diagnostic of the integrity of the protein tertiary structure. If the protein of interest does not have any tryptophan, it is possible to label it with extrinsic fluorophores that can be used to report the local environment of the labeled position, for example by studying fluorescence quenching, or to measure the distance between two specific protein locations from the Förster resonant energy transfer (FRET) efficiency between two suitable extrinsic fluorophores.

FRET arises from a dipole–dipole interaction between a donor fluorophore in its excited state and an acceptor fluorophore in its ground state. The main requirement is that the fluorescence spectrum of the donor overlaps with the absorption spectrum of the acceptor (i.e., the overlap integral should be non-null). If this condition is accomplished, radiation-less transfer of excitation from the donor to the acceptor occurs via a dipole–dipole interaction. The transfer efficiency depends on the spectral overlap, on the relative orientation of the transition dipoles of the two dyes and on their distance. The distance dependence is of particular interest for biophysical applications. Due to the dipole–dipole nature of the interaction, the FRET efficiency scales as the inverse of the sixth power of the inter-dye distance according to the expression:

$$E = \frac{R_0^6}{R_0^6 + r^6} \quad (1)$$

where E is the FRET efficiency, r is the inter-dye distance and R_0 is the characteristic Förster radius, i.e., the distance at which the efficiency is 0.5 for a specific donor/acceptor pair. Thanks to this relation, by experimentally measuring E, it is possible to determine the average distance between the two fluorophores. For typically used fluorophores pairs, the accessible distance range is 2–8 nm, making FRET a particularly suitable "spectroscopic ruler" to detect protein conformational transitions.

The implementation of fluorescence in laser-induced temperature-jump experiments requires a probe beam tuned to excite either the intrinsic fluorescence of the protein (i.e., the tryptophan absorption maximum at 280 nm) or of the donor fluorophore for FRET applications.

2 Materials

2.1 Instrumentation

A typical setup for an IR absorption laser-induced temperature-jump instrument is shown in Fig. 1. The pump beam originates from a pulsed infrared laser run at a low frequency (1–10 Hz) (*see* **Note 1**). Typically a Q-switched Nd:YAG is used for this purpose. Because the fundamental of this laser (1064 nm) is not in resonance with the vibrational modes of D_2O (the solvent used for protein IR spectroscopy), the beam is converted to longer wavelengths taking advantage of the stimulated Raman emission effect (a Raman cell filled with high pressure H_2 is used for this purpose). The output from the Raman cell contains the fundamental plus different Stokes and anti-Stokes conversions. This heterogeneous output is

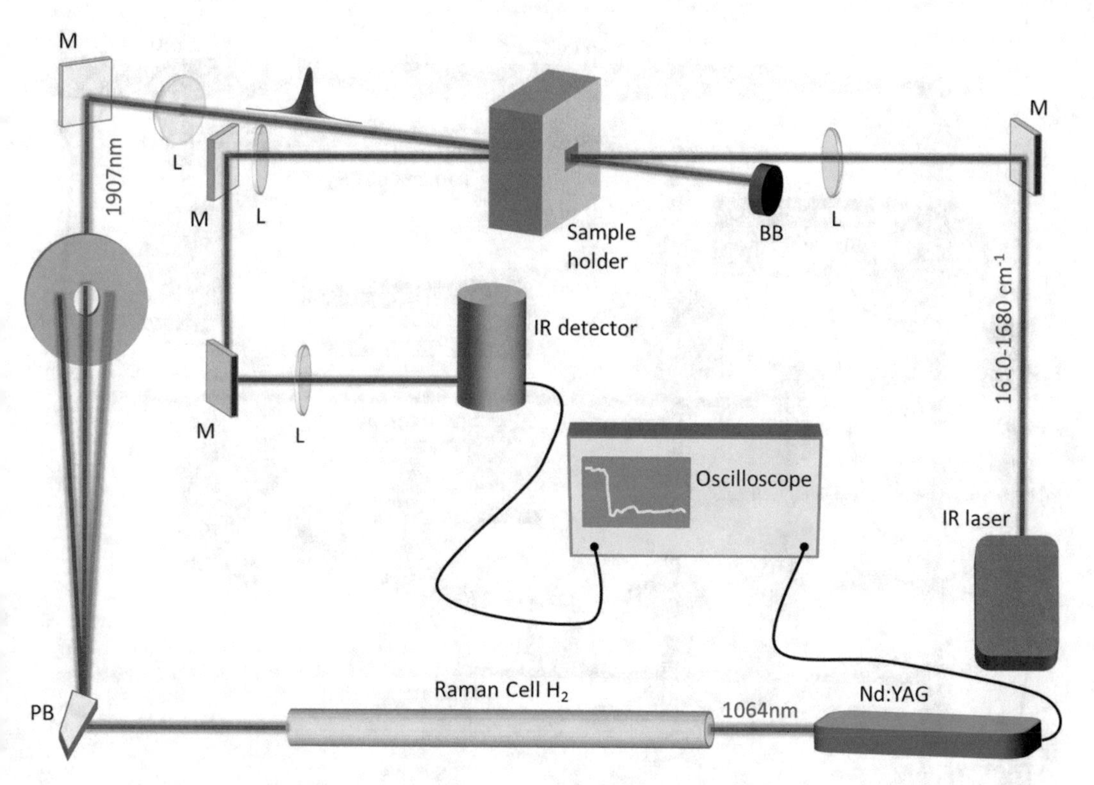

Fig. 1 Schematic representation of the apparatus for infrared T-jump measurements installed at IMDEA Nanociencia in Madrid (abbreviations used for the components: mirror (M), lens (L), Pellin–Broca prism (PB), beam blocker (BB))

dispersed by a Pellin–Broca prism, and the beam corresponding to the first Stokes beam at 1907 nm is chosen for D_2O excitation. The protein sample dissolved in D_2O is placed on an IR transmission spectroscopic cell (e.g., two $CaCl_2$ windows separated by a Teflon spacer) of typically 50 μm pathlength. The 1907 nm pump beam is then focused onto the sample (with a spot of about 1 mm diameter), where it induces the rapid increase in temperature (the thin spacer thickness facilitates relatively uniform heating along the pathlength with only one pump beam coming from the front). The probe comes from a continuous wave quantum cascade laser that is tunable within the range of the amide I band. This beam is focused on the center of the heated volume to maximize the magnitude and homogeneity of the temperature jump. The transmitted probe beam is collimated and then focused on a fast mercury-cadmium-telluride (MCT) detector. The output of the detector is analyzed by an oscilloscope that is synchronized with the pump beam.

A typical setup for a spectrally resolved fluorescence laser-induced temperature-jump instrument is shown in Fig. 2. The configuration of the pump beam is similar, with only two differences: (1) the Raman converter is filled with D_2 in order to obtain a

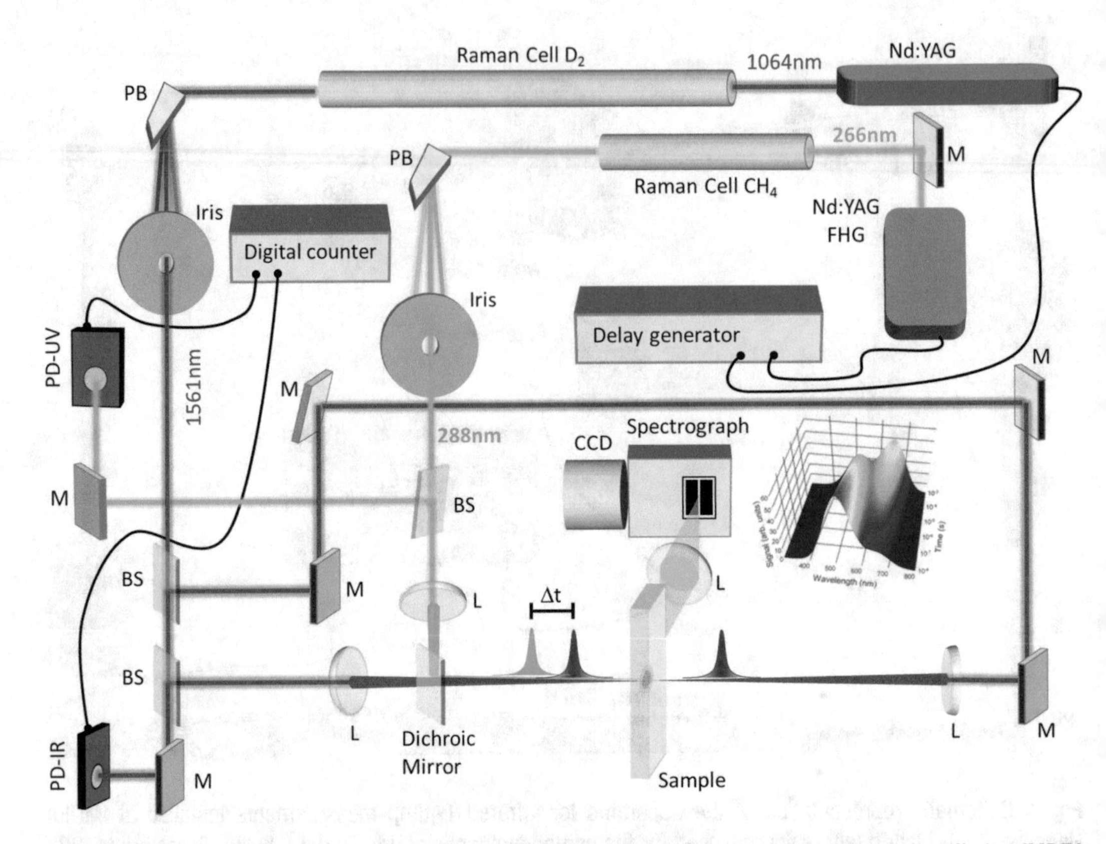

Fig. 2 Schematic representation of the apparatus for fluorescence T-jump measurements installed at IMDEA Nanociencia in Madrid (abbreviations used for the components: mirror (M), lens (L), Pellin–Broca prism (PB), beam blocker (BB), beam splitter (BS), photodiode (PD))

first Stokes beam at 1561 nm that is in resonance with the absorption spectrum of regular H_2O; (2) the pump beam is separated in two halves, using one beam to reach the sample from the front and the other one from the back (*see* **Note 2**). The probe beam is emitted by a laser with an appropriate wavelength to excite the singlet state of the fluorophore in the protein of interest. In the scheme depicted in Fig. 2, the fourth-harmonic of a Nd:YAG at 266 is used to directly excite tryptophan (*see* **Note 3**). If the protein is labeled with extrinsic fluorophores, other excitation sources are needed, e.g., the second or third harmonic of a Nd:YAG laser or other kind of pulsed lasers. In this setup, the pulsed probe laser is run at the same frequency of the pump laser and triggered using a pulse delay generator, so that the time delay between the pump and probe pulses can be carefully set and changed automatically following a predefined sequence. The sequence of time delays generates the full kinetic experiment (each probe pulse delayed by a specific time relative to the pump pulse provides a given time point in the kinetic decay of the protein). The probe beam is then coupled to the front pump beam with a dichroic filter, making them collinear, and is focused to a diffraction limited spot on the sample and in the center of the pump beams. The fluorescence emitted by the sample is collected at 90° (*see* **Note 4**), and focused to the entrance slit of a spectrograph connected to a CCD camera that records the spectrum. A kinetic experiment in this type of setup is conducted by performing a sequence of pump-probe pulses in which the time between pump and probe is varied between 1 ns and 10 ms by the time delay generator. After the probe pulse is fired, fluorescence emission is collected, resolved spectrally via a spectrograph and recorded by the CCD camera. Each recorded fluorescence spectrum corresponds to a snapshot of the properties of the protein at a specific time in the reaction, as determined by the delay between the two pulses (*see* **Note 5**). The system is run at a constant rate (1–10 Hz) set so that it is sufficiently slow to ensure full heat dissipation and return to the initial temperature before the next cycle. A programmed sequence of time delays is then performed at that rate to generate a set of time-resolved fluorescence spectra (each spectrum could be from a single shot or from an average of multiple shots).

2.2 Chemicals and Reagents

Buffer solutions should be prepared with the highest possible purity: MilliQ water or HPLC-grade water should be used. For the infrared experiments, NMR-grade D_2O should be used. All the chemicals needed to prepare the buffers should be at the highest available purity, especially for the fluorescence experiments where the presence of impurities can give unexpected background. N-acetyl-tryptophanamide (Sigma), or another reference fluorophore that can be excited at the probe wavelength and has strongly temperature-dependent quantum yield, is needed to calibrate the temperature jump in the fluorescence experiments. Samples for tryptophan fluorescence temperature-jump experiments should be prepared in a 2% v/v solution of carbon disulfide (CS_2, Sigma), which acts as an efficient tryptophan triplet quencher to minimize triplet buildup that can result on decreased and/or delayed fluorescence.

2.3 Other Materials

Laser burn paper is used to record the position of the pump laser spot.

Temperature-sensitive liquid crystal sheets are used to locate the far-infrared probe beam of the infrared experiments.

Sample cuvette holders with Peltier temperature controllers.

Optical power and energy meters.

3 Methods

- *Protein production.* Different strategies can be used to obtain the protein object of the study. The most common technique uses

3.1 Sample Preparation for Infrared Absorption Measurements

cell culture (bacteria, yeast, or eukaryotic cells) to over express recombinant proteins [10]. This approach renders high yields and is generally applicable to any protein. Other synthetic approaches are solid-phase peptide synthesis [11] (which is typically limited to protein of less than 100 amino acids) or cell-free in vitro protein synthesis [12], which may be preferable for proteins difficult to express recombinantly, but at the expense of lower production yields.

- *Protein purification.* Depending on the specific protein and on the method used to produce it, different purification strategies can be used. For example, his-tag, ionic exchange, reverse phase, size-exclusion, or other types of affinity chromatography methods are available. If the final concentration after purification is insufficient, a step in which the protein sample is concentrated by ultrafiltration might be required.
- *Protein deuteration.* H_2O has strong absorption in the IR region overlapping with the amide I band. This is a major limitation for measuring the amide I band of proteins in solution. However, the absorption of D_2O is shifted to lower wavenumbers relative to H_2O, so that it does not overlap with the amide I band anymore. For IR measurements in aqueous solution (as opposed to dry films), it is therefore necessary to substitute H_2O by D_2O as solvent for sample preparation. The labile amide protons of the protein must be deuterated, and suitably deuterated buffers should also be used. To achieve uniform deuteration of protein amide protons, it is necessary to repeat several cycles of lyophilization to remove H_2O followed by dilution in pure D_2O (*see* **Note 6**). Typically, three cycles are enough to achieve the complete deuteration of the protein. As a final step, the effectiveness of the deuteration process should be evaluated by measuring the IR absorption spectrum of the sample resuspended in D_2O with the final deuterated buffer.
- *Sample preparation.* The final sample should be prepared in deuterated buffers using chemicals which do not have any absorption in the IR region of interest. The final concentration of the protein should be in the range 0.5–2.5 mM, depending on the amount of secondary structure of the actual protein. Not being able to reach these high protein concentrations is often times the main limitation for IR temperature-jump measurements. On the other hand, the required volume of protein solution needed for a single measurement is very small (less than 20 μL), then the sample can be prepared at the minimal volume required to measure its pH using an ultrathin pH electrode (similar to those use for NMR sample preparation).
- *pH adjustment.* For these experiments it is important to realize that the readout of a glass electrode used to measure pH is

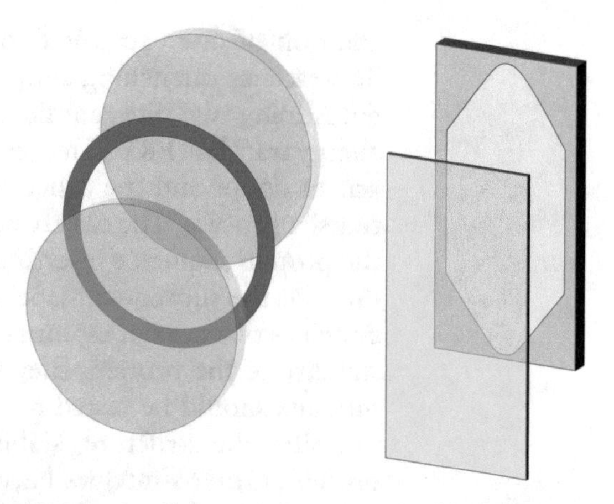

Fig. 3 Typical types of sample cells used to perform T-jump experiments in the infrared (left) and in fluorescence (right). For the IR experiments, the sample solution is sandwiched between two round windows of a material transparent to both the pump and probe beams (e.g., CaF_2) with a 50-μm Teflon spacer. For the fluorescent experiments, an all-glass (UV-silica to be transparent at the wavelengths of both the lasers) cuvette is filled with the sample solution

affected by D_2O, resulting in an isotopic shift relative to electrode reading in pure water [13]. The addition of salts and/or denaturants in very high concentrations also affect electrode-based measurements of pH [14]. These effects are usually accounted for by applying a simple linear correction to the glass electrode reading.

- Mounting the sample in the cell holder. The sample cell is constituted by two windows of CaF_2 or of any other material that does not absorb in the range of the pump and probe wavelengths. Typical windows are round windows with a diameter of 12 mm and a thickness of 2 mm. The sample solution is sandwiched between the windows with a Teflon spacer with thickness that defines the optical path length, as depicted in Fig. 3, left. Typical spacer thickness is about 50 μm to minimize the decay of the temperature jump along the path length due to the solution's absorbance of the pump beam. The cell should be mounted manually, clamped together (*see* **Note 7**), and held in a customized sample holder that is temperature controlled (e.g., using a Peltier thermoelectric system) to accurately define the initial temperature of the experiment (before the temperature jump).

3.2 Sample Preparation for Fluorescence Measurements

- *Protein production.* If the protein to be studied is intrinsically fluorescent (i.e., it has tryptophan residues), the same production strategies described in Subheading 3.1 can be used. If the protein is not fluorescent, fluorescent labels need to be attached to the protein using standard protein chemistry procedures. The addition of one extrinsic fluorophore enables measurement of fluorescence quenching and/or photon-induced electron transfer. Adding two different fluorophores enables Förster resonant energy transfer (FRET) measurements in which one fluorophore acts as donor and the other as acceptor of the resonant energy transfer process. The easiest way to achieve this goal is to modify the protein sequence inserting one or two cysteines that will be used for the subsequent labeling [15]. If the protein has natural cysteines that are not essential for the biological function and/or stability of the protein, they should be mutated out. The new mutants should be tested to verify that the point mutations do not alter the structure, stability, and activity of the wild-type protein. Other strategies have been proposed, for example, the incorporation of unnatural amino acids that allow click chemistry with orthogonal reactivities to achieve site selectivity and specificity [16]. The labeling with the fluorescent dyes should be done following the instructions of the manufacturer for the specific reactive group. *See* chapter on FRET labeling for more technical details on how to label proteins with extrinsic fluorophores.

- *Protein purification.* The same purification protocols cited in Subheading 3.1 are to be used for obtaining proteins for fluorescence temperature-jump experiments. For samples for FRET experiments, the protocols should be optimized to ensure that the sample only contains protein molecules simultaneously labeled with both donor and acceptor (*see* chapter on FRET labeling).
- *Sample preparation.* The sample should be prepared in the required buffer at a final concentration in the range 1–10 μM. For tryptophan measurements, 2% v/v C_2S can be added to the sample to minimize tripled buildup.
- Mounting the sample in the cell holder. For fluorescence experiments, since the detection path is perpendicular to the excitation path, the whole sample cell needs to be completely transparent both in the UV-VIS (for the probe beam and the fluorescence spectrum) and in the NIR (for the pump beam). Practically, one can use UV-silica grade cuvettes with a sample thickness in the range of 100–500 μm. These cuvettes are constituted by two rectangular plates. One is carved at the required pathlength thickness to contain the sample. The other plate is set on top of the carved plate filled with the sample solution (using 1.5 times the void volume), starting from one end and setting the plate carefully (by slowly dropping the other end of the top plate until it touches the bottom plate) to avoid formation of bubbles. The top plates are then pushed together applying gentle pressure so that the excess liquid is removed and the two plates become

stuck by capillarity. *See* for example the sample cell depicted in Fig. 3, right.

3.3 Infrared Temperature-Jump Kinetic Measurements

- *Setup optimization.* The probe beam needs to be aligned onto the MCT detector, which is normally done by maximizing the readout signal. This can be done by inserting a mechanical chopper on the probe beam path, so that an easily recognizable pattern can be observed on the oscilloscope and maximized in amplitude. Once the probe beam is aligned onto the detector, the pump beam must be aligned by adjusting the last mirror to steer the pump beam until the probe beam is centered onto the larger probe beam. To maximize the overlap of the two beams, the position of the pump beam should be scanned along the *X*–*Y* axes looking to maximize the change in transmission of a sample with only buffer (no protein) taking place concomitantly to the pump pulse (about 8 ns). This sudden change in transmission reflects the jump in temperature induced by the pump beam on the sample (Fig. 4, left). Therefore, this sudden change in transmission can be used to optimize the alignment since it will be maximal when the probe beam is exactly placed in the center of the pump beam (which has a Gaussian beam profile). This step in transmission is also used to determine the magnitude of the temperature jump.
- *Temperature-jump calibration.* Before making the actual measurements, the amplitude of the temperature jump must be calibrated. As first step, one must measure what is the conversion factor between the amplitude of the detector readout (which depends on the specific setup and alignment) and the transmitted intensity as a function of temperature. The sample cell should be filled with just the buffer, and the magnitude of the step in transmission needs to be measured at different temperatures covering the required temperature range, as this change in transmission depends on the initial temperature. The signal measured before the pump pulse is fired is proportional to the intensity of the transmitted light at the initial temperature (base temperature at which the whole sample holder is kept). A plot of the magnitude of the step in transmission as a function of the base temperature will give rise to a linear correlation in which the slope provides the conversion factor from buffer transmission to temperature. One can then set the pump pulse energy to achieve the desired temperature jump. For example, using 8 ns pump pulses of 30 mJ it is usually possible to obtain temperature jumps of about 10–12 K.
- *Data collection and pre-analysis.* The actual experiment consists of measuring a series of time traces at different initial temperatures. The buffer and the sample containing the protein must be

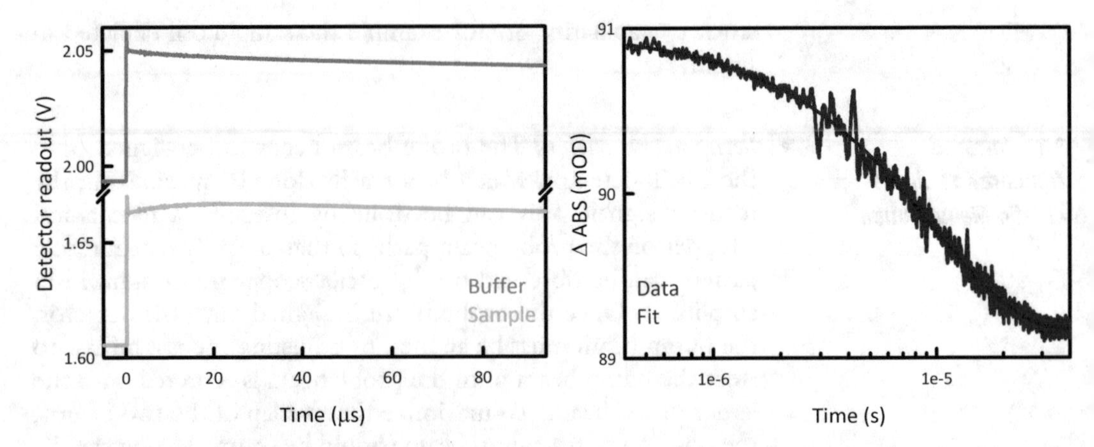

Fig. 4 Typical data obtained for a kinetic experiment in the infrared. In the left panel, the time traces of the protein sample and of the buffer needed as a reference are reported. In the right panel, the resulted transient absorption decay is represented, with a fit to an exponential decay function

measured in exactly the same temperature conditions. A single time trace is normally very noisy, so that it is usually necessary to acquire hundreds of traces and average them. This can be simply done by setting the oscilloscope to acquire directly the average of the appropriate number of traces (the instrument is run at a constant rate of 1–10 Hz). One can also acquire single traces directly, which can then be averaged at the time of analysis. This second strategy is more cumbersome since it requires longer acquisition times and produces much larger datasets that need to be post-processed, but it allows for a more accurate data analysis. This is because the pulsed energy of the pump laser is not constant and thus the temperature jump exhibits pulse-to-pulse variability that introduce uncertainty in the experiment. The availability of single traces allows to select traces within a certain temperature-jump threshold and discard the rest (rather than average all of them). Recording single traces also allows to identify traces with clear signs of cavitation, which is the major source of artifacts in laser-induced temperature-jump experiments, especially at higher temperatures (*see* **Note 8**).

- *The time-resolved IR absorption kinetic decays.* The kinetic trace reflecting the time-dependent changes in amide I band absorption of the protein in response to the jump in temperature starting from the initial temperature are obtained from the averaged time traces of the sample with only buffer and the sample containing the protein as:

$$\Delta A(t) = -\mathrm{Log}\frac{V_{\text{sample}}}{V_{\text{buffer}}} \tag{2}$$

where V_{sample} and V_{buffer} are the measured traces for the sample and the buffer, respectively. The resulting decay can be fitted with the appropriate decay function that describes the system under study. An example of a typical infrared T-jump kinetic decays is shown in Fig. 4.

3.4 Fluorescence Temperature-Jump Kinetic Measurements

- *Setup optimization.* In these experiments, the most critical aspect of the alignment is to ensure the required overlap of the three excitation beams (the front and back pump beams and the probe beam) onto the sample. In this case, the optimal alignment of the three beams needs to be carried out daily because the exact position of this multi-focal point is strongly sensitive to minor changes in the detailed alignment of all components. The position is defined by the probe excitation beam that should be focused to approximately the center of the sample cell. The front and back pump beams need to be aligned, so that their center coincides with the probe beam (which is typically diffraction limited, and thus much smaller in diameter at the focal point than the ~1 mm diameter of the pump beams). A first coarse alignment of the pump beams is carried out using burn paper to signal the position of the infrared pump beam and adjusting the last mirror in the beam path to set its approximate position relative to the probe beam. After the coarse alignment is done for both front and back pump beams, it needs to be optimized by maximizing the amplitude of the temperature jump (*see* below). The position of the fluorescence collection optics needs to be optimized as well. The collection optics alignment is performed by placing a solution of a reference dye in the sample cuvette, recording its spectrum with the spectrograph/CCD camera, and maximizing the overall intensity as a function of the optics position.
- *Temperature-jump calibration.* The temperature jump is determined from the decrease in fluorescence intensity induced by the pump probe in a reference sample with a well-characterized temperature-dependent quantum yield. For example, the fluorescence intensity of tryptophan decreases by about 1% for each degree of increase in temperature [17]. The temperature jump is determined from the differences in intensity between two spectra: one acquired right after the temperature jump is completed (maximal temperature), which practically means exciting the sample at a time delay slightly longer than the duration of the pump pulse (e.g., 15 ns); and a second spectrum corresponding to the base temperature, which is recorded by exciting the reference sample at a time previous to the pump pulse period, and thus previous to the temperature jump.
- *Data collection.* In this type of fluorescence temperature-jump instrument, a kinetic experiment consists of a series of fluorescence spectra recorded after firing the ~5 ns pulsed probe laser

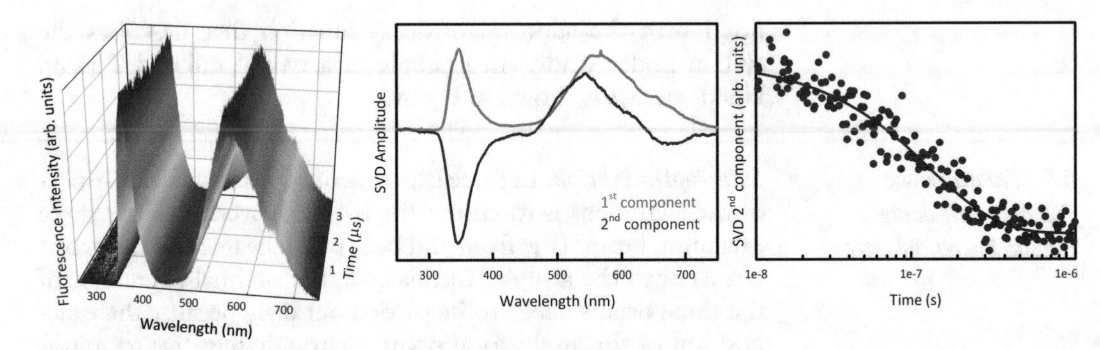

Fig. 5 Typical data obtained for a fluorescence experiment in which FRET is studied. The experimental data are a series of fluorescence spectra recorded at different delay times (left panel). Performing a singular value decomposition of the experimental data, it is possible to identify the components that describe the overall experiment (central panel) and to analyze the temporal evolution of the component reporting for the FRET efficiency (right panel)

(here the probe is a pulsed UV laser rather than the CW laser used for infrared experiments) at different times relative to the pump pulse. The spectrum at each time delay corresponds to a snapshot of the fluorescence properties of the protein sample under study at that given time from the heating pulse. A complete experiment consists of cycling through a series of suitable time delays and recording a matrix of fluorescence spectra as a function of time (*see* **Note 9**). Similarly to the IR experiments, it is usually convenient to average several spectra for each delay time to minimize shot-to-shot variability and fluctuations in fluorescence intensity. In addition, care must be exerted to ensure that the pulse energy of the probe excitation laser is kept low enough to avoid photobleaching and blinking of the fluorophores. Alternatively, one can use photoprotection cocktails added to the sample solution to minimize the photodegradation of the sample, or employ a customized cuvette fitted with entry and an exit ports to permit the circulation of fresh sample during the experiment.

- *Data analysis.* The resulting three-dimensional dataset (fluorescence intensity vs. wavelength vs. time) is best analyzed using singular value decomposition (SVD) procedures [18], which is an efficient method to further reduce noise (using the entire matrix of time-resolved fluorescence spectra) and identify relevant components that represent the temporal evolution of specific spectral features. The first SVD component (the one with highest singular value) corresponds to the average spectrum, and its amplitude reflects its variation as a function of time (e.g., changes in quantum yield). Additional non-random SVD components (usually the very first few as the SVD components are ranked according to their singular values, or contributions) report on spectral changes and their amplitudes on how they change with time. These spectral changes can be either shifts of an emission maximum or anti-correlated changes in intensity such as the ones that take place for a donor and an acceptor FRET pair. Some non-random components may also correspond to artefacts with a specific time dependence (e.g., a slow oscillation of the probe excitation laser energy output). Components representing well-understood signal artifacts can be easily discarded, and keep only those components that represent the changes in fluorescence properties of the sample that are of interest. The changes in amplitude as a function of time of each of the selected components can then be fitted to an appropriate time decay function (i.e., single, double, multi-, or stretched exponential). Figure 5 shows an example of fluorescence temperature-jump kinetic data obtained with this type of instrument and a sample labeled with a donor/acceptor FRET pair. In this case the second component is an anti-correlation between the emission intensity of donor and acceptor fluorophores, which is proportional to the changes in FRET efficiency.

4 Notes

1. The pulse frequency should be low enough to allow for the temperature perturbation to be completely quenched when the following pulse arrives, avoiding accumulation effects that can result in a drift in the experimental conditions.
2. For the fluorescence temperature-jump experiment the sample is normally placed in a UV-grade quartz cell of 0.2–0.5 mm pathlength to increase the number of molecules that are excited at a given protein concentration. The unwanted implication of the longer pathlength is that the heating profile from a single beam decays exponentially along the optical path due to the absorption of the solvent according to Lambert–Beer's law. A simple way to remedy this problem is to split the pump beam into two and bring them to the sample cell from its both sides in an counter propagating configuration (dual front-back excitation). The key here is to ensure that the two beams are as collinear as possible and focused to the center of the sample. This symmetrical excitation setup reduces the gradient of temperature along the optical path, which is non-negligible for fluorescence temperature-jump experiments. In our hands, the dual front–back excitation allows to achieve an almost constant temperature jump along the whole sample thickness.
3. Laser pulses at 266 nm are not optimal for tryptophan excitation because its maximum absorption is at 280 nm. An even more serious problem is that the high energy associated with

266 nm facilitates crossing the photoionization barrier from the singlet state, thus causing significant photodegradation of the sample during the experiment. There are no many existing options of ns pulsed lasers in the 280–290 nm range that is ideal for repeated excitation of tryptophan fluorescence, and the ones available are very expensive. However, it is possible to convert the fourth harmonic of a Nd:YAG laser from 266 to 288 nm using stimulated Raman emission. This can be achieved by selecting with a Pellin-Broca prism the first Stokes from the combined beam exiting from a Raman cell filled with CH_4 at high pressure (500 psi).

4. The collection of fluorescence is carried out at a 90° angle despite the short pathlength of the cell (0.2–0.5 mm) to minimize the amount of stray light from the excitation probe beam as well as the two pump beams that reaches the detector. Any residual stray light can be easily removed using bandpass optical filters.

5. To improve the precision of the kinetic measurements, it is better to directly measure the delay between the pump and probe pulses rather that rely on the time set by the delay generator because the electronics controlling the laser triggering contain a certain degree of jitter in the nanosecond timescale, which is particularly problematic for the shortest time delays in the experiment. One can easily achieve this by inserting a microscope slide at a 45° angle in the beam path (for the pump beam before it is split into front and back beams and for the probe beam before the focusing lens) so that a small fraction of the light is reflected and sent to a fast (ns) photodiode that is sensitive in the required spectral range (IR or UV). The electrical signals from the two photodiodes are then sent to a digital counter that measure the actual delay between the two pulses. The photodiodes should be placed in a position such that the length of the optical path between the beam splitter and the photodiode is identical to the length between the beam splitter and the sample (e.g., a 10 cm longer optical path corresponds to an extra delay of 0.333 ns).

6. To reach a complete deuteration of the protein amide bonds for IR absorption measurements, it could be necessary either to heat the sample solution above the melting point of the protein to expose the hydrophobic core of the protein and make it more accessible to the solvent (this strategy is viable only for protein that do not suffer from misfolding or aggregation) or to perform even more cycles of lyophilization.

7. The cell assembly process starts with one window and the Teflon spacer placed on top. A sufficient volume of the sample solution is placed on top of the window's center (added volume

should be about twice as much as the void volume of the mounted cell, defined by the cylinder with radius and length as determined by the spacer) to avoid the appearance of bubbles. Finally, the second window is placed on top, and the assembly is placed on the sample holder with two plastic O-rings on top and below to protect the windows from scratches or breaking. Finally, the whole assembly is clamped together by applying gentle and uniform pressure by tightening the set screws in the cell holder. A relatively high pressure needs to be applied to seal the cell assembly and limit the evaporation of the sample during the experiments (especially at the higher temperatures).

8. One of the major sources of artefacts in laser-induced temperature-jump experiments is the production of photo-acoustic (shock) waves that lead to cavitation [19]. The sudden (<10 ns) and steep raise in temperature induced by the pump laser pulse produces a shock wave that propagates centrifugally from the focal point of the pump in the sample. The shockwave propagates through the sample at 343 m/s, resulting in a pressure drop from the center of the heated volume. If the pressure drop reaches a certain magnitude, the liquid cavitates producing a bubble that results in a void in the solution that strongly deflects the probe beam, resulting in an artifactual time trace. This phenomenon is stochastic, but it increases in probability as the initial temperature increases (the water density decreases) or as the pump robe energy increases (this is what practically restricts the temperature magnitude to about 15° or lower), it becomes more problematic in experiments at higher initial temperatures. Tracers with signs of cavitation should not be used for further analysis, as the trace distortion is very marked. When traces are averaged directly with the oscilloscope, the presence of a few cavitation events may go unnoticed, resulting in distortions that affect the entire measurement. The acquisition of single traces permits to eliminate those with signs of cavitation during post-acquisition analysis. In parallel, it is good practice to try to minimize photoacoustic cavitation effects by removing the possible nucleation centers for cavitation from the solution, for example, by thoroughly degassing the sample before the experiments.

9. In this type of instrument, the delay times are set by the delay generator that triggers the probe laser, which enables the programming of sophisticated acquisition schemes for the kinetic experiments. The delay generator is controlled by a computer in which it is possible to set up the experiment, so that the delays are extended in a logarithmic rather than linear timescale. This strategy enables the full observation of complex kinetic relaxations that extend over many time decades in a

single experiment. It also allows to randomize the order of the time delays during the experiment and then reorder them to get the relevant time series. This feature is extremely useful to minimize contributions from artifactual fluctuations in fluorescence emission, such as the shot-to-shot drift of most pulsed lasers, or photodamage induced in the sample after so many excitations.

References

1. Bryngelson JD, Onuchic JN, Socci ND, Wolynes PG (1995) Funnels, pathways, and the energy landscape of protein folding: a synthesis. Proteins 21:167–195
2. Cerminara M, Muñoz V (2016) When fast is better: protein folding fundamentals and mechanisms from ultrafast approaches. Biochem J 473:2545–2559
3. Jones CM, Henry ER, Hu Y, Chan CK, Luck SD, Bhuyan A, Roder H, Hofrichter J, Eaton WA (1993) Fast events in protein folding initiated by nanosecond laser photolysis. Proc Natl Acad Sci U S A 90:11860–11864
4. Chen E, Kumita JR, Woolley GA, Kliger DA (2003) The kinetics of helix unfolding of an azobenzene cross-linked peptide probed by nanosecond time-resolved optical rotatory dispersion. J Am Chem Soc 125:12443–12449
5. Lu HSM, Volk M, Kholodenko Y, Gooding E, Hochstrasser RM, DeGrado WF (1997) Aminothiotyrosine disulfide, an optical trigger for initiation of protein folding. J Am Chem Soc 119:7173–7180
6. Ballew RM, Sabelko J, Reiner C, Gruebele M (1996) A single-sweep, nanosecond time resolution laser temperature-jump apparatus. Rev Sci Instrum 67:3694–3699
7. Dyer RB, Gai F, Woodruff WH, Gilmanshin R, Callender RH (1998) Infrared studies of fast events in protein folding. Acc Chem Res 31:709–716
8. Krimm S, Bandekar J (1986) Vibrational spectroscopy and conformation of peptides, polypeptides and proteins. Adv Protein Chem 38:181–364
9. Lakowicz JR (2006) Principles of fluorescence spectroscopy. Springer, New York
10. ROsano GL, Ceccarelli EA (2014) Recombinant protein expression in Escherichia coli: advances and challenges. Front Microbiol 5:172.1–172.17
11. Hojo H (2014) Recent progress in the chemical synthesis of proteins. Curr Opin Struct Biol 26:26–23
12. Carlson ED, Gan R, Hodgan CE, Jewett MC (2012) Cell-free protein synthesis: applications come of age. Biotechnol Adv 30 (5):1185–1194
13. Covington AK, Paabo M, Robinson RA, Bates RG (1968) Use of the glass electrode in deuterium oxide and the relation between the standardized pD (paD) scale and the operational pH in heavy water. Anal Chem 40:700–706
14. Garcia-Mira M, Sanchez-Ruiz JM (2001) pH corrections and protein ionization in water/guanidinium chloride. Biophys J 81:2489–3502
15. Obermaier C, Gabriel A, Westermeier R (2015) Principles of protein labelling techniques. Methods Mol Biol 1295:153–165
16. Lang K, Chin JW (2014) Bioorthogonal reactions for labelling proteins. ACS Chem Biol 9:16–20
17. Thompson PA (1997) Laser temperature jump for the study of early events in protein folding. Techn Protein Chem 8:735–743
18. Henry ER, Hofrichter J (1992) Singular value decomposition – application to analysis of experimental data. Methods Enzymol 210:129–192
19. Wai WO, Aida T, Dyer RB (2002) Photoacoustic cavitation and heat transfer effects in the laser-induce temperature jump in water. Appl Phys B Lasers Opt 74:57–66

Chapter 7

Measurement of Submillisecond Protein Folding Using Trp Fluorescence and Photochemical Oxidation

David Witalka and Lisa J. Lapidus

Abstract

Observation of protein folding on submillisecond time scales requires specialized ultra-rapid mixers coupled to optical or chemical probes. Here we describe the protocol for employing a microfabricated mixer with a mixing time of 8 μs coupled to a UV confocal microscope. This instrument can detect Trp fluorescence and also excite hydroxyl radicals that label the folding protein which can be detected by mass spectrometry.

Key words Protein folding, Microfluidic mixer, Trp fluorescence, Fast photochemical oxidation of proteins (FPOP)

1 Introduction

Protein folding remains one of the fundamental unsolved problems of modern science for a number of reasons. Two of the most significant reasons are (1) the wide range of time scales (from femtoseconds to seconds) over which folding processes occur and (2) the difficulty in observing all atoms of the protein over such a wide range of time scales. Considering the first problem, the time dimension, conventional experimental methods of observing refolding of proteins in "stopped flow" mixers are on time scales longer than 1 ms due to turbulence in the mixing process that requires that long to subside. Work over the past two decades in ultra-rapid mixers have pushed this "dead time" lower by shrinking the volume of the mixer to control or eliminate turbulence [1–4]. The mixers described in this chapter operate in the laminar flow regime, experiencing no turbulence even for linear flow rates as high as 1 m/s [5]. Mixing occurs by diffusion and can be as fast as 8 μs under the conditions described here. This mixer is continuous flow so measurements at various locations within the mixer correspond to distinct slices of time during folding. Another advantage of this mixer is that it uses very small volumes of protein solution, ~

Victor Muñoz (ed.), *Protein Folding: Methods and Protocols*, Methods in Molecular Biology, vol. 2376,
https://doi.org/10.1007/978-1-0716-1716-8_7,

1 nL/h. These mixers are fabricated in a clean room using a protocol described previously [6].

The second problem in understanding protein folding is that proteins are rather large irregular molecular objects, and no techniques are available to observe the motion of all parts of the protein on all time scales. Therefore, we must use multiple probes to observe the folding of different parts of the protein in parallel. In this chapter, we describe two probes, tryptophan fluorescence and fast photochemical oxidation of proteins (FPOP) [7–9]. The first technique is commonly used in observing protein folding and reports on the local environment of tryptophan side chains, including both solvent accessibility and quenching by other side chains. The second technique is fairly novel but is similar to hydrogen exchange in that it measures the solvent accessibility of various amino acids. In FPOP, hydroxyl radicals are created by a UV laser which rapidly (within 1 μs) attach to certain amino acid side chains. The change in molecular weight of the protein due to the labeling can be detected by mass spectrometry. In contrast to hydrogen exchange, the labeling is covalent. Therefore, the mass spectrometry can occur at any time after the folding reaction (Figs. 1 and 2).

2 Materials

1. Lysozyme solution: 500 μM, 6 M guanidine HCl, 20 mM glutamine. For 1 mL of solution dissolve 57.3 mg of stock guanidine HCl in 1 mL of deionized water. Add 2.92 mg of stock glutamine. Finally add 7.15 mg of Lysozyme from chicken egg whites to achieve final solution.
2. Buffer solution: 100 mM potassium phosphate buffer (pH 7) with 20 mM glutamine and 15 mM hydrogen peroxide. Will use approximately 2 mL.
3. Denaturant solution: 6 M guanidine HCl, 20 mM glutamine. Will use approximately 2 mL.
4. Collection solution: 100 nM catalase and 70 mM methionine. Will use 20 μL per collection point.
5. Microfluidic mixer: The fused silica mixing chips were fabricated at the University of Michigan Lurie Nanofabrication facility and the Michigan State University Keck Microfabrication Facility, as described below. A more detailed protocol is described in ref. 6.

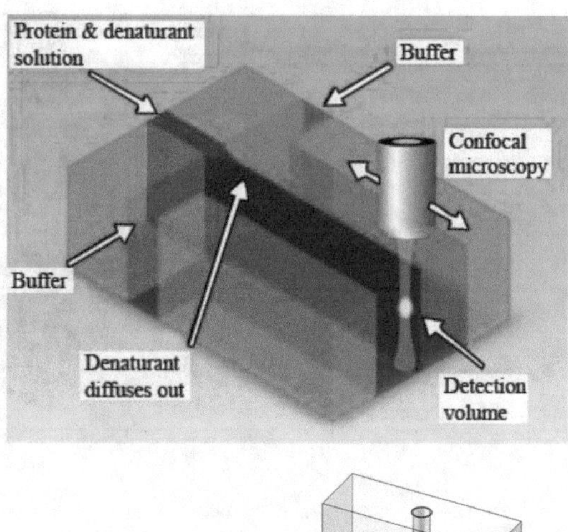

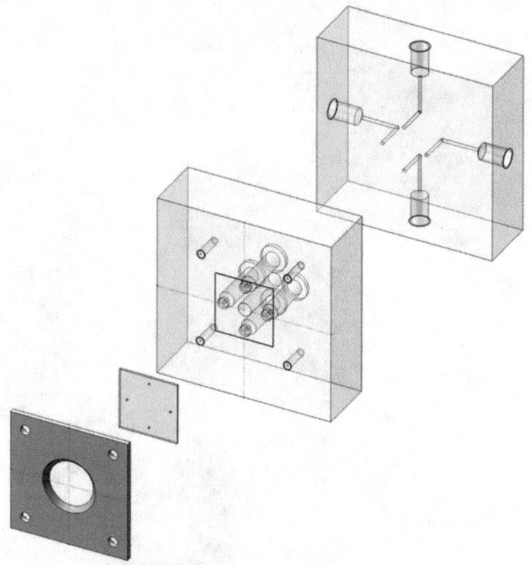

Fig. 1 (Top) Schematic of the mixing region of the microfluidic chip. Protein and denaturant enter from the top left and is met by buffer flowing from either side at a much higher flow rate, constricting the protein jet. The denaturant diffuses out quickly, and the protein remains in the jet as it flows down the exit channel to be excited by fluorescence or labeled by radicals. (Bottom) Exploded view of the chip (small gray square) mounted to the manifold (middle large square) which holds the solutions. For FPOP, the manifold does not contain an exit reservoir but a hose to direct solutions to the collection tubes. The chip is secured to the manifold by the face plate (medium square) which allows access by the microscope objective. The pressure manifold (large square at top right) applies programmable pressures to each solution well. (Figures taken from ref. 10)

3 Methods

3.1 Mixer Fabrication

- A fused silica wafer (525 μm thick, 4 in. in diameter) is cleaned and coated with KMPR1005 photoresist.
- Using a negative photolithography mask (which contains patterns for 12 mixers) in high pressure contact with the wafer, the

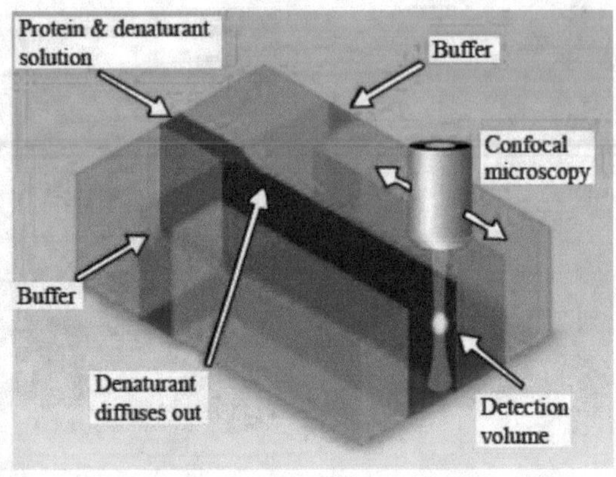

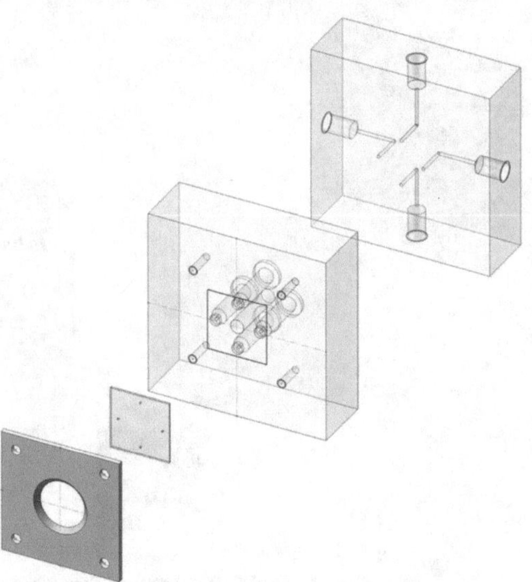

Fig. 2 (Left) Fluorescence within the mixing region and exit channel of the microfluidic chip. The protein enters from the top and is mixed with buffer at ~75 μm, and continues down the mixing region. The circles indicate positions where sample was collected for FPOP, with corresponding mass spectra of hen egg lysozyme on the right. (*Note*: positions on the left are not to scale). (Spectra taken from ref. 11)

wafer is exposed to UV light for 3.5 s and then baked at 100 °C for 7 min.

- The pattern is developed using AZ-300 MIF developer for 135 s to expose the channels that will be etched.
- The channels are created by reactive ion etching optimized for oxide substrates. 10 μm deep channels require 20 min of etching. The remaining photoresist was removed by organic solvent.
- Holes in the mixer, which allow the entrance and exit of solutions, are created by sandblasting with 27.5 μm particles.

- To seal the channels, the wafer is bonded with another fused silica wafer (170 μm thick, 4 in. in diameter). Both wafers are cleaned to remove any surface residues and then directly bonded by finger pressure. The wafer is then baked at 1000 °C for 2 h.
- The wafer is diced into individual chips.

3.2 Setup

- The mixing manifold holds the microfluidic mixer and contains reservoirs for the solutions to be mixed which will flow under the control of air pressure. Side and center reservoir pressures can be controlled separately.
- Place the mixing manifold atop a three-axis piezoelectric scanner with the x-axis aligned perpendicular to the exit channel of the chip, the y-axis in parallel with the exit channel, and the z-axis vertical. The scanning range is 100 μm in each direction. The piezoelectric scanner is attached to a motorized microscope stage used for coarse grain adjustments while the piezoelectric scanner allows for fine grain adjustments.
- An argon-ion laser produces a collimated beam at 257 nm and enters an inverted microscope where it is focused by a 0.5 NA UV objective to a 1-μm diameter inside the exit channel.
- The power of the laser should be adjusted to achieve ~5 mW power after the objective.
- The fluorescence is collected by the objective and travels through a dichroic mirror, focusing lens and pinhole before hitting a photon counter. In order to collect fluorescence spectra, the beam is diverted into a spectrograph and CCD camera.

3.3 Aligning Equipment

- Placing the protein and buffer solutions into their corresponding reservoirs, adjust channel air pressure so that the center channel is 4× that of the side channels.
- Scan the x–y plane of the exit channel using the photon counter in order to locate the fluorescent jet of the protein near the mixing region. Scan 100 μm across in both directions (*see* **Note 1**).
- Once the channel has been located, fix the y-value of the piezoelectric scanner and do another scan in the x–z plane. An hourglass pattern should appear with a bright spot in the middle, indicating the z-value where the beam is focused in the exit channel.
- Adjust the channel pressures so the side channels achieve a flow rate of 1 m/s (testing conditions) and the center and side pressures are equal (optimal mixing conditions) and scan again in the x–y plane, now scanning just 10 μm in the x direction around the jet. Repeat the process of fixing y and scanning over the x–z plane to achieve the ideal laser focus.

3.4 Taking Fluorescence Data

- Add protein solution to center channel reservoir and buffer solution to side channel reservoir (*see* **Note 2**).
- Follow the setup procedure every time a new solution is placed in the reservoir.
- Starting at the mixing region, scan across the exit channel going 10 μm in the *x* direction and 100 μm in the *y* direction (*see* **Notes 3** and **4**).
- Once the scan is complete use the motorized microscope stage to move the chip 80 μm down the exit channel and repeat the scan. Repeat this process for desired range of data (*see* **Note 5**).
- Empty the side channel reservoir and replace the buffer solution with denaturant and repeat data collection steps. This will serve as the control because the protein will remain unfolded throughout the exit channel (*see* **Notes 6** and **7**).

3.5 FPOP

- Place 20 μL of the collection solution in each Eppendorf tube, the number of tubes corresponding to the number of collection points desired. Also include a tube for flow waste (*see* **Note 8**).
- Set the side channel pressure to achieve a 1 m/s flow rate and an equal pressure on the center channel. Move to your desired collection point down the exit channel and realign the laser so that its focus is in the middle of the jet.
- Hold the laser at this point for 10 min, letting the flow in the exit channel drain into the waste tube (*see* **Note 9**).
- Switch over the flow to a collection tube containing the 20 μL of collection solution. Hold this position for an additional 20 min. Roughly 100 μL of solution should have been collected.
- Move the laser to the next data collection point and again let the exit channel drain to waste for 10 min before final collection.
- Samples should be left at room temperature for 10 min before storing at 5 °C to allow the complete decomposition of the hydrogen peroxide by the catalase.
- Prior to mass spec analysis, samples should be filtered and desalted.

3.6 Analyzing with Mass Spec

- Here we make use of a Waters Quattro Premier mass spectrometer coupled to an Acquity HPLC system.
- Keeping solutions at room temperature, inject 10–20 μL with an HPLC flow rate of 0.1 mL/min using the gradient from 2% acetonitrile (98% water) to 75% acetonitrile over 12 min and then switch back to 2% acetonitrile for 3 min as a re-equilibration step.

- 0.1% Formic acid is added to the water mobile phase to ensure protein solubility.
- Using electrospray ionization in the positive ion mode, we collect the mass spectra for our sample (*see* **Note 10**). The capillary voltage, extractor voltage, and cone voltage should be set to 3.17 kV, 5 V, and 25 V, respectively. Source temperature and desolvation temperature were 120 and 350 °C, respectively.

4 Notes

1. The exit channel of the chip is 500 μm long, giving a time range of 500 μs at 1 m/s. Flow rates lower than 1 m/s can be used to extend the time range. Fluorescence data is collected in 100 μm increments due to the range of the piezo electric scanner. Each scan is displaced by only 80 μm, so that fluorescence data from two different scans can be overlaid to make a continuous trace. Often a small offset is applied to each scan to match up with the previous scan.
2. Protein aggregation is quite commonly observed because of the close proximity of the chip surfaces. If protein is aggregating inside the mixer consider lowering the concentration.
3. If the jet is slightly crooked you can collect data as normal. However, if the jet is pressed against one of the walls of the exit channel, one side channel is clogged and a new chip should be used.
4. The exit channel may not be perfectly parallel with the stage motor axis. If the exit channel drifts out of the scan area adjust the x motor axis to bring it back to center.
5. Fluorescent data can be collected at multiple flow rates and overlaid to determine if any kinetic features are due to imperfections in the chip. Data that does not match up at different flow rates should be discarded. This will not include data within the mixing time, which depends on flow rate.
6. It is a good idea to clean the manifold with deionized water before and after experiments to prevent residue build up.
7. Chips clogged by protein aggregation can be cleaned using a piranha method ((H_2SO_4:H_2O_2 = 3:1, heated to 90 °C), and reused.
8. Glutamine acts as a scavenger in the FPOP experiment, controlling the time the protein is exposed to the radicals. Vary the amount of glutamine depending on protein properties in order to extend/shorten the labeling time. For open and unstructured proteins, more glutamine can be added to reduce the amount of labels added.

9. If bubbles form and block the exit channel, reduce the concentration of hydrogen peroxide.
10. The volume of sample collected can be varied depending on properties of mass spectrometer used.

References

1. Kathuria SV, Guo L, Graceffa R, Barrea R, Nobrega RP, Matthews CR, Irving TC, Bilsel O (2011) Minireview: Structural insights into early folding events using continuous-flow time-resolved small-angle X-ray scattering. Biopolymers 95(8):550–558. https://doi.org/10.1002/bip.21628
2. Hertzog DE, Michalet X, Jager M, Kong XX, Santiago JG, Weiss S, Bakajin O (2004) Femtomole mixer for microsecond kinetic studies of protein folding. Anal Chem 76 (24):7169–7178
3. Kane AS, Hoffmann A, Baumgartel P, Seckler R, Reichardt G, Horsley DA, Schuler B, Bakajin O (2008) Microfluidic mixers for the investigation of rapid protein folding kinetics using synchrotron radiation circular dichroism spectroscopy. Anal Chem 80(24):9534–9541. https://doi.org/10.1021/ac801764r
4. Shastry MCR, Luck SD, Roder H (1998) A continuous-flow capillary mixing method to monitor reactions on the microsecond time scale. Biophys J 74(5):2714–2721
5. Yao S, Bakajin O (2007) Improvements in mixing time and mixing uniformity in devices designed for studies of protein folding kinetics. Anal Chem 79:5753–5759
6. Izadi D, Nguyen T, Lapidus L (2017) Complete procedure for fabrication of a fused silica ultrarapid microfluidic mixer used in biophysical measurements. Micromachines 8(1):16
7. Chen J, Rempel DL, Gau BC, Gross ML (2012) Fast photochemical oxidation of proteins and mass spectrometry follow submillisecond protein folding at the amino-acid level. J Am Chem Soc 134(45):18724–18731. https://doi.org/10.1021/ja307606f
8. Chen J, Rempel DL, Gross ML (2010) Temperature jump and fast photochemical oxidation probe submillisecond protein folding. J Am Chem Soc 132(44):15502–15504. https://doi.org/10.1021/ja106518d
9. Gau BC, Sharp JS, Rempel DL, Gross ML (2009) Fast photochemical oxidation of protein footprints faster than protein unfolding. Anal Chem 81(16):6563–6571. https://doi.org/10.1021/ac901054w
10. Waldauer SA (2010) Early events in protein folding investigated through ultrarapid microfluidic mixing. Michigan State University, East Lansing
11. Wu L, Lapidus LJ (2013) Combining ultrarapid mixing with photochemical oxidation to probe protein folding. Anal Chem 85 (10):4920–4924. https://doi.org/10.1021/Ac3033646

Chapter 8

Native State Hydrogen Exchange-Mass Spectrometry Methods to Probe Protein Folding and Unfolding

Pooja Malhotra and Jayant B. Udgaonkar

Abstract

Native state hydrogen exchange (HX) methods provide high-resolution structural data on the rare and transient opening motions in proteins under native conditions. Mass spectrometry-based HX methods (HX-MS) have gained popularity because of their ability to delineate population distributions, which allow a direct determination of the mechanism of inter conversion of the partially folded states under native conditions. Various technological advancements have provided further impetus to the development of HX-MS-based experiments to study protein folding. Classical HX-MS studies use proteolytic digestion to produce fragments of the protein subsequent to HX in solution, in order to obtain structural data. New chemical fragmentation methods, which achieve the same result as proteolysis and cause minimal change to the HX pattern in the protein, provide an attractive alternative to proteolysis. Moreover, when used in conjunction with proteolysis, chemical fragmentation methods have significantly increased the structural resolution afforded by HX-MS studies, even bringing them at par with the single amino acid resolution observed in NMR-based measurements. Experiments based on one such chemical fragmentation method, electron transfer dissociation (ETD), are described in this chapter. The ETD HX-MS method is introduced using data from a protein which is inherently resistant to proteolytic digestion as example of how such an experiment can provide high-resolution structural data on the folding-unfolding transitions of the protein under native conditions.

Key words Mass spectrometry, Electron transfer dissociation, Protein folding, Backbone amide hydrogen, Native-state studies, Proteolysis

1 Introduction

Proteins have dynamic structures which undergo multiple folding and unfolding transitions in solution [1]. Unfolding reactions may involve either partial opening of only some parts of the protein structure to form partially folded/unfolded states or a complete opening to form the globally unfolded state (U) [2]. Even under native conditions, which favor the population of the native state (N) at >99%, unfolding and partial opening reactions take place continuously, albeit with a very slow rate constant compared to that found in the denaturing conditions that favor unfolding reactions.

Victor Muñoz (ed.), *Protein Folding: Methods and Protocols*, Methods in Molecular Biology, vol. 2376,
https://doi.org/10.1007/978-1-0716-1716-8_8,

Native state hydrogen exchange (HX) methods can monitor these rare unfolding transitions that take place under native conditions with high structural resolution [3]. In an HX reaction, a protium or deuterium, present at an exchangeable backbone amide site in the protein, exchanges with a deuterium or a protium in the surrounding solution [4]. The methodology makes use of the fact that exchange occurs at a backbone amide site only subsequent to a partial or global unfolding event which exposes the given site to the surrounding solvent. Native state HX is, therefore, silent to the large population of the N state present under native conditions and monitors only rare and transient unfolding/opening events [3].

For several decades, HX reactions have been monitored in protein folding studies using nuclear magnetic resonance (NMR) spectroscopy [3, 5–7], which provides single amino acid resolution. Despite the high structural resolution afforded by NMR methods, mass spectrometry (MS) has emerged as an attractive alternative to measure HX reactions [8] in recent years. This is, in part, due to the low sample concentrations required for MS studies. The more important advantage afforded by MS experiments is that they yield population distributions [9]. The number of species populated during the transient unfolding events under native conditions as well as their mechanism of inter-conversion can be directly deduced from the mass distributions obtained in hydrogen exchange mass spectrometry experiments (HX-MS) [8]. The wealth of information obtained from HX-MS experiments has further triggered technological advancements which have helped to improve the methodology. At present, mass spectrometers are designed with an 'HX module', which is a refrigerated chamber attached to the liquid chromatography (LC) system, online with the MS. The HX module significantly reduces the loss of information in the HX pattern due to back exchange taking place during the chromatographic step, making the experiment more automated as well as reproducible.

Another notable advancement in mass spectrometric methods, which has benefitted HX-MS studies tremendously, is the development of chemical fragmentation techniques [10]. A determination of the HX pattern in fragments of the protein, generated subsequent to HX in solution, increases the structural resolution of HX-MS studies. In most cases, the protein is fragmented by proteolytic digestion under acidic, quenched conditions [11, 12] (*see* **Notes 1** and **2**). Chemical fragmentation provides two big advantages in HX-MS experiments. First, it can be used as an alternative to proteolytic digestion in those cases where the latter may not be feasible [13]. Second, in conjunction with proteolytic cleavage, chemical fragmentation may even provide single amino acid resolution in HX-MS studies [14]. The most commonly used fragmentation technique in mass spectrometry is a high-energy fragmentation method called collision-induced dissociation

(CID). This, however, results in significant "scrambling," i.e., intramolecular migration of protium or deuterium along the protein backbone, thus altering the HX pattern in the protein molecule [10, 15]. The development of chemical fragmentation methods other than CID has, therefore, proved to be very useful for HX-MS experiments [16, 17].

One such recently developed methodology is electron transfer dissociation (ETD) [18]. In this technique, radical anions of an "ETD reagent" are produced within the mass spectrometer and allowed to collide with the positively charged ions of the protein. The transfer of electrons from the radical anion to the protein results in fragmentation at the N-Cα bonds, generating c ions (on the N terminal end) and z ions (on the C terminal end). In addition to being a milder fragmentation method, and therefore resulting in minimal scrambling of the HX pattern, ETD, unlike CID, also has the advantage of being able to fragment longer peptides or even intact proteins [17]. In this chapter, we describe an example of the use of ETD, as an alternative to proteolytic digestion of a protein in HX-MS studies [13]. The same principles can also be applied for further fragmentation of peptides initially generated by proteolysis, in order to achieve even higher structural resolution [14].

Single-chain monellin (MNEI) is a small 97 residue protein which is resistant to digestion by acid proteases [13]. Here, we discuss an HX-MS study of MNEI in which the transient unfolding reactions of the protein are studied under native conditions. The use of ETD enables a determination of the unfolding events with a high degree of structural resolution. The focus of the present chapter is on the practical aspects of performing such ETD-HX-MS experiments.

2 Materials

2.1 Buffers for the HX Reaction

All buffers used in these experiments were of the highest purity grade, from Sigma Aldrich. Guanidine hydrochloride (GdnHCl) of the highest purity grade was procured from United States Biochemicals. All buffers were made in Milli-Q water.

- Deuteration buffer: 10 mM Tris in D_2O. 1.2 mg of Tris base was dissolved in 10 ml of D_2O. The pH of the solution was adjusted to 8 using 1 N deuterated hydrochloric acid (DCl).
- 1 N deuterated sodium hydroxide (NaOD) and 1 N DCl were used to adjust the pH of the protein during deuteration.
- Exchange buffer: 20 mM Tris (H_2O) at pH 8, 25 °C. 24.22 g of Tris base was dissolved in 100 ml of Milli-Q water, to make a 2 M stock of Tris buffer. The pH of the solution was adjusted to

8 using concentrated hydrochloric acid (HCl). 500 μl of the 2 M stock was added to 49.5 ml of Milli-Q water to achieve a concentration of 20 mM Tris buffer (*see* **Note 3**). The solution was filtered through a 0.22 μm filter before use.

- Quench buffer: 100 mM glycine hydrochloride, containing 8 M guanidine hydrochloride (GdnHCl), at pH 2.2 on ice. 1.12 g of glycine hydrochloride and 76.42 g of GdnHCl were weighed out and dissolved in 100 ml of Milli-Q water (*see* **Note 4**). After complete dissolution of the salts, the solution was allowed to come to room temperature before adjusting the pH to 2.2, with concentrated sodium hydroxide (NaOH) (*see* **Note 5**). The solution was filtered through a 0.22 μm filter before use.

2.2 Sample Desalting

- Sephadex G-25, 5 ml, Hi-trap column from GE.
- ÄKTA basic HPLC system.
- Desalting solvent: Milli-Q water at pH 2.6, on ice. 1 L of Milli-Q water was kept at 4 °C overnight. Keeping the solution on ice, the pH was adjusted to 2.6, using concentrated HCl (*see* **Note 6**). The solution was degassed and stored at 4 °C prior to use.

2.3 LC-MS

- NanoACQUITY UPLC system, with HDX module (Waters Corp.), coupled to a Synapt G2 HD mass spectrometer (Waters Corp.), with an ESI (electrospray ionization) source, a Q-TOF (quadrupole-time of flight) mass analyzer, and the ETD (electron transfer dissociation) option.
- ACQUITY UPLC BEH C18 VanGuard Pre-column, 130 Å, 1.7 μm, 2.1 mm × 5 mm, for the UPLC system.
- Solvents used in LC-MS: Milli-Q water and acetonitrile, each containing 0.1% formic acid, were used as the solvents in the UPLC system attached to the mass spectrometer. 500 μl of formic acid was added to 500 ml of water/acetonitrile.
- Manual syringe pump for direct injection of the sample into the mass spectrometer.
- Glass Hamilton syringe for direct injection of the sample into the mass spectrometer (*see* **Note 7**).
- ZipTip for desalting small volumes of the protein solution.

2.4 ETD

- ETD reagent: 1,4 dicyanobenzene.

3 Methods

Figure 1 shows a schematic representation of the ETD-HX-MS experiment.

3.1 Deuteration of the Protein

- In the case of MNEI, the protein was deuterated by the administration of pH jumps which resulted in unfolding and refolding of the protein. However, the method of deuteration may vary from protein to protein (*see* **Notes 8** and **9**). The lyophilized protein was dissolved, at a concentration of 500–700 μM, in the deuteration buffer (10 mM Tris, D_2O, pH 8). The pH of the solution was increased to12.6 by the addition of a fixed volume of 1 N NaOD.
- After exactly 5 min, the pH was dropped to 1.6 by the addition of a fixed volume of 1 N DCl. The pH of the solution was then slowly readjusted to 8 by the drop wise addition of 1 N NaOD (*see* **Notes 10–12**).
- For checking the extent of deuteration, the fully deuterated protein was desalted using a Ziptip C18 filter and eluted in 50% acetonitrile, 0.1% formic acid solution. This was diluted further, in the same solvent, to a final concentration of 200 nM protein and injected directly into the mass spectrometer using a manual syringe pump and a glass Hamilton syringe. The observed mass of the deuterated protein was compared to the theoretically expected value for MNEI (*see* **Note 13**).

3.2 Hydrogen Exchange (HX) Reaction

- The HX reaction was initiated by diluting 500 μM of the deuterated protein 15-fold into the protonated exchange buffer (20 mM Tris, H_2O, pH 8) (*see* **Note 14**).
- After different time intervals of exchange at 25 °C, the reaction was quenched by adding 125 μl of the exchange reaction to 375 μl of the quench buffer, on ice (*see* **Notes 15–17**). The final quenched reaction contained 75 mM glycine hydrochloride, 6 M GdnHCl and 8 μM protein at a pH of 2.6, on ice.
- The quenched reaction was incubated on ice for 1 min (*see* **Note 18**).
- 400 μl of the quenched solution was then injected into the ÄKTA basic HPLC system, for desalting using a Sephadex G-25 Hi-trap column (*see* **Note 19**). The protein was eluted in Milli-Q water, pH 2.6. The column was kept surrounded by ice throughout the experiment in order to minimize back exchange due to an increase in temperature (*see* **Notes 20** and **21**).

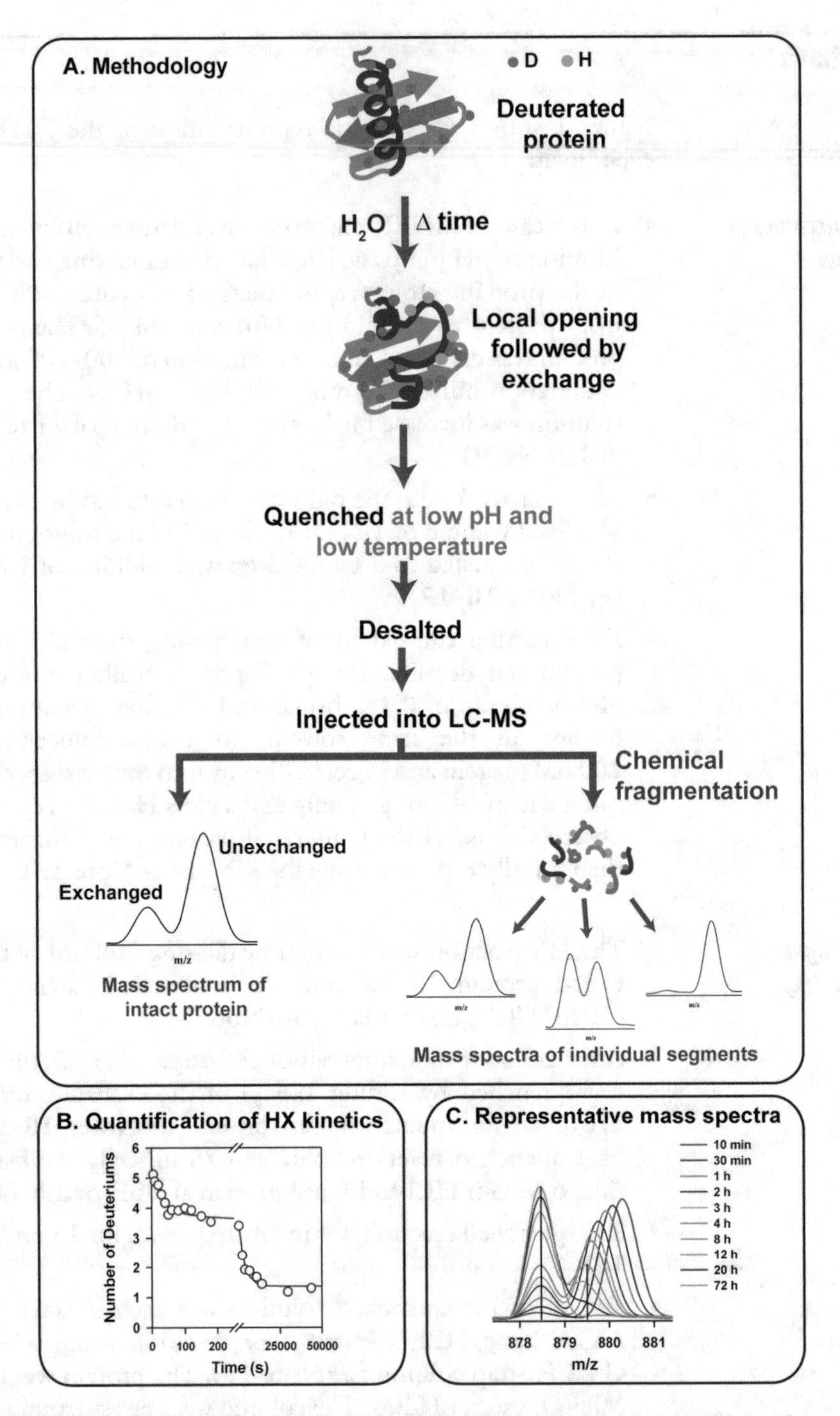

Fig. 1 Overview of the HX-ETD-MS experiment. (**a**) The methodology involves the HX reaction, resulting in exchanged protein, and its quenching after different time intervals, followed by desalting on an HPLC and final injection into an LC-MS system. The sample is further desalted by an online UPLC before entering the mass spectrometer where it is analyzed either in the intact form or fragmented by ETD and subsequently analyzed for mass determination. (**b**) A plot of the number of deuterium versus exchange time which can be used to

- The eluted protein was injected immediately into the LC-MS system. The time required for the injection to be executed was calibrated carefully, in order to maintain a constant level of back exchange (*see* **Note 20**).

3.3 Mass Spectrometry of the Intact Protein

- The quenched and desalted reaction was injected into the HDX module of the NanoACQUITY UPLC system (*see* **Note 22**).
- The protein was further desalted on the ACQUITY UPLC BEH C18 VanGuard Pre-column connected to the UPLC system. The protein was loaded onto the column for 1 min at a flow rate of 100 μl/min, in 0.05% formic acid. The protein eluted in a 3-min chromatographic run, between 35% and 95% acetonitrile, in 0.1% formic acid (*see* **Note 23**). The UPLC system was online with the mass spectrometer and the eluted protein enters the MS directly. The parameters used for the ionization of the protein in the MS were: capillary voltage, 3 kV; source temperature, 80 °C; desolvation temperature, 200 °C.
- For the intact protein analysis, a cumulative ion count of $>10^6$ was obtained by averaging 40 scans in the TIC (total ion chromatogram). The averaged spectrum was background subtracted (polynomial order 25; below curve 2% and tolerance 0.01; flatten edges) and smoothed (smooth window: 25 channels; number of smooths: 5; Savitzky Golay algorithm) in the Masslynx v4.1 software. The centroids were generated and the average *m/z* as well as charged state of the multiple peaks of the protein were used by the software to generate the observed mass of protein.
- The observed mass of the protein was used to calculate the number of deuteriums incorporated into the protein at each time (t) of the exchange reaction as: Observed mass of the exchanged protein at time (t) – Observed mass of the protonated protein. The number of deuterium was plotted as a function of exchange time, as shown in Fig. 1, in order to quantify the exchange kinetics. The resulting curve was fit to a single exponential or a sum of exponentials to obtain rate constants and amplitudes of the kinetic phases of exchange.
- For plotting HX spectra, the peak list was exported to SigmaPlot after zooming into the peak of interest (*see* **Note 24**). For visualizing changes in the mass spectra at different exchange

Fig. 1 (continued) quantify exchange kinetics. The solid line through the data is a double exponential fit which yields the rate constants and amplitudes associated with the two phases of exchange. (**c**) Representative mass spectra plotted for each time of exchange after normalizing to the area under the curve (*see* Methods, Subheading 3.3). The vertical solid lines mark the beginning and end of a kinetic phase. It should be noted that these spectra correspond to a DHX reaction and hence move towards lower *m/z* values with exchange time

times, the intensities at each m/z value were normalized to the total area under the curve at that exchange time and plotted in a single graph as shown in Fig. 1.

3.4 Electron Transfer Dissociation (ETD)

- Radical anions of 1,4-dicyanobenzene were generated using a glow discharge current of 35 μA. A total of 10^6 counts of the reagent were obtained per scan using a makeup gas (nitrogen) flow rate of 25 ml/min. Lower counts of the ETD reagent reduced the extent of fragmentation.
- Fragmentation was observed to be optimum for the +11 charged state of MNEI (*see* **Note 25**) (Fig. 2).
- The instrument parameters used for ETD fragmentation were: sample cone voltage, 30 V; extraction cone voltage, 4 V; trap wave velocity, 300 m/s; wave height, 0.35 V; transfer collision energy, ramped from 10 to 14 eV (*see* **Note 26**).
- The centroid spectra were generated as mentioned above for the intact protein and analyzed using the BioLynx software to identify the individual c and z ions, shown in Fig. 3. The intensity weighted isotopic abundances were used to determine the average mass of each c and z ions.
- The deuterium retention in each ion was determined by comparing the mass of the ion obtained from an exchanged sample to that of the ion obtained from the fully protonated protein.
- Multiple subtractions of the masses of consecutive c and z ions were used to determine the extent of exchange in the different sequence segments of the protein (Fig. 3). For example, the average mass of the c39 ion (spanning residues 1–40) minus the average mass of the c4 ion (spanning residues 1–5) yielded the number of deuterium retained at the backbone amide sites in the sequence segment spanning residues 6–40 (*see* **Note 27**).
- The number of deuteriums retained in each sequence segment of the protein was plotted with respect to the time of exchange in order to determine the exchange kinetics of each sequence segment of the protein (Fig. 4) (*see* **Note 28**).
- The percentage of exchange was calculated as: (Number of deuterium retained/number of deuterium at the first time point of exchange) × 100 for each sequence segment of the protein. This was then mapped on to the structure of the protein as shown in Fig. 5, in order to visualize the loss of structure resulting in exchange in different parts of the protein structure.

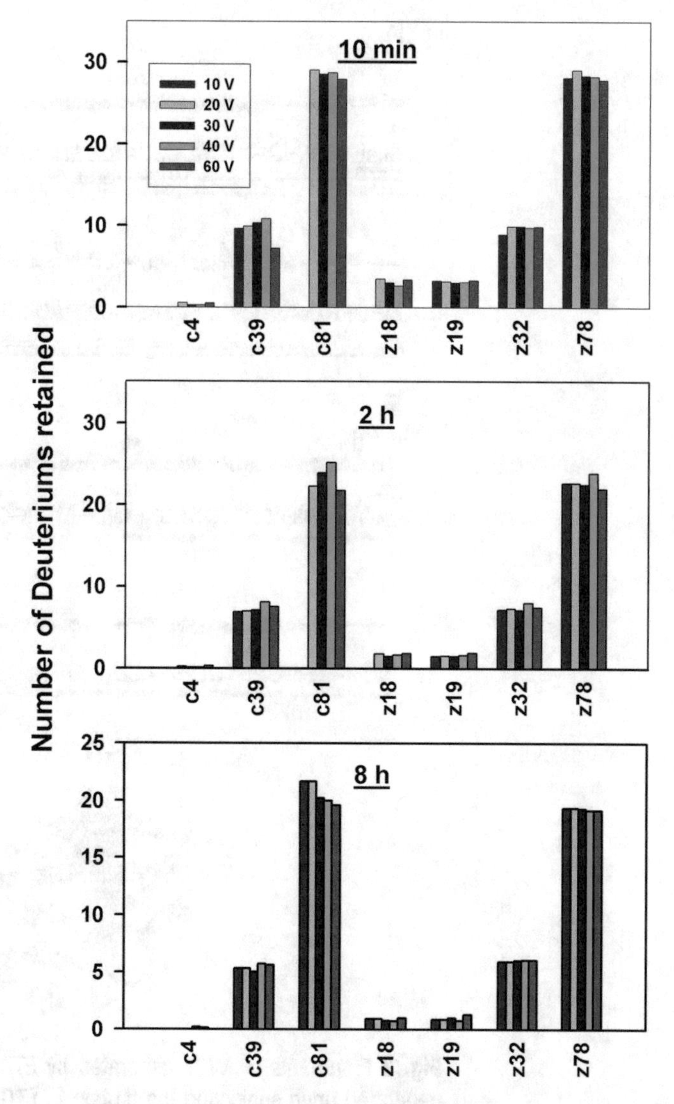

Fig. 2 Scrambling in ETD experiments. The number of deuterium measured after three different times of exchange while varying the cone voltage between 0 and 60 V and keeping all other instrument parameters constant. A constant deuterium content at all values of the cone voltage, for each individual c or z ion, confirms that fragmentation by ETD does not cause intramolecular migration of deuterium/protium which results in scrambling of the HX pattern in protein molecules. (Reprinted from ref. 13, with permission from American Chemical Society)

4 Notes

1. HX-MS experiments usually rely on proteolytic cleavage to obtain site-specific information on the HX pattern in the protein [11]. Typically, the exchange reaction is quenched at a low

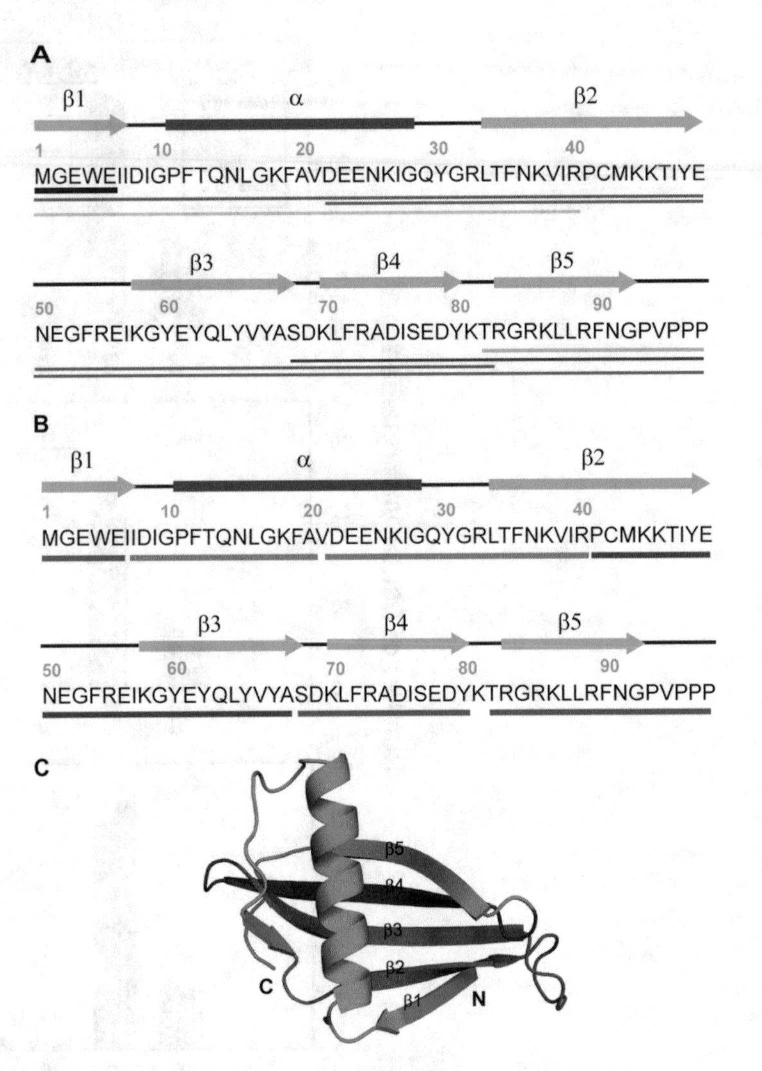

Fig. 3 Fragments of MNEI generated by ETD. (**a**) The individual c and z ions generated upon subjecting the protein to ETD are shown as solid colored lines beneath the secondary structure and sequence of the protein. The β sheets are shown as light blue arrows and the single helix is shown as a solid dark blue bar. (**b**) The solid vertical lines beneath the secondary structure and the sequence of the protein represent the sequence segments of the protein obtained from multiple subtractions of the c and z ions as described in the Methods, Subheading 3.4. (**c**) The sequence segments shown in (**b**) are mapped on to the three dimensional structure of MNEI (PDB ID: 1IV7) using the PyMol software. (Reprinted from ref. 13, with permission from American Chemical Society)

pH and temperature, followed by enzymatic digestion by proteases which are active under the acidic conditions of the quenched reaction. One of the most commonly used enzymes for proteolytic cleavage in HX-MS experiments is pepsin—an acid protease which is active in the pH range of 2–4, even at low temperature. Pepsin cleaves non-specifically at peptide bonds

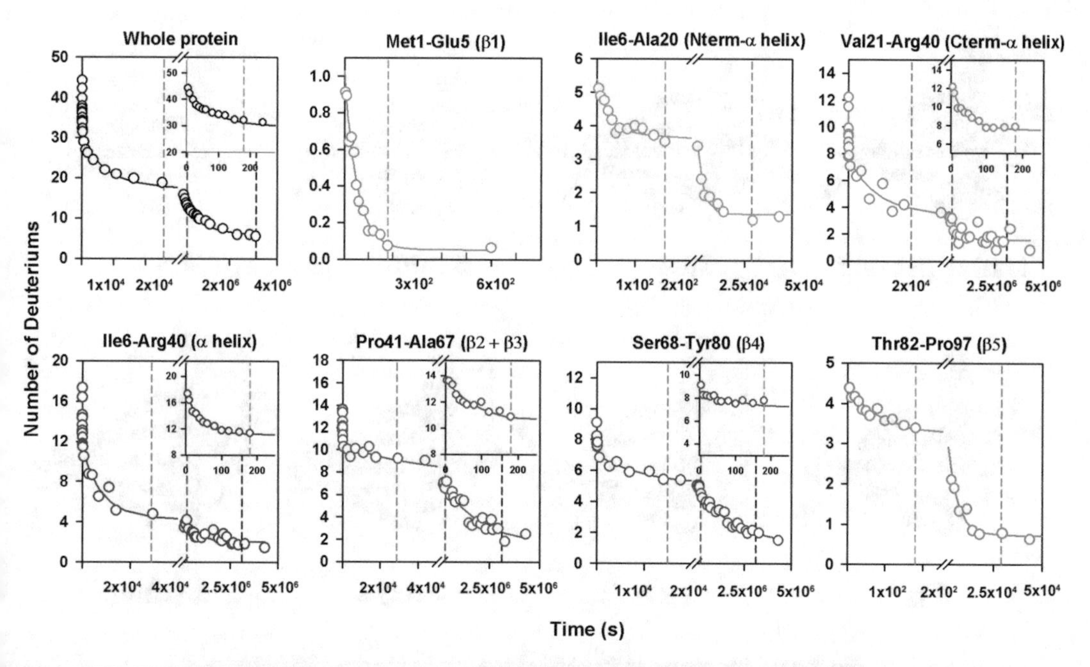

Fig. 4 HX-MS kinetics of individual sequence segments of MNEI. The number of deuteriums versus time of exchange plotted for each sequence segment of the protein (indicated above each panel) as well as for the intact protein (first panel). The solid lines through the data represent fits to a single exponential, or a sum of multiple exponentials. The dashed vertical lines denote the end of each kinetic phase of exchange. The insets show a zoomed in view of the initial kinetic phase. (Reprinted from ref. 13, with permission from American Chemical Society)

and usually yields peptide fragments which are 5–30 amino acid residues long. Other proteases such as protease type XVIII from *Rhizhopus* species and protease type XIII from *Aspergillus saitoi* have also been found to be compatible with the quenched conditions of HX-MS reactions [19]. In fact, the use of multiple proteases in the same experiment has proven very useful in generating multiple overlapping fragments and improving sequence coverage in HX-MS studies.

2. Experimentally, proteolytic cleavage of the protein under study can be accomplished in various ways. The protease may be directly added to the quenched reaction and subsequently separated chromatographically in the LC system attached to the MS [11]. Another convenient method for removal of the protease from the quenched solution is the use of pepsin beads which can be spun down and easily removed from solution [20]. A more common practice, however, is to use pepsin cartridges in which the enzymes have been immobilized on a hydrophobic/hydrophilic surface [21]. The cartridge is an enzyme column which can be easily incorporated into the LC-MS system for online digestion of the quenched reaction.

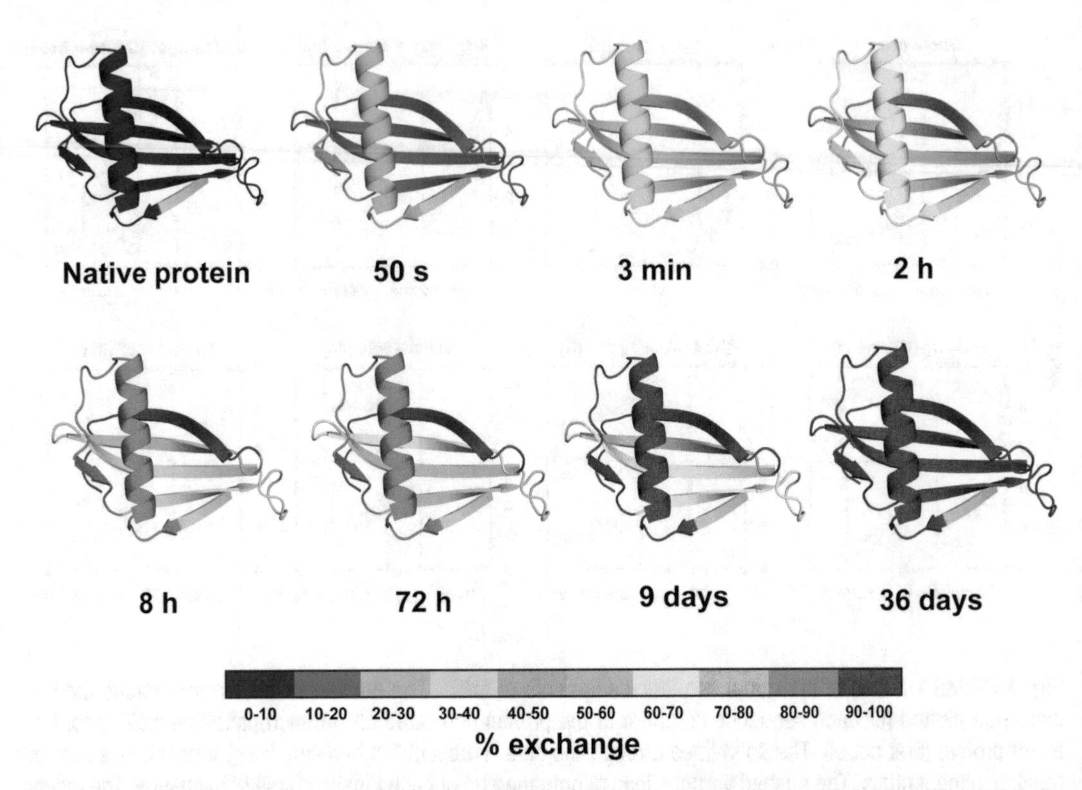

Fig. 5 Structural information from HX kinetics. A combination of the information obtained from the HX kinetic curves and the structural localization of a given sequence segment generated by ETD allows the HX kinetics to be mapped to each individual structural element of the protein, providing a clear visualization of the kinetics of loss of structure in different parts of the protein molecule. The % of exchange was calculated as explained in Methods, Subheading 3.4 and represented by the color bar shown at the bottom of the figure. (Reprinted from ref. 13, with permission from American Chemical Society)

The pepsin column is usually a part of the refrigerated HX module (*see* below) of the UPLC system.

3. Since the stock solution of Tris is being diluted 100-fold, the pH of the final buffer may deviate from 8. Hence, the pH of the 20 mM Tris buffer was checked and readjusted, if required, to 8.
4. GdnHCl is highly hygroscopic. Care should be taken to add only half the volume i.e. 50 ml of water to the salts. The volume should be adjusted to 100 ml only after the salts have dissolved entirely. The dissolution of GdnHCl in water is an endothermic reaction, and the solution may need to be kept in a hot water bath for the GdnHCl to dissolve completely at such a high concentration.
5. The quench buffer cannot be kept on ice for a prolonged period because at a high concentration of 8 M, GdnHCl will precipitate out of solution at a low temperature. Hence, in order to achieve a final pH of 2.6 under quenched conditions

on ice, the pH of the quench buffer was adjusted to 2.2 at room temperature. It was checked in multiple experiments that upon quenching the exchange reaction on ice, the final pH was indeed 2.6.

6. It is important to adjust the pH of the Milli-Q water used for desalting while on ice since pH is significantly affected by changes in the temperature of the solution.
7. The solutions which are injected directly into the mass spectrometer usually contain a significant percentage of organic solvents, and it is therefore advisable to use glass, and not plastic, syringes for these injections.
8. To accomplish complete deuteration at all exchangeable sites, a protein is typically denatured in deuterated solvent, thus exposing all exchangeable sites to the surrounding solvent, and then refolded back to the native structure. For some proteins, the native state is dynamic enough that complete deuteration can be achieved by incubation in D_2O at room temperature [20]. In other cases, the denaturation in D_2O is achieved by an increase in temperature [22] or pH jumps [23].
9. In the case of MNEI, pH jumps were used since the protein does not exchange out completely upon incubation in D_2O at room temperature, even after several days, and it precipitates upon heating. For the pH jumps, the fixed volumes of 1 N NaOD and 1 N DCl required to execute the changes in pH were carefully calibrated before every experiment.
10. It is important to minimize the time for which the protein is exposed to a high pH because, at alkaline pH, proteins are usually prone to deamidation, i.e., conversion of asparagine residues to aspartic acid residues. It is important to therefore establish that the time for which the protein has been exposed to a high pH does not result in significant deamidation. This can be checked by measuring the mass of the protein before and after the jump to a high pH (>10). A deamidation reaction results in an increase in the mass of the protein by 1 Da [24].
11. The pH of the protein solution was first dropped to 1.6 and then readjusted to 8 because a direct drop to pH 8 resulted in precipitation of the protein [23]. This is possibly because the p*I* of MNEI is 8.56 and refolding around pH 8 may have caused the protein to come out of solution. It should be noted here, again, that deuteration protocols need to be established for each protein on a case by case basis.
12. The HX reaction can be monitored as a DHX reaction in which the deuterium at the backbone amide sites in the protein exchange with the hydrogen in solution. In these cases, as also described here for MNEI, the reaction has to begin with a deuterated protein. However, the HX reaction can also be

carried out as a HDX reaction in which the protein is protonated to begin with, and the deuterium in solution are incorporated into the protein backbone amide sites as the exchange reaction progresses.

13. Hydrogens attached to electronegative atoms such as N, O, and S are prone to exchange in a protein. However, the exchangeable hydrogens in the side chains of amino acid residues exchange too rapidly to be measured. Hence, in practice, the HX experiment measures only the comparatively slower exchange of the backbone amide hydrogens in a protein. However, upon deuteration, the increase in mass of the protein corresponds to exchange at *all* exchangeable sites in the protein including the side chains. The theoretical value for the number of exchangeable backbone amide sites in a protein can be calculated as follows:

    Number of exchangeable hydrogens = (Number of amino acid residues − Number of proline residues in the protein) + Number of hydrogens attached to C, O, N in the amino acid side chains (at the pH at which the experiment has been performed).

14. It is important that for the initiation of the HX reaction, the deuterated or protonated protein is diluted at least 15- to 20-fold into the exchange buffer (which would be protonated for a DHX reaction and deuterated for a HDX reaction). This ensures that the deuterium/protium in the solution is >95%, thus facilitating complete exchange at all backbone amide sites in the protein.

15. If the quench buffer does not contain a salt like GdnHCl, which precipitates at the required high concentrations and temperatures slightly above freezing, the entire solution can be kept on ice for the duration of the experiment. However, in this case, to prevent precipitation of GdnHCl, aliquots of 375 μl of the quench buffer were kept on ice for ~5 min prior to quenching the exchange reaction.

16. It is important to use an acidic buffer instead of using concentrated acid alone for quenching the exchange reaction, in order to accurately control the changes in the pH of the solution. Intrinsic exchange rates in an HX reaction are minimum only at a pH of 2.6 and increase sharply on either side of this pH value [25]. It is therefore imperative to control the pH of the quenched solution accurately and consistently in each of the exchange and quenched reactions.

17. A fourfold dilution of the exchange reaction into the quench buffer further ensures that the pH of the reaction not only drops to 2.6 but is also maintained at 2.6 till the sample is injected into the mass spectrometer.

18. The incubation of the quenched reaction in 6 M GdnHCl was found to facilitate subsequent fragmentation of the protein by ETD [13]. This step could likewise be included in experiments in which the protein is being enzymatically digested as well.
19. Although the LC system online with the mass spectrometer also desalts the sample, an additional desalting step prior to this is very useful, particularly in cases where the salt concentration of the sample is >6 M. If the salt concentration of the quenched sample is 50–100 mM, it may be directly injected into the LC-MS system.
20. In all HX reactions, back exchange, i.e., a reverse exchange of the protium which are incorporated into the deuterated protein during the exchange reaction back into the solution, poses a major problem [3]. Under ideal circumstances, quenching the exchange reaction at a pH of 2.6 and 4 °C should completely halt any further exchange. However, since back exchange cannot be completely avoided in practice, it is imperative to control the pH of the quenched solution accurately and consistently in each of the exchange and quenched reactions, in order to maintain a constant level of back exchange for all the HX reactions. A failure to reproducibly control both these parameters would result in significant variations in the extents of exchange in multiple HX reactions.
21. It is also important to do a "back exchange control" in all exchange reactions to ascertain the level of back exchange taking place. For this purpose, the deuterated protein is diluted into deuterated exchange buffer (resulting in no exchange) and then immediately diluted into the quench buffer. The remaining steps are exactly the same as in an exchange reaction. The observed mass of the protein in the "back exchange control" provides a measure of the loss in information due to back exchange that occurs during the processing of a sample. For example, for MNEI, 192 deuterium were incorporated into the fully deuterated protein. The back exchange control, however, retained only 72 of these deuterium, even though the number of exchangeable backbone amide sites in this protein is 90. Hence 18 backbone amide deuterium were lost due to back exchange during the processing of the sample [23].
22. The HX module is a refrigerated chamber of the UPLC system which maintains the desalting column as well as the connecting tubing at 4 °C. This enables better control on the temperature of the quenched reaction during the experiment.
23. The chromatographic run has to be standardized such that the protein can be desalted within the shortest possible time to minimize back exchange.

24. The charged state with the highest intensity is usually used for depicting the changes in the mass spectra with time of exchange.
25. The charged state of the protein used for ETD fragmentation needs to be optimized carefully. Although a higher charged state of the protein would result in more efficient fragmentation [26], it may have significantly lower intensity than a lower charged state. Hence fragmentation should initially be carried out with multiple charged states of the protein in order to ascertain which charged state gives optimum fragmentation.
26. To be able to fix the ETD instrument parameters, each of the parameters were systematically varied in order to ensure that no scrambling of the hydrogen–deuterium pattern took place during fragmentation. It has been reported previously using the same mass spectrometer and ionization source that increasing the cone voltage has the maximum effect on scrambling [27]. The deuterium retention in MNEI was measured by varying the cone voltage in the range of 0–60 V, keeping all other parameters fixed. As shown in Fig. 2, a constant deuterium retention observed in MNEI at all the values of the cone voltage in the range of 0–60 V confirms a lack of scrambling during ETD. This is an important control that needs to be carried out for all ETD experiments.
27. None of the subtractions used to arrive at the mass of a given sequence segment of the protein involved overlapping c and z ions. The mass of a smaller ion segment was subtracted from that of a larger one only when the smaller ion segment was completely contained within the larger segment.
28. The mass spectra of the individual sequence segments were also plotted similar to the intact protein in order to visualize changes in the spectral pattern of each part of the protein with exchange time.

References

1. Woodward C, Simon I, Tuchsen E (1982) Hydrogen exchange and the dynamic structure of proteins. Mol Cell Biochem 48:135–160
2. Bai Y, Sosnick TR, Mayne L, Englander SW (1995) Protein folding intermediates: native-state hydrogen exchange. Science 269:192–197
3. Englander SW (2000) Protein folding intermediates and pathways studied by protein folding. Annu Rev Biophys Biomol Struct 29:213–238
4. Hvidt A, Nielsen SO (1966) Hydrogen exchange in proteins. Adv Protein Chem 21:287–386
5. Chamberlain AK, Handel TM, Marqusee S (1996) Detection of rare partially folded molecules in equilibrium with the native conformation of RNaseH. Nat Struct Mol Biol 3:782–787

6. Udgaonkar JB, Baldwin RL (1990) Early folding intermediate of ribonuclease A. Proc Natl Acad Sci U S A 87:8197–8201
7. Bhuyan AK, Udgaonkar JB (1998) Two structural subdomains of barstar detected by rapid mixing NMR measurement of amide hydrogen exchange. Proteins 30:295–308
8. Malhotra P, Udgaonkar JB (2016) How cooperative are protein folding and unfolding transitions? Protein Sci 25:1924–1941
9. Ferraro DM, Lazo ND, Robertson AD (2003) EX1 hydrogen exchange and protein folding. Biochemistry 43:587–594
10. Elviri L (2012) ETD and ECD mass spectrometry fragmentation for the characterization of protein post translational modifications. In: Tandem mass spectrometry – applications and principles. IntechOpen, London, pp 163–178
11. Zhang Z, Smith DL (1993) Determination of amide hydrogen exchange by mass spectrometry: a new tool for protein structure elucidation. Protein Sci 2:522–531
12. Hu W, Walters BT, Kan Z-Y, Mayne L, Rosen LE, Marqusee S, Englander SW (2013) Stepwise protein folding at near amino acid resolution by hydrogen exchange and mass spectrometry. Proc Natl Acad Sci U S A 110:7684–7689
13. Malhotra P, Udgaonkar JB (2016) Secondary structural change can occur diffusely and not modularly during protein folding and unfolding reactions. J Am Chem Soc 138:5866–5878
14. Rand KD, Zehl M, Jensen ON, Jørgensen TJD (2009) Protein hydrogen exchange measured at single-residue resolution by electron transfer dissociation mass spectrometry. Anal Chem 81:5577–5584
15. Wysocki VH, Resing KA, Zhang Q, Cheng G (2005) Mass spectrometry of peptides and proteins. Methods 35:211–222
16. Rand KD, Adams CM, Zubarev RA, Jørgensen TJD (2008) Electron capture dissociation proceeds with a low degree of intramolecular migration of peptide amide hydrogens. J Am Chem Soc 130:1341–1349
17. Zehl M, Rand KD, Jensen ON, Jorgensen TJD (2008) Electron transfer dissociation facilitates the measurement of deuterium incorporation into selectively labeled peptides with single residue resolution. J Am Chem Soc 130:17453–17459
18. Syka JEP, Coon JJ, Schroeder MJ, Shabanowitz J, Hunt DF (2004) Peptide and protein sequence analysis by electron transfer dissociation mass spectrometry. Proc Natl Acad Sci U S A 101:9528–9533
19. Cravello L, Lascoux D, Forest E (2003) Use of different proteases working in acidic conditions to improve sequence coverage and resolution in hydrogen/deuterium exchange of large proteins. Rapid Commun Mass Spectrom 17:2387–2393
20. Wani AH, Udgaonkar JB (2009) Native state dynamics drive the unfolding of the SH3 domain of PI3 kinase at high denaturant concentration. Proc Natl Acad Sci U S A 106:20711–20716
21. Singh J, Sabareesan AT, Mathew MK, Udgaonkar JB (2012) Development of the structural core and of conformational heterogeneity during the conversion of oligomers of the mouse prion protein to worm-like amyloid fibrils. J Mol Biol 423:217–231
22. Hamid Wani A, Udgaonkar JB (2006) HX-ESI-MS and optical studies of the unfolding of thioredoxin indicate stabilization of a partially unfolded, aggregation-competent intermediate at low pH. Biochemistry 45:11226–11238
23. Malhotra P, Udgaonkar JB (2015) Tuning cooperativity on the free energy landscape of protein folding. Biochemistry 54:3431–3441
24. Lehmann WD, Schlosser A, Erben G, Pipkorn R, Bossemeyer D, Kinzel V (2000) Analysis of isoaspartate in peptides by electrospray tandem mass spectrometry. Protein Sci 9:2260–2268
25. Bai Y, Milne JS, Mayne L, Englander SW (1993) Primary structure effects on peptide group hydrogen exchange. Proteins 17:75–86
26. Sterling HJ, Williams ER (2010) Real-time hydrogen/deuterium exchange kinetics via supercharged electrospray ionization tandem mass spectrometry. Anal Chem 82:9050–9057
27. Rand KD, Pringle SD, Morris M, Engen JR, Brown JM (2011) ETD in a traveling wave ion guide at tuned Z-spray ion source conditions allows for site-specific hydrogen/deuterium exchange measurements. J Am Soc Mass Spectrom 22:1784–1793

Chapter 9

Multi-Probe Equilibrium Analysis of Gradual (Un)Folding Processes

Ginka S. Kubelka and Jan Kubelka

Abstract

Studies of small proteins that exhibit noncooperative, gradual (un)folding can offer unique insights into the rarely accessible intermediate stages of the protein folding processes. Detailed experimental characterization of these intermediate states requires approaches that utilize multiple site-specific probes of the local structure. Isotopically edited infrared (IR) spectroscopy has emerged as a powerful methodology capable of providing such high-resolution structural information. Labeling of selected amide carbonyls with ^{13}C results in detectable side-bands of amide I′ vibrations, which are sensitive to local conformation and/or solvent exposure without introducing any significant structural perturbation to the protein. Incorporation of isotopically labeled amino acids at specific positions can be achieved by the chemical synthesis of the studied proteins. We describe the basic procedures for synthesis of ^{13}C isotopically edited protein samples, experimental IR spectroscopic measurements and analysis of the site-specific equilibrium thermal unfolding of a small protein from the temperature-dependent IR data.

Key words Isotopic editing, Infrared spectroscopy, Amide I′, Solid phase peptide synthesis, Thermal unfolding, Secondary structure, Tertiary structure

1 Introduction

Small proteins which fold and/or unfold gradually, i.e., through a continuum of partially folded states [1–3], are extremely valuable models for studying protein folding. Gradual (un)folding means that the folding pathways are in principle experimentally accessible, in contrast to proteins that fold and unfold cooperatively in an all-or-none fashion [4]. To take advantage of this behavior in order to uncover the sought-after details of the protein folding process, it is necessary to employ experimental methods capable of providing maximum amount of structural detail about the partly folded, intermediate states of the protein at all stages of folding. One of such experimental methods is ^{13}C isotopically edited infrared (IR) spectroscopy.

Victor Muñoz (ed.), *Protein Folding: Methods and Protocols*, Methods in Molecular Biology, vol. 2376,
https://doi.org/10.1007/978-1-0716-1716-8_9,

IR spectroscopy is a well-established experimental method for probing structural changes in peptides and proteins during folding or unfolding, which generally relies on the sensitivity of the amide I′ band (predominantly amide C=O stretch, usually measured in D_2O-based solutions for N-deuterated amides) to the secondary structure [5]. Traditionally, only the average secondary structural information can be obtained due to the inherently low resolution of the IR. However, this fundamental limitation can be overcome by introducing isotopically labeled amino acids on the amide C=O at specific locations within the peptide or protein sequence [6, 7]. Amide I′ vibrations of ^{13}C-labeled residues are decoupled from ^{12}C vibrations and result in a separate signal shifted to lower wavenumbers (Fig. 1), which contains the information about the local structure within the labeled protein segment.

Chemical synthesis of peptides and small proteins offers the possibility to insert ^{13}C=O labeled amino acids into selected, known positions. Incorporation of two or three ^{13}C amino acids in neighboring or alternating positions provides enough spectral signal as well as specific coupling patterns to report on the backbone conformation (secondary structure) of the labeled region [8]. Furthermore, changes in tertiary structure can be probed through the solvent exposure of labeled amides [9–12]. In this case, labeling of a single buried amide is often sufficient [11], and the isotopic shift can be further enhanced by ^{13}C=^{18}O substitution [13]. The ^{13}C amide I′ signal is also a very sensitive probe of the β-sheet structure [14] and has been used in studies of polypeptide aggregation [15–17]. Other isotopic labeling schemes are possible as well: for example, deuteration of specific side chain residues can be used to probe the burial of the hydrophobic core in proteins [18].

Isotopic editing therefore provides a versatile and minimally perturbing experimental probe of the local protein secondary and tertiary structure. In combination with the inherent fast time-scale of IR spectroscopy, it also allows direct measurements of protein conformational dynamics with site-specific resolution [19, 20]. However, thus far, the applications of this methodology have focused predominantly on short model peptides. Only recently has systematic incorporation of isotopically labeled amino acids at multiple positions within a protein been used to draw a detailed picture of its folding mechanism [21–23].

One of the challenges has been the analysis of site-specific structural transitions from the ^{13}C side-bands, which is complicated by several factors. First, the ^{13}C amide I′ bands is obscured by overlap with the ^{12}C (unlabeled) amide I′ as well as with the IR signals of several of the amino acid side chains (Fig. 1). This is of particular concern in proteins that are larger than model peptides, and consequently, there is generally lower relative proportion of ^{13}C-labeled amides to the ^{12}C-unlabeled ones and to the number

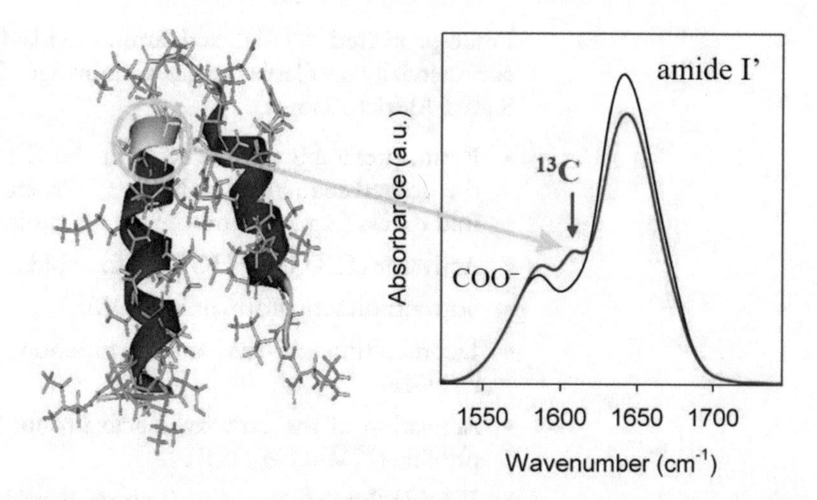

Fig. 1 Site-specific resolution of protein structure by means of ^{13}C isotopically edited IR spectroscopy. Amide I′ signal of the model protein selectively labeled in the highlighted region with three ^{13}C amino acids (red) overlaid with the amide I′ of the same, unlabeled protein (black)

of interfering side chain groups. Second, since the most commonly used denaturants, urea and guanidinium, also strongly absorb in the amide I′ IR region [24], temperature is almost exclusively used to induce unfolding. Unfortunately, we [25, 26] as well as others [9, 10] have shown that the amide I′ IR exhibits a strong intrinsic temperature dependence, which can severely frustrate the interpretation of the spectral changes in terms of structural transitions [27].

In order to correct for the nonstructural temperature effects, we have developed a novel methodology for analysis of the temperature-dependent amide I′ IR bands that allows considering the frequency, intensity, and band-shape changes together [27]. The application of this analysis to ^{13}C IR thermal unfolding data of a small protein is presented below as an example. For analysis of the solvent exposure of buried amide groups, which is indicative of the unfolding of the protein hydrophobic core, the shift of the ^{13}C peak maximum to the lower frequency can be utilized [11].

2 Materials

2.1 Peptide Synthesis

Synthetic peptides can be easily synthesized on a peptide synthesizer applying standard solid phase peptide synthesis (SPPS) techniques [28]. Amino acids with 9-fluorenylmethoxycarbonyl (Fmoc) protection on the amino group and acid labile side chain protecting groups are commercially available from a number of suppliers (EMD Novabiochem, Anaspec, Sigma-Aldrich, etc.), as are preloaded resins and all reagents for the peptide synthesis.

Fmoc-protected ^{13}C-labeled amino acids for synthesis are also commercially available (e.g., Cambridge Isotope Laboratories, Sigma-Aldrich/Isotec).

- Resin, preferably preloaded with the C-terminal amino acid of the desired sequence (*see* **Note 1**). Protected amino acids: fourfold excess (e.g., 0.2 mmole for 50μmole scale synthesis).
- Activator (HBTU or HCTU): fourfold excess.
- Solvent: dimethylformamide (DMF).
- Deprotection of the amino function: 20% v/v piperidine in DMF.
- Activation of the carboxylic acid group: 0.4 M *N*-methylmorpholine (NMM) in DMF.
- Peptide Synthesizer (e.g., Tribute Peptide Synthesizer, Protein Technologies, Inc.).
- Cleavage: 81.5% v/v trifluoroacetic acid (TFA), 5% v/v thioanisole, 5% w/v phenol, 1% v/v triisopropylsilane, 5% v/v H_2O, 2.5% v/v ethanedithiol (EDT) (*see* **Note 2**).
- 13-mL filter tubes (*see* **Note 3**), 50-mL centrifuge tubes.
- Precipitation: methyl *tert*-butyl ether (MTBE) or diethyl ether.
- Lyophilizer.
- Low-temperature (0 °C) centrifuge.
- Purification: HPLC, 0.1% v/v TFA in H_2O (buffer A), 0.1% v/v TFA in acetonitrile (buffer B), C18 column.

2.2 FTIR Spectroscopy

- D_2O (e.g., Cambridge Isotope Laboratories), sodium phosphate salts (monobasic, dibasic).
- FTIR spectrometer containing a constantly purged sample compartment with cell holder, equipped with a sample temperature control via a water recirculator bath and temperature jacket or a Peltier device, liquid sample cell, CaF_2 windows, gaskets, Teflon spacer (50μm) (*see* **Note 4**).

3 Methods

3.1 Peptide Synthesis

- Use standard procedures for the Fmoc solid phase peptide synthesis [28].
- Fill vials with amino acids and activator (HBTU, HCTU).
- Place the resin into the reaction vessel.
- Fill the solvent and reagent bottles and connect to the peptide synthesizer.
- Load the vials into the synthesizer, starting with the C-terminus.

- Start the synthesis.
- When the synthesis is complete, wash the resin with the peptide into a filter tube using DMF.

3.2 Post-Synthesis Work-Up and Cleavage

- Remove DMF under reduced pressure and wash once again with DMF.
- Wash three times with dichloromethane (DCM).
- Wash twice with methanol and dry under reduced pressure for at least ½ h.
- Add 10 mL of the cleavage mixture for every 100 mg resin/peptide and nutate the mixture for 2–3 h, then let the liquid drop into pre-chilled (−20 °C) ether (30 mL ether/8 mL cleavage solution), where it precipitates (*see* **Note 5**).
- Wash the resin twice with 1 mL cleavage solution or TFA, add wash solutions to the ether.
- Vortex the ether mixture well for 3 min, then chill for 1 h at −20 °C.
- Centrifuge the mixture for 5 min at 2540 × *g* and 0 °C, and discard the ether carefully.
- Add 20 mL chilled ether to the peptide, vortex thoroughly, centrifuge again, discard ether.
- Repeat **step 8** two more times.
- Remove the remaining ether under vacuum.
- Dissolve the dry peptide in a small amount of water, freeze, and lyophilize.
- Purify by reversed phase HPLC by dissolving the peptide in 100% buffer A, run a linear gradient over 50 min from 0% to 50% buffer B (*see* **Note 6**).
- Freeze the collected peaks in liquid N_2, lyophilize, and analyze by MS, if available (*see* **Note 7**).

3.3 FTIR Sample Preparation

- H/D exchange of the peptide: Dissolve 3–4 mg of peptide in 2 mL D_2O, nutate for 1 h, heat to ~45 °C for 10 min, add 2μL HCl (conc.) to remove TFA (*see* **Note 8**), nutate for another 45 min, freeze in liquid N_2, and lyophilize.
- Buffer: Exchange the sodium phosphate buffer salts by dissolving in D_2O, nutating for 1 h, freezing, and lyophilization. Mix to get ~0.1 M deuterated buffer with the desired pH.
- Protein sample: Dissolve 1 mg of the exchanged peptide in 60μL deuterated phosphate buffer while trying to minimize air contact with the sample.

3.4 FTIR Spectroscopy

- Measure the FTIR spectrum of the water vapor with the empty sample compartment (*see* **Note 9**). Usually, quickly opening and closing the sample compartment cover before starting the scan will introduce enough water vapor.
- Fill the (clean and dry) sample cell with ~30μL of the deuterated phosphate buffer solution, while trying to minimize air contact with the sample (*see* **Note 10**).
- Measure the buffer spectrum or spectra (e.g., at preset temperatures).
- Disassemble the sample cell, clean, and dry all components thoroughly, in particular the windows and the spacer.
- Fill ~30μL of the peptide solution into the IR sample cell (*see* **Note 10**).
- Measure the protein solution FTIR spectrum or spectra.

3.5 FTIR Data Processing

The processing of raw FTIR data follows standard, established procedures [5, 29].

- Subtract the FTIR spectra of the buffer and of the residual water vapor from the corresponding protein solution spectra (*see* **Note 11**).
- Truncate the corrected protein spectra to contain the region of interest (e.g., amide I′ ~ 1500–1750 cm^{-1}, *see* Fig. 1).
- If necessary, correct the residual spectral baseline, e.g., by a second-order (quadratic) polynomial.

3.6 Example Data Analysis: Thermal Unfolding

The analysis of ^{13}C data depends on the objective of the particular study, choice of isotopic labels, etc. We present an example of the analysis of thermal unfolding of the secondary structure in a model protein [22].

- Normalize the ^{13}C-labeled and -unlabeled amide I′ FTIR spectra to the integral amide I′ intensity at the highest measured temperature.
- Subtract the unlabeled amide I′ IR spectrum from the ^{13}C-labeled one to obtain a set of difference spectra as a function of temperature (*see* Fig. 2).
- Truncate the difference spectra to keep only the positive part, which corresponds to the ^{13}C-labeled amide I′ signal, e.g., by setting all negative intensity values to zero (*see* Fig. 2).
- Use the parametric *S*hifted *M*ultivariate *S*pectra *A*nalysis (SMSA) [22, 27] with each of the truncated ^{13}C amide I′ spectral sets to obtain the unfolding thermodynamics for the corresponding isotopically labeled segment of the protein (*see* Fig. 2, **Note 12**). The MATLAB code is available upon request.

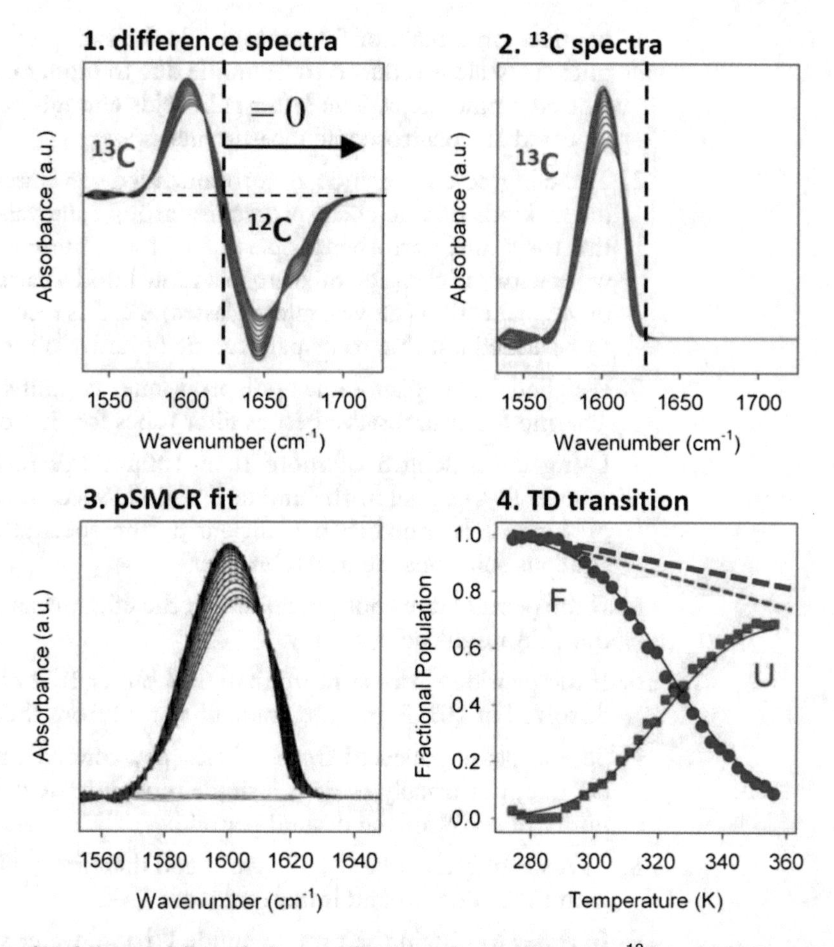

Fig. 2 Example of the modeling of site-specific thermal unfolding from ^{13}C isotopically edited amide I′ spectrum using parametric Shifted Multivariate Spectra Analysis (pSMSA) method assuming a two state model. (1) Temperature-dependent difference spectra (^{13}C-labeled minus unlabeled) are truncated to keep only the positive (^{13}C) part. (2) The ^{13}C experimental differences are fitted with the pSMSA model. (3) The pSMSA fit (black) to the experimental data (red points) with the residuals in green. (4) Fractional populations of the two states: folded (blue) and unfolded (red). The values contain thermodynamic populations and intrinsic temperature-dependent intensity changes shown as dashed lines (blue—folded, red—unfolded). Solid lines are the predictions of the model, while the symbols correspond to the experimental data points given the model prediction of two basis spectra for states F and U (not shown)

4 Notes

1. The first amino acid on the resin defines the success of the whole synthesis. Therefore, use of preloaded resins with a known loading density of the first amino acid is preferred. Low-loaded resins are beneficial for longer peptides or more difficult sequences, since high loading density facilitates aggregation during the synthesis. We usually synthesize unlabeled

peptides on a scale of 50μmole, while for isotopically labeled ones the scale is reduced to 25μmole due to higher cost of the labeled amino acids. The latter still yields enough peptide for repeated IR spectroscopic measurements.

2. These ingredients include all recommended scavengers for different kinds of side chain protecting group removals. Due to the toxic and corrosive properties of these ingredients only work with the cleavage mixture in a fume hood wearing appropriate protection (gloves, safety glasses). EDT is recommended to be added last due to its particularly offensive odor.
3. We find that disposable polypropylene columns (Pierce, Thermo Scientific) serve best as filter tubes for that purpose.
4. Using a pathlength of more than 100μm will result in an intense D_2O signal in the amide I′ region. Spectral resolution of 4 cm^{-1} is more than sufficient as the spectral bands in aqueous solutions are much broader.
5. If the peptide does not precipitate in the ether, evaporation of the TFA might be necessary.
6. If the peptide elutes at more than 30% buffer B, it can also be dissolved in 10% B and the gradient started from there.
7. Several peaks collected from HPLC may contain the desired peptide, but usually there is a single most intense peak with a purity of >95% of the desired peptide.
8. TFA strongly absorbs at 1673 cm^{-1} and therefore will interfere with the amide I′ band if not removed.
9. In order to obtain the protein amide I′ band, water vapor and buffer spectra have to be subtracted from that of the protein, which all have to be collected under conditions as close to identical as possible. Furthermore, since the analysis of the site-specific unfolding relies on the difference spectra of the ^{13}C-labeled and -unlabeled protein, high degree of consistency is required between the experimental measurements of the isotopically edited and unlabeled samples. For thermal unfolding experiments, the spectra are collected at temperatures from 0 to 90 °C at steps of 3–5 °C. Preferably, the temperatures are changed automatically, through a software interfaced with the spectrometer control. Prior to the spectral scan, a delay of 3–5 min is applied after reaching each desired temperature to allow for full sample equilibration. The standard experimental parameters are 4 cm^{-1} spectral resolution and 256 or 512 repeated scans.
10. Make sure the cell is completely filled, and no air bubbles are visible after closing the cell. Use more solution if necessary.
11. The buffer spectrum is subtracted from that of the protein solution (protein plus buffer) by minimizing the absorbance

signal in the spectral region with no contribution from the protein, e.g., 1800–1900 cm^{-1}. The water vapor is subtracted to eliminate the sharp features (either positive or negative) associated with the atmospheric H_2O spectrum.

12. The *S*hifted *M*ultivariate *S*pectra *A*nalysis (SMSA) is a generalization of the standard multivariate curve resolution (MCR) scheme:

$$\mathbf{D}_{\nu,T} = \sum_{d} \mathbf{S}_{\nu-\boldsymbol{\delta\nu}_{d,T},d}\mathbf{C}_{d,T} + \mathbf{E}_{\nu,T} \tag{1}$$

where **D** is the matrix of the experimental spectra, **S** is the matrix of the spectra of individual components, **C** of their concentrations, and **E** the errors, ν denotes frequency, T temperature, and d indexes the particular component (species, structure with distinct spectroscopic signature). The difference from the common MCR, where **S** is temperature independent, is that the **S** components can change with temperature via frequency shifts $\delta\nu_{d,T}$. The frequency shifts of amide I′ with temperature are linear [25]:

$$\boldsymbol{\delta\nu}_{d,T} = \boldsymbol{a}_d(T - T_0) \tag{2}$$

where $\boldsymbol{a}_d$ is a vector of d temperature-dependent slopes, one for each component. The model is optimized by finding the matrices **S**, **C** (Eq. 1) and the frequency slopes in vector $\boldsymbol{a}$ (Eq. 2) that minimize the least squares error:

$$\begin{aligned}\chi^2 &= \sum_{\nu,T}\left\|\mathbf{D}_{\nu,T} - \sum_{d}\mathbf{S}_{\nu-\boldsymbol{\delta\nu}_{d,T},d}\mathbf{C}_{d,T} + \mathbf{E}_{\nu,T}\right\|_F^2 \\ &= N\sum_{t,T}\left\|\widetilde{\mathbf{D}}_{t,T} - \sum_{d}\widetilde{\mathbf{S}}_{t,d}e^{i2\pi(t-1)\boldsymbol{\delta\nu}_{d,T}/N}\mathbf{C}_{d,T}\right\|_F^2\end{aligned} \tag{3}$$

where subscript F denotes Frobenius norm, and the equality follows from Parseval identity with the following constraints: (1) both **S** and **C** are non-negative, (2) the individual component spectra (columns of **S**) and their concentration profiles (rows of **C**) are unimodal, and the slopes $\boldsymbol{a}$ are non-negative (amide I frequencies shift higher with temperature) and, for this particular application, bounded by the maximum slope measured for a model amide, N-methylacetamide (0.078 cm^{-1} K^{-1}) [25].

The parametric version (pSMSA) further considers **C** (in Eq. 1) to conform to the simple thermodynamic model for the protein unfolding:

$$\mathrm{A}_1 \rightleftharpoons \mathrm{A}_2 \rightleftharpoons \mathrm{A}_3 \cdots \rightleftharpoons \mathrm{A}_n \tag{4}$$

where A_i represents the individual thermodynamic states (e.g., folded, intermediates, and unfolded). The fractional populations of the individual states can be expressed as:

$$X_i = \frac{\Pi^{i}_{j=1} K_{j-1}}{\sum_{k=1}^{d} \Pi^{k}_{j=1} K_{j-1}} \quad (5)$$

where K_i is the equilibrium constant for the $i \rightleftharpoons i+1$ transition:

$$K_i = e^{-\frac{\Delta G_i}{RT}} = e^{-\frac{\Delta H_i}{R}\left(\frac{1}{T} - \frac{1}{T_{mi}}\right)} \quad (6)$$

with $K_0 \equiv 1$. In (6) ΔG_i is the (standard) free energy for the particular step, usually expressed in terms of the enthalpy ΔH_i and the midpoint transition temperature $T_{mi} = \Delta H_i/\Delta S_i$. The temperature dependence of the spectral intensities is taken into account by linear baselines, as the amide I′ intensity decreases linearly with temperature [22]. The effective component concentrations, which reflect the contribution of the particular component spectrum to the experimental data, are therefore expressed as:

$$C_i = [1 - b_i(T - T_0)]X_i \quad (7)$$

where b_i is the linear slope of the intensity decrease with temperature and X_i is given in (5). The effective "concentrations" of the species C_i are therefore described by $4d$—2 parameters ($d - 1$ enthalpies, $d - 1$ temperatures, d frequency slopes, and d intensity slopes). The intensity slopes b_i are constrained to be positive (i.e., the intensity can only decrease with temperature) and to not exceed the value of 1×10^{-2} K^{-1} [26, 27].

The details of the implementation are described in ref. 23, and MATLAB (MathWorks, Inc., Mattick, MA) code is available upon request.

Acknowledgments

This work was supported by the National Science Foundation (NSF) grant CAREER 0846140.

References

1. Garcia-Mira MM, Sadqi M, Fischer N, Sanchez-Ruiz JM, Muñoz V (2002) Experimental identification of downhill protein folding. Science 298:2191–2195
2. Sadqi M, Fushman D, Muñoz V (2006) Atom-by-atom analysis of global downhill protein folding. Nature (442):317–321
3. Udgaonkar JB (2008) Multiple routes and structural heterogeneity in protein folding. Annu Rev Biophys 37:489–510
4. Eaton WA (1999) Searching for "downhill scenarios" in protein folding. Proc Natl Acad Sci U S A 96:5897–5899

5. Barth A, Zscherp C (2002) What vibrations tell us about proteins. Q Rev Biophys 35:369–430
6. Tadesse L, Nazarbaghi R, Walters L (1991) Isotopically enhanced infrared spectroscopy: a novel method for examining secondary structure at specific sites in conformationally heterogeneous peptides. J Am Chem Soc 113:7036–7037
7. Decatur SM (2006) Elucidation of residue-level structure and dynamics of polypeptides via isotope-edited infrared spectroscopy. Acc Chem Res 39:169–175
8. Huang R, Kubelka J, Barber-Armstrong W, Silva RA, Decatur SM, Keiderling TA (2004) Nature of vibrational coupling in helical peptides: an isotopic labeling study. J Am Chem Soc 126:2346–2354
9. Manas ES, Getahun Z, Wright WW, Degrado WF, Vanderkooi JM (2000) Infrared spectra of amide groups in alpha-helical proteins: evidence for hydrogen bonding between helices and water. J Am Chem Soc 122:9883–9890
10. Walsh STR, Cheng RP, Wright WW, Alonso DOV, Daggett V, Vanderkooi JM, DeGrado WF (2003) The hydration of amides in helices; a comprehensive picture from molecular dynamics, IR, and NMR. Protein Sci 12:520–531
11. Brewer SH, Song BB, Raleigh DP, Dyer RB (2007) Residue specific resolution of protein folding dynamics using isotope-edited infrared temperature jump spectroscopy. Biochemistry 46:3279–3285
12. Fesinmeyer RM, Peterson ES, Dyer RB, Andersen NH (2005) Studies of helix fraying and solvation using C-13 isotopomers. Protein Sci 14:2324–2332
13. Marecek J, Song B, Brewer S, Belyea J, Dyer RB, Raleigh DP (2007) A simple and economical method for the production of C-13, O-18-labeled FMOC-amino acids with high levels of enrichment: applications to isotope-edited IR studies of proteins. Org Lett 9:4935–4937
14. Kubelka J, Keiderling TA (2001) The anomalous infrared amide I intensity distribution in ^{13}C isotopically labeled peptide β-sheets comes from extended, multiple-stranded structures. An ab initio study. J Am Chem Soc 123:6142–6150
15. Petty SA, Decatur SM (2005) Intersheet rearrangement of polypeptides during nucleation of beta-sheet aggregates. Proc Natl Acad Sci U S A 102:14272–14277
16. Welch WRW, Kubelka J, Keiderling TA (2013) Infrared, VCD and Raman spectral simulations for β-sheet structures with various isotopic labels, inter-strand and stacking arrangements using Density Functional Theory. J Phys Chem B 117:10343–10358
17. Welch WRW, Keiderling TA, Kubelka J (2013) Structural analyses of experimental ^{13}C edited amide I' IR and VCD for peptide β-sheet aggregates and fibrils using DFT-based spectral simulations. J Phys Chem B 117:10359–10369
18. Sagle LB, Zimmermann J, Dawson PE, Romesberg FE (2004) A high-resolution probe of protein folding. J Am Chem Soc 126:3384–3385
19. Huang CY, Getahun Z, Zhu YJ, Klemke JW, DeGrado WF, Gai F (2002) Helix formation via conformation diffusion search. Proc Natl Acad Sci U S A 99:2788–2793
20. Hauser K, Krejtschi C, Huang R, Wu L, Keiderling TA (2008) Site-specific relaxation kinetics of a tryptophan zipper hairpin peptide using temperature jump IR spectroscopy with isotopic labeling. J Am Chem Soc 130:2984–2992
21. Amunson KE, Ackels L, Kubelka J (2008) Site-specific unfolding thermodynamics of a helix-turn-helix protein. J Am Chem Soc 130:8146–8147
22. Kubelka GS, Kubelka J (2014) Site-specific thermodynamic stability and unfolding of a de novo designed protein structural motif mapped by C-13 isotopically edited IR spectroscopy. J Am Chem Soc 136:6037–6048
23. Lai JK, Kubelka G, Kubelka J (2015) Sequence, structure and cooperativity in folding of elementary protein motifs. Proc Natl Acad Sci U S A 112:9890–9895
24. Fabian H, Naumann D (2004) Methods to study protein folding by stopped-flow FT-IR. Methods 34:28–40
25. Amunson KE, Kubelka J (2007) On the temperature dependence of amide I frequencies of peptides in solution. J Phys Chem B 111:9993–9998
26. Ackels LA, Stawski P, Amunson KE, Kubelka J (2009) On the temperature dependence of the amide I intensities of peptides in solution. Vibr Spectrosc 50:2–9
27. Kubelka J (2013) Multivariate analysis of spectral data with frequency shifts: application to temperature dependent infrared spectra of peptides and proteins. Anal Chem 85:9588–9595
28. Chan WC, White PD (2000) FMOC solid phase peptide synthesis: a practical approach. Oxford University Press, Oxford, UK
29. Goormaghtigh E, Cabiaux V, Ruysschaert JM (1994) Determination of soluble and membrane protein structure by Fourier transform infrared spectroscopy. II. Experimental aspects, side chain structure, and H/D exchange. In: Hilderson HJ, Ralston GB (eds) Subcellular biochemistry. Plenum, New York, pp 363–403

Chapter 10

NMR Analysis of Protein Folding Interaction Networks

Eva de Alba

Abstract

Theory and experimental evidence unequivocally indicate that protein folding is far more complex than the two-state (all-or-none) model that is usually assumed in the analysis of folding experiments. Proteins tend to fold hierarchically by forming secondary structure elements, followed by supersecondary arrangements, and other intermediate states that ultimately adopt the native tertiary fold as a result of a delicate balance between interatomic interactions and entropic contributions. However, small proteins with simple folds typically follow downhill folding, characterized by very small energetic barriers (<3 RT) that allow multiple protein conformations to be populated along the folding path down the free energy landscape, reaching the native fold at the lowest energy level.

Here we describe the use of solution-state nuclear magnetic resonance (NMR) for the analysis of protein folding interaction networks at atomic resolution. The assignment of NMR spectra acquired at different unfolding conditions provides hundreds of atomic unfolding curves that are analyzed to infer the network of folding interactions. The method is particularly useful to study small proteins that fold autonomously in the sub-millisecond timescale. The information obtained from the application of this method can potentially unveil the basic relationships between protein structure and folding.

Key words Protein folding, Nuclear magnetic resonance, Interaction networks, Minimal cooperativity

1 Introduction

The complexity of protein folding results from the intricate network of weak interactions that are responsible for stabilizing protein three-dimensional structures. These interaction networks allow for individual proteins to explore different pathways during the folding process [1, 2]. Moreover, we know now that proteins do not necessary fold following a two-state mechanism, despite the fact that the pervasive use of simple analytical methods to interpret low-resolution experiments make it difficult to resolve the underlying complexity [3, 4]. In fact, small proteins have been shown to fold downhill, sampling a large conformational space without encountering significant free energy barriers [5–11].

Victor Muñoz (ed.), *Protein Folding: Methods and Protocols*, Methods in Molecular Biology, vol. 2376,
https://doi.org/10.1007/978-1-0716-1716-8_10,

Downhill folding of small proteins and archetypal folds (simple combinations of α-helices, β-hairpins, β-turns, and short loops) is characterized by broad, minimally cooperative unfolding transitions [12–14]. In addition, these transitions exhibit different features depending on the probe used to monitor unfolding. Here, it is described in detail a method that exploits the probe-dependence of minimally cooperative unfolding transitions to monitor the equilibrium thermal unfolding of fast folding proteins at atomic resolution using NMR [14–16]. The method also involves the computational analysis of the hundreds of atomic unfolding curves obtained from NMR data to calculate cross-correlations that lead to the characterization of the folding interaction network [15].

2 Materials

2.1 Suitable Proteins

2.1.1 Proteins That (Un)Fold in the Microsecond Timescale

One requirement for the NMR analysis of protein folding at atomic resolution is that proteins under study fold faster than the NMR chemical shift timescale. Therefore, the folding–unfolding process needs to be in fast conformational exchange relative to the chemicals shift timescale. The bioinformatics tool PREFUR predicts the protein folding–unfolding rates based on protein size and structural classification [17].

2.1.2 Structural Motifs

Small proteins and fold archetypes (supersecondary structural elements that fold autonomously) are particularly well suited as they usually fold in the microsecond time regime [18]. Some examples are: α-helix bundles, β-hairpins, small three-stranded antiparallel β-sheets, helix-loop-helix (e.g., EF-hands), helix-hairpin motifs (e.g., zinc fingers), minimal α–β parallel folds (Figs. 1 and 2a). Suitable examples can be found in the protein structure database (PDB), or designed de novo using bioinformatic tools such as Modeller [19] and I-TASSER [20].

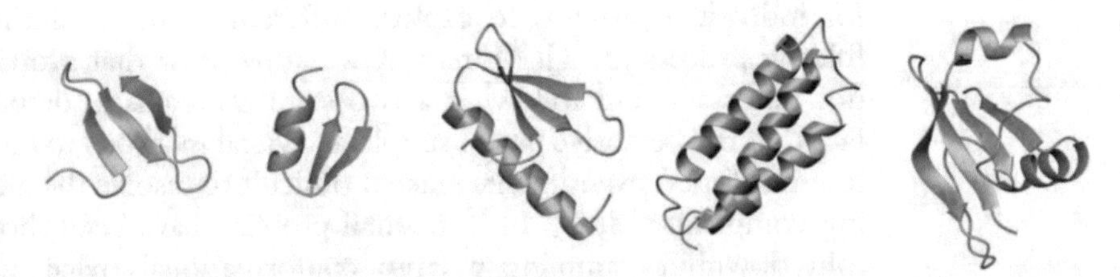

Fig. 1 Examples of small proteins that fold autonomously into elementary structure motifs that can be used to build a catalog of folding archetypes. From left to right: three-stranded antiparallel β-sheet, zinc finger, α + β minimal motif, three-helix bundle, α,β-sandwich

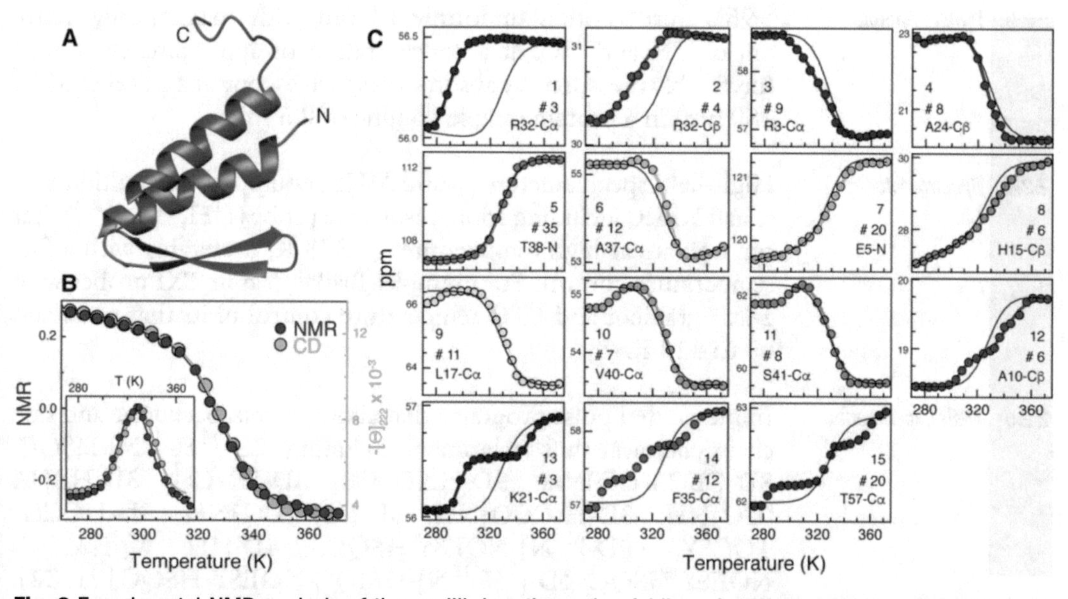

Fig. 2 Experimental NMR analysis of the equilibrium thermal unfolding of gpW at atomic resolution [15]. (**a**) Ribbon diagram of the lowest energy structure from the NMR ensemble obtained at pH 6.5 (PFB ID 2L6Q) [16]. (**b**) The global NMR thermal unfolding behavior represented by the second component from the singular value decomposition (SVD) of the 180 atomic unfolding curves (blue, left scale) compared to the unfolding curve from circular dichroism (green, right scale). The inset shows the derivative of the curves. (**c**) The 15 different types of atomic unfolding behaviors obtained for gpW from cluster analysis. Clusters 1–8 are 2SL (two state-like), clusters 9–13 are 3SL (three state-like), and clusters 14 and 15 show curves with complex patterns (CP group). Each panel shows a representative experimental unfolding curve as example (colored circles), the number of cluster elements, and the expected global behavior for reference (black curve). The latter was calculated by fixing the thermodynamic parameters to those of the two-state fit to the blue curve in panel B ($T_m = 329$ K and $\Delta H = 133$ kJ/mol) and fitting the "native" and "unfolded" baselines for each probe

2.2 NMR Samples and Experiments

2.2.1 Type of NMR Tubes

Pressure valve NMR tubes with medium wall thickness (0.77 mm) that stand pressures of up to ~20 bar to avoid solvent evaporation at high temperature. Pressure valve NMR tubes are connected to a gas manifold. Once the desired pressure is reached, the valve is closed, the gas line is disconnected from the valve, and the tube/sample is ready for use.

2.2.2 Chemicals

Ethylene glycol: 100%, for temperature calibration in NMR tube with regular thickness. Sodium trimethylsilyl propanesulfonic acid: 0.01 mM concentration, for chemical shift referencing. Deuterium-enriched buffers and salts to alleviate interference from the NMR signals of the selected buffer. Deuterium oxide at 5–10% as reference signal for the NMR spectrometer. Helium or other inert gas to generate pressure inside the thick-wall NMR tube.

2.2.3 Protein Sample

>95% pure protein uniformly labeled with magnetically active nuclei ^{15}N and ^{13}C, at a concentration of approximately 1 mM (preferably measured by absorption spectroscopy at a wavelength of 280 nm) in a protein sample volume of 0.5 ml.

2.2.4 Spectrometer

High-field spectrometers (>600 MHz) equipped for multidimensional NMR, including triple-resonance probes (^{1}H, ^{15}N, ^{13}C) that can operate at high temperatures (>373 K) preferably with a fine temperature control. For example, Bruker 5 mm TXI probes with z-axis gradient and BTO temperature control units that can stand up to 423 K.

2.2.5 NMR Experiments

Implemented pulse programs that allow protein backbone and side chain chemical shift assignment including: 2D-[^{1}H-^{15}N]-HSQC, 3D-CBCA(CO)NH, 3D-HNCACB, 3D-HNCO, 3D-HBHA (CO)NH, 3D-H(CCO)NH, 3D-(H)C(CO)NH, 3D-HCCH-TOCSY, 3D-[^{15}N]-NOESY-HSQC, 4D-[^{1}H-^{13}C]-HMQC-NOESY-HSQC, 4D-[^{1}H-^{15}N]-HMQC-NOESY-HSQC [21, 22].

2.2.6 NMR Software

Software for NMR experiment processing and analysis, for example, NMRPipe [23] for processing, PIPP [24] and SPARKY [25] for analysis.

2.3 Computational Analysis

Data analysis programs such as Grace, SigmaPlot, Origin, and MATLAB. The MATLAB package for numerical data analysis, complemented with custom-made programs, was used for the computational analyses described in this work. The network graphs were created with the graph plotting software visone [26] (available at http://www.visone.info).

3 Methods

3.1 Conditions Necessary for the Atomic-Resolution Analysis

3.1.1 Reversibility of the Equilibrium Thermal Unfolding

The atom-by-atom analysis described here requires a fully reversible thermal unfolding process so that it can be considered at thermodynamic equilibrium. The reversibility of thermal unfolding is checked by preparing the sample at the typical NMR concentration (sub-mM to mM). The sample is heated up to 373 K for a period of 2 h and then cooled down. Two "fingerprint" NMR experiments (typically, 2D-[^{1}H-^{15}N]-HSQC) are acquired before and after heating. Reversibility of the unfolding process will result in superimposable signals of both spectra (*see* **Notes 1** and **2**).

3.1.2 Folding Kinetics in Fast Conformational-Exchange

Folding should fall within the fast conformational-exchange regime in NMR. Two-dimensional [^{1}H-^{15}N]-HSQC spectra are acquired at low temperature, at the approximate global denaturation midpoint (determined previously with a low-resolution probe, such as circular dichroism, in a thermal unfolding experiment), and at the highest temperature. For a suitable protein system, the 2D-[^{1}H-^{15}N]-HSQC spectrum at the denaturation midpoint will show a single set of signals without significant line broadening, and with chemical shifts that fall approximately halfway between those observed at low and high temperatures (*see* **Note 3**).

3.2 NMR Experiments to Monitor Protein Unfolding at Atomic Resolution

3.2.1 Temperature Calibration of the NMR Probe

The NMR probe is calibrated at each temperature of the study using an air flow rate of 535 L/h, and following the ethylene glycol chemical shift temperature dependence as described in Amman et al. [27] for the 272–416 K range. Standard one-pulse 1D ^{1}H-NMR experiments are used to monitor chemical shift changes with temperature, after the equilibration of the ethylene glycol sample for 30 min at each temperature. The typical range for temperature calibration is from 273 to 370 K.

3.2.2 Chemical Shift Referencing

One-pulse 1D ^{1}H-NMR experiments are used to calibrate the position of the water signal relative to DSS (0.0 ppm) at each temperature. The resonance frequency of the water is used as reference for ^{1}H chemical shifts. The references for ^{13}C and ^{15}N are obtained from the gyromagnetic ratios relative to the ^{1}H reference.

3.2.3 Chemical Shift Assignment

Protein backbone amide ^{15}N, $^{13}C_{\alpha}$, and $^{13}C_{\beta}$ chemical shifts can be obtained with a basic set of double- and triple-resonance experiments: 2D-[^{1}H-^{15}N]-HSQC, 3D-CBCA(CO)NH, and 3D-HNCACB. Side chain ^{1}H and ^{13}C chemical shifts are assigned with the experiments: 3D-HBHA(CO)NH, 3D-H(CCO)NH, 3D-(H)C(CO)NH, and 3D-HCCH-TOCSY. All assignments are confirmed with the experiments: 3D-[^{15}N]-NOESY-HSQC, and 4D-[^{1}H-^{13}C]-HMQC-NOESY-HSQC (*see* **Notes 4** and **5**).

3.2.4 Monitoring NMR Signal Changes with Temperature

The protein sample is kept at a He pressure of 8 bar inside the pressure valve NMR tube to minimize buffer evaporation at high temperature. The stepwise change in sample temperature needs to be small enough to unambiguously follow chemical shift changes of the NMR signals. These intervals are typically from 3 to 5 K (Fig. 2b, c), although they can be reduced when changes in chemical shifts are large or result in signal overlap (*see* **Note 4**). Significantly broader signals are typically observed at low temperatures and will need special treatment for chemical shift determination (*see* **Note 4**).

3.3 Analysis of Individual Atomic Equilibrium Unfolding Curves

3.3.1 Classification of Atomic Unfolding Curves

All experimental unfolding curves obtained from chemical shift changes with temperature (*see* **Note 5**) are classified in three groups according to the unfolding behavior: (a) two-state-like (2SL) transition curves that display a distinctly sigmoidal curve characterized by a single transition and a single peak in their derivative; (b) three-state-like (3SL) transition curves that display two apparent transitions (a bi-sigmoidal shape) and two separated peaks in the derivative; and (c) complex pattern (CP) curves that include all curves that do not fit any of the former two categories (multiple transitions or no obvious transition). The curves classified as 2SL and 3SL are fitted to the thermodynamic two-state or three-state unfolding model equation, respectively, using the procedures described below. A final check based on the comparison of fitting residuals is performed. Namely, the atomic curves classified in groups 2SL and 3SL are fitted to both two-state and three-state models (vide infra) and the best fit using a Fisher-test criterion (considering number of degrees of freedom and number of parameters) defines the final adscription of the atomic curve to either group.

3.3.2 Analysis of Two-State-Like Unfolding Curves

The atomic unfolding curves of the 2SL group are fitted to a two-state model (*see* **Note 6**). The average chemical shift $\langle cs \rangle$ for 2SL curves is given by the equation:

$$\langle cs \rangle = \left(cs_f + cs^s_f(T - T_o)\right)\cdot p_f + \left(1 - p_f\right)\cdot\left(cs_u + cs^s_u\,(T - T_o)\right)$$

cs_f and cs^s_f are the intercept and slope of the pre-transition baseline, and $cs_u$and cs^s_u are the intercept and slope of the post-transition baseline. Both baselines, $cs_f$and cs_u, are assumed to have linear temperature dependence. p_f and p_u represent the probabilities of the folded and unfolded states such that $p_u = (1 - p_f)$; T_o is an arbitrary reference temperature (chosen within the experimental range for convenience).

The native state is used as thermodynamic reference, thus, $p_f$is given by:

$$P_f = 1/[1 + \exp\left((-\Delta H + T\Delta H/T_m)/RT\right))]$$

ΔH is the enthalpy change upon unfolding and T_m is the midpoint of the thermal unfolding curve (*see* **Note 7**). The analysis involves fitting six free floating parameters: the four parameters of the two baselines, T_m and ΔH. For simplicity, this model and the three-state model below ignore changes in heat capacity upon unfolding.

3.3.3 Analysis of Three-State-Like Unfolding Curves

The atomic unfolding curves from the 3SL group are fitted to a three-state model. The average chemical shift $\langle cs \rangle$ of a 3SL curve at any given temperature is given by:

$$\langle cs \rangle = \left(cs_f + cs^s_f(T - T_o)\right)\cdot p_f + cs_i\cdot p_i + \left(cs_u + cs^s_u\,(T - T_o)\right)\cdot P_u$$

cs_f and cs^s_f represent the intercept and slope of the pre-transition, cs_i is the intercept for the intermediate state (for simplicity it is assumed to be temperature independent since this baseline is typically not well resolved); cs_u and cs^s_u represent the intercept and slope of the post-transition. p_f, p_i, and p_u are the folding, intermediate and unfolding probabilities such that $p_f + p_i + p_u = 1$. The probabilities are calculated with the relationships:

$$p_f = 1/\left[1 + \exp\left(\left(-\Delta H_1 + \frac{T\Delta H_1}{T_{1m}}\right)/RT\right) + \exp\left(\left(-\Delta H_2 + \frac{T\Delta H_2}{T_{2m}}\right)/RT\right)\right]$$

$$p_i = \exp\left(\left(-\Delta H_1 + \frac{T\Delta H_1}{T_{1m}}\right)/RT\right)/\left[1 + \exp\left(\left(-\Delta H_1 + \frac{T\Delta H_1}{T_{1m}}\right)/RT\right) + \exp\left(\left(-\Delta H_2 + \frac{T\Delta H_2}{T_{2m}}\right)/RT\right)\right]$$

$$p_u = 1 - p_f - p_i$$

ΔH_1 and T_{1m} are the parameters for the intermediate relative to the native state, and ΔH_2 and T_{2m} are the parameters for the unfolded state relative to the native state.

3.4 Clustering of Atomic Unfolding Curves and Network Analysis

3.4.1 Average Atomic Unfolding Behavior from NMR Compared to the Global Unfolding Process

A matrix composed of the chemical shift values versus temperature for all atoms used in the study is created to obtain the average atomic unfolding behavior using singular value decomposition (SVD) procedures (*see* **Note 8**). This average unfolding curve is compared with that obtained by a low-resolution collective probe such far UV circular dichroism (Fig. 2b).

3.4.2 Data Clustering

The data set containing all of the atomic unfolding curves is analyzed globally using data clustering methods such as *K*-means or hierarchical clustering (*see* **Note 9**). Atomic unfolding curves belonging to the 2SL and 3SL groups were clustered according to both the thermodynamic parameters obtained in their fits (T_m and ΔH for 2SL and T_{m1}, T_{m2}, ΔH_1, and ΔH_2 for 3SL) or according to the probability of the native state as a function of temperature obtained from the two-state and three-state fits. In the case of the 3SL, the calculation of the probability is the compounded $p_f + p_i$ curve assuming that the signal of the intermediate is 50% of that of the native state. Both methods should render comparable results.

Atomic unfolding curves included in the CP group are clustered according to their similarity after normalization of the signal using the *Z*-score procedure to provide a common frame of reference (*see* **Note 10**). *Z*-scored unfolding curves are calculated using the following expression:

$$z(T) = (x(T) - \mu)/\sigma$$

μ is the mean of all the data being clustered together and σ is the standard deviation.

3.4.3 Calculation of the Thermodynamic Coupling Index Matrix (TCI)

The thermodynamic coupling index (TCI) for each pair of residues in a protein is calculated from the pairwise comparison between all the atomic unfolding curves of the first residue in the pair with all the curves of the second residue. The thermodynamic coupling index of residues *x* and *y* is then calculated by summing all the possible pairwise comparisons of atomic folding curves as reported in the following equation:

$$\mathrm{TCI}_{x,y} = \sum_i \sum_j \ln\left(\frac{\left\langle\sqrt{\delta(p_m - p_n)(p_m - p_n)^T}\right\rangle}{\left\langle\sqrt{\left(p_{x,i} - p_{y,j}\right)^2}\right\rangle}\right)$$

where p_m and p_n are row vectors from a matrix including all the *z*-scored atomic unfolding curves. The denominator corresponds to the root-mean-square deviation (RMSD) for all chemical shift curves in the data set, and $p_{x,i}$ and $p_{y,j}$ are the vectors from the same matrix that correspond to the chemical shift curves of residues *x* and *y*, respectively (*i* runs over all curves of residue *x* and *j* over all curves of residue *y*). The TCI for residues *x* and *y* is positive when the average Euclidean distance between all their cross-pairs is smaller than the mean RMSD for all curves in the protein and negative otherwise (*see* **Note 11**). The TCI matrix is constructed by repeating the same procedure over all possible residue pairs (Fig. 3).

3.4.4 Comparing Thermodynamic Coupling Index Matrix and Native Three-Dimensional Structure

The contact map is a two-dimensional presentation of the protein's native three-dimensional (3D) structure. It is obtained by identifying all residue pairs in the 3D structure with atoms at distances indicative of a close contact (e.g., 0.5 nm). The resulting contact map is a matrix of the same rank as the TCI (*see* **Note 12**). A graphical comparison can be obtained by overlaying the two matrices in different colors (For example; for TCI, the highest coupling shown in red, the lowest in dark blue, and black circles for the contact map matrix). The command "image" of MATLAB generates this type of graph by placing the contact map and the TCI on the first and second layers, respectively (Fig. 3).

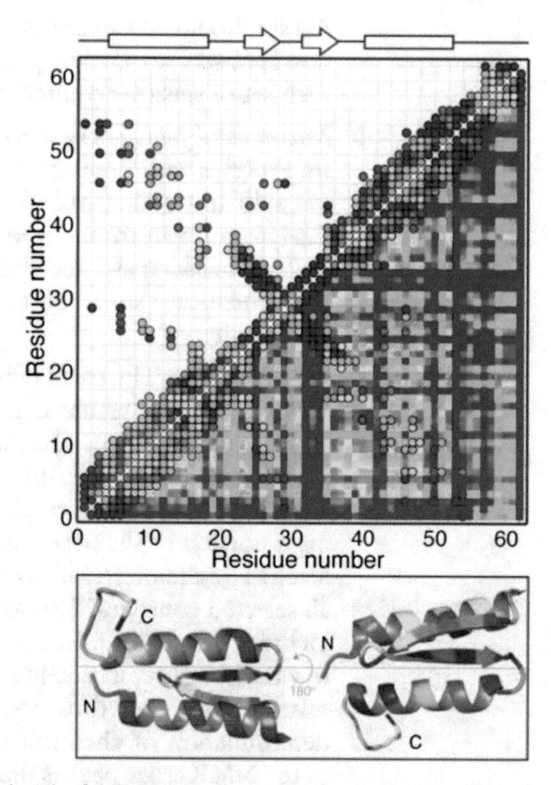

Fig. 3 Mapping the folding interaction network of gpW from the residue–residue thermodynamic coupling matrix [15]. Upper panel shows the experimental residue–residue thermodynamic coupling matrix (highest coupling in red, lowest in dark blue), with native contacts identified using hollow black circles. The bottom right triangle shows the entire coupling matrix, whereas the upper left triangle shows only primary couplings (interacting residues). The structural and sequence distribution of the overall degree of coupling (defined as the sum of the couplings of a given residue to all other residues) is shown in the bottom panel. Residues were classified in six levels of coupling (from low to high: dark blue, light blue, cyan, yellow, orange, red) using data clustering tools

4 Notes

1. *Use of fast acquisition techniques*: Fast NMR acquisition techniques such as SOFAST-HMQC [28] can be used to obtain preliminary information on protein stability and behavior with temperature, thus avoiding the need of uniform isotopic labeling.
2. *Checking protein sample status*: Some proteins undergo degradation and/or oligomerization at high temperature during the time required to acquire triple-resonance NMR experiments. Thus, it is advisable to check the status of the protein by periodically acquiring 2D-[^{1}H-^{15}N]-HSQC spectra. At each temperature, the HSQC spectra acquired before and after the

3D set should be superimposable. If degradation/oligomerization is observed, a newly prepared sample needs to be used for each set of spectra acquired at each temperature.

3. *Microsecond-folding times required for the atom-by-atom analysis*: Proteins with folding times comparable to the NMR chemical shift timescale (~0.1–5 ms) will be difficult to study with the atom-by-atom method because the NMR signals will severely broaden at intermediate degrees of unfolding, making the chemical shift determination highly inaccurate or impossible.
4. *Signal overlap*: The small range of ^{1}H chemical shifts in protein NMR (~12 ppm) results in frequent signal overlap that worsens at high temperature. Overlapping signals that give rise to a broad cross-peaks without distinguishable maxima should be analyzed by contour averaging. This analysis is based on selecting a portion of the peak with a reasonable number of contour levels. The chemical shift is obtained by averaging the centers of all selected contours. The software PIPP for peak picking [24] includes this feature. In addition, solvent viscosity increases at low temperature, increasing in turn the protein rotational correlation time and thus resulting in signal broadening. The determination of chemical shifts from the shallow maximum of the NMR cross-peak is therefore less accurate at low temperature, and hence the use of contour averaging is also advised.
5. *Chemical shift temperature dependence of protein ^{1}H nuclei*: Amide ^{15}N, $^{13}C_{\alpha}$, and $^{13}C_{\beta}$ chemical shifts are ideal choices for the atom-by-atom analysis of protein folding because their values are directly related to structural changes. In contrast, chemical shifts of protein ^{1}H nuclei, specifically amide $^{1}H^{N}$, strongly depend on temperature even in the absence of secondary or tertiary structural modifications. Therefore, it is more difficult to derive information on the protein unfolding process from ^{1}H chemical shifts changes with temperature.
6. *Characteristics of the unfolding atomic curve*: The steepness of the atomic unfolding curve (ΔH), the denaturation temperature (T_{m}), together with the group adscription are the defining characteristics of any given atomic unfolding curve.
7. *Additional methods to obtain denaturation midpoints from atomic unfolding curves*: Model-free denaturation midpoints can also be obtained by calculating the derivative of the atomic equilibrium unfolding curve using numerical methods. The curve derivative is analyzed with an algorithm that estimates the position of the maximum in the derivative data, by finding the point that divides the area under the curve into two equal halves [29]. This maximum in the derivative is then taken as the denaturation midpoint [29]. For curves with two transitions (SL2 group), a similar procedure is followed where two

maxima are identified. This method serves the purpose of cross-validating the denaturation midpoints obtained with the two- and three-state fits, which could be affected by poorly defined baselines [30].

8. *Average normalized unfolding curve*: The first component of the matrix multiplied by the first singular value from the SVD provides the temperature averaged chemical shifts for the whole dataset. The amplitude of the first component (the first row of the V matrix) provides the averaged normalized equilibrium unfolding curve for the whole dataset.
9. *Comparison between different unfolding curves*: The changes in chemical shift values upon unfolding vary broadly, thus, it is very important to define a reference to compare heterogeneous atomic unfolding curves. In addition to using the basic thermodynamic parameters obtained from the fits (T_m and ΔH), the native signal calculated from the native probability generated by the two-state and three-state fits can be used. In the latter case, it can be assumed that the intermediate provides a fraction (e.g., 50%) of the native signal. For curves that do not belong to either the SL2 or SL3 groups, or for comparing curves from different groups, the best procedure is to compare directly the chemical shift versus temperature curves once they have been *z*-scored to provide a common frame of reference.
10. *Identifying the optimal number of clusters*: Clustering should be performed individually for each of the three groups of atomic unfolding curves. The number of clusters to use depends on the total number of curves within each group and their heterogeneity. Clustering methods tend to gather the most dissimilar unfolding curves into clusters containing only one or two atoms and, group all others into a few highly populated clusters. This problem is minimized by beginning the clustering with approximately five times fewer clusters than curves in each group, and iteratively decreasing the number of clusters until only a few clusters are produced, including only one or two atomic curves.
11. *Interpretation of TCI matrix*: The thermodynamic coupling index (TCI) is positive when the sum of atomic couplings for two residues is stronger than the mean coupling for the entire dataset (all atoms of the protein under study) and negative when it is weaker. Residues with positive TCI values reflect highly coupled atomic unfolding behaviors.
12. *The TCI matrix* versus *the contact map*: The red/orange dots in Fig. 3 define the critical network of residue–residue contacts that are responsible for the global unfolding process of the autonomously folding gpW [15], whether they are or not in spatial contact in the 3D structure. These residues are typically

structurally connected by secondary or tertiary contacts. That is, by mutual coupling to another residue that is in contact with both for secondary contacts. Finally, black circles show contacts in the 3D structure.

Acknowledgments

The author acknowledges support from the NSF-CREST Center for Cellular and Biomolecular Machines at the University of California, Merced (NSF-HRD-1547848).

References

1. Bryngelson JD, Onuchic JN, Socci ND, Wolynes PG (1995) Funnels, pathways, and the energy landscape of protein-folding - a synthesis. Proteins Struct Funct Genet 21:167–195
2. Pande VJ (2008) Computer simulations of protein folding. In: Muñoz V (ed) Protein folding, misfolding and aggregation: classical themes and novel approaches. RSC, Cambridge, pp 161–187
3. Jackson SE (1998) How do small single-domain proteins fold? Fold Des 3:R81–R91
4. Kubelka J, Hofrichter J, Eaton WA (2004) The protein folding 'speed limit'. Curr Opin Struct Biol 14:76–88
5. Muñoz V (2007) Conformational dynamics and ensembles in protein folding. Annu Rev Biophys Biomol Struct 36:395–412
6. Naganathan AN, Doshi U, Fung A, Sadqi M, Muñoz V (2006) Dynamics, energetics, and structure in protein folding. Biochemistry 45:8466–8475
7. Yang WY, Gruebele M (2003) Folding at the speed limit. Nature 423:193–197
8. Muñoz V, Sanchez-Ruiz JM (2004) Exploring protein-folding ensembles: a variable-barrier model for the analysis of equilibrium unfolding experiments. Proc Natl Acad Sci U S A 101:17646–17651
9. Naganathan AN, Sanchez-Ruiz JM, Muñoz V (2005) Direct measurement of barrier heights in protein folding. J Am Chem Soc 127:17970–17971
10. Naganathan AN, Perez-Jimenez R, Sanchez-Ruiz JM, Muñoz V (2005) Robustness of downhill folding: guidelines for the analysis of equilibrium folding experiments on small proteins. Biochemistry 44:7435–7449
11. Muñoz V (2002) Thermodynamics and kinetics of downhill protein folding investigated with a simple statistical mechanical model. Int J Quant Chem 90:1522–1528
12. Garcia-Mira MM, Sadqi M, Fischer N, Sanchez-Ruiz JM, Muñoz V (2002) Experimental identification of downhill protein folding. Science 298:2191–2195
13. Sadqi M, Fushman D, Muñoz V (2006) Atom-by-atom analysis of global downhill protein folding. Nature 442:317–321
14. Fung A, Li P, Godoy-Ruiz R, Sanchez-Ruiz JM, Muñoz V (2008) Expanding the realm of ultrafast protein folding: gpW, a midsize natural single-domain with $\alpha+\beta$ topology that folds downhill. J Am Chem Soc 130:7489–7495
15. Sborgi L, Verma A, Piana S, Lindorff-Larsen K, Cerminara M, Santiveri CM, Shaw DE, de Alba E, Muñoz V (2015) Interaction networks in protein folding via atomic-resolution experiments and long-time-scale molecular dynamics simulations. J Am Chem Soc 137:6506–6516
16. Sborgi L, Verma A, Muñoz V, de Alba E (2011) Revisiting the NMR structure of the ultra-fast downhill folding protein gpW from bacteriophage λ. PLoS One 6:e26409
17. de Sancho D, Muñoz V (2011) Integrated prediction of protein folding and unfolding rates from only size and structural class. Phys Chem Chem Phys 13:17030–17043
18. de Alba E, Santoro J, Rico M, Jiménez MA (1999) De novo design of a monomeric three-stranded antiparallel β-sheet. Protein Sci 8:854–865

19. Marti-Renom MA, Stuart A, Fiser A, Sánchez R, Melo F, Sali A (2000) Comparative protein structure modeling of genes and genomes. Annu Rev Biophys Biomol Struct 29:291–325
20. Roy A, Kucukural A, Zhang Y (2010) I-TASSER: a unified platform for automated protein structure and function prediction. Nat Protoc 5:725–738
21. Bax A, Grzesiek S (1993) Methodological advances in protein NMR. Acc Chem Res 26:131–138
22. Cavanagh J, Fairbrother WJ III, Palmer AG, Rance M, Skelton NJ (1995) Chemical exchange effects in NMR spectroscopy. In: Protein NMR spectroscopy: principles and practice. Academic Press, San Diego, pp 391–404
23. Delaglio F, Grzesiek S, Vuister GW, Zhu G, Pfeifer J (1995) NMRPipe: a multidimensional spectral processing system based on UNIX pipes. J Biomol NMR 6:277–293
24. Garret DS, Powers R, Gronenborn AM, Clore GM (1991) A common sense approach to peak picking in two-, three-, and four-dimensional spectra using computer analysis of contour diagrams. J Magn Reson 95:214–220
25. Goddard TD, Kneller DG (2000) SPARKY 3. University of California, San Francisco
26. Brandes U, Wagner D (2004) Visone – analysis and visualization of social networks. In: Juenger M, Mutzel P (eds) Graph drawing software. Springer-Verlag, New York, pp 321–340
27. Amman C, Meier P, Merbach AE (1982) A simple multinuclear NMR thermometer. J Magn Reson 46:319–321
28. Schanda P, Brutscher B (2005) Very fast two-dimensional NMR spectroscopy for real-time investigation of dynamic events in proteins on the time scale of seconds. J Am Chem Soc 127:8014–8015
29. Naganathan AN, Muñoz V (2008) Determining denaturation midpoints in multiprobe equilibrium protein folding experiments. Biochemistry 47:6752–6761
30. Sadqi M, Fushman D, Muñoz V (2007) Structural biology – analysis of protein-folding cooperativity – reply. Nature 445:E17–E18

Chapter 11

NMR Relaxation Dispersion Methods for the Structural and Dynamic Analysis of Quickly Interconverting, Low-Populated Conformational Substates

Sivanandam Veeramuthu Natarajan, Nicola D'Amelio, and Victor Muñoz

Abstract

Most biomolecular processes involve proteins shuttling among different conformational states, particularly from highly populated ground states to the lowly populated excited states that determine the interconversion rates and biological function, and which are invisible to most structural biology techniques. These structural transitions are rare and relatively fast: happen in the millisecond–microsecond timescale (ms–μs). NMR spectroscopy can access these timescales via relaxation dispersion techniques (RD-NMR). The exchange parameters extracted from RD-NMR experiments provide pivotal information on these otherwise invisible states that reports on key properties of the high free energy, reactive regions of the protein's energy landscape, including the mechanisms of folding/unfolding and of the interconversion between active and inactive states. Here, we describe a simple, step-by-step protocol to carry out RD-NMR experiments on proteins to detect the existence of such conformational substates and characterize their structural properties (chemical shifts).

Key words NMR, Protein folding, Excited states, Conformational substates relaxation dispersion, CPMG, Conformational dynamics, Chemical exchange

1 Introduction

Biomolecules in general, and proteins in particular, operate as cellular nanomachines thanks to their ability to spontaneously self-assemble into defined 3D structures and change shape on cue. The native structure determines the biological function, but the dynamics that drive the changes in biomolecular conformation is the key to understanding their mechanisms of action. With the relatively recent advent of single-molecule spectroscopic techniques to monitor the stochastic fluctuations of individual molecules, it is now possible to experimentally watch biomolecules at work. However, the structural resolution of single-molecule techniques is still insufficient to characterize the mechanisms in depth, and thus NMR is still the leading technique for the study of dynamic

Victor Muñoz (ed.), *Protein Folding: Methods and Protocols*, Methods in Molecular Biology, vol. 2376,
https://doi.org/10.1007/978-1-0716-1716-8_11,

biomolecular processes. This is due to the unrivaled capability of NMR to provide information at the atomic level, in solution, and across virtually all relevant timescales, ranging from picoseconds to minutes. NMR provides detailed information on each individual atom of the protein in its natural environment, and specific experiments can be designed that are able to separate the contributions from different timescales to the overall dynamics of the system under study.

Nuclear-spin relaxation is the physical phenomenon that provides the richest source of dynamic information because its physics is directly linked to anisotropic tensors, which are for most cases directly modulated by the molecular motions of the protein. Each nucleus type and each value of the static magnetic field gives opportunities to probe different frequencies of biomolecular motion. At the commonly used magnetic fields for biomolecular NMR, the motions that typically give rise to nuclear-spin relaxation in solution are in the pico- to nanoseconds timescale. The upper limit (about 10 ns for proteins amenable to NMR studies in solution) is determined by the overall tumbling of the protein, which produces motions unable to promote nuclear transitions (and therefore relaxation) and, most importantly, defines the timescale at which the magnetization decays due to spin relaxation. Analysis of nuclear-spin relaxation data provides the value of the order parameter S^2, which informs on the local disorder of the protein backbone and ranges from 1 (mostly rigid) to 0 (completely disordered), and its associated correlation time [1, 2].

Motions in the range from nanoseconds to microseconds cannot be probed directly by a specific NMR experiment, but they can be probed by indirect measurements based on residual dipolar couplings (RDC) [3, 4]. In isotropic tumbling conditions, the dipolar coupling between pairs of active NMR nuclei is averaged to zero. However, RDCs can be measured as a contribution to the overall scalar coupling when the system is weakly aligned. Given the angular nature of this information, data are somewhat ambiguous and require measurements in multiple alignment media to accurately determine the alignment tensor [5, 6]. Once this is done, the structural and dynamical parameters can be disentangled to provide the RDC order parameter S^2_{RDC}, which describes the degree of local order. In contrast to the order parameter obtained by nuclear relaxation, S^2_{RDC} provides information in a window from picoseconds to milliseconds (since RDCs are based on the chemical shift, the upper limit is determined by the acquisition time). Therefore, the comparison of S^2_{RDC} and S^2 highlights motions that are slower than the overall tumbling time of the protein [7].

Slower motions going up to the timescale defined by the NMR signal detection (microseconds–milliseconds) are the most important ones to the folding and functional dynamics of proteins. In this range, NMR can take advantage from the analysis of the chemical exchange phenomenon, which arises from the transverse relaxation that occurs upon the interconversion between conformers with differences in chemical shift taking place during the experimental NMR timescale. Exchange phenomena that occur during the acquisition time change the frequency of the signal (the NMR signal is recorded over a total duration of hundreds of milliseconds). These effects alter the shape of the NMR peaks, resulting in linewidth broadening that can have devastating consequences on the signal to noise ratio. Nonetheless, the effects on the signal linewidth and frequency of the affected resonances provide important information on the exchanging species and their interconversion dynamics. This approach is the most important one to probe the majority of biological phenomena because its sub-millisecond to hundreds of milliseconds time window overlaps with the typical timescales of protein conformational changes, (un)folding transitions, biomolecular recognition and binding events, and allosteric transitions.

Therefore, NMR, which has the additional advantage of monitoring biomolecules in their functional environment (in aqueous solution), can probe the microsecond to milliseconds conformational conversions of proteins at the atomic level. There are many NMR experimental approaches that have been developed to probe such motions based on two kinds of experiments: relaxation dispersion (spin lock relaxation and Carr-Purcell-Meiboom-Gill (CPMG)) and exchange spectroscopy (CEST, EXSY). Relaxation dispersion NMR (RD-NMR) monitors fast processes (microseconds to a few milliseconds), whereas EXSY [8–11] or CEST can be used for processes longer than a few milliseconds [12–16]. Among different relaxation dispersion techniques, those based on spinlock (measuring rotating frame relaxation) offer the possibility to quantify exchange rates as fast as 10^5 s^{-1} (tens of microseconds), although in this case the information on the chemical shift of the exchanging site is lost.

RD-NMR techniques are the best tool for the investigation of protein conformational dynamics due to their optimal overlap in timescale. They also allow insight in the nature of the exchanging conformers and their rates. For example, the exchange between two species has contributions to the transverse relaxation time (R_2) that depend on the population of the two states (p_A and p_B), the forward and backward rate constants k_+ and k_-, and the difference in chemical shift between the two species ($\Delta\omega$). Therefore, RD-NMR can resolve the chemical shifts of minimally populated conformations that are normally invisible to any other bulk measurement. For instance, the technique can efficiently detect populations in solution that are as low as a fraction of 1%. In this chapter, we focus on RD-NMR experiments applied to the study of protein folding using the fast folder gpW as example. We illustrate how these techniques can be applied to reveal the chemical shift (thus the structure) of low populated, excited conformational substates such as those that determine the rates of folding, folding upon binding, catalysis, and allosteric transitions. In addition, we describe how to prepare the samples, perform the basic RD-NMR experiments, and analyze the results in terms of a two-site exchange process.

2 Materials

2.1 NMR Samples

1. Uniformly ^{15}N labeled and selectively $^{13}C\alpha$, ^{13}CO, or methyl ^{13}C-labeled 1 mM samples of the protein of interest (here gpW) are required to measure, ^{15}N, $^{1}H^{N}$ $^{13}CH_3$, $^{13}C\alpha$, and ^{13}CO relaxation dispersion profiles (*see* **Note 1**).
2. All the protein samples contained 30μM DSS (4,4-dimethyl-4-silapentane-1-sulfonic acid) for referencing and 10% D_2O for deuterium lock.

2.2 NMR Tubes

1. Regular high-resolution 5 mm NMR tubes.
2. In case of probing dynamics at low temperature or even below zero (in the case of gpW, the experimental temperature was very close to 0 °C), high pressure NMR tubes are used to prevent freezing.

3 Methods

3.1 Theory of Relaxation Dispersion NMR

3.1.1 CPMG Relaxation: The Pulse Sequence

The CPMG relaxation dispersion experiment (Fig. 1) is basically an R_2 spin-relaxation pulse sequence repeated for different frequencies of 180° pulse repetitions (*see* **Note 2**). All these subsequent experiments have in common a fixed total length for the pulse train (T_{CPMG}). After bringing the magnetization into the plane orthogonal to the magnetic field, the CPMG block (τ_{CPMG}–180°–τ_{CPMG}) is repeated N times for the duration of the total time T_{CPMG} (typically 20–40 ms). The pulse sequence block is set after each CPMG block to measure the relevant signals, and each 180° pulse refocuses the magnetization on the transverse plane for the next measurement. The CPMG pulse repetition frequency ($\nu_{CPMG}/2$) determines the timescale at which the relaxation contributions from conformational exchange can be detected: the higher the frequency, the smaller the contributions to R_{2eff} (line broadening) that come from conformational exchange. Plotting R_{2eff} as a function of the ν_{CPMG} produces the relaxation dispersion profile [17] (Fig. 2, *see* **Note 3**). An important part of setting-up all relaxation dispersion experiments is the choice of the appropriate values of ν_{CPMG} as defined in Eqs. (1)–(4). This can be confusing because certain researchers use τ_{CP} (the distance between two consecutive

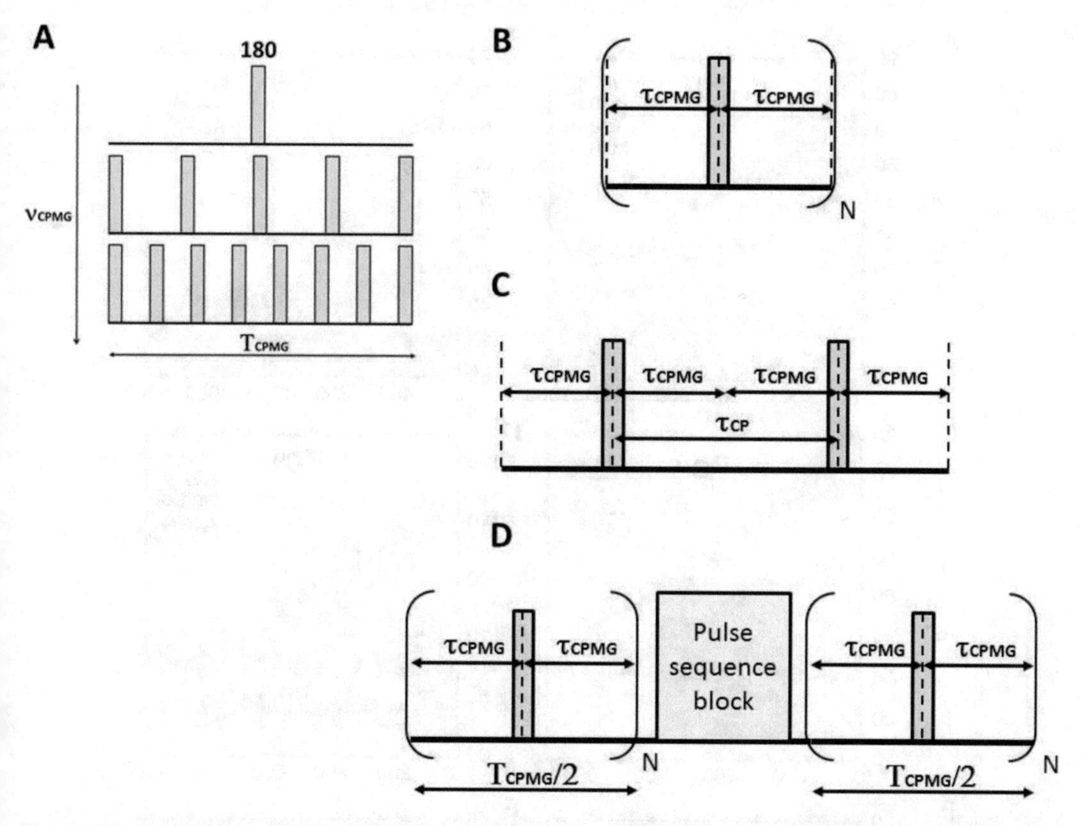

Fig. 1 (**a**) Basic CPMG experiment scheme. Increasing the number of pulses during the relaxation delay T_{CPMG} increases ν_{CPMG}. Definition of the delays τ_{CPMG} (**b**) and τ_{CP} (**c**) in a CPMG relaxation dispersion basic block. (**d**) The CPMG sequence can be separated into two parts to include other pulse sequence blocks, as in relaxation compensated CPMG [19, 39]

180° pulses) and define $\nu_{\mathrm{CPMG}} = 1/(2\tau_{\mathrm{CP}})$, whereas others prefer to use τ_{CPMG} (*see* Fig. 1b) which corresponds to $\tau_{\mathrm{CP}}/2$ (*see* Fig. 1c) and define ν_{CPMG} as $1/(4\tau_{\mathrm{CPMG}})$. In other cases, τ_{CP} is used to denote τ_{CPMG} (*see* **Note 4**). Pulse sequences commonly used for the measurement of relaxation dispersion of ^{1}H [18], ^{15}N [19], and ^{13}C [20] can be found in the literature.

3.1.2 RD Experiments: Mathematical Description

In CPMG relaxation experiments, a train of 180° pulses is applied during acquisition. Defining ν_{CPMG} as the frequency at which a 360° rotation is applied to the magnetization, the dependence of the exchange contribution to R_2 can be described as [21, 22]:

$$R_{2,\mathrm{ex}} = \frac{k_{\mathrm{ex}}}{2} - 2\nu_{\mathrm{CPMG}} \sinh^{-1}\left(\frac{k_{\mathrm{ex}}}{\xi} \sinh \frac{\xi}{4\nu_{\mathrm{CPMG}}}\right) \quad (1)$$

where $k_{\mathrm{ex}} = k_{+}/p_{\mathrm{A}} = k_{-}/p_{\mathrm{B}}$ and

$$\xi = \sqrt{\left(k_{\mathrm{ex}}^{-2} - 4p_{\mathrm{A}}p_{\mathrm{B}}\Delta\omega^{2}\right)} \quad (2)$$

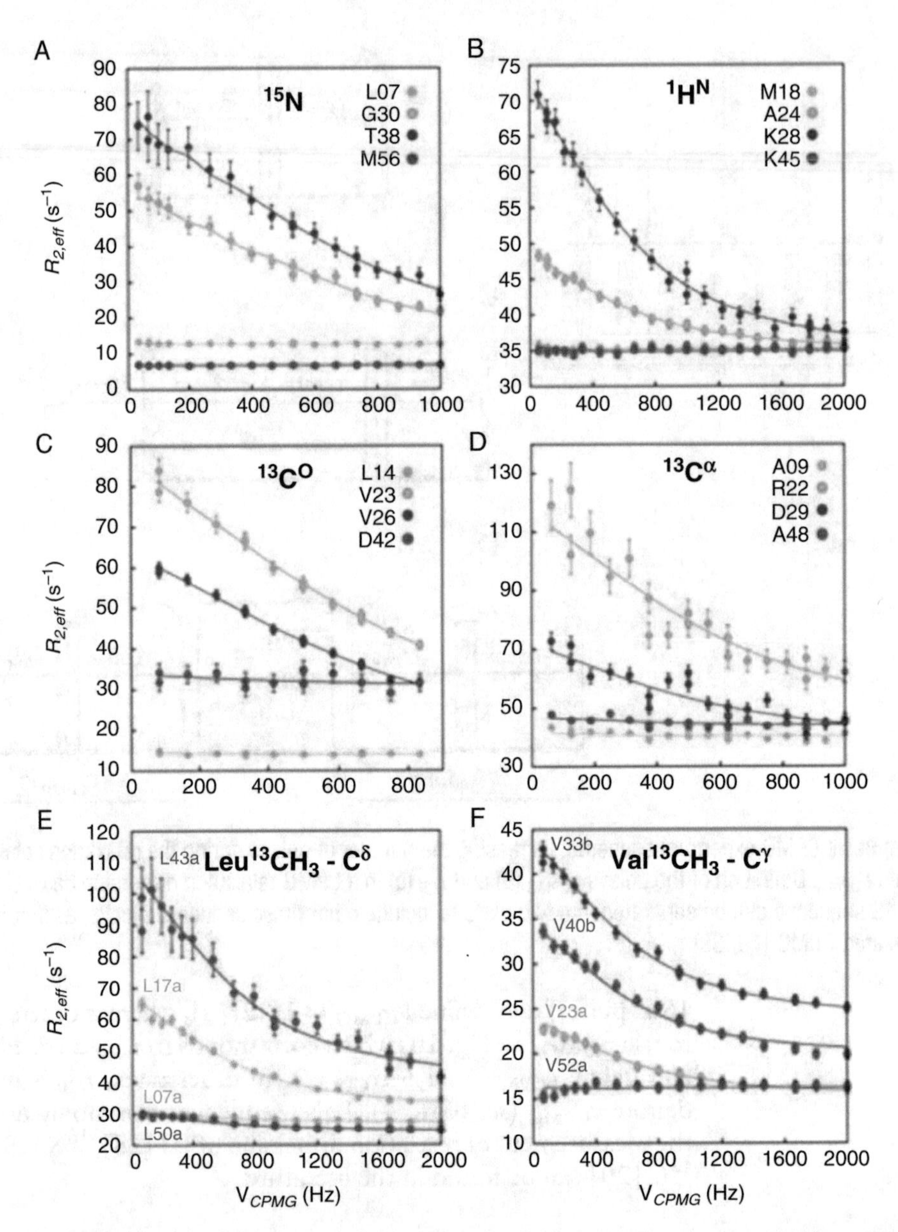

Fig. 2 Exemplary relaxation dispersion profiles for gpW measured on (**a**) ^{15}N, (**b**) ^{1}HN, (**c**) ^{13}C$_{O}$, and (**d**) ^{13}Cα probes, recorded at 1 °C, 11.7 T, showing the existence of sub-millisecond timescale conformational fluctuations in gpW. A flat profile means no detectable dispersion. Panels (**E**) and (**F**) provide examples of relaxation dispersion profiles of methyl groups. Reprinted with permission from *J. Am. Chem. Soc.* 2014, 136, 20, 7444–7451. Copyright 2014 American Chemical Society

In cases where $\nu_{\text{CPMG}} < 0.1k_{\text{ex}}$, the equation reduces to [23]:

$$R_{2,\text{ex}} = \left[1 - \frac{4\nu_{\text{CPMG}}}{k_{\text{ex}}}\right] \frac{p_{\text{A}} p_{\text{B}} \Delta\omega^2}{k_{\text{ex}}} \quad (3)$$

and to:

$$R_{2,\mathrm{ex}} = \frac{p_{\mathrm{A}} p_{\mathrm{B}} \Delta\omega^2}{k_{\mathrm{ex}}} \tag{4}$$

for $\nu_{\mathrm{CPMG}} \ll 0.1k_{\mathrm{ex}}$.

Experimentally, the upper limit for the pulse repetition rate is 10^3 Hz, and this might practically limit measurements to exchange processes of up to 10^4 Hz [24]. In practice, nuclei such as ^{13}C and ^{1}H allow slightly faster repetition rates than ^{15}N (up to 2 kHz vs 1 kHz).

3.1.3 Determining the Exchange Parameters by Fitting CPMG Relaxation Dispersion Curves

The relaxation profile curves depend on several interrelated parameters (pb, $\Delta\omega$, and k_{ex}) and thus are not mathematically well defined. It is thus essential to limit some of the parameters during the fit to ensure proper convergence. The range of each parameter can be established based on empirical values determined independently, or on reasonable physical assumptions. For instance, Eq. (3) (used for very fast conformational exchange) shows clearly that pb, $\Delta\omega$, and k_{ex} cannot be extricated as they are all part of a product. However, global fitting of multiple relaxation dispersion curves arising from signals of different residues within a protein (or different nuclei) can permit the effective determination of the three parameters. This is because one can assume that p_{B} (and consequently p_{A}) and k_{ex} are protein specific, and hence should have the same values for all signals whereas the exchange regime is signal specific because it also depends on the difference in chemical shift between the two exchanging conformations. There are several useful guidelines that one can follow to evaluate the general parameters p_{B} and k_{ex} using composite data from multiple signals (*see* **Notes 5–7**).

The values of $\Delta\omega$ are possibly the most interesting parameters to derive from RD-NMR experiments. This is so because they provide the chemical shift values of the minimally populated excited species that is in exchange with the ground state of the protein. The chemical shifts can then be used to derive structural information of the excited conformational substate (which is otherwise invisible). Because the signal-specific $\Delta\omega$ is obtained from the product $p_A{\cdot}p_B{\cdot}\Delta\omega^2/k_{\mathrm{ex}}$ (Eqs. 1–4), the reliability of their determination depends on their absolute magnitude (should be at least 25 Hz [25]) and on the accuracy of the estimated values of p_{B} and k_{ex}. In cases where $\Delta\omega$ is very small, it is possible to increase their magnitude (and hence their detectability) using multiple quantum versions of the relaxation dispersion experiments (*see* **Note 8**).

The last piece of relevant information is the sign of $\Delta\omega$ for each signal. The sign is required to determine the chemical shift of each signal in the excited state (B) and thus to structurally characterize it. This information cannot be extracted from the fitting because the relaxation dispersion curves depend on $\Delta\omega^2$ (*see* Eqs. 1–4). Hence, the sign for $\Delta\omega$ must be evaluated independently and provided by the user to the fitting program for the analysis (*see* **Note 9**).

3.2 Relaxation Dispersion in the Rotating Frame

For those cases where the conformational exchange is faster than 10^4 Hz, an experiment based on measurements of the off-resonance relaxation rate $R_{1\rho}$ can be attempted. In this case, the presence of a transverse spin-locking field of frequency ω_1, applied off-resonance by Ω, aligns the spins to an effective field B_{eff} $(=\omega e/\gamma)$ oriented at an angle θ from the laboratory frame. Provided that $\Delta\omega/k_{\mathrm{ex}} \ll 1$, the relaxation of the spins along this effective field is given by [26, 27]:

$$R_{1\rho} = R_1 \cos^2\theta + \left(R_2 + R_{2,\mathrm{ex}}\right) \sin^2\theta \tag{5}$$

where

$$R_{2,\mathrm{ex}} = p_{\mathrm{A}} p_{\mathrm{B}} \Delta\omega^2 \frac{k_{\mathrm{ex}}}{k_{\mathrm{ex}}^2 + \omega_{\mathrm{e}}^2} \tag{6}$$

Since Ω depends on the angle θ ($\Omega = \omega_1/\sin\theta$), Eqs. (5) and (6) show that a change in offset makes it possible to resolve faster exchange rates using the same applied radiofrequency power. Practically, the sine scaling of the exchange term in Eq. (5) limits the maximum value of detectable exchange rates to 10^5 Hz.

3.3 Basic NMR Setup

3.3.1 Temperature Calibration of the NMR Probe

Temperature calibration is a must for relaxation dispersion measurements. Often the discrepancy between the real sample temperature and the set temperature can be of several degree Celsius. As a procedure for temperature calibration, we use sealed neat ethylene glycol samples and run a 1D ^{1}H-NMR experiment with constant airflow of 400 L/h. The chemical shift difference between the hydroxyl and methylene protons is linearly dependent on the temperature. 1D proton spectra are measured at a set of temperatures ranging from 265 to 370 K. The chemical shifts as a function of temperature result on a calibration curve that relates the temperature set on the instrument to the actual temperature of the sample. This procedure is described in detailed by Ammann et al. [28].

3.3.2 ^{15}N and ^{3}C Hard Pulse Calibration

^{15}N and ^{13}C pulse widths should be calibrated before the RD-NMR experiments. This can be done using the Bruker standard pulse calibration sample: 100 mM ^{13}C methanol and 100 mM ^{15}N urea in DMSO-d6. A procedure for that is as follows:

- Insert the sample into the magnet and equilibrate at a temperature of 298 K. After tuning and matching, shim the sample using the DMSO deuterium lock.

- Run *decp90* and *decp90f3* pulse programs in the Bruker spectrometer console with single scan and d1 = 5 s. Set the delay d2 to 1/2 J (CH) and 1/2 J (NH) for ^{13}C and ^{15}N, respectively, to produce an antiphase doublet signal.
- For ^{13}C pulse calibration, the methyl proton antiphase doublet signal at 3.2 ppm is calibrated to null point by increasing the pulse *p3* in the *decp90* pulse sequence.
- For ^{15}N pulse calibration, the proton antiphase doublet signal at 5.4 ppm is calibrated to null point by increasing the pulse *p21* in the *decp90f3* pulse sequence.

3.3.3 Chemical Shift Referencing

Proton spectra of the protein samples are referenced to the DSS signal (DSS added to the sample, *see* Subheading 2.1, **item 2**), which is then calibrated to 0 ppm. The DSS residual methyl protons can be used for ^{13}C chemical shift calibration. Once determined, the referencing is applied to two- and three-dimensional spectra for chemical shift assignments.

3.3.4 Chemical Shift Assignments

Relaxation dispersion measurements are often performed on backbone (^{15}N, ^{1}HN,^{13}Cα, ^{13}CO) or methyl (^{1}H or ^{13}C) signals. In the case of the protein gpW, their assignment was achieved from the following experiments: 2D [^{15}N-^{1}H]-HSQC, 2D [^{13}C-^{1}H]-HSQC, 3D-HNCA, 3D-HNCACB, 3D-CBCA(CO)NH, and 3D-HNCO. The ^{1}H and ^{13}C assignment of side chains was achieved by 3D-C(CCO)NH, 3D-[^{15}N]-NOESY-HSQC, 3D-[^{13}C]-NOESY-HSQC, and 3D-HCCH-TOCSY experiments on a uniformly ^{13}C- and ^{15}N-labeled protein.

3.4 The RD-NMR Experiment

The protocol presented here is an extension of the original one developed for ^{15}N-CPMG relaxation dispersion experiments [29]. We include multiple probes (such as ^{15}N, ^{1}H^{N},^{13}C$^{\alpha}$, ^{13}C$_{O}$, and methyl groups) for the study of protein conformational substates.

3.4.1 Setting Up an RD-NMR Experiment

As described in a previous protocol [29], we employ a ^{1}H continuous wave (CW) decoupling sequence [19] for ^{15}N RD-NMR experiments and a sensitivity improved ^{13}C RD-NMR experiment for methyl groups [19, 20] and other ^{13}C nuclei [30]. The protocol is similar for different nuclei. The only differences are the total time T_{CPMG} and the number of CPMG repetitions (N).

- The protein sample is prepared as described in Subheading 2.1 and placed in the magnet to reach the desired temperature.
- Subsequently, the NMR probe is tuned and matched (^{13}C, ^{15}N, and ^{1}H channels), and shimming is performed on the deuterium lock signal of D_2O (10%).

- A simple ^{1}H proton spectrum is used to calibrate the ^{1}H 90° hard pulse. In the Bruker spectrometers, the command *pulsecal* does it automatically.
- 2D [^{15}N-^{1}H]-HSQC and 2D [^{13}C-^{1}H]-HSQC spectra with a number of scans equal to 4 or more, depending on simple concentration, are usually acquired before starting the CPMG RD-NMR experiments.
- A series of pulses have to be calibrated as follows:
  - For ^{1}H, ^{15}N, and ^{13}C relaxation dispersion measurements, CPMG pulses should have a power resulting in a 180° pulse duration not longer than: 20μs for ^{1}H, 30μs for ^{13}C, and 80μs for ^{15}N.
  - For ^{15}N relaxation dispersion measurements, ^{1}H decoupling should have a power equivalent to a hard 90° pulse of 15μs (*see* **Note 10**). ^{15}N must sometimes be decoupled from ^{13}C (even when labeling methyl groups by using ^{13}C glucose labeled in position 1, as the Cα carbon can be labeled). This is obtained by using a power level equivalent to a 90° hard pulse of about 22μs.
  - For ^{1}H relaxation dispersion measurements, ^{15}N must be decoupled during acquisition with a power level equivalent to a 90° hard pulse of about 250μs (*see* **Note 11**).
  - Finally, water suppression is achieved by selective 90° pulses on the water resonance whose typical duration is 1500μs.
- Delays and scans should be set as follows:
  - For the RD-NMR experiments, the delay *d1* is typically set to 2.5 s.
  - For samples of concentrations about 0.5 mM, the number of scans is usually set to either 16 or 32.
  - Dummy scans are set to 256–512, to allow temperature equilibration in case of heating caused by the pulse sequence.
  - Setting up the number of repetitions N (vc_list). The ν_{CPMG} is internally calculated by the pulse sequence from the provided list called "vc_list" in the Bruker spectrometers. The first data point is the reference spectrum with $\nu_{CPMG} = 0$ ($N = 0$), followed by a series of spectra with different ν_{CPMG} (according to the values N in the vc_list). As N is an integer multiplying the minimum number of pulses in the pulse sequence within the total relaxation time T_{CPMG}, the frequency can be calculated as $\nu_{CPMG} = N/(T_{CPMG})$ which implies $\tau_{CPMG} = (T_{CPMG})/4N$ (*see* **Note 12**). For nuclei as ^{15}N, where the pulses are significantly longer than that of ^{1}H (about 80μs, due to ^{15}N's low gyromagnetic ratio), pulse repetition rate is limited by sample heating and instrumental

limitations (*see* **Note 13**). An example of values for the vc_list is 0, 40, 1, 36, 2,32, 3, 28, 4, 26, 5, 24, 6, 22, 7, 20, 8, 18, 9, 16, 10, 14, 11, 12, 22, 2, 40. A few data points are repeated in order to include in the statistical error estimation during the data analysis. Data points are given in sparse order with the intent to average the effect of phenomena which might arise during the long acquisition times of the complete experiment (sample heating due to the CPMG 180° pulses, partial precipitation).

- After acquisition, the split AU program in the Topspin software (from Bruker BioSpin, Rheinstetten, Germany) is used to split the individual 2D planes that correspond to different values of the vc_list.

3.4.2 Data Analysis

All the RD-NMR spectra are processed together with the same number of data points and processing parameters. For this purpose, either Topspin or NMRPipe [31] software can be used. Here we describe the general steps to be taken for data analysis.

- Evaluate peak intensities in each 2D spectrum. NMRPipe uses shell scripts to process the multiple data together and also has built-in functions (such as *seriesTab*) to integrate the peak intensities with respect to the reference spectrum. Peak intensities can also be obtained from Topspin.
- Relaxation dispersion measurements are based on the quenching of exchange contributions in transversal relaxation times. Therefore, a spectrum acquired without any quenching effects must be used as reference. This spectrum is simply obtained by using $N = 0$ in the list of repetitions (vc_list in bruker pulse sequences), which corresponds to skipping the CPMG pulse train (*see* **Note 14**).
- The list of R_2 values as a function of ν_{CPMG}, each labeled with their assignments, is used as the input for the RD analysis that aims to estimate k_{ex}, $\Delta\omega$, and p_{B} using one of the existing RD-NMR data analysis software (*see* **Note 15**).

4 Notes

1. $^{13}CH_3$ selective methyl labeling and selective ^{13}Cα, ^{13}CO labeling can be achieved using [1-^{13}C]glucose and [2-^{13}C]glucose [32], respectively, as the sole carbon source during the overexpression of the recombinant protein in the host (e.g. *E. coli)*. $^{15}NH_4Cl$ is used as the sole nitrogen source in both cases. Selecting an appropriate labeling strategy for the biomolecular system is key for NMR structure and dynamics studies. Several

selective labeling methods have been reported [33] for various types of NMR experiments ranging from sequential assignments to stereo-specific assignments and selective labeling for relaxation experiments [34]. Traditionally, a ^{15}N labeling scheme is used in protein NMR because it is relatively inexpensive and allows direct monitoring of the protein backbone. However, more nuclear probes are often required to overcome signal overlap or single probe ambiguity in determining the exchange parameters. The presence of homonuclear coupling complicates the interpretation of the relaxation dispersion profiles (its effect is not taken into account in Eq. (1)). Therefore, selective labeling schemes have been developed to obtain samples in which the nucleus of interest is isolated from the other scalar coupling partners. In this way, samples with isolated $^{13}C^{\alpha}$, $^{1}H^{\alpha}$, ^{13}CO, and $^{13}CH_3$ have been produced [32, 35, 36]. There are exclusive isotope labeling schemes that can be found in the literature [34]; however, they have to be adapted to the biomolecular system under investigation.

2. In these experiments, the frequency ν_{CPMG} is calculated every two 180° pulses for an easier comparison with rotating frame relaxation experiments.

3. In a typical relaxation dispersion profile as that shown in Fig. 2, the signals that experience a change in chemical shift between the interconverting conformational substates produce a dispersion profile whereas the signals that have an unchanged chemical shift display a flat profile (independent on the CPMG frequency). There are several pulse sequences that are used by the NMR community to perform these experiments. Depending on the biomolecular system under study (large proteins or small-medium size proteins) and the accessibility of the timescales that are in the intermediate to fast exchange regime, one can adopt single quantum or multiple quantum techniques [19, 37, 38] to acquire data. In general, the common practice is to run the experiment using the maximum number of 180° pulse repetitions that can be accommodated into the CPMG block.

4. Using our nomenclature, the two definitions are equivalent and they both define the frequency of 360° rotations in the total relaxation delay T_{CPMG} ($2\tau_{CP}$ or $4\tau_{CPMG}$). This definition allows for direct comparison with relaxation dispersion in the rotating frame, where the frequency corresponds to rotations of 360° and implies that the minimum frequency in a CPMG experiment must contain at least two 180° pulses. A minimum of two pulses allows the creation of two CPMG periods of duration $T_{CPMG}/2$. This can be useful for the design of some pulse sequences: for example, the two CPMG periods can be separated by a block (Fig. 1d) that allows the evolution of

heteronuclear J coupling (as in the case of relaxation compensated sequences [19, 39]). In other cases, a 180° pulse can be introduced to refocus the evolution of unwanted *J* couplings when working without protein perdeuteration (as in the case of ^{1}H amide relaxation dispersion, where a selective 180 REBURP pulse refocuses the evolution of the coupling with Hα [18]). The evolution of *J* couplings can complicate significantly the dispersion profiles and their interpretation, especially when the scalar couplings are of the same order of magnitude as the pulse repetition frequency. This is due to the creation of operators that have very different relaxation behavior. Continuous wave decoupling during the CPMG period can in principle avoid the evolution of J coupling, even if it does not fully simplify the complexity of relaxation [40]. Alternative pulse sequences have been developed, where the contributions from in-phase and antiphase magnetization created from the heteronuclear coupling are quantifiable [19, 39].

5. NMR signals experiencing the largest variations in chemical shift between the interconverting species are likely to be unaffected by low frequency CPMG repetitions since they require fast repetition rates to be refocused. As a consequence, these signals will tend to display relaxation dispersion curves with an initial plateau (corresponding to low values of CPMG pulse repetition rates). Once the value of k_{ex} is known, the presence of an initial plateau on the profile allows for the accurate determination of p_B (if the population is of at least 0.5%) and consequently, $\Delta\omega$, which also contributes to scale the curve [38].
6. Obtaining an accurate value of k_{ex} from the global fitting of multiple signals requires that the relaxation dispersion curves reach a plateau at the highest CPMG frequencies. Practically this means that the pulse repetition is fast enough to completely refocus the exchanging signals. This issue sets the limit for the k_{ex} values that are accessible to the technique, which must be lower than the maximal repetition rate. In cases where k_{ex} is too fast to determine the fast exchange regime even for signals that experience large chemical shift changes upon exchange, the relaxation dispersion curves will not be able to provide accurate values of the parameters.
7. A simple procedure to assess the reliability of the fitted p_B and k_{ex} values involves performing multiple global fits on a grid of fixed values for one of the two parameters (p_B or k_{ex} but not both) and floating the rest. Analysis of the chi-squared (χ^2) of the fit as a function of the fixed parameter will indicate if the parameter can be determined from the given data (the curve produces a clear minimum at the exact value) and of its statistical significance (how deep the minimum is). A good example of

this analysis can be found in the study of the protein gpW, whose very fast k_{ex} was determined using this procedure [17].

8. Multiple quantum coherences have intrinsically higher frequencies than single quantum ones. Therefore, it is possible to increase the sensitivity to detect $\Delta\omega$ by measuring multiple quantum relaxation dispersion. A simple way to increase the magnitude of ^{1}H $\Delta\omega$ is to measure the relaxation dispersion of the double quantum ^{1}H,^{15}N coherence, in which the much larger $\Delta\omega$ of ^{15}N contributes to the overall $\Delta\omega$ [37, 41]. The combination of single quantum and double quantum relaxation dispersion experiments at multiple fields results in the robust determination of $\Delta\omega$ and exchange parameters [37, 41]. Another example is the use of double or triple quantum ^{1}H relaxation dispersion applied to methyl groups, which enhances the $\Delta\omega$ magnitude by a factor of 9 [25]. Generally, enhancement of $\Delta\omega$ (for example, by using triple quantum instead of single quantum experiments) not only scales the relaxation dispersion curves but it is also likely to change the exchange regime, thus providing multiple curves from which more robust estimates of the set of parameters can be extracted [25].
9. Two methods are commonly used for the determination of the sign of $\Delta\omega$. One method is based on the comparison of the chemical shifts measured by two different experiments performed at the same field; for example, HSQC and HMBC spectra can provide the sign of ^{15}N $\Delta\omega$. The other method is based on the comparison of spectra measured at different fields. Both methods are described in detail in [42].
10. Here we recommend using a pseudo 3D version of the ^{15}N-CPMG sequence with ^{1}H CW decoupling [19] to avoid the evolution of coupling during the CPMG period.
11. For measurements on amide protons of non-perdeuterated proteins, coupling to Hα can be quenched by a selective 180° REBURP pulse centered on the ^{1}H amide region. This pulse should be rather selective to avoid water irradiation, and with a duration of at least 1500μs.
12. For example, given a constant relaxation time T_{CPMG} of 20 ms, the minimum frequency ($N = 1$) is $1/0.02 = 50$ Hz; this corresponds to introducing two 180° pulses during 20 ms and a τ_{CPMG} of 5 ms (minus the half duration of a 180° pulse). As N increases, the delay between two consecutive pulses decreases as $(T_{CPMG})/4N$-N$p_{180}/2$, where p_{180} is the duration of the CPMG 180° pulse.
13. In practice, τ_{CPMG} cannot be shorter than 200μs, resulting in maximum frequencies for ^{15}N of about 1 kHz that are smaller than for ^{1}H (2 kHz) or ^{13}C (1.5 kHz). The upper value of the

frequency is determined by T_{CPMG}. Longer values allow sampling slower dynamics, but the gain in long timescale information is counterbalanced by signal loss due to transverse relaxation. Typical T_{CPMG} for ^{15}N relaxation dispersion experiments in proteins are 20–40 ms.

14. As the intensity of the signal is inversely proportional to R_2, the dispersion curve can be obtained by comparing the intensities of spectra acquired with increasing CPMG frequencies with the reference spectrum [40, 43]:

$$R_2(\tau_{CP}) = -\frac{1}{T_{CPMG}} \ln\left(\frac{I(\tau_{CP})}{I_0}\right) \quad (7)$$

where $I(\tau_{CP})$ represents peak intensity at each τ_{CP} delay, and I_0 is the intensity in the reference spectrum. The error for each point is calculated as [43, 44]:

$$\sigma(\tau_{CP}) = \frac{1}{T_{CPMG}}\left(\frac{\varepsilon_I}{I(\tau_{CP})}\right) \quad (8)$$

15. There are several software programs available for the analysis of the RD-NMR experiments: CHEMEX (https://github.com/gbouvignies/chemex/releases), CATIA (https://www.ucl.ac.uk/hansen-lab/catia/), CPMGfit (http://www.palmer.hs.columbia.edu/software/cpmgfit.html), relax [45], GLOVE [43], and NESSY [46]. These programs fit the data assuming different exchange models (mostly exchange between two-states) to extract the relevant exchange parameters. Often the relaxation dispersion profiles from a single probe (^{15}N or ^{13}C) are not sufficient to distinguish between two-state or complex exchange models [38]. To circumvent this problem, multiple probes (^{15}N, $^{1}H_N$, $^{13}C_\alpha$, $^{13}C_O$, and $^{13}C^\beta$ of the side chains like methyl groups) and/or fields are used in a global fit. Generally, including multiple experimental datasets results in a more robust analysis, and hence in more accurate values of the optimal exchange parameters [17, 47].

Acknowledgments

This work was funded by Advanced Grant ERC-2012-ADG-323059 from the European Research Council to V.M. V.M. also acknowledges support from the W.M. Keck foundation and the National Science Foundation (grants NSF-MCB-1616759 and NSF-CREST-1547848).

References

1. Lipari G, Szabo A (1982) Model-free approach to the interpretation of nuclear magnetic resonance relaxation in macromolecules. 1. Theory and range of validity. J Am Chem Soc 104:4546–4559
2. Lipari G, Szabo A (1982) Model-free approach to the interpretation of nuclear magnetic resonance relaxation in macromolecules. 2. Analysis of experimental results. J Am Chem Soc 104:4559–4570
3. Ban D, Sabo TM, Griesinger C et al (2013) Measuring dynamic and kinetic information in the previously inaccessible supra-τ(c) window of nanoseconds to microseconds by solution NMR spectroscopy. Molecules 18:11904–11937
4. Kleckner IR, Foster MP (2011) An introduction to NMR-based approaches for measuring protein dynamics. Biochim Biophys Acta 1814:942–968
5. Tolman JR (2002) A novel approach to the retrieval of structural and dynamic information from residual dipolar couplings using several oriented media in biomolecular NMR spectroscopy. J Am Chem Soc 124:12020–12030
6. Meiler J, Prompers JJ, Peti W et al (2001) Model-free approach to the dynamic interpretation of residual dipolar couplings in globular proteins. J Am Chem Soc 123:6098–6107
7. Lange OF, Lakomek N-A, Farès C et al (2008) Recognition dynamics up to microseconds revealed from an RDC-derived ubiquitin ensemble in solution. Science 320:1471–1475
8. Jeener J, Meier BH, Bachmann P et al (1979) Investigation of exchange processes by two-dimensional NMR spectroscopy. J Chem Phys 71:4546–4553
9. Palmer AG 3rd, Kroenke CD, Loria JP (2001) Nuclear magnetic resonance methods for quantifying microsecond-to-millisecond motions in biological macromolecules. Methods Enzymol 339:204–238
10. Fejzo J, Westler WM, Macura S et al (1990) Elimination of cross-relaxation effects from two-dimensional chemical-exchange spectra of macromolecules. J Am Chem Soc 112:2574–2577
11. Fejzo J, Westler WM, Macura S et al (1991) Strategies for eliminating unwanted cross-relaxation and coherence-transfer effects from two-dimensional chemical-exchange spectra. J Magn Reson 92:20–29
12. Bouvignies G, Vallurupalli P, Kay LE (2014) Visualizing side chains of invisible protein conformers by solution NMR. J Mol Biol 426:763–774
13. Forsén S, Hoffman RA (1963) Study of moderately rapid chemical exchange reactions by means of nuclear magnetic double resonance. J Chem Phys 39:2892–2901
14. Fawzi NL, Ying J, Ghirlando R et al (2011) Atomic-resolution dynamics on the surface of amyloid-β protofibrils probed by solution NMR. Nature 480:268–272
15. Rennella E, Huang R, Velyvis A et al (2015) 13CHD2–CEST NMR spectroscopy provides an avenue for studies of conformational exchange in high molecular weight proteins. J Biomol NMR 63:187–199
16. Vallurupalli P, Bouvignies G, Kay LE (2012) Studying “invisible” excited protein states in slow exchange with a major state conformation. J Am Chem Soc 134:8148–8161
17. Sanchez-Medina C, Sekhar A, Vallurupalli P et al (2014) Probing the free energy landscape of the fast-folding gpW protein by relaxation dispersion NMR. J Am Chem Soc 136:7444–7451
18. Ishima R, Torchia DA (2003) Extending the range of amide proton relaxation dispersion experiments in proteins using a constant-time relaxation-compensated CPMG approach. J Biomol NMR 25:243–248
19. Hansen DF, Vallurupalli P, Kay LE (2008) An improved 15N relaxation dispersion experiment for the measurement of millisecond time-scale dynamics in proteins. J Phys Chem B 112:5898–5904
20. Lundström P, Vallurupalli P, Religa TL et al (2007) A single-quantum methyl 13C-relaxation dispersion experiment with improved sensitivity. J Biomol NMR 38:79–88
21. Bloom M, Reeves LW, Wells EJ (1965) Spin echoes and chemical exchange. J Chem Phys 42:1615–1624
22. Mulder FA, van Tilborg PJ, Kaptein R et al (1999) Microsecond time scale dynamics in the RXR DNA-binding domain from a combination of spin-echo and off-resonance rotating frame relaxation measurements. J Biomol NMR 13:275–288. https://doi.org/10.1023/A:1008354232281
23. Luz Z, Meiboom S (1963) Nuclear magnetic resonance study of the protolysis of

trimethylammonium ion in aqueous solution—order of the reaction with respect to solvent. J Chem Phys 39:366–370

24. Farber PJ, Mittermaier A (2015) Relaxation dispersion NMR spectroscopy for the study of protein allostery. Biophys Rev 7:191–200
25. Yuwen T, Vallurupalli P, Kay LE (2016) Enhancing the sensitivity of CPMG relaxation dispersion to conformational exchange processes by multiple-quantum spectroscopy. Angew Chem Int Ed Engl 55:11490–11494
26. Davis DG, Perlman ME, London RE (1994) Direct measurements of the dissociation-rate constant for inhibitor-enzyme complexes via the T1 rho and T2 (CPMG) methods. J Magn Reson B 104:266–275
27. Deverell C, Morgan RE, Strange JH (1970) Studies of chemical exchange by nuclear magnetic relaxation in the rotating frame. Mol Phys 18:553–559
28. Ammann C, Meier P, Merbach A (1982) A simple multinuclear NMR thermometer. J Magn Reson 46:319–321
29. Ishima R (2013) CPMG relaxation dispersion. In: Methods in molecular biology. Humana, New York, pp 29–49
30. Lundström P, Hansen DF, Kay LE (2008) Measurement of carbonyl chemical shifts of excited protein states by relaxation dispersion NMR spectroscopy: comparison between uniformly and selectively (13)C labeled samples. J Biomol NMR 42:35–47
31. Delaglio F, Grzesiek S, Vuister GW et al (1995) NMRPipe: a multidimensional spectral processing system based on UNIX pipes. J Biomol NMR 6:277–293
32. Lundström P, Teilum K, Carstensen T et al (2007) Fractional 13C enrichment of isolated carbons using [1-13C]- or [2-13C]-glucose facilitates the accurate measurement of dynamics at backbone Cα and side-chain methyl positions in proteins. J Biomol NMR 38:199–212
33. Tugarinov V, Kanelis V, Kay LE (2006) Isotope labeling strategies for the study of high-molecular-weight proteins by solution NMR spectroscopy. Nat Protoc 1:749–754
34. Lundström P, Ahlner A, Blissing AT (2012) Isotope labeling methods for relaxation measurements. In: Advances in experimental medicine and biology. Springer Nature, Switzerland, pp 63–82
35. Lundström P, Hansen DF, Vallurupalli P et al (2009) Accurate measurement of alpha proton chemical shifts of excited protein states by relaxation dispersion NMR spectroscopy. J Am Chem Soc 131:1915–1926
36. Goto NK, Gardner KH, Mueller GA et al (1999) A robust and cost-effective method for the production of Val, Leu, Ile (delta 1) methyl-protonated 15N-, 13C-, 2H-labeled proteins. J Biomol NMR 13:369–374
37. Korzhnev DM, Kloiber K, Kay LE (2004) Multiple-quantum relaxation dispersion NMR spectroscopy probing millisecond time-scale dynamics in proteins: theory and application. J Am Chem Soc 126:7320–7329
38. Korzhnev DM, Salvatella X, Vendruscolo M et al (2004) Low-populated folding intermediates of Fyn SH3 characterized by relaxation dispersion NMR. Nature 430:586–590
39. Loria JP, Patrick Loria J, Rance M et al (1999) A relaxation-compensated Carr–Purcell–Meiboom–Gill sequence for characterizing chemical exchange by NMR spectroscopy. J Am Chem Soc 121:2331–2332
40. Mulder FAA, Skrynnikov NR, Hon B et al (2001) Measurement of slow (μs–ms) time scale dynamics in protein side chains by15N relaxation dispersion NMR spectroscopy: application to Asn and Gln residues in a cavity mutant of T4 lysozyme. J Am Chem Soc 123:967–975
41. Millet O, Patrick Loria J, Kroenke CD et al (2000) The static magnetic field dependence of chemical exchange linebroadening defines the NMR chemical shift time scale. J Am Chem Soc 122:2867–2877
42. Skrynnikov NR, Dahlquist FW, Kay LE (2002) Reconstructing NMR spectra of "invisible" excited protein states using HSQC and HMQC experiments. J Am Chem Soc 124:12352–12360
43. Sugase K, Konuma T, Lansing JC et al (2013) Fast and accurate fitting of relaxation dispersion data using the flexible software package GLOVE. J Biomol NMR 56:275–283
44. Ishima R, Torchia DA (2005) Error estimation and global fitting in transverse-relaxation dispersion experiments to determine chemical-exchange parameters. J Biomol NMR 32:41–54
45. Morin S, Linnet TE, Lescanne M et al (2014) relax: the analysis of biomolecular kinetics and thermodynamics using NMR relaxation dispersion data. Bioinformatics 30:2219–2220
46. Bieri M, Gooley PR (2011) Automated NMR relaxation dispersion data analysis using NESSY. BMC Bioinformatics 12:421
47. Hansen DF, Vallurupalli P, Lundström P et al (2008) Probing chemical shifts of invisible states of proteins with relaxation dispersion NMR spectroscopy: how well can we do? J Am Chem Soc 130:2667–2675

Part III

Single-Molecule Spectroscopy Techniques

Chapter 12

Labeling of Proteins for Single-Molecule Fluorescence Spectroscopy

Franziska Zosel, Andrea Holla, and Benjamin Schuler

Abstract

Single-molecule fluorescence spectroscopy has become an important technique for studying the conformational dynamics and folding of proteins. A key step for performing such experiments is the availability of high-quality samples. This chapter describes a simple and widely applicable strategy for preparing proteins that are site-specifically labeled with a donor and an acceptor dye for single-molecule Förster resonance energy transfer (FRET) experiments. The method is based on introducing two cysteine residues that are labeled with maleimide-functionalized fluorophores, combined with high-resolution chromatography. We discuss how to optimize site-specific labeling even in the absence of orthogonal coupling chemistry and present purification strategies that are suitable for samples ranging from intrinsically disordered proteins to large folded proteins. We also discuss common problems in protein labeling, how to avoid them, and how to stringently control sample quality.

Key words Labeling, Maleimide, Cysteine, RP-HPLC, Ion-exchange chromatography, Fluorescent dye, Fluorescence spectroscopy, Single-molecule detection, Förster resonance energy transfer, FRET, Intrinsically disordered proteins, Protein folding

1 Introduction

Single-molecule Förster resonance energy transfer (FRET) has become a valuable tool for studying biomolecular structures, dynamics, and interactions [1]. As a single-molecule technique, it has the power to detect rare events, dynamics, and conformations that might otherwise be hidden in the ensemble average. FRET relies on labeling the biomolecules of interest with two fluorescent dyes with suitable spectral properties, so that radiationless energy transfer from the donor fluorophore to the acceptor fluorophore can occur. The transfer process relies on the coupling of the transition dipoles of the two fluorophores, and the resulting efficiency of energy transfer, $E(r)$, depends on the inverse sixth power of the distance, r, between the dye molecules, $E(r) = 1/(1 + (r/R_0)^6)$. This dependence makes FRET ideally suited for interrogating even

Victor Muñoz (ed.), *Protein Folding: Methods and Protocols*, Methods in Molecular Biology, vol. 2376, https://doi.org/10.1007/978-1-0716-1716-8_12,

small distance changes around the Förster radius, R_0, the distance where the transfer efficiency E equals ½ (*see* **Note 1**). Typical values of R_0 for single-molecule dyes are in the range of 4–7 nm, which makes FRET a sensitive reporter for processes on the biomolecular length scale [2]. Correspondingly, a wide range of questions have been addressed with single-molecule FRET spectroscopy [3], including the mechanisms of molecular machines [4], protein–nucleic acid interactions [5, 6], enzymatic reactions [7], protein and RNA folding [8–10], as well as the distance distributions and dynamics of unfolded and intrinsically disordered proteins (IDPs) [11–13]. Single-molecule experiments provide access to a wealth of information on biomolecular dynamics over a wide range of timescales; useful techniques are correlation spectroscopy, the analysis of broadening and exchange between subpopulations in FRET efficiency histograms, and kinetics from fluorescence trajectories of immobilized molecules or from microfluidic mixing [9]. By combining wavelength-selective photon detection with fluorescence polarization sensitivity and time-correlated single-photon counting, a multitude of complementary observables become accessible that turn single-molecule FRET into a versatile spectroscopic toolbox [14].

However, recording high-quality single-molecule data requires high-quality samples. Quality here primarily refers to labeling specificity and sample purity. Although spectroscopic approaches such as alternating excitation [15] enable a lot of sample heterogeneity to be resolved, single-molecule FRET experiments greatly benefit from the preparation of pure samples to avoid experimental artifacts, to reduce measurement time, and to be able to detect and resolve even small subpopulations unequivocally. Fortunately, sample consumption is exceedingly low, so very stringent purification procedures can be applied to accomplish this goal. To achieve site-specific labeling, many strategies have been described [16], including the labeling of cysteine residues [17, 18], the introduction of unnatural amino acids [19], or the use of enzymatic tags [20]. With the same attachment chemistry for both fluorophores, it is often not possible to resolve labeling permutants (each dye can be on either labeling site). This lack of complete site-specificity may not be essential if FRET is used as a spectroscopic ruler with a single pair of donor and acceptor, since the distance between the two dyes is the same for both permutants. However, variations in the local environment of the fluorophores can cause site-specific variations in quantum yields [21], and in advanced applications, such as 3-color-FRET [22] or FRET in combination with a fluorescence quencher [23], it is crucial to know and control the exact location of both dyes.

In spite of the wide range of sophisticated strategies for labeling with orthogonal chemistry that are now available, the most commonly used approach is still the derivatization of two reactive cysteine residues with maleimide-functionalized dyes, primarily owing to its simplicity. In the following protocol, we describe a reliable strategy for this method that we use routinely in our laboratory and that (starting from purified unlabeled protein) usually yields samples with excellent quality in a few days. Our implementation of the method offers several advantages:

1. The strategy is simple and can be applied to a wide range of proteins, from IDPs to larger folded proteins.
2. Maleimide coupling can be carried out in a short time (<2 h) and with high yields (>90%).
3. By engineering the reactivity of the two cysteine residues and/or stringent chromatographic purification steps, even labeling permutants can often be separated.
4. Sample loss is reduced by avoiding concentration steps.
5. When RP-HPLC is used as purification method, the labeled protein is obtained in lyophilized form, simplifying storage and use in downstream applications.

We note that some of the procedures described here can also be useful for experiments closely related to single-molecule FRET, such as photoinduced electron transfer (PET) [24], or for ensemble methods requiring site-specific labeling with fluorophores, spin labels, or other chemical moieties.

2 Materials

Prepare all solutions using ultrapure water and analytical-grade reagents and test them for fluorescence impurities before use. Prepare and store all reagents at room temperature (unless indicated otherwise).

2.1 Instruments

1. HPLC system, equipped with a diode array detector and fluorescence module (e.g., Agilent 1200 series, Agilent Technologies). A multi-wavelength absorbance detector is the minimum requirement.
2. FPLC system, equipped with two pumps, a sample pump, a mixer, and a multi-wavelength absorbance detector (e.g., ÄKTA avant, Cytiva).
3. C18-functionalized reversed-phase chromatography columns (e.g., Reprosil Gold 200 5 μm, Dr. Maisch; Xterra RP18 5 μm, Waters; Sunfire 3.5 μm, Waters), semi-preparative scale (4.6 × 250 mm, column volume 4.15 mL). For larger proteins, C4 or C8 columns can be useful. A 2-μm inline metal frit should be used to prevent larger particles from entering and damaging the column.

4. Desalting columns (e.g., HiTrap, 5 mL, Cytiva) and ion-exchange chromatography columns (e.g., Mono Q/S, 1 mL, Cytiva).
5. UV-Vis spectrophotometer.
6. Ultrasonic bath.
7. Vacuum filtration system for buffers.
8. Lyophilizer or centrifugal evaporator with cold trap ("SpeedVac").
9. Optional: Anaerobic atmosphere (e.g., nitrogen glove box).

2.2 Reagents

1. 1 mL of a 1 M solution of Tris(2-carboxyethyl)phosphine (TCEP) in water. Adjust with sodium hydroxide to pH 7. Store in aliquots at −20 °C.
2. 1 mL of a 1 M solution of dithiothreitol (DTT) in water. Store in aliquots at −20 °C and minimize room temperature exposure to avoid oxidation.
3. RP-HPLC buffer A: 0.5 L of water with 0.1% (v/v) trifluoroacetic acid (TFA). Prepare freshly, do not use for more than 2 days. Stir to mix components properly (*see* **Note 2**).
4. RP-HPLC buffer B: 0.5 L of 100% acetonitrile (ACN).
5. Ion-exchange binding buffer (low salt): 0.5 L of 20 mM potassium phosphate pH 7.2, 20 mM potassium chloride, filtered and degassed. Optional: Add 0.001% Tween 20 after filtration (*see* **Note 3**).
6. Ion-exchange elution buffer (high salt): 0.5 L of 20 mM potassium phosphate pH 7.2, 1 M potassium chloride, filtered and degassed. Optional: Add 0.001% Tween 20 after filtration (*see* **Note 3**). Ideally, the ion-exchange buffer system should fulfill all the criteria listed under *Labeling buffer*, so the protein can be used for labeling directly as it elutes from the column.
7. Dimethyl sulfoxide, anhydrous (DMSO): Store desiccated.
8. Dimethyl formamide, anhydrous (DMF): Store desiccated.
9. Dissolve 1 mg of maleimide-functionalized dye (*see* Subheading 3.1.2) in 200 μL DMSO and sonicate in ultrasonic bath for 10 min to monomerize the dye. Split into 10–20 μL aliquots and lyophilize to obtain 50–100 μg aliquots. Store at −20 °C. Lyophilized dye can be stored for >1 year at −20 °C without significant loss in reactivity.
10. Labeling buffer (*for RP-HPLC*): e.g., 0.1 M potassium phosphate, pH 7.0. Prepare 10 mL by mixing 0.615 mL 1 M K_2HPO_4 and 0.385 mL 1 M KH_2PO_4 with 9 mL of water, followed by degassing. Store at 4 °C. Addition of salt or glycerol is possible, as well as the use of other buffer systems (e.g., HEPES). The pH should be between 7.0 and 7.4 in

order that a sufficient fraction of cysteine residues ($pK_a = 8.3–8.5$) are in the reactive thiolate anion form but no cross-reactivity with amines occurs [25].

The following chemicals should be avoided: (1) primary amines in the buffer (e.g., Tris), as they exhibit some reactivity towards maleimides; (2) thiols (DTT, 2-mercaptoethanol), as they compete with the cysteine for the maleimide-functionalized dye; and (3) TCEP, as it can react with maleimides [17, 26] and undergo other side reactions [27].

For ion-exchange chromatography, labeling can be performed directly in the elution buffer if the buffer system meets the abovementioned criteria.

11. Denaturing labeling buffer: 10 mL of 6 M GdmCl and 0.1 M potassium phosphate, pH 7.0. The true pH of GdmCl-containing solutions can be obtained according to Ref. 28. This buffer should only be used if efficient refolding of the protein of interest is possible and labeling under native conditions is not feasible, e.g., because of protein aggregation or surface interactions.
12. Liquid nitrogen.
13. 10 mg/mL of trypsin protease in 1 mM HCl. Store in aliquots at −80 °C.
14. Trypsin digest buffer: 50 mM HEPES, pH 7.5, and 100 mM NaCl. Store at 4 °C.

3 Methods

3.1 Selection of Labeling Sites and Fluorophores

3.1.1 Design Considerations for Introducing Cysteine Residues

1. Remove solvent-accessible cysteine residues from the protein by site-directed mutagenesis, e.g., by replacing them with serine residues. Check the stability and/or functionality of the modified protein if necessary.
2. Introduce two cysteine residues at suitable sites (e.g., replacing serine) by site-directed mutagenesis. For folded proteins, labeling sites should be solvent-exposed and separated by a distance that maximizes the change in transfer efficiency during the process of interest. A computational tool is available to identify labeling sites for best resolution [29]. In IDPs and unfolded proteins, labeling sites should be spaced ~40–100 amino acids apart, depending on the amino acid sequence and residual structure [30].
3. Avoid labeling sites where dyes can come into direct contact to prevent self-quenching [31]. Tryptophan residues in close proximity to the dyes can also quench fluorescence by photo-induced electron transfer [21, 32] and can complicate the quantitative analysis of transfer efficiencies (*see* Fig. 1c) and fluorescence correlation experiments [21]. In IDPs, such quenching effects can modulate the signal up to a sequence separation of ~20 amino acids between dye and quencher

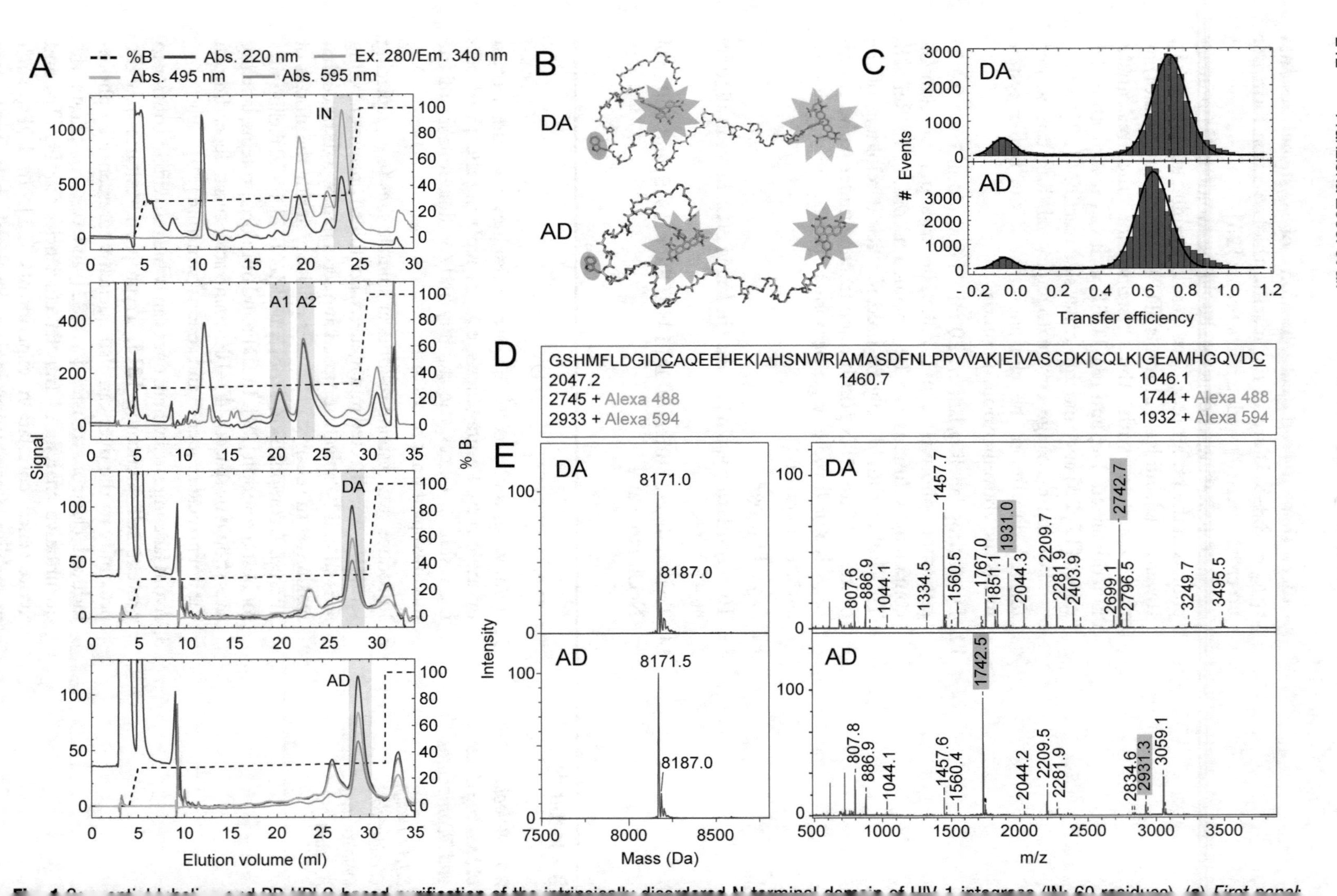

Fig. 1 Sequential labeling and RP-HPLC based purification of the intrinsically disordered N-terminal domain of HIV-1 integrase (IN; 60 residues). (a) *First panel*

(20 and 23 mL) and double-labeled protein (31 mL). *Third panel*: Single-labeled protein from peak A1 was reacted with an equimolar amount of Alexa 488, yielding sample DA. *Fourth panel*: Single-labeled protein from peak A2 was reacted with an equimolar ratio of Alexa 488, yielding sample AD. (**b**) Backbone representation of IN DA and AD products, with dyes attached (Alexa 488: green, Alexa 594: orange). A Trp residue (brown) can quench the dye at position 11. (**c**) Single-molecule measurements of IN DA and AD. The transfer efficiency peak of AD is shifted to lower values owing to quenching of the acceptor dye in this construct [21]. The peak below zero transfer efficiency arises from molecules lacking an active acceptor dye (the shift to $E < 0$ is caused by the correction for direct excitation of the acceptor). (**d**) Sequence of IN, with tryptic fragments indicated. Additionally, the expected masses upon reaction with either Alexa 488 (+698 Da) or Alexa 594 (+886 Da) are given. The cysteine residues used for labeling are underlined. The reactivity of the other cysteine residues was suppressed by the complexation with Zn^{2+} ions. (**e**) *Left*: ESI-MS spectra of both donor–acceptor-labeled IN variants from (**a**) (expected mass: 8170.5 Da). Additional peaks with a mass of +16 suggests methionine oxidation. *Right*: Corresponding MALDI-MS spectra after trypsin digest. In the sample from peak DA, only the N-terminal fragment is detected with an Alexa 488 modification (green), and only the C-terminal fragment with an Alexa 594 modification (orange). The peaks from the other permutation are absent. In the sample from peak AD, this pattern is switched, confirming site-specific labeling in both cases

[33]. If necessary, replace tryptophan by phenylalanine by site-directed mutagenesis. Generally, rhodamine and oxazine dyes are more susceptible to quenching than cyanine dyes [34].

4. To maximize site-specific labeling, the reactivity of the individual cysteine residues can be tuned by modulating the pK_a of the thiol group [18, 35]. Positive charges in the vicinity will lower the pK_a and increase the population of the reactive sulfhydryl anion at pH 7. Conversely, negative charges near the sulfhydryl group decrease its reactivity by increasing its pK_a. Prediction tools for cysteine reactivity, pK_a, and solvent accessibility exist for folded proteins [36, 37]. For unfolded proteins, a cysteine residue at the negatively charged C-terminus will have a lower reactivity than a cysteine at the N-terminus. The difference in reactivity can be further increased by introducing charged residues as immediate neighbors. Under these conditions, adding sub-stoichiometric concentrations of dye usually results in the preferential modification of one cysteine residue and increases the overall yield of site-specifically labeled protein.

3.1.2 Choice of Dye Pair

Fluorophores suitable for single-molecule FRET spectroscopy are characterized by high photostability and quantum yield. Purchase maleimide-functionalized dyes to label the sulfhydryl groups of cysteine residues within the protein of interest. Among the best-established labels are the Alexa Fluor dye series (Thermo Fisher Scientific), particularly Alexa Fluor 488, 594, and 647; the ATTO dye series (ATTO-TEC); and the cyanine dyes (Cytiva, Thermo Fisher Scientific). The CF dye series (Biotium) includes suitable and very hydrophilic far-red dyes, in particular CF640R, CF660R, and CF680R. Some "self-healing" cyanine dyes are also commercially available (Lumidyne Technologies), offering higher photostability than their unmodified precursors [38]. Select the dyes according to the required spectral properties (e.g., excitation and detection wavelengths) and the expected inter-dye distance. The Förster radius of the dye pair should match the expected distances to yield maximum transfer efficiency resolution. When working in a high-fluorescence background environment (e.g., inside live cells), using longer excitation wavelengths has proven useful to reduce problems with background luminescence or cellular autofluorescence [39, 40]. Furthermore, take into account dye charge in the selection of the FRET pair: when charge interactions are governing the behavior of a protein, select a dye pair that perturbs the overall charge distribution as little as possible. The retention of dyes on RP-HPLC columns can be used as an indicator for the hydrophobicity of fluorophores and can thus aid the selection for hydrophilicity [41].

3.2 Preparing the Protein for Labeling

Purify the protein of interest (with two cysteine residues for labeling) according to established protocols. Include reducing agent during the purification to avoid disulfide-mediated adduct formation and to maintain the reduced state of the cysteine residues. The requirements for efficient labeling are: (1) maintaining the reactive reduced state of the cysteine residues; (2) removing impurities and reducing agents that interfere with the labeling reaction; and (3) obtaining the protein at high concentration (>10 μM) to maximize the labeling rate (*see* **Note 4**). With the following protocols, protein samples meeting these requirements can be prepared efficiently in a single chromatography step (based on RP-HPLC or ion-exchange chromatography). The respective chromatography can also be used for further purification steps of the dye-labeled protein, using similar elution gradients throughout. The protocols are optimized for initial protein amounts of 0.1–2 mg (RP-HPLC) or 3–50 mg (ion-exchange chromatography).

3.2.1 RP-HPLC–Based Purification

RP-HPLC separates molecules by hydrophobicity. After chromatography, the protein is lyophilized and can readily be dissolved in any buffer to any concentration, which eliminates the need for buffer exchange or concentration steps and greatly enhances the overall yield (*see* **Note 5**). The low pH (≈2) of the mobile phase and the absence of oxygen during lyophilization keep the cysteine residues reduced until the protein is dissolved for labeling. As the protein is usually denatured during HPLC, this approach is useful for proteins that can be refolded or do not require refolding (peptides, IDPs) and works best for peptides and proteins with a mass <15 kDa.

The RP-HPLC gradient is designed to separate the protein of interest from impurities in a minimum amount of time. A useful starting point for the purification of most proteins is given below.

Step 1: 1 column volume (CV) at 5% B (*see* **Note 6**)—elution of all molecules that do not interact with the column (salts, reducing agents, etc.).

Step 2: 1 CV at 10% B—elution of all molecules that interact weakly with the column.

Step 3: 6 CVs with a gradient from 10% to 60% B—elution of the protein of interest (*see* **Notes 7** and **8**).

Step 4: 1 CV at a 100% B—elution of tightly bound molecules.

Estimate the percentage of buffer B where the protein elutes (%B_{el}) by calculating %B 1 CV before the detection of the protein in the absorbance channel (include the volume of the injection loop in the calculation). Adjust the gradient by increasing %B at step 2 and making the gradient in step 3 shallower. As a rule of thumb, a gradient from %B_{el}−15% to %B_{el}+5% over 5 CVs yields good separation; the protein should elute at the end of the gradient to achieve a narrow peak. If there are impurities eluting close to the protein of interest, a shallower gradient of 10% over 5 CVs is recommended.

The semi-preparative columns used here are operated with a flow rate of 1 mL/min.

A standard RP-HPLC purification protocol is as follows:

1. Use a sample that is already enriched in the protein of interest (>50%, e.g., by immobilized metal ion-affinity chromatography), ideally with protein concentrations >50 μM. The semi-preparative HPLC columns used here have a binding capacity of roughly 1 mg protein.
2. Add 5 mM TCEP or DTT to the sample. Incubate for 10 min to fully reduce the protein (*see* **Note 9**).
3. Perform a test run to identify at what percentage of buffer B the protein elutes. For that purpose, dilute 20 μL (>1 nmol) of the protein in HPLC buffer A and inject. Optimize the gradient accordingly. Detect the peptide bond absorbance at 220 nm (*see* **Note 10**), and the absorbance of aromatic amino acids at 280 nm. Excite and detect fluorescence according to the spectral properties of the aromatic residues in the protein of interest (Phe: 256/280 nm, Tyr: 275/305 nm, Trp: 280/350 nm) to distinguish it from impurities.
4. Inject 0.5–1 mg of the sample and collect the fraction(s) of interest. If desired, remove 50 μL of the elution peak for mass spectrometry. Perform multiple runs for larger amounts of protein to maintain good peak resolution (*see* **Note 11**).
5. Flash-freeze the collected fractions in liquid nitrogen immediately after collection.
6. Lyophilize samples in a lyophilizer or centrifugal evaporator and store dry at −80 °C until use. In this state, the sulfhydryl groups on the protein typically remain reactive for years.

3.2.2 Ion-Exchange–Based Purification

If a protein is too large for HPLC or its folded state must be maintained during labeling, ion-exchange chromatography is the purification method of choice. The method separates molecules by surface charge. The protein binds to the column and is eluted with a salt gradient, which concentrates dilute samples in the process.

Ion-exchange chromatography requires the choice of a suitable ion-exchange column, a corresponding working pH and an initial salt concentration. Depending on the net charge of the protein, either an anion- or cation-exchange column is selected for purification. For proteins with an isoelectric point (pI) < 7.0, we recommend using an anion-exchange column (Mono Q); for proteins with a pI > 7.5, a cation-exchange column (Mono S). The pH of the binding and elution buffers should be between 7.0 and 7.5, and at least 0.5 units higher (anion-exchange chromatography) or lower (cation-exchange chromatography) than the pI of the protein. In this way, the protein binds to the column at low salt concentrations and elutes in a buffer that is suitable for labeling. Accordingly, we recommend the use of phosphate or HEPES buffers for ion-exchange chromatography (*see* also Subheading 2.2, **Note 12**). The optimal salt concentration in the initial binding buffer strongly depends on the net and surface charge of the protein; a typical buffer system is described in Subheading 2.2. For more information, refer to Ref. 42.

A standard ion-exchange purification protocol is as follows:

1. Buffer-exchange the sample by dialyzing against binding buffer (supplied with 1 mM TCEP or DTT) or using a desalting column (e.g., HiTrap).
2. Add 5 mM TCEP or DTT to the sample. Incubate for 10 min to completely reduce the protein (*see* **Note 9**).
3. Equilibrate the ion-exchange column with 10 CVs binding buffer, followed by 10 CVs elution buffer and 10–20 CVs binding buffer. Equilibrating with elution buffer is necessary to saturate the functional groups of the column with the chosen counter ion and to obtain a stable, linear salt gradient for elution. Record the absorbance at 280 nm (or at 220 nm if the protein does not contain aromatic residues, *see* **Note 10**).
4. Apply 3–50 mg of protein, either using a sample pump or by injection (if applying a small volume) (*see* **Note 13**).
5. Wash the column with binding buffer until the absorbance at 280 nm returns to baseline.
6. Elute the protein with a gradient of 0–40% elution buffer in 40 CVs. Optimize the gradient if required (*see* **Note 14**).
7. Use the collected fractions immediately for labeling or flash-freeze in liquid nitrogen and store at −80 °C.

3.3 Labeling Reaction

The starting material for labeling is the protein purified either by RP-HPLC or ion-exchange chromatography (*see* Subheading 3.2). Depending on the desired purity and yield of the product, different labeling strategies are possible, as outlined below. If you purify the protein with RP-HPLC, you can choose between native and denaturing labeling conditions (*see* Subheading 2.2). We recommend labeling under denaturing conditions if the protein is aggregation-prone or sticky, i.e., interacts strongly with surfaces. For purification with ion-exchange chromatography, use the dye with the highest net charge in the first labeling step to maximize resolution during purification.

There are several strategies for labeling depending on the order of addition of the two dyes to the reaction mixture. In the sequential labeling strategy, the protein is first labeled with one dye at a sub-stoichiometric ratio of dye to protein. Single-labeled protein is purified and then reacted with the second dye. The first dye will primarily couple to the more reactive cysteine residue, so choose the dye accordingly. If both cysteine residues have similar reactivities, a separation of labeling permutants can often be achieved chromatographically (*see* Figs. 1a and 2a). This approach typically

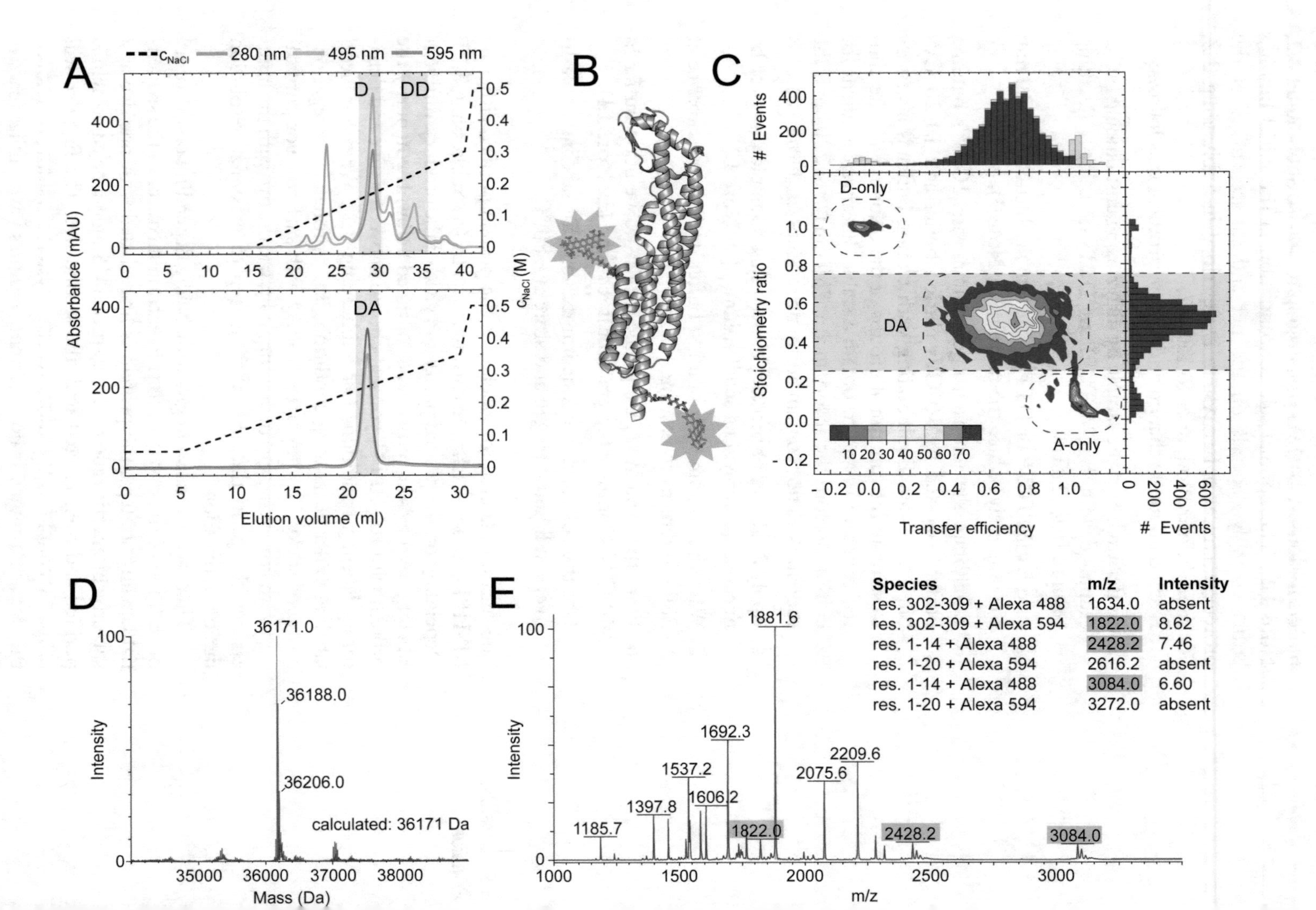
cNaCl
280 nm
495 nm
595 nm
A
D
DD
DA
Absorbance (mAU)
cNaCl (M)
Elution volume (ml)
B
C
Events
D-only
DA
A-only
Stoichiometry ratio
Transfer efficiency
Events
D
36171.0
36188.0
36206.0
calculated: 36171 Da
Intensity
Mass (Da)
E
Species | m/z | Intensity
res. 302-309 + Alexa 488 | 1634.0 | absent
res. 302-309 + Alexa 594 | 1822.0 | 8.62
res. 1-14 + Alexa 488 | 2428.2 | 7.46
res. 1-20 + Alexa 594 | 2616.2 | absent
res. 1-14 + Alexa 488 | 3084.0 | 6.60
res. 1-20 + Alexa 594 | 3272.0 | absent
1185.7
1397.8
1537.2
1606.2
1692.3
1822.0
1881.6
2075.6
2209.6
2428.2
3084.0
Intensity
m/z

chromatography on a MonoQ column. Unlabeled protein (24 mL) elutes before the two single-labeled permutants (29 and 31 mL) and double-labeled protein (34 mL). *Bottom panel*: Anion exchange chromatography after the second labeling reaction. Single-labeled protein from peak D at 29 mL was incubated overnight with a threefold molar excess of Alexa 594. Before ion exchange chromatography, uncoupled dye was removed from the sample on a HiTrap desalting column. (**b**) Structure of ClyA (PDB 1QOY [47]) with dyes attached (Alexa 488: green, Alexa 594: orange). (**c**) Single-molecule measurement of donor–acceptor-labeled ClyA. Fluorescence was collected after pulsed-interleaved excitation, so the relative fractions of donor-only, donor–acceptor, and acceptor-labeled protein can be identified from the transfer efficiency vs. stoichiometry ratio plot. The transfer efficiency histogram (*top*) is depicted before (gray bars) and after (blue bars) selecting bursts with a stoichiometry ratio between 0.25 and 0.75 (blue-shaded area). The projected stoichiometry ratio histogram is shown to the right. (**d**) ESI-MS spectrum of donor–acceptor-labeled ClyA with peak masses indicated. The expected mass of donor–acceptor-labeled ClyA is 36,171 Da. (**e**) MALDI-MS spectrum of ClyA after trypsin digest. Only the N-terminal fragment is detected with an Alexa 488 modification (green), and only the C-terminal fragment with an Alexa 594 modification (orange). The peaks from the other permutation are absent. The remaining peaks correspond to ClyA peptides without reactive cysteine residues

yields the highest quality samples. Advantages are that site-specific labeling is possible; the product usually has the highest purity; and correctly labeled fractions are easily identified from the chromatogram. A disadvantage is that two chromatography runs are required, which might decrease the overall yield.

An alternative is to perform both labeling reactions simultaneously. In this case, both dyes are added at the same time at equimolar concentrations. This approach is particularly useful for preliminary tests of the suitability of labeling positions or if only small amounts of protein are available. An advantage is that only one chromatography run is required. The disadvantages are that no site-specific labeling is possible; the elution profile is more heterogeneous since it contains all possible labeled species, so the correctly labeled fraction might be difficult to identify from the chromatogram; and additional quality control (FRET, mass spectrometry) of several fractions is often required to identify the correctly donor–acceptor–labeled protein.

Finally, one can perform the labeling reaction in semi-sequential fashion. In this case, the first dye is added at a sub-stoichiometric ratio of dye to protein and incubated to react with the protein to completion. Afterwards, an excess of the second dye is added to the same reaction. This approach is useful if only small amounts of protein are available but site-specific labeling is important. Advantages are that only one chromatography run is required and that site-specific labeling is possible if the cysteine residues have sufficiently different reactivity. Disadvantages are that the second cysteine residue might lose reactivity due to oxidation during incubation with the first dye, especially at high protein concentrations; the elution profile is more heterogeneous, so the correctly labeled fraction might be difficult to identify; and additional quality control (FRET, mass spectrometry) of several fractions is often required to identify the donor–acceptor–labeled protein.

A standard labeling protocol is outlined here:

The following steps are carried out in a low-humidity and oxygen-free atmosphere (e.g., in a nitrogen glove box), if available.

1. Dissolve the dye in DMSO to a concentration of 10 μg/μL or in DMF to a concentration of 5 μg/μL. The solubility of dyes in DMF is generally lower than in DMSO. Use DMF for proteins prone to methionine oxidation (*see* Subheading 3.6.5). Protect the dissolved dye from light.
2. Sonicate the dissolved dye in an ultrasonic bath for 15 min to completely dissociate dye oligomers, which can increase the fraction of undesired double-donor- or double–acceptor–labeled protein.

3. Determine the protein concentration. *RP-HPLC*: Dissolve the lyophilized protein in labeling buffer (native or denaturing) to reach a protein concentration of 10–200 μM (*see* **Notes 4, 15,** and **16**). *Ion-exchange chromatography*: Determine the protein concentration in the elution fraction. If the protein is already coupled to the first dye, use the dye absorbance to calculate the protein concentration. For an example calculation, *see* **Note 17**.

   Add the dye to the protein solution according to the respective labeling scheme and mix the labeling reaction gently. When working with aggregation-prone proteins, add the protein solution to the dissolved dye to ensure rapid mixing. Keep the percentage of organic solvent below 5% for maximum labeling efficiency. Unused dye dissolved in organic solvent can be lyophilized and reused. For an example calculation of how much dye to add to a labeling reaction, *see* **Note 17**.

   Sequential labeling with first dye: Add a 0.7:1 molar ratio of dye to the protein solution.

   Sequential labeling with second dye: Add a 1.2–3-fold molar excess of dye to the protein solution.

   Simultaneous labeling: Add both dyes at an equimolar ratio to the protein (i.e., 1:1:1).

   Semi-sequential labeling: Add a 0.7:1 molar ratio of the first dye to the protein solution. Incubate the reaction protected from light for 1–2 h at room temperature. Add the second dye in a 1.2–3-fold molar excess to the protein solution.

4. Incubate the reaction protected from light for 2–3 h at room temperature or overnight at 4 °C (which can increase the labeling yield).

5. Quench unreacted dye (to reduce its affinity to the column) and reduce oxidized cysteines by addition of 10 mM DTT. Add the labeling reaction to the DTT solution to ensure fast mixing. Incubate for 10 min at room temperature and either purify the reaction directly or flash-freeze it in liquid nitrogen and store at −80 °C until further purification.

3.4 Purification of Dye-Labeled Protein

In this step, donor–acceptor-labeled protein is purified from the labeling reaction. The most common impurities that interfere with single-molecule experiments are: (1) dye not coupled to the protein and (2) protein carrying only donor dye(s). The purification procedure is independent of the labeling strategy and needs to be carried out twice for sequential labeling.

3.4.1 RP-HPLC-Based Purification

Since dyes usually contain large aromatic moieties, dye-coupled proteins often have a larger retention time than the unmodified protein, and single-labeled protein elutes before double-labeled protein (*see* **Notes 18** and **19**; Figs. 1a and 3b). In many cases, it

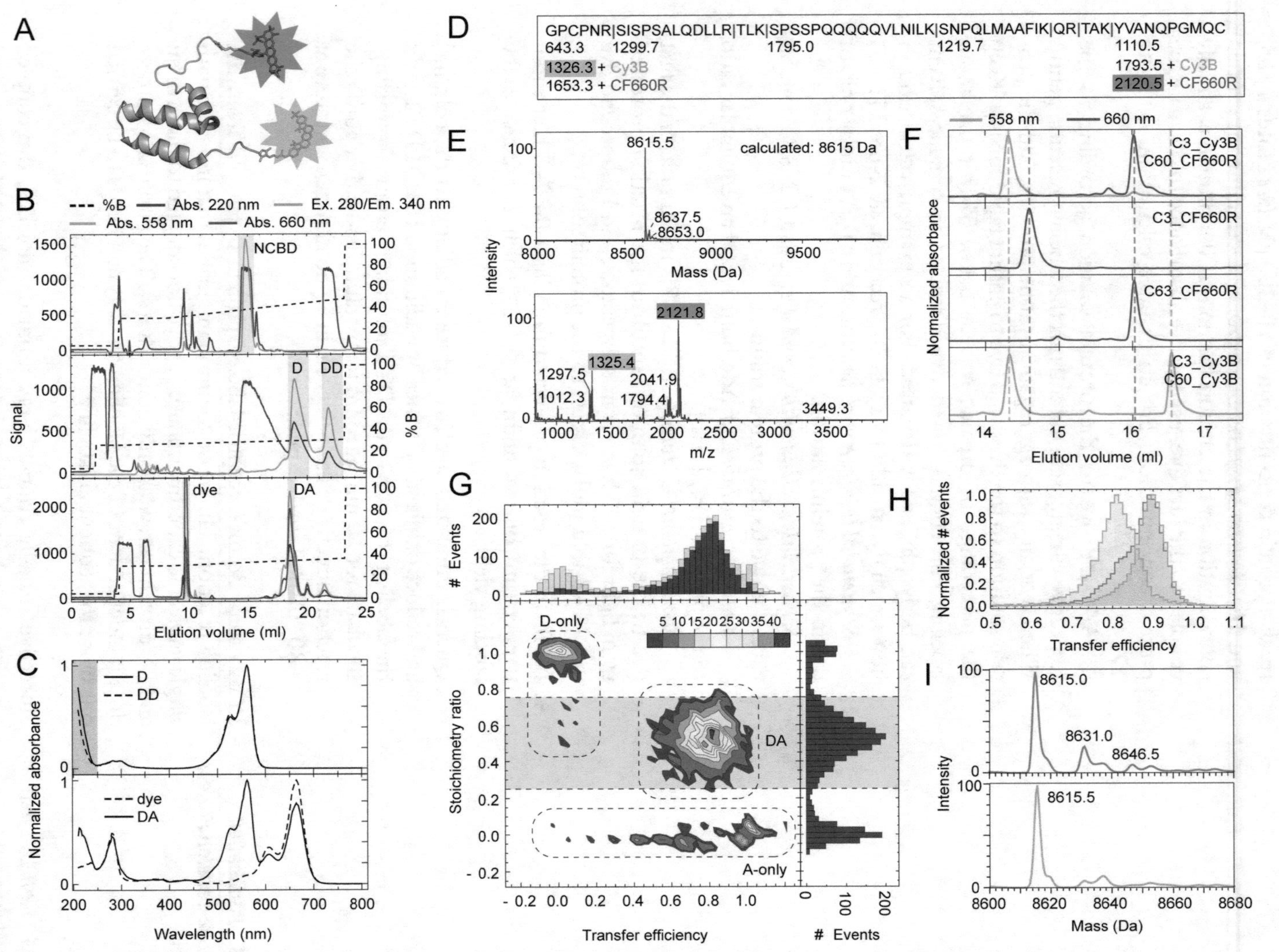

Fig. 3 Site-specific labeling, RP-HPLC-based purification, and characterization of the molten-globule-like nuclear coactivator binding domain of CBP/p300 (NCBD;

donor–acceptor-labeled protein. The fractions of interest are shaded in gray. *Top*: NCBD was co-expressed with its binding partner ACTR and can be distinguished from it and other impurities by its fluorescence peak at 340 nm (arising from a single Tyr residue). *Middle*: NCBD was incubated for 3 h with a 0.7:1 molar ratio of Cy3B to protein. Unlabeled protein elutes before single-labeled (D) and double-labeled protein (DD). *Bottom*: The single-labeled protein was incubated for 3 h with a 1.5-fold molar excess of Biotium CF660R and purified (DA). Unreacted dye elutes before the labeled protein, reflecting its hydrophilicity. (**c**) Absorbance spectra recorded during the HPLC runs in (**a**), with the corresponding elution peaks indicated. *Top*: Single-and double-Cy3B-labeled protein can be distinguished by their 220/558 nm absorbance ratios. The absorbance spectrum of the single-labeled protein (black line) has a higher relative absorbance in the peptide bond region (<250 nm, blue-shaded area). *Bottom*: Donor–acceptor-labeled protein (DA) can be distinguished from free dye by its absorbance spectrum. In the elution peak DA, the absorbance ratio at the emission maxima of both dyes matches the ratio of the extinction coefficients provided by the suppliers (Cy3B: 130.000 M^{-1} cm^{-1}, CF660R: 100.000 M^{-1} cm^{-1}). (**d**) NCBD sequence with trypsin cleavage sites and the masses of the individual fragments indicated. The N- and C-terminal fragments contain the cysteine residues for labeling (C3/C63). The expected molecular weights of these fragments after coupling of Cy3B (+683 Da) or CF660R (+1010 Da) are shown. (**e**) *Top*: The ESI-MS spectrum of donor–acceptor-labeled NCBD shows a dominant peak at the expected mass. *Bottom*: MALDI-MS spectrum of NCBD after trypsin digest. The N-terminal fragment is only detected with a Cy3B modification (green), the C-terminal fragment only with a CF660R modification (red). (**f**) Analytical RP-HPLC elution profiles of NCBD after trypsin digest confirm site-specific labeling close to 100%. By digesting different NCBD variants, we are able to determine the elution times of all possible labeled peptides. *First panel*: donor–acceptor-labeled NCBD; *second panel*: single-cysteine NCBD variant labeled with CF660R at C3; *third panel*: single-cysteine NCBD variant labeled with CF660R at C63; *fourth panel*: double-Cy3B-labeled NCBD. Thus, in the Cy3B/CF660R variant, we find C3 exclusively modified with Cy3B, and C63 exclusively modified with CF660R. (**g**) Single-molecule measurement of donor–acceptor-labeled NCBD. Fluorescence was collected after pulsed-interleaved excitation, so the relative fractions of donor-only, donor–acceptor, and acceptor-labeled protein can be identified from the transfer efficiency vs. stoichiometry ratio plot. The transfer efficiency histogram (*top*) is depicted before (gray bars) and after (blue bars) selecting bursts with a stoichiometry ratio between 0.25 and 0.75 (blue-shaded area). The projected stoichiometry ratio histogram is shown to the right. (**h**) Methionine oxidation in NCBD. Transfer efficiency histograms of reduced (orange), partly oxidized (blue), and fully oxidized (green) NCBD. Protein oxidation causes a pronounced shift to lower transfer efficiencies. (**i**) ESI-MS spectra of partly oxidized (blue) and reduced (orange) NCBD. In the upper MS spectrum, oxidation products with additional masses +16 and +31.5 are present, suggesting the addition of one or two oxygen atoms. This observation can be attributed to the oxidation of two methionine residues in the protein (sequence in (**d**))

is possible to separate the two labeling permutants, so site-specific labeling can be achieved (Fig. 1a). In some cases, dye isomers can lead to different elution times of the labeled protein, resulting in multiple peaks for protein carrying a single dye. When labeling with donor and acceptor dye simultaneously, these effects can lead to very complex chromatograms. We recommend a shallow gradient (a change of 5–10% B over 5 CVs) to resolve the different labeled species. Shallower gradients usually improve separation. The peak corresponding to unlabeled protein can be collected and used in another labeling reaction. If you label a large amount of protein (>0.5 mg), it is advisable to purify the labeling reaction in two runs to maintain peak separation. If the donor–acceptor-labeled protein does not have the desired quality (e.g., donor-only–labeled contamination), another RP-HPLC run on a column with different properties usually helps to isolate the desired species.

A standard RP-HPLC purification protocol is outlined here:

1. Perform a test run on the HPLC to optimize the elution gradient (the gradient from the initial purification run is a good starting point). For that purpose, dilute 5 μL of the quenched labeling reaction in RP-HPLC buffer A and inject. Detect protein absorption at 220 nm and dye absorption at the appropriate wavelength (*see* **Note 20**). If using fluorescence detection, excite at 280 nm; detect aromatic residues at 340 nm and the dye at its emission wavelength. When labeling with donor and acceptor simultaneously, excite fluorescence at the donor absorbance maximum and detect at the donor and acceptor emission maxima to identify the correctly labeled species via FRET.
2. Inject the labeling reaction and collect the appropriate peaks. If desired, take a 50 μL sample for mass spectrometry (*see* **Note 11**).
3. Flash-freeze the appropriate fractions in liquid nitrogen immediately after collection.
4. Lyophilize samples and store dry at −80 °C until use (or proceed analogously to labeling with the second dye).

3.4.2 Ion-Exchange Chromatography-Based Purification

Many commonly available dyes have a nonzero net charge to increase their hydrophilicity, so that labeling often alters the charge of the protein. Ion-exchange chromatography then allows the separation of unlabeled, single-labeled, and double-labeled reaction products. As in the case of RP-HPLC, the single-labeled permutants often elute as two separable peaks from the column, thus enabling site-specific labeling. In ion-exchange chromatography, the elution profile strongly depends on the net charge of the protein and the dyes. For example, when labeling a positively charged protein with a negatively charged dye, the double-labeled species is expected to elute first from a cation-exchange column, followed by single- and unlabeled protein. Optimization of the gradient can lead to better separation, but for ion-exchange chromatography this is not as straightforward as for RP-HPLC. Very shallow salt gradients usually lead to peak broadening rather than improved separation. pH and counter ions can be altered systematically to optimize the separation of labeling products [42].

A standard ion-exchange purification protocol is outlined here:

1. Equilibrate the ion-exchange column with binding buffer (*see* Subheading 3.2.2). Record the absorbance at 280 nm and at the wavelength of maximum absorbance of the dyes. If using fluorescence detection, proceed analogously to RP-HPLC.
2. Dilute the labeling reaction ~3-fold in the binding buffer or exchange the buffer with a desalting column to enable binding of the labeled protein to the ion-exchange column. When labeling with an excess of dye, remove it with a desalting column prior to ion-exchange chromatography.
3. Inject the labeling reaction and wash the column with binding buffer until the absorbance at 280 nm returns to baseline.
4. Elute the protein with a gradient of 0–40% elution buffer in 40 CVs (*see* **Note 14**).
5. Use the collected fractions immediately for labeling with the second dye or flash-freeze in liquid nitrogen and store them at −80 °C until use.

3.5 Preparation of Samples for Single-Molecule Spectroscopy

1. *Lyophilized protein from RP-HPLC*: Dissolve the protein in a suitable buffer (e.g., 50 mM sodium phosphate, pH 7) at a concentration of 1–100 μM. Since the protein will be diluted >10,000-fold for single-molecule measurements, it is possible to use additives to increase protein stability or solubility in the stock solutions if necessary (e.g., glycerol, GdmCl) without interfering with the final solution conditions of the measurements. *Ion-exchange–purified protein:* Determine the concentration via the absorbance of the dyes. Add stabilizing agents if necessary. Do not include thiols in storage buffers as they can promote hydrolysis of the maleimide-sulfhydryl bond over time [43].
2. Split the protein sample into small aliquots for single use to avoid repeated freeze–thaw cycles. Flash-freeze aliquots in liquid nitrogen and store at −80 °C.
3. For more dilute aliquots (1–100 nM protein), include 0.001% Tween 20 (v/v) in the buffer to avoid loss of protein to surfaces.

3.6 Quality Control

3.6.1 Protein Identity and Impurities

The identity of donor–acceptor-labeled protein is most reliably assessed with electrospray ionization mass spectrometry (ESI-MS), as depicted in Figs. 1e, 2d, and 3e. Common contaminants are double-donor and double-acceptor–labeled species. If 2-mercaptoethanol was used as a reducing agent, peaks with an additional mass of +76 Da point towards the formation of 2-mercaptoethanol-cysteine adducts. Note that the molecular weight of the dye given by the supplier often does not match the observed shift in mass, since it includes counter ions that are usually removed during purification.

3.6.2 UV-Vis Absorption Spectrum

Record a UV-Vis absorption spectrum of the labeled protein under native and denaturing (e.g., in 6 M GdmCl) conditions. Note that the peaks might shift slightly (<10 nm) compared to free dye.

1. Check the absorption maxima of both dyes. Do their relative amplitudes match the ratio of the published extinction coefficients? If not, this might be an indication of double-donor or double-acceptor labeled species in the sample (*see* **Notes 20** and **21**).
2. Check the shape of the recorded spectra. Are they similar under native and denaturing conditions? If not, this might indicate a stacking interaction of the dyes, which will perturb their spectroscopic properties [21, 31].

3.6.3 Single-Molecule Spectroscopy

A FRET efficiency histogram from a confocal recording of freely diffusing donor–acceptor-labeled protein is a straightforward way to assess sample purity (or a camera-based recording of immobilized samples if total internal reflection fluorescence is used). Ideally, use alternating or pulsed-interleaved excitation [15, 44] to determine the amount of donor-only, donor-acceptor, and acceptor-only species in the sample (*see* Fig. 2c and 3g). Donor-only and acceptor-only peaks can even be present if the donor–acceptor-labeled protein is very pure according to chromatography or mass spectrometry, e.g., owing to background fluorescence or the inactivation of dyes in a previous passage through the confocal volume. Note that in the presence of excessive amounts of single-labeled protein, even alternating excitation may not suffice to fully isolate the signal from donor–acceptor-labeled protein.

3.6.4 Site-Specific Labeling

Site-specific labeling can be confirmed by tryptic digest of the protein, either in combination with mass spectrometry (MS) or analytical RP-HPLC (*see* Fig. 3f).

A protocol for tryptic digestion is as follows:

1. Dilute 0.1–1 nmol of donor–acceptor-labeled protein in 50 μL trypsin digest buffer. The higher the sensitivity of the fluorescence detector, the less protein is required. If unlabeled cysteine residues are present in the protein, add 1 mM TCEP to the buffer.

2. Add a sub-stoichiometric amount of trypsin protease (1:10 molar ratio of trypsin:protein) and incubate overnight in the dark at 37 °C. For IDPs, a digestion time of 1 h is usually sufficient.
3. Analyze peptide fragments with analytical RP-HPLC on a C18 column. A gradient from 5% to 50% buffer B over 15 CVs usually yields good separation. Detect absorbance at 220 nm as well as at the absorbance maxima of the dyes. Excite fluorescence at 280 nm, detect at 340 nm and at the emission maxima of the dyes of interest.
4. Analyze the chromatogram. For a site-specifically labeled protein, only one peak should appear with the absorbance and fluorescence of each dye (*see* Fig. 3f). If the protein is not site-specifically labeled, two peaks will appear for each dye. The area under these peaks (dye absorbance) is proportional to the amount of dye coupled to each site. The identity of the labeling site can often be determined using additional information about the corresponding tryptic peptide, e.g., the predicted extinction coefficient at 220 nm or the fluorescence of aromatic amino acids.

Otherwise, site-specificity can be tested by MS. Again, the protein is digested with trypsin or other proteases (sometimes a service provided at the MS facility), and the mass of the tryptic peptides is analyzed. Depending on the dye modification, the tryptic peptides will have different masses. If only the mass of one peptide with donor dye is detected, and the mass of the peptide with acceptor dye is absent from the spectrum (and vice versa), then it is a good indication that site-specific labeling was achieved (*see* Figs. 1d, e, 2e, and 3d, e). Note that the relative peak areas in the MS spectrum are not quantitative because of different mobility of the different ions. The large negative charge of many commonly used fluorophores can reduce the mobility of labeled protein in positive ion mode substantially.

3.6.5 Methionine Oxidation

Oxidation of methionine to methionine sulfoxide is a common side reaction, e.g., when incubating proteins in the presence of DMSO for an extended time [45], so DMF should be used for labeling oxidation-prone proteins. Methionine oxidation is easily recognized by a +16 shift in the mass spectrum (or multiples of it in the case of several methionine residues, *see* Fig. 3i). It can have a significant effect on protein structure and activity, as illustrated in Fig. 3h. Methionine sulfoxide can be reduced again to methionine either enzymatically (using methionine sulfoxide reductases) or with dimethyl sulfide in the presence of 7 M hydrochloric acid [45], a harsh procedure that might be unsuitable for larger proteins.

4 Notes

1. R_0 can be calculated from the spectral overlap integral, J, between the fluorophores, the orientational factor, κ^2, the quantum yield of the donor, Q, and the refractive index, n, of the medium: $R_0 = \left(\frac{9000(\ln 10)}{128\pi^5 N_A}\frac{Q\kappa^2 J}{n^4}\right)^{1/6}$, with N_A being Avogadro's number. If the fluorophores rotate freely and the relative orientation of the fluorophore averages on a timescale shorter than the donor excited state lifetime, κ^2 averages to a value of 2/3. The validity of this approximation can be tested by fluorescence anisotropy measurements [14]. J is calculated from the donor emission spectrum, $f_D(\lambda)$ (normalized to an area of 1), and the wavelength-dependent molar extinction coefficient of the acceptor, $\varepsilon_A(\lambda)$, according to $J = \int f_D(\lambda)\varepsilon_A(\lambda)\lambda^4 d\lambda$.
2. Incomplete mixing can lead to shifts in the retention time of proteins. If used infrequently, it is advisable to purchase TFA in 1-mL glass ampules. For frequent use, buy TFA in a 1-L glass bottle for maximum shelf life. Use filter tips to handle the corrosive liquid.
3. The addition of the mild detergent Tween 20 to the buffer is not strictly required but significantly reduces protein–column interactions and increases the yield of eluted protein. The stability and function of proteins is usually not affected by low concentrations of Tween 20, but suitable tests should be performed for every protein.
4. This concentration range usually gives the best results. Lower concentrations lead to a lower labeling efficiency due to hydrolysis of the dye before it reacts with the protein. Higher protein concentrations can be less efficient because of the formation of intermolecular disulfide bridges competing with the maleimide addition.
5. We do not recommend using any membrane-based concentration or filtration steps during labeling and subsequent purification, since dye-coupled proteins often have the tendency to stick to ultrafiltration membranes, leading to loss of sample.
6. Use at least 5% ACN at all times to avoid collapse of the stationary phase of the column.
7. Especially when purifying IDPs, RP-HPLC allows the separation of the full-length protein from truncations (*see* Fig. 1a). Usually, the full-length protein elutes at higher percentage of buffer B.
8. If the protein of interest does not elute in the gradient (**step 3**) or at 100% B (**step 4**), it might have excessive affinity to the column. To elute it, inject 1 mL of 6 M guanidinium chloride

while flowing a mixture of 95% A/5% B over the column, followed by an elution gradient (without injecting sample).

9. Do not use 2-mercaptoethanol as a reducing agent in this step, as it can form cysteine adducts.
10. The wavelength for detecting peptide bond absorption can be adjusted from 200 to 235 nm, depending on the amount of protein, the required sensitivity, and the absorption of the buffer.
11. To maximize column lifetime, run an elution gradient without injecting sample after the purification run to remove residual bound protein (and dye).
12. Even though phosphate buffer is not recommended for anion-exchange chromatography by some suppliers (as phosphate ions strongly interact with the anion-exchange resin), it is frequently used in this application [42]. We recommend the use of phosphate buffer because it is ideal for subsequent labeling.
13. The volume applied to the ion-exchange column only matters if the protein does not strongly bind to the resin. In this case, concentrating and applying the protein in a small volume might be necessary. Alternatively, the salt concentration in the binding buffer can be reduced to enhance binding (e.g., using only 10 mM phosphate buffer). On the other hand, proteins that bind strongly to the ion-exchange resin will be highly concentrated on the column and therefore might aggregate. Lowering the affinity by increasing the salt concentration in the binding buffer can help to prevent aggregation.
14. To increase column lifetime, take care to remove any precipitated protein before injection, e.g., by centrifugation. Ion-exchange columns should be cleaned after each use by injections of 1–3 CV 1 M sodium hydroxide and 6 M GdmCl (both filtered), each followed by a wash with 10 CVs binding buffer.
15. Minimize the amount of time between dissolving the protein and starting the labeling reaction (ideally <15 min), especially at high protein concentrations, to minimize oxidation. The shorter the time, the higher the labeling efficiency.
16. For the first labeling reaction, it is crucial to accurately determine the protein concentration. The correct dye-to-protein ratio increases labeling efficiency and avoids excessive formation of double-labeled protein (with two identical dyes). If the protein does not contain aromatic residues that absorb at 280 nm, it is advisable to determine the extinction coefficient at 225 nm, either by dissolving a known amount of protein in a certain volume or by quantifying the protein concentration via

a bicinchoninic acid (BCA) or Bradford assay. The calculated extinction coefficient can subsequently be used for the rapid quantification of protein concentration with UV spectroscopy. Once the protein is coupled to a fluorescent dye, use the dye absorbance peak and published extinction coefficient to determine the protein concentration. Note that many dyes strongly absorb at 280 nm, which affects concentration determination via ε_{280} (cf. **Note 21**).

17. Use the Lambert-Beer law to determine the protein concentration (c_{Protein}) and total amount of protein (n_{Protein}) in a sample of volume V_{Protein}:

$$c_{\text{Protein}} = \frac{A_\lambda}{\varepsilon_\lambda \cdot d} \text{ and } n_{\text{Protein}} = c_{\text{Protein}} \cdot V_{\text{Protein}}$$

here, λ is the wavelength used for concentration determination, A_λ is the measured absorbance, d is the length of the optical path, and ε_λ the extinction coefficient at λ.

Example: Measured $A_{280} = 0.082$, with an optical path of 1 mm and ε_{280}=5500 M^{-1} cm^{-1}. The total volume of the sample is 400 μL. The dye (molecular weight 721 g/mol) is dissolved in DMSO at a concentration of 10 μg/μL and is to be reacted with the protein at a 0.7-fold molar ratio.

$$c_{\text{Protein}} = \frac{A_\lambda}{\varepsilon_\lambda \cdot d} = \frac{0.082}{5500\text{M}^{-1}\text{cm}^{-1} \cdot 0.1 \text{ cm}} = 149 \text{ μM}$$

$$n_{\text{Protein}} = c_{\text{Protein}} \cdot V_{\text{Protein}} = 149 \frac{\text{nmol}}{\text{mL}} \cdot 0.4 \text{ mL} = 60 \text{ nmol}$$

$$c_{\text{Dye}} = \frac{10 \text{ gL}^{-1}}{721 \text{ g mol}^{-1}} = 0.0139 \text{ mol L}^{-1} = 13.9 \text{ nmol μL}^{-1}$$

$$V_{\text{Dye}} = \frac{n_{\text{Protein}} \cdot (\text{dye-to-protein ratio})}{c_{\text{Dye}}} = \frac{60 \text{ nmol} \cdot 0.7}{13.9 \text{ nmol μL}^{-1}} = 3.02 \text{ μL}$$

Add the 400 μL of protein solution to 3 μL of dissolved dye and mix rapidly, e.g., by pipetting or inverting the tube.

18. Usually, single- and double-labeled species can be distinguished by the ratio of peptide bond absorption (measured at 220 nm or 235 nm) to dye absorption (Fig. 3c). Judging protein absorption at 280 nm is not advisable if protein and dye have similar extinction coefficients at this wavelength. If a diode-array detector is present, record the absorbance spectra over the whole chromatography run to facilitate post-run analysis.

19. If there is insufficient separation between unlabeled and single-labeled protein, try another column with different properties (e.g., switch from a 5-μm to a 3.5-μm pore size).

20. Protein labeled correctly with two FRET dyes can usually be identified by the ratio of the absorbance peak intensities, which should match the ratio of the extinction coefficients of the dyes. Store absorbance spectra over the entire chromatography run to facilitate analysis. In the calculation, account for acceptor dye absorbance at the absorbance maximum of the donor dye.
21. Calculate labeling efficiencies by determining the protein concentration via absorbance at 280 nm and the dye concentration at its absorbance maximum. Account for dye absorbance at 280 nm (usually provided by the supplier as "correction factor"). This approach is not reliable for proteins with a low ε_{280} (<20,000 M^{-1} cm^{-1}), since the dye then dominates absorbance at 280 nm.

Acknowledgments

We thank Dr. Serge Chesnov and the Functional Genomics Center Zurich for expert mass spectrometry and MS data analysis, Fabian Dingfelder for contributing chromatograms and single-molecule data on ClyA, Erik D. Holmstrom for providing single-molecule data on NCBD, and Daniel Nettels for providing data analysis software and for helpful discussions.

References

1. Lerner E, Cordes T, Ingargiola A, Alhadid Y, Chung S, Michalet X, Weiss S (2018) Toward dynamic structural biology: two decades of single-molecule Förster resonance energy transfer. Science 359(6373):eaan1133
2. Lakowicz JR (2006) Principles of fluorescence spectroscopy, 3rd edn. Springer, New York
3. Selvin PR, Ha T (2008) Single-molecule techniques: a laboratory manual. Cold Spring Harbor Laboratory Press, New York
4. Dunkle JA, Cate JHD (2010) Ribosome structure and dynamics during translocation and termination. Annu Rev Biophys 39:227–244
5. Kapanidis AN, Strick T (2009) Biology, one molecule at a time. Trends Biochem Sci 34 (5):234–243
6. Ha T, Kozlov AG, Lohman TM (2012) Single-molecule views of protein movement on single-stranded DNA. Annu Rev Biophys 41:295–319
7. Smiley RD, Hammes GG (2006) Single molecule studies of enzyme mechanisms. Chem Rev 106(8):3080–3094
8. Zhuang XW (2005) Single-molecule RNA science. Annu Rev Biophys Biomol Struct 34:399–414
9. Schuler B, Hofmann H (2013) Single-molecule spectroscopy of protein folding dynamics-expanding scope and timescales. Curr Opin Struct Biol 23(1):36–47
10. Muñoz V, Cerminara M (2016) When fast is better: protein folding fundamentals and mechanisms from ultrafast approaches. Biochem J 473(17):2545–2559
11. Brucale M, Schuler B, Samori B (2014) Single-molecule studies of intrinsically disordered proteins. Chem Rev 114(6):3281–3317
12. Schuler B, Hofmann H, Soranno A, Nettels D (2016) Single-molecule FRET spectroscopy and the polymer physics of unfolded and intrinsically disordered proteins. Annu Rev Biophys 45:207–231
13. Ferreon AC, Moran CR, Gambin Y, Deniz AA (2010) Single-molecule fluorescence studies of intrinsically disordered proteins. Methods Enzymol 472:179–204

14. Sisamakis E, Valeri A, Kalinin S, Rothwell PJ, Seidel CAM (2010) Accurate single-molecule FRET studies using multiparameter fluorescence detection. Methods Enzymol 475:455–514
15. Kapanidis AN, Laurence TA, Lee NK, Margeat E, Kong X, Weiss S (2005) Alternating-laser excitation of single molecules. Acc Chem Res 38(7):523–533
16. Kapanidis AN, Weiss S (2002) Fluorescent probes and bioconjugation chemistries for single-molecule fluorescence analysis of biomolecules. J Chem Phys 117 (24):10953–10964
17. Kim Y, Ho SO, Gassman NR, Korlann Y, Landorf EV, Collart FR, Weiss S (2008) Efficient site-specific labeling of proteins via cysteines. Bioconjug Chem 19(3):786–791
18. Ratner V, Kahana E, Eichler M, Haas E (2002) A general strategy for site-specific double labeling of globular proteins for kinetic FRET studies. Bioconjug Chem 13(5):1163–1170
19. Lemke EA (2011) Site-specific labeling of proteins for single-molecule FRET measurements using genetically encoded ketone functionalities. Methods Mol Biol 751:3–15
20. Popp MW (2015) Site-specific labeling of proteins via sortase: protocols for the molecular biologist. Methods Mol Biol 1266:185–198
21. Haenni D, Zosel F, Reymond L, Nettels D, Schuler B (2013) Intramolecular distances and dynamics from the combined photon statistics of single-molecule FRET and photoinduced electron transfer. J Phys Chem B 117 (42):13015–13028
22. Hohng S, Joo C, Ha T (2004) Single-molecule three-color FRET. Biophys J 87 (2):1328–1337
23. Zosel F, Haenni D, Soranno A, Nettels D, Schuler B (2017) Combining short- and long-range fluorescence reporters with simulations to explore the intramolecular dynamics of an intrinsically disordered protein. J Chem Phys 147(15):152708
24. Doose S, Neuweiler H, Sauer M (2009) Fluorescence quenching by photoinduced electron transfer: a reporter for conformational dynamics of macromolecules. ChemPhysChem 10 (9–10):1389–1398
25. Hermanson GT (2013) Bioconjugate Techniques, 3rd edn, pp 1–1146
26. Shafer DE, Inman JK, Lees A (2000) Reaction of Tris(2-carboxyethyl)phosphine (TCEP) with maleimide and alpha-haloacyl groups: anomalous elution of TCEP by gel filtration. Anal Biochem 282(1):161–164
27. Liu P, O'Mara BW, Warrack BM, Wu W, Huang Y, Zhang Y, Zhao R, Lin M, Ackerman MS, Hocknell PK, Chen G, Tao L, Rieble S, Wang J, Wang-Iverson DB, Tymiak AA, Grace MJ, Russell RJ (2010) A tris (2-carboxyethyl) phosphine (TCEP) related cleavage on cysteine-containing proteins. J Am Soc Mass Spectrom 21(5):837–844
28. Garcia-Mira MM, Sanchez-Ruiz JM (2001) pH corrections and protein ionization in water/guanidinium chloride. Biophys J 81 (6):3489–3502
29. Kalinin S, Peulen T, Sindbert S, Rothwell PJ, Berger S, Restle T, Goody RS, Gohlke H, Seidel CA (2012) A toolkit and benchmark study for FRET-restrained high-precision structural modeling. Nat Methods 9(12):1218–1225
30. Müller-Späth S, Soranno A, Hirschfeld V, Hofmann H, Rüegger S, Reymond L, Nettels D, Schuler B (2010) Charge interactions can dominate the dimensions of intrinsically disordered proteins. Proc Natl Acad Sci U S A 107(33):14609–14614
31. Chung HS, Louis JM, Eaton WA (2010) Distinguishing between protein dynamics and dye photophysics in single-molecule FRET experiments. Biophys J 98(4):696–706
32. Chen HM, Ahsan SS, Santiago-Berrios MB, Abruna HD, Webb WW (2010) Mechanisms of quenching of Alexa fluorophores by natural amino acids. J Am Chem Soc 132(21):7244
33. Soranno A, Holla A, Dingfelder F, Nettels D, Makarov DE, Schuler B (2017) Integrated view of internal friction in unfolded proteins from single-molecule FRET, contact quenching, theory, and simulations. Proc Natl Acad Sci U S A 114(10):E1833–E1839
34. Marme N, Knemeyer JP, Sauer M, Wolfrum J (2003) Inter- and intramolecular fluorescence quenching of organic dyes by tryptophan. Bioconjug Chem 14(6):1133–1139
35. Lutolf MP, Tirelli N, Cerritelli S, Cavalli L, Hubbell JA (2001) Systematic modulation of Michael-type reactivity of thiols through the use of charged amino acids. Bioconjug Chem 12(6):1051–1056
36. Jacob MH, Amir D, Ratner V, Gussakowsky E, Haas E (2005) Predicting reactivities of protein surface cysteines as part of a strategy for selective multiple labeling. Biochemistry 44 (42):13664–13672
37. Rostkowski M, Olsson MH, Sondergaard CR, Jensen JH (2011) Graphical analysis of pH-dependent properties of proteins predicted using PROPKA. BMC Struct Biol 11:6

38. Altman RB, Zheng Q, Zhou Z, Terry DS, Warren JD, Blanchard SC (2012) Enhanced photostability of cyanine fluorophores across the visible spectrum. Nat Methods 9 (5):428–429
39. Koenig I, Zarrine-Afsar A, Aznauryan M, Soranno A, Wunderlich B, Dingfelder F, Stüber JC, Plückthun A, Nettels D, Schuler B (2015) Single-molecule spectroscopy of protein conformational dynamics in live eukaryotic cells. Nat Methods 12(8):773–779
40. Aigrain L, Crawford R, Torella J, Plochowietz A, Kapanidis A (2012) Single-molecule FRET measurements in bacterial cells. FEBS J 279:513–513
41. Borgia A, Zheng W, Buholzer K, Borgia MB, Schuler A, Hofmann H, Soranno A, Nettels D, Gast K, Grishaev A, Best RB, Schuler B (2016) Consistent view of polypeptide chain expansion in chemical denaturants from multiple experimental methods. J Am Chem Soc 138 (36):11714–11726
42. Ion Exchange Chromatography Principles and Methods (2021) Cytiva. https://cdn.cytivalifesciences.com/dmm3bwsv3/AssetStream.aspx?mediaformatid=10061&destinationid=10016&assetid=13101. Accessed 08.07.2021
43. Fontaine SD, Reid R, Robinson L, Ashley GW, Santi DV (2015) Long-term stabilization of maleimide-thiol conjugates. Bioconjug Chem 26(1):145–152
44. Hendrix J, Lamb DC (2013) Pulsed interleaved excitation: principles and applications. Methods Enzymol 518:205–243
45. Shechter Y (1986) Selective oxidation and reduction of methionine residues in peptides and proteins by oxygen exchange between sulfoxide and sulfide. J Biol Chem 261(1):66–70
46. Benke S, Roderer D, Wunderlich B, Nettels D, Glockshuber R, Schuler B (2015) The assembly dynamics of the cytolytic pore toxin ClyA. Nat Commun 6:6198
47. Wallace AJ, Stillman TJ, Atkins A, Jamieson SJ, Bullough PA, Green J, Artymiuk PJ (2000) E. coli hemolysin E (HlyE, ClyA, SheA): X-ray crystal structure of the toxin and observation of membrane pores by electron microscopy. Cell 100(2):265–276
48. Kjaergaard M, Teilum K, Poulsen FM (2010) Conformational selection in the molten globule state of the nuclear coactivator binding domain of CBP. Proc Natl Acad Sci U S A 107(28):12535–12540

Chapter 13

Single-Molecule Fluorescence Spectroscopy Approaches for Probing Fast Biomolecular Dynamics and Interactions

Zifan Wang, Nivin Mothi, and Victor Muñoz

Abstract

Single-molecule fluorescence spectroscopy, and particularly its Förster resonance energy transfer implementation (SM-FRET), provides the opportunity to resolve the stochastic conformational fluctuations undergone by individual protein molecules while they fold–unfold, bind to their partners, or carry out catalysis. Such information is key to resolve the microscopic pathways and mechanisms underlying such processes, and cannot be obtained from bulk experiments. To fully resolve protein conformational dynamics, SM-FRET experiments need to reach microsecond, and even sub-microsecond, time resolutions. The key to reach such resolution lies in increasing the efficiency at which photons emitted by a single molecule are collected and detected by the instrument (photon count rates). In this chapter, we describe basic procedures that an end user can follow to optimize the confocal microscope optics in order to maximize the photon count rates. We also discuss the use of photoprotection cocktails specifically designed to reduce fluorophore triplet buildup at high irradiance (the major cause of limiting photon emission rates) while improving the mid-term photostability of the fluorophores. Complementary strategies based on the data analysis are discussed in depth by other authors in Chap. 14.

Key words Single-molecule fluorescence spectroscopy, Förster resonance energy transfer, Time resolution, Photon count rate, Shot noise, Triplet buildup, Photoblinking and photobleaching

1 Introduction

Single-molecule fluorescence spectroscopy offers the opportunity to resolve the stochastic conformational fluctuations of biomolecules [1] and has been widely used as a technique to measure the dynamics and kinetics of biomolecular processes [2]. In this regard, the implementation of Förster resonance energy transfer (FRET) single-molecule measurements is particularly attractive, as the FRET efficiency (E) allows to probe nanoscale changes in the distance between two macromolecules, or two segments of one macromolecule [3], thus providing an efficient method to detect both intramolecular conformational changes and binding reactions of proteins, RNA, and DNA [4–17]. Here we focus on the applications of this technique to the investigation of fast (sub-millisecond)

Victor Muñoz (ed.), *Protein Folding: Methods and Protocols*, Methods in Molecular Biology, vol. 2376, https://doi.org/10.1007/978-1-0716-1716-8_13,

protein conformational dynamics, a problem for which attaining single-molecule resolution is essential.

The FRET efficiency indicates the probability that an excited donor fluorophore transfers its excitation energy to an acceptor fluorophore. Practically, in single-molecule experiments E is estimated from the ratio between the number of photons emitted by the acceptor and the total number of emitted photons (donor plus acceptor). The temporal resolution of a standard SM-FRET experiment is hence limited by the time it takes to collect enough photons from an individual molecule as to determine E with the desired accuracy. Such accuracy is limited by shot noise statistics according to the equation:

$$\sigma = \sqrt{\frac{E(1-E)}{N}} \tag{1}$$

where the standard deviation (σ) in the determination of E decreases proportionally to the square root of the number of photons used in the determination (N). Equation (1) makes it apparent that the time resolution of SM-FRET experiments is a direct function of the photon count rate. For instance, 50 measured photons lead to $\pm 7\%$ error for $E = 0.5$. Given the maximum collection efficiency of current confocal microscopes, and the emission properties (lifetimes, quantum, yields, triplet buildup, and photon budgets) of the organic fluorophores that are used for these experiments, the theoretical maximally achievable count rate is about 2 MHz, which would result in an effective time resolution of ~25μs (for $N = 50$). New statistical methods of analysis of individual photons (rather than binning photons in bunches) can increase the resolution down to twice the mean inter-photon time (~1μs). These analytical methods are discussed in Chap. 14 of this volume.

However, regardless of the method of analysis, the actual photon count rates that are typically obtained in SM-FRET experiments are often much lower than this theoretical maximum, i.e., in the 50–100 kHz range, due to a combination of factors. Some of these factors can be optimized by the end user to increase the effective count rates to values closer to the theoretical maximum. In this regard, there are two complementary approaches: (1) The optimization of the instrument components, which involves the laser excitation source, the properties of the objective (particularly its numerical aperture, N.A.), and the number of elements (lenses, filters, and mirrors) that are introduced in the optical path; (2) the implementation of procedures to reduce/minimize photophysical processes of the fluorophores (blinking and bleaching) in conditions of high irradiance. Both approaches can be attempted independently, or in combination for optimal results. In this chapter, we describe some of the basic procedures and components to be used in order to maximize the photon count rates of SM-FRET experiments performed on a standard confocal single-molecule fluorescence microscope. For the optimization of the instrument, we refer all the specific details to our customized confocal fluorescence microscope. However, the components and procedures that we discuss here are easily transferred to any other confocal SM-FRET microscope, whether custom-built or commercial.

2 Materials

2.1 Confocal Microscope

1. Inverted microscope: Nikon, Eclipse Ti-U.
2. Oil-immersion objective: Nikon standard immersion oil with N.A. = 1.49 and glass coverslips with matched refractive index (*see* **Note 1**).
3. Beam collimator: OZ optics, calibrated from 400 nm to 700 nm.
4. Polarization maintaining single-mode optical fiber: Thorlabs, single-mode transmission from 400 to 680 nm.
5. Avalanche single photon detector (APD): Excelitas SPCM, with 180μm active area (*see* **Note 2**).
6. Optics (*see* **Note 3**):
   - (a) Dichroic mirror: Chroma, cutting edge at 490 nm.
   - (b) Long-pass filters: Chroma, cutting edge at 505 nm.
   - (c) Long-pass filters: Chroma, cutting edge at 590 nm.
   - (d) Band-pass filter: Chroma, 97% transmission from 502 nm to 545 nm.
7. Time-correlated single photon counting electronics: HydraHarp 400 from PicoQuant GmbH.
8. Excitation laser: Coherent, continuous wave laser system working at 488 nm (*see* **Note 4**).
9. Fluospheres: Lifetechnologies, FluoSpheres, 0.1μm, yellow-green colors (*see* **Note 5**).

2.2 Photoprotection Cocktail

All solutions are prepared using Milli-Q water and HPLC-grade reagents.

1. Trolox: Sigma-Aldrich, HPLC grade.
2. Cysteamine: Sigma-Aldrich, HPLC grade.
3. Activated charcoal: Sigma-Aldrich, 12–20 mesh.

2.3 Sample Well

1. Sample holder: Adams & Chittenden Scientific Glass, 3.5″ × 1.17″ × 3/8″ thick w/3/8″ hole in center, borosilicate.
2. Coverslips: Fisher Scientific, optical borosilicate glass, with uniform thickness and size.

3. UV curing optical adhesives: Thorlabs, fast UV curing adhesive with refractive index of 1.56.
4. Vectabond reagent: Vector laboratories, reactive Vectabond.
5. Polyethylene glycol electrolyte (PEG): Iris Biotech GmbH, conjugating NHS-ester polyethylene glycol, 5000 Dalton.

3 Methods

3.1 Optimizing Confocal Microscope Setup for High Count Rates

3.1.1 Setup of the Confocal Microscope for High Count Rates

A simple, customized single-molecule fluorescence confocal microscope setup specifically designed to reduce the number of optical elements in the emission path to the barebones minimum is shown in Fig. 1. In this instrument, the small active area of the APD detectors (200μm) is directly used as confocal pinhole. The confocal microscope is constructed based on an inverted optical microscope (which could be of any of the common brands: Nikon, Zeiss, Olympus). The excitation source is a continuous wave laser working at 488 nm (for excitation of the donor fluorophore Alexa 488) whose beam is delivered to the microscope through a dichroic mirror, D1, and then focused to the sample by a 100×

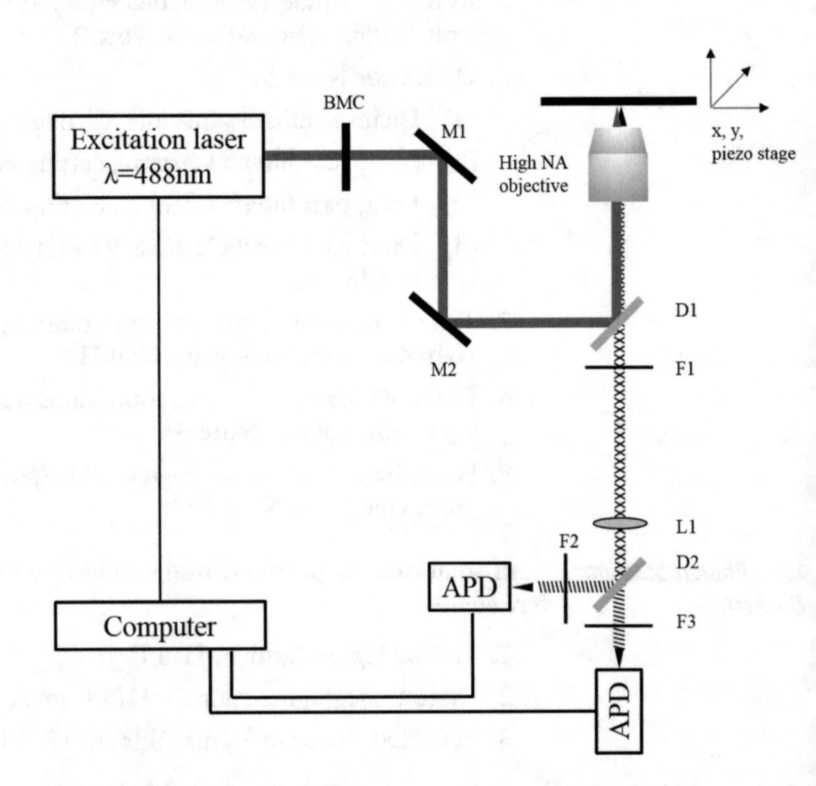

Fig. 1 Schematic of confocal fluorescence microscope setup for high count rate single-molecule FRET experiments. D1 and D2, Dichroic mirrors; F1, emission filter; F2 and F3, band pass filter and long pass filters, respectively; L1, tube lens; APD, avalanche photo diode. BMC, beam collimator

oil-immersion objective (N.A 1.49). The fluorescence emitted from the sample is back collected by the same objective, using the dichroic mirror D1 and a long pass filter, F1, to separate the fluorescence from any scattered excitation light. The collected fluorescence is then focused by a tube lens (included in the inverted microscope) and then split into two detection channels (donor and acceptor) by a second dichroic mirror, D2. Additional band-pass filters, F2, and long-pass filters, F3, are placed in front of the detectors to reduce leak through from the other channel and any leftover scattered excitation light. The path is designed so that the two APD detectors are placed exactly at the focal distance to act as effective pinhole (their active area is only ~200 μm) and reject the out of focus light without the need of adding a separate pinhole and lenses to subsequently focus the light onto the APD detectors (*see* **Note 6**).

3.1.2 Alignment of the Excitation Pathway

1. Attach the optical fiber from the excitation laser into the FC adaptor on the beam collimator (*see* **Note 7**).
2. Turn on the laser and steer the excitation beam (using the steering mirror) toward the dichroic mirror. The excitation laser beam should be aligned parallel to the surface of the optical table (*see* **Note 8**). The dichroic mirror is used to steer the beam to the light port of the microscope. In the absence of any sample, the laser beam spot should be visible on the ceiling above the objective as a blue, speckled circle. Looking at the laser spot, continue to adjust the position of the beam collimator and the steering mirror until you see a perfect circle projected on the ceiling (Fig. 2, *see* **Note 9**).
3. Place an alignment mirror upside down on the stage to reflect the light back towards the objective and onto the dichroic mirror.
4. Place an alignment pinhole in front of the collimator to analyze the reflected beam (*see* **Note 10**).

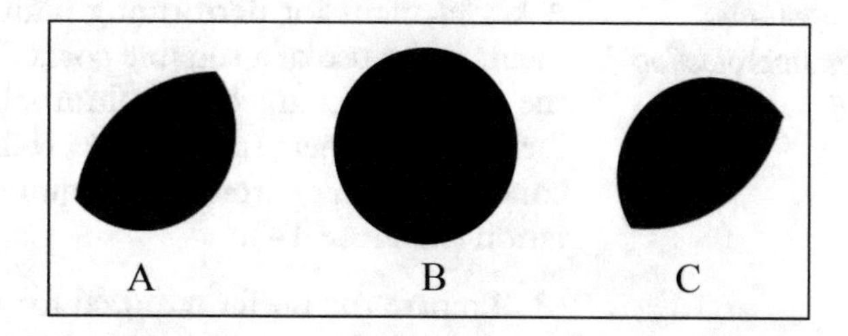

Fig. 2 Excitation beam coming out of the objective. The figure shows the shape of the laser beam as it comes out of the objective in an inverted fluorescence microscope, and observed on a screen kept above the microscope: (**a**) Top-left clipped; (**b**) fine alignment (**c**) bottom-right clipped

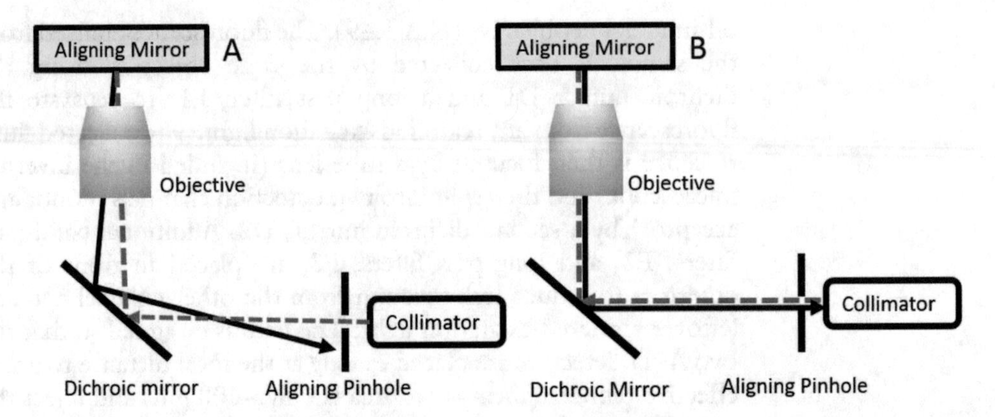

Fig. 3 Schematic of the optical path during alignment of the excitation laser onto the sample using an alignment mirror. (**a**) The excitation and reflected beams do not overlap, and hence, the optical path is not properly aligned. (**b**) The excitation and reflected beams perfectly overlap, indicating that the system is well aligned

5. Move the objective one focal length away from the alignment mirror, so that the profiles of the excitation and reflected beams are similar in size.
6. Adjust the position of collimator and the mirror until the incoming excitation light and the reflected light overlap completely (Fig. 3).

3.1.3 Alignment of the Emission Pathway

1. Remove the alignment mirror from the microscope piezo stage and place a coverslip. Focus the excitation light to the interface of the coverslip, rotate the objective fine focus knob forward ~15 to 20μm, and place 1 drop of the fluorospheres aligning solution on the coverslip (*see* **Note 11**).
2. Cover the two APD detectors with a blackout cloth or curtain, turn off the room lights and then turn on the detectors (*see* **Note 12**).
3. Maximize the photon count rate by adjusting the *X*, *Y*, and *Z* positions of the detector (*see* **Note 13**).

3.2 Preparation of the Photoprotection Cocktail

A key element for performing high count rate SM-FRET experiments is the use of a suitable cocktail of photoprotectors to reduce the photobleaching of the fluorophores and quickly quench the formation of their triplet states (blinking), particularly under the conditions of high irradiance required to saturate fluorophore excitation (*see* **Note 14**).

1. Prepare the buffer required for the experimental system at the desired pH.
2. Prepare the stock solutions of trolox and cysteamine:
   (a) Dissolve trolox in absolute methanol (~5.5 mg/100μL).
   (b) Dissolve cysteamine in HPLC water or filtered Milli-Q water (~40 mg/500μL).

3. Add the required amount of the stock solutions of trolox and cysteamine to the final protein sample so that the final concentrations of trolox and cysteamine are 1 mM and 10 mM, respectively.
4. Add active charcoal to the solution and keep under mild shaking overnight. On the next day, filter the solution with 0.2μm syringe filter followed by 0.05μm syringe filter (3x) to completely eliminate the activated charcoal.

3.3 Preparation of the Sample Well for Small Volume Samples

3.3.1 Cleaning the Sample Holder and Coverslips

1. Rinse the coverslips and the sample holders with acetone/methanol/water three times.
2. Immerse the coverslips and the sample holders in a 1 M KOH solution and sonicate for 30 min.
3. Rinse thoroughly the coverslips and the sample holders with Milli-Q water.
4. Store the coverslips and the sample holders in acetone.

3.3.2 PEGylation of the Sample Holder

1. Prepare the Vectabond reagent solution by diluting 500μL reactive Vectabond in 25 mL of acetone.
2. Remove the coverslips and the sample holders from the acetone and dry them by blowing nitrogen onto them.
3. Attach the coverslips to the sample holder using UV-curing optical glue (Fig. 4).
4. Load the Vectabond reagent solution prepared in the first step to the sample wells and incubate for 5 min at room temperature.
5. Remove the Vectabond reagent solution from the sample wells and rinse them with Milli-Q water.
6. Prepare the PEG solution by dissolving 160 mg of NHS-PEG in 800μL of 100 mM sodium borate at pH 8.5.

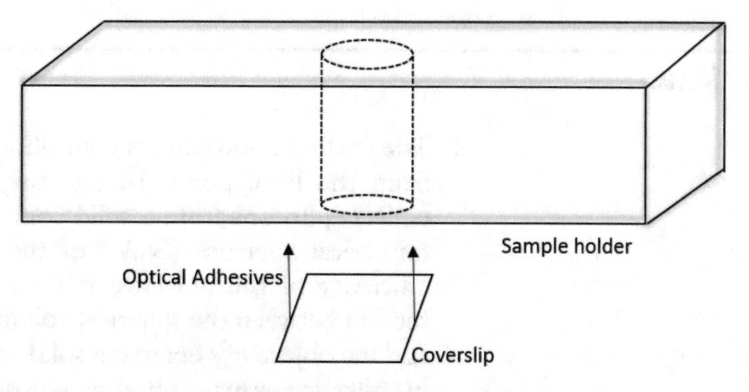

Fig. 4 Schematic of a microscope thick slide containing a cylindrical hole that will constitute the sample well for the SM-FRET experiments once a glass coverslip is glued to the bottom of the slide

7. Add 100μL of the PEG solution on each of the Vectabond-treated sample wells and incubate at room temperature for 3 h.
8. Remove the PEG solution from the sample wells and rinse the sample wells with the photoprotection cocktail prepared in Subheading 3.2.

3.3.3 Long-Term Storage

1. Rinse the unused sample wells with Milli-Q water and dry them by blowing with nitrogen.
2. Place the sample wells in an airtight box and fill the box with nitrogen gas.
3. Store the sample wells at −80 °C (*see* **Note 15**).

3.4 Single-Molecule Fluorescence Measurements with the Confocal Microscope

1. Carefully load 1 drop (~10μL) of optical oil on top of the oil-immersion objective (*see* **Note 16**).
2. Place the sample well slide on top of the microscope XYZ piezo stage.
3. Direct the reflection beam to the eyepiece channel of the microscope and move the objective towards the sample well to focus the excitation light at the interference between the coverslip and air (*see* **Note 17**).
4. Prepare your sample by serial dilutions using the photoprotection cocktail solution prepared in Subheading 3.2 in a dark room (*see* **Note 18**).
5. Load the sample solution at the required concentration (about 25 pM) into the sample well and shift the microscope focus into the solution (move the objective fine focus knob forward ~15 to 20μm).
6. Redirect the emission beam towards the detection channel of the microscope.
7. Turn off the room lights, turn on the APD detectors, and start collecting data (*see* **Note 19**).

4 Notes

1. The excited fluorophores emit photons that spread spherically from the focal point. In contrast, the confocal microscope collects photons from a solid cone of size determined by the numerical aperture (N.A.) of the objective. The collection efficiency of the objective is then determined by the cross-section between the spherical volume of fluorescence emission and the objective's detection solid cone. When one is interested in collecting as many photons as possible an objective should be chosen that is suitable for confocal microscopy and has the highest possible N.A [18]. In our experiments, we use a 100x

oil-immersion objective with N.A. =1.49. This objective contains a correction ring to account for temperature-induced changes. This ring should be adjusted before performing the experiment.

2. Excelitas SPCM APD detectors have the highest quantum efficiency in the market (about 75%). Dark counts (<1000 cps) and time resolution (250 ps) are other important factors to consider when choosing the APD detectors for SM-FRET measurements.
3. Each element placed in the optical path, including lenses, dichroic mirrors, band-pass filters, and long-pass filters decreases the collection efficiency due to transmission losses of about 5–10% per element. Therefore, for high count rate measurements having fewer elements in the path is always preferable. The optical setup shown in Fig. 1 has been specifically designed with this idea in mind and for the SM-FRET pair composed of Alexa 488 as donor and Alexa 594 as acceptor. A different customized optical setup should be considered for other fluorophores pairs.
4. For high count rate experiments, it is very important to use a continuous wave laser as excitation source rather than a pulsed laser. The pulsed lasers typically used in these experiments produce picosecond pulses at 80, 40, or 20 MHz repetition rates. Because the fluorescence lifetimes of the fluorophores used for SM-FRET measurements are <5 ns, and hence significantly shorter than the pulse repetition rate, these lasers are unable to minimize the time the fluorophore spends in the ground state between excitation events, and hence tend to produce lower count rates. Moreover, for the same total excitation power, pulsed lasers concentrate all that power into the very short duration of the pulse, which results in much higher rates of photophysical artifacts (triplet buildup and photobleaching) than for continuous wave lasers, particularly at the high irradiance required for high count rate measurements [1].
5. The fluorospheres yellow and green recapitulate the emission spectra of Alexa 488 and Alexa 594. Fluorospheres, which are extremely bright compared to single fluorophore molecules, are hence an excellent choice to use as reference in the optimization of the optical path and the optical alignment. A different set of fluorospheres should be considered for other fluorophore pairs.
6. With this design, the active area of the APD detectors functions as an effective pinhole, thereby reducing the number of optical elements present in the emission path, and hence the losses in collection efficiency.

7. The ends of the optical fiber are covered with a plastic tip to protect it from dust. To remove the tip, unscrew the threaded barrel that connects the FC connector and adaptor. Before screwing the barrel into the adaptor, ensure that the FC connector key has slid into the FC adaptor notch. Typically, this can be done by rotating the FC connector until you feel the key slip into the notch or hear a click.
8. An aligning target placed at certain height should be used, changing the mirror and making sure the laser beam stays in the same position (does not drift) when the alignment target is moved along the excitation path (the best way to do this is placing a rail on the optical table and setting the alignment target on a rail holder to slide it back and forth).
9. If the excitation beam is completely off, i.e., you do not see the beam reflected on the ceiling, please rotate the microscope nosepiece to the empty position (no objective position) and start adjusting the beam's position from here until you see the light passing through. Be very careful when rotating the objective wheel because the 100x objective is bulky and hence it is easy to scratch it against the moving stage.
10. The power of the excitation laser should be reduced to a safe level (minimal intensity) before attempting to place the aligning pinhole.
11. The fluorospheres aligning solution can be prepared by diluting 1μL of the carboxylate-modified fluorospheres (Thermo Fisher) stock solution into 1 mL of Milli-Q water. Depending on the chosen FRET fluorophore pairs, select appropriate fluorospheres to recapitulate the entire emission spectrum of the FRET dye pairs. For instance, for the FRET pair Alexa 488 and Alexa 594, the best fluorophores selection will be FluoSpheres™ carboxylate-modified microspheres, 0.1μm, yellow-green fluorescent, 2% solids (Thermo Fisher, F8803).
12. APD detectors are extremely sensitive to light, it is important to make sure the room lights are off before you turn on the current on the detectors. Also, make sure the emission filter is set in place before you turn on the detectors to avoid damaging them due to excessive light.
13. The detectors should be placed on *XYZ* stages so that their position can be finely adjusted. Checking the counts registered by the detector, adjust *X* and *Y* at a fixed value of Z to maximize the counts, then change *Z* to a new position, and repeat the process; keep on repeating this procedure until you find the position that results in the highest counts, which will correspond to the optimal alignment.
14. The most important events negatively impacting fluorescence emission are photophysical artifacts of the fluorophore,

including triplet buildup (photoblinking) and irreversible photobleaching. Photoblinking results from the transition of the fluorophore in the excited state to a triplet state by intersystem crossing processes. The triplet is a long-lasting (micro to milliseconds), nonfluorescent state that stalls the excitation-emission cycle until after the fluorophore has relaxed back to the ground state, thereby greatly decreasing the effective count rate. Photobleaching is the irreversible loss of fluorescence emission of the fluorophore, which usually happens by the formation of radical states or other type of nonradiative intermediates. Oxygen is one of the most efficient triplet quenchers available, operating via internal conversion or energy transfer [19]. However, oxygen can also induce photobleaching. A photoprotection cocktail for high count rates has a different formulation from that of one that seeks to maintain the photostability for the longest times under low irradiance. For high count rates under high irradiance, it becomes essential to minimize triplet buildup, and thus ensure that the dissolved oxygen levels are kept high. One solution to this issue is to avoid any components that reduce the levels of dissolved oxygen and use efficient triplet quenchers (e.g., trolox) and oxygen radical scavengers, such as cysteamine. Such cocktails have proved to be extremely efficient for high count rates measurements using rhodamine derivatives [20].

15. If a −80 °C freezer is not available, a −20 °C freezer can be used instead. The unused PEGylated sample well should be good for at least 1 month.
16. Be careful while adding the oil to the objective to make sure it does not overflow or spill through as this can damage the objective.
17. The observation of a typical diffraction pattern with concentric rings provides easy indication that the excitation light is focused exactly at the interference between the coverslip and air. For safety reasons, we recommend to use a microscope camera on the eyepiece to observe the diffraction pattern.
18. To reduce photophysical damage of the fluorophores before starting the single-molecule measurements, it is best to perform all the serial dilutions and store the solution in the dark until the time of measurement.
19. Check the APD detectors' dark counts before starting to collect data. If the dark counts are significantly higher than the technical specifications from the factory certificate, check the instrument light insulation for any possible light leaks.

References

1. Wang Z, Campos LA, Muñoz V (2016) Chapter fourteen—Single-molecule fluorescence studies of fast protein folding. Methods Enzymol 581:417–459
2. Weiss S (1999) Fluorescence spectroscopy of single biomolecules. Science 283 (5408):1676–1683
3. Stryer L, Haugland RP (1967) Energy transfer: a spectroscopic ruler. Proc Natl Acad Sci U S A 58(2):719–726
4. Seidel R, Dekker C (2007) Single-molecule studies of nucleic acid motors. Curr Opin Struct Biol 17(1):80–86
5. Zhuang X (2005) Single-Molecule RNA Science. Annu Rev Biophys Biomol Struct 34 (1):399–414
6. Goodson KA, Wang Z, Haeusler AR, Kahn JD, English DS (2013) LacI-DNA-IPTG loops: equilibria among conformations by single-molecule FRET. J Phys Chem B 117 (16):4713–4722
7. Chris A, Ferreon M, Deniz AA (2012) Protein folding at single-molecule resolution. Biochim Biophys Acta 1814(8):1021–1029
8. Haran G (2012) How, when and why proteins collapse: the relation to folding. Curr Opin Struct Biol 22(1):14–20
9. Schuler B, Eaton WA (2008) Protein folding studied by single molecule FRET. Curr Opin Struct Biol 18(1):16–26
10. Michalet X, Weiss S, Jäger M (2006) Single-molecule fluorescence studies of protein folding and conformational dynamics. Chem Rev 106:1785–1813
11. Aznauryan M, Nettels D, Holla A, Hofmann H, Schuler B (2013) Single-molecule spectroscopy of cold denaturation and the temperature-induced collapse of unfolded proteins. J Am Chem Soc 135 (38):14040–14043
12. Schuler B, Soranno A, Hofmann H, Nettels D (2016) Single-molecule FRET spectroscopy and the polymer physics of unfolded and intrinsically disordered proteins. Annu Rev Biophys 45(1):207–231
13. Hofmann H (2014) Single-molecule spectroscopy of unfolded proteins and chaperonin action. Biol Chem 395(7–8):689–698
14. Smiley RD, Hammes GG (2006) Single molecule studies of enzyme mechanisms. Chem Rev 106(8):3080–3094
15. Chung HS, Eaton WA (2013) Single-molecule fluorescence probes dynamics of barrier crossing. Nature 502(7473):685–688
16. Chung HS, Louis JM, Eaton WA (2010) Distinguishing between protein dynamics and dye photophysics in single-molecule FRET experiments. Biophys J 98(4):696–706
17. Campos LA, Sadqi M, Liu J, Wang X, English DS, Muñoz V (2013) Gradual disordering of the native state on a slow two-state folding protein monitored by single-molecule fluorescence spectroscopy and NMR. J Phys Chem B 117(42):13120–13131
18. Michalet X et al (2014) Silicon Photon-Counting Avalanche Diodes for Single-Molecule Fluorescence Spectroscopy. IEEE J Sel Top Quantum Electron 20(6):248–267
19. Grewer C, Brauer H-D (1994) Mechanism of the triplet-state quenching by molecular oxygen in solution. J Phys Chem 98 (16):4230–4235
20. Campos LA, Liu J, Wang X, Ramanathan R, English DS, Muñoz V (2011) A photoprotection strategy for microsecond-resolution single-molecule fluorescence spectroscopy. Nat Methods 8(2):143–146

Chapter 14

Theory and Analysis of Single-Molecule FRET Experiments

Irina V. Gopich and Hoi Sung Chung

Abstract

Inter-dye distances and conformational dynamics can be studied using single-molecule FRET measurements. We consider two approaches to analyze sequences of photons with recorded photon colors and arrival times. The first approach is based on FRET efficiency histograms obtained from binned photon sequences. The experimental histograms are compared with the theoretical histograms obtained using the joint distribution of acceptor and donor photons or the Gaussian approximation. In the second approach, a photon sequence is analyzed without binning. The parameters of a model describing conformational dynamics are found by maximizing the appropriate likelihood function. The first approach is simpler, while the second one is more accurate, especially when the population of species is small and transition rates are fast. The likelihood-based analysis as well as the recoloring method has the advantage that diffusion of molecules through the laser focus can be rigorously handled.

Key words FRET efficiency histograms, Maximum likelihood, Recoloring

1 Introduction

Conformational structure and dynamics of a single protein can be studied using single-molecule Förster resonance energy transfer (FRET) [1–5]. The experimental output is a sequence of photons emitted by the donor and acceptor fluorophores (dyes) attached to a molecule (*see* Fig. 1). The probability that a photon is emitted by the donor or by the acceptor depends on the distance between them. Photon sequences contain information about the inter-dye distance that keeps changing because of conformational dynamics.

Single molecules can be immobilized (i.e., attached to a surface, embedded in gels, or trapped in lipid vesicles) [6–8], or freely diffuse in solution [9, 10]. The fluorescence signal from diffusing molecules is a sequence of short bursts generated by single molecules traversing the laser spot. A molecule may enter the laser spot several times before it leaves the spot forever; then, another molecule enters the spot [11]. The conformational dynamics of diffusing molecules are not affected by potential interactions with a surface in the case of immobilization, but the duration of a burst is limited by

Victor Muñoz (ed.), *Protein Folding: Methods and Protocols*, Methods in Molecular Biology, vol. 2376,
https://doi.org/10.1007/978-1-0716-1716-8_14,

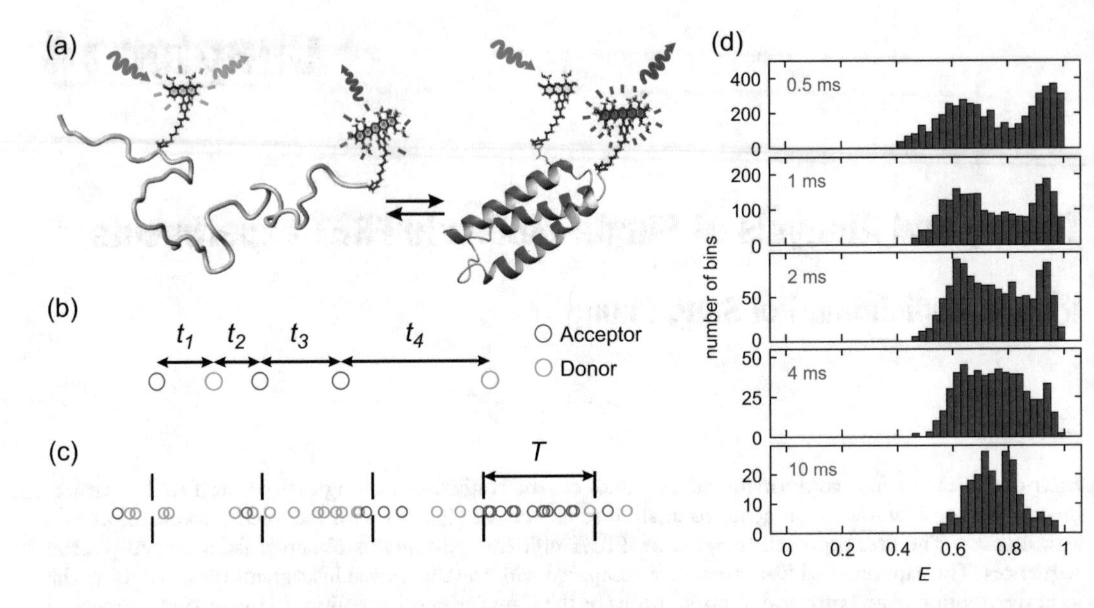

Fig. 1 (**a**) Illustration of a protein with attached donor and acceptor dyes that can exist in folded and unfolded states. After laser excitation, the donor can emit a photon or the excitation energy can be transferred to the acceptor, which then can emit a photon. (**b**) A sequence of acceptor (red) and donor (green) photons with recorded photon arrival times. (**c**) A sequence of photons divided into bins of duration *T*. FRET efficiency in each bin is calculated as the fraction of acceptor photons. (**d**) FRET efficiency histograms of an immobilized protein, α_3D, at different bin times. The histograms show collapse to a single peak as the bin time increases caused by transitions between folded and unfolded states. (Panel (**d**) adapted from Ref. 19)

the time a molecule spends inside the confocal volume, usually on the order of a few milliseconds. On the other hand, photons emitted by an immobilized molecule are recorded for an extended period of time on the order of many seconds, until fluorophores photobleach.

The traditional method of analyzing photon sequences uses FRET efficiency histograms. To construct a histogram, photons are first divided into time bins (*see* Fig. 1c), or bursts of photons from freely diffusing molecules are selected using a search criterion [12, 13]. Then a FRET efficiency in each bin or burst is calculated as the ratio of acceptor photons to the total number of photons. In the histogram, the position of the peak is related to the distance between the two fluorophores [10, 14], and the histogram shape can provide information about conformational dynamics [15–19] (*see* Fig. 1d).

Constructing FRET efficiency histograms is a convenient way to visualize single-molecule data and an excellent tool for the selection of conformational states [20, 21]. By changing experimental conditions (e.g., by adding denaturant to solution), one can readily see the variation in the population of conformational states. Considerable progress in the methods of analyses of FRET efficiency histograms has been made [11, 19, 22–28]. However, extracting quantitative information about conformational dynamics from the histograms is still challenging.

Photons in the above histogram analysis are time-binned. In an alternative photon-by-photon analysis, photon sequences are analyzed without binning. The parameters of the proposed model are found by maximizing an appropriate likelihood function that is calculated using the colors and arrival times of photons [29]. The maximum likelihood analysis is especially advantageous when the photon count rate is too low, and transitions between conformational states are too fast to assign a bin of photons to a single state [30]. This photon-by-photon method can also be used to study short-lived intermediate states and transition paths [31, 32].

Likelihood-based analyses have been used in a variety of contexts, including single-molecule fluorescence spectroscopy [33–40]. These analyses have been often performed using Hidden Markov Models (HMM) for binned photon sequences, in which the numbers of photons in a bin are recorded instead of individual photon information [41–46]. In standard HMM, time is discrete, the observables are photon counts or FRET efficiencies in a bin, and all photons in a bin are emitted from a single conformational state. This method is applicable when conformational dynamics are slow compared to the bin time [47], although the bin size can be reduced so that each bin contains either zero or one photon [48]. The photon-by-photon analysis that we discuss here differs from HMM since the observables are inter-photon times instead of photon counts in a bin. Given a correct model and a sufficient amount of data, one can determine the transition rates that are comparable to and even faster than the photon count rate [30, 49].

In this chapter, we summarize the results of the theory of two-color photon counting in single-molecule measurements. We focus on the FRET efficiency histograms and likelihood-based analysis, which can be exploited to study fast conformational dynamics of both immobilized and diffusing molecules. Both FRET efficiency histograms and the likelihood functions are presented in the same conceptual framework. We devote much attention to the theory underlying each method. In Subheading 2, we discuss several factors that influence photon emission and introduce the basic parameters involved in the theory. In Subheading 3, we discuss the various methodologies for the analysis of two-color single-molecule measurements, including FRET efficiency histograms and maximum likelihood analysis of photon sequences; how to validate model parameters using recolored photon sequences; the application of maximum likelihood method and recoloring to detect a fast process that cannot be probed by the histogram method; and we finish by discussing some practical issues in the two modes of single-molecule FRET experiments.

2 Photon Statistics and Basic Parameters

Conformational dynamics of a protein lead to variations of the inter-dye distance. These variations affect the intensity of donor and acceptor photon emission. Statistics of photon counts depend on how fast the fluctuations of the inter-dye distance are. An important parameter that characterizes a sequence of photons is the mean time between consecutively detected photons, which is usually of the order of 10–100 microseconds (due primarily to low detection efficiency). If the dynamics are faster than this inter-photon time, then photons are uncorrelated and statistics of photons are essentially Poissonian (*see* **Note 1**). For the photons of one color, this means that the distribution of inter-photon times t is exponential, $n\exp(-nt)$, and the probability of detecting N photons in a time bin of duration T is $(nT)^N e^{-nT}/N!$. Here n is the photon count rate (or the fluorescence intensity), which is the mean number of photons detected per unit time. For the acceptor and donor photons, the corresponding photon count rates are the only parameters that characterize Poisson statistics.

Stochastic processes that influence photon statistics occur on a wide range of time scales. Fast processes compared to the photon count rate include photophysical ones (excitation, decay, energy transfer, etc.) as well as changes in the dye orientations and inter-dye distances that occur on the sub-microsecond time scale. The statistics of photons in this case are Poissonian, and all parameters of fast processes as well as background noise are included into the donor and acceptor count rates. On the other hand, the dynamic processes on a time scale that is comparable to or slower than the mean time between photons change the statistics of photon counts. These processes include conformational dynamics on the microsecond time scale and slower and translational diffusion through the laser spot.

When analyzing experimental sequences of photons using FRET efficiency histograms or likelihood functions, one can extract the transition rates and the photon count rates (or the apparent FRET efficiencies defined in Subheading 2.2) associated with the conformational states that interconvert on a time scale comparable to or longer than the inter-photon time. At this stage, the photon count rates and the apparent FRET efficiencies can be regarded as phenomenological parameters. Background noise and other complications such as spectral crosstalk (a fraction of donor photons detected in the acceptor channel) and dye orientational dynamics on the sub-microsecond time scale are included into these count rates and do not affect the interconversion rates. To interpret the extracted FRET efficiencies in terms of the inter-dye distances, these complications need to be carefully considered.

The photon count rates and their dependence on the conformational and translational coordinates are the input of the theory considered in Subheading 3. Next, we show how the count rates are related to the parameters of the simplest scheme describing FRET.

2.1 Photon Count Rates

Consider a donor-acceptor pair (DA) with a fixed inter-dye distance. The donor is excited to form D*A (*see* Fig. 2). The excitation rate constant, k_{ex}, is proportional to the laser intensity and, therefore, depends on the location of the molecule in the laser spot. The donor can either emit a photon or decay non-radiatively, or the excitation can be transferred to the acceptor to form DA*. The energy transfer rate, k_{ET}, depends on the inter-dye distance r and the dye orientations. Dye rotation is often assumed to be fast compared to the donor lifetime, which is usually on the order of a few nanoseconds [50]. Consequently, dye orientations are averaged out (*see* **Note 2**) so that the energy transfer rate is:

$$k_{ET} = k_D \left(\frac{R_0}{r} \right)^6 \tag{1}$$

where k_D is the decay rate constant of the donor in the absence of acceptor and R_0 is the Förster radius.

The donor and acceptor photon count rates (i.e., the mean numbers of photons detected per unit time) can be presented as the product of the radiative rate, the population of the corresponding excited state, and the detection efficiency, since not every emitted photon is detected:

$$\begin{aligned} n_A &= \zeta_A k_A^{rad} p_{ss}(DA^*) = \zeta_A \varphi_A k_A p_{ss}(DA^*) \\ n_D &= \zeta_D k_D^{rad} p_{ss}(D^* A) = \zeta_D \varphi_D k_D p_{ss}(D^* A) \end{aligned} \tag{2}$$

Here $k_{A,D}^{rad}$ are the acceptor and donor radiative decay rates, k_A and k_D are the decay rates of the acceptor and donor excited states, which include both radiative and non-radiative decays, $\phi_A = k_A^{rad}/k_A$ and $\phi_D = k_D^{rad}/k_D$ are the acceptor and donor fluorescence quantum yields, ζ_A and ζ_D are the corresponding detection efficiencies (i.e., the fraction of photons that is observed), and $p_{ss}(DA^*)$ and $p_{ss}(D^*A)$ are the steady-state populations of the acceptor and donor excited states that are normalized as $p_{ss}(DA) + p_{ss}(D^*A) + p_{ss}(DA^*) = 1$. These populations satisfy the steady-state rate equations that correspond to the kinetic scheme in Fig. 2:

$$\begin{aligned} &-k_{ex} p_{ss}(DA) + k_D p_{ss}(D^* A) + k_A p_{ss}(DA^*) = 0 \\ &k_{ex} p_{ss}(DA) - (k_D + k_{ET}) p_{ss}(D^* A) = 0 \\ &k_{ET} p_{ss}(D^* A) - k_A p_{ss}(DA^*) = 0 \end{aligned} \tag{3}$$

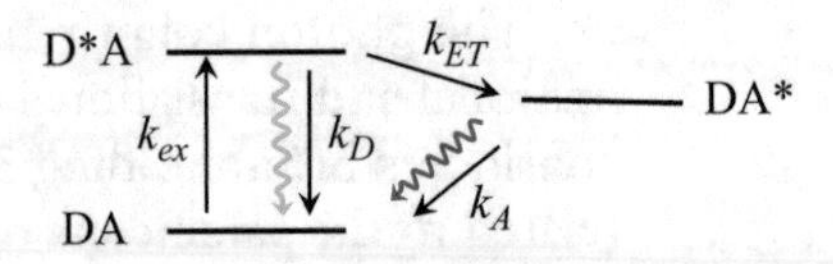

Fig. 2 The kinetic scheme with three electronic states (ground, DA, excited donor, D*A, and excited acceptor, DA*) that describes FRET. The decay rate constants k_A and k_D include both the radiative and nonradiative decay, k_{ex} is the excitation rate that is proportional to the light intensity, and k_{ET} is the rate of energy transfer that depends on the distance between the fluorophores

Solving these equations in the limit $k_{ex} \ll k_{A,D}$ (so that $p_{ss}(DA) \approx 1$) and using Eq. 2, we find the count rates in the limit of low laser intensity:

$$\begin{aligned} n_A &= \zeta_A \varphi_A k_{ex} \frac{k_{ET}}{k_D + k_{ET}} \\ n_D &= \zeta_D \varphi_D k_{ex} \frac{k_D}{k_D + k_{ET}} \end{aligned} \tag{4}$$

The procedure for calculating the count rates can be applied to kinetic schemes of any complexity [51–54] to include various factors such as spectral crosstalk, dye quenching, direct acceptor excitation, donor re-excitation, as well as sub-microsecond fluctuations of the inter-dye distance and orientation. Examples of how to include these factors are described in detail in Ref. 55. The count rates are first expressed in terms of the steady-state probabilities of the excited states that emit photons. These probabilities are then obtained by solving the steady-state rate equations.

Finally, we note that background noise can be readily handled when background photons are Poissonian and independent of the photons emitted by the molecule. In this case, the photon count rate is the sum of the count rate of the photons emitted by the molecule and the background count rate. Thus, Poissonian background noise shifts the count rates but does not alter photon statistics.

2.2 Apparent FRET Efficiency

The photon count rates in Eq. 4 depend on both the inter-dye distance (through the energy transfer rate, k_{ET}) and the translational coordinate (through the excitation rate, k_{ex}, and the detection efficiencies, $\zeta_{A,D}$). In addition to the photon count rates, we will also use alternative parameters related to the count rates, namely, the total count rate, n, and the *apparent* FRET efficiency, ε, which is the probability that the detected photon is emitted by the acceptor fluorophore:

$$\mathcal{E} = \frac{n_A}{n_A + n_D} \tag{5a}$$

$$n = n_A + n_D \tag{5b}$$

Using Eqs. 1 and 4 in Eq. 5a, we find that the apparent FRET efficiency is related to the inter-dye distance by:

$$\mathcal{E} = \frac{\gamma k_{ET}}{k_D + \gamma k_{ET}} = \frac{\gamma}{\gamma + (r/R_0)^6} \tag{6}$$

where the gamma factor is defined as the product of the ratios of quantum yields and detection efficiencies, $\gamma = \zeta_A \phi_A / \zeta_D \phi_D$. Note that the apparent FRET efficiency does not involve the excitation rate, k_{ex}, which depends on the location in the laser spot. On the other hand, using Eq. 4, one can show that:

$$n_A + \gamma n_D = \zeta_A \phi_A k_{ex} \tag{7}$$

This does not involve the energy transfer rate, k_{ET}, which depends on the inter-dye distance. Thus, when $\gamma = 1$, the apparent FRET efficiency, ε, depends only on the inter-dye distance, whereas the total count rate, n, depends only on the location in the laser spot.

The apparent FRET efficiency is different from the gamma-corrected FRET efficiency defined as $n_A/(n_A + \gamma n_D)$. In the simplest case considered above,

$$\frac{n_A}{n_A + \gamma n_D} = \frac{k_{ET}}{k_D + k_{ET}} = \frac{1}{1 + (r/R_0)^6} \tag{8}$$

The gamma-corrected FRET efficiency for the scheme in Fig. 2 is equal to the "true" FRET efficiency, i.e., the probability that the excited state D*A transfers its energy before returning to the ground state [50]. The relation between the gamma-corrected efficiency and the inter-dye distance is particularly simple, which is why the gamma-corrected definition for the FRET efficiency is often used [56]. However, as will be shown later, the statistics of photons are more naturally expressed in terms of the total count rates and the apparent efficiency defined in Eqs. 5a, 5b.

In general, simple relations between the FRET efficiency and the inter-dye distance in Eqs. 6 and 8 do not hold. These relations can be affected by linker dynamics, polymer chain dynamics, spectral crosstalk, donor re-excitation, background noise, etc. [55–60]. In order to extract the inter-dye distance from the measured FRET efficiency, these factors should be carefully taken into account.

3 Methods

3.1 FRET Efficiency Histograms

A common method of analyzing photon sequences involves FRET efficiency histograms. Photon sequences are divided into equally spaced time bins. In the case of diffusing molecules, the vast majority of these bins contain only noise (e.g., detector dark counts). The bins with noise are discarded by imposing a threshold on the total number of photons. Each bin contains different numbers of acceptor, N_A, and donor, N_D, photons. A FRET efficiency is defined as the ratio of the number of acceptor photons to the total number of photons in a bin (*see* **Note 3**):

$$E = \frac{N_A}{N_A + N_D} \tag{9}$$

This FRET efficiency, which is also referred to as the "proximity ratio" [23], fluctuates even when the inter-dye distance is fixed, because the emission of photons is a stochastic process.

The notations introduced here are used throughout the chapter. Capital N_A and N_D denote photon counts that are random integers, small n_A and n_D are the mean numbers of photons per unit time (i.e., the photon count rates). Analogously, the FRET efficiency, E, is a random rational number related to photon counts by Eq. 9, and $\boldsymbol{\varepsilon}$ is the apparent FRET efficiency related to the photon count rates by Eq. 5a. The photon counts, N_A and N_D, and the FRET efficiency, E, (Eq. 9) are the observables, whereas photon count rates, n_A and n_D, and the apparent FRET efficiency, $\boldsymbol{\varepsilon}$, are the parameters that are extracted in data analysis.

We shall review several quantitative approaches to the analysis of FRET efficiency histograms. The first one focuses on the joint distribution of acceptor and donor photon counts, $P(N_A, N_D)$, and its dependence on model parameters [16, 17, 22, 23, 26, 61]. The joint distribution is used to calculate the FRET efficiency histogram, which is then compared to that obtained from data. In the second approach, the FRET efficiency histogram is directly approximated by a set of Gaussians with mean, variance, and amplitude that are not adjustable but are explicit functions of model parameters [18, 19]. Finally, as shown in Subheading 3.3, FRET efficiency histograms can be constructed from the photon sequences that are recolored using model parameters [19, 29, 62].

3.1.1 Single Conformational State

We start by considering an immobilized molecule with a single conformational state. This implies that all fluctuations of fluorescence intensity are fast compared to the mean time between photons. In this case, the donor and acceptor photons are uncorrelated, and the photon statistics are Poissonian. Specifically, the joint probability of detecting N_A acceptor and N_D donor photons in a time bin T is:

$$P(N_A, N_D) = \frac{(n_A T)^{N_A}}{N_A!} \mathrm{e}^{-n_A T} \frac{(n_D T)^{N_D}}{N_D!} \mathrm{e}^{-n_D T} \tag{10}$$

where n_A and n_D are the acceptor and donor photon count rates (*see* Eq. 4). In general, these count rates involve the parameters of all processes that influence photon emission.

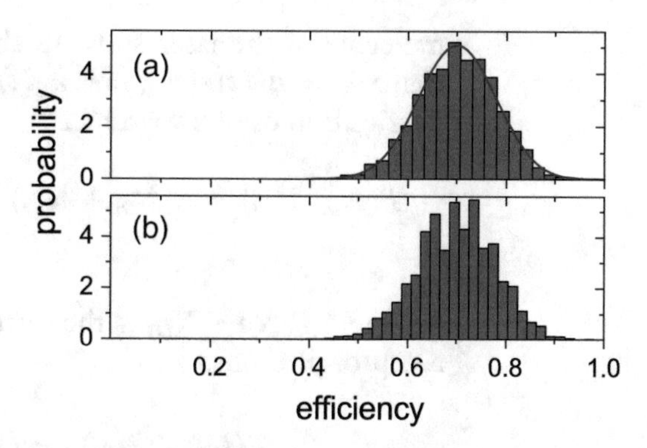

Fig. 3 FRET efficiency histograms for a single conformational state. The histograms are calculated using the Poisson joint distribution, Eq. 11. The parameters are: (**a**) apparent FRET efficiency $\varepsilon = 0.7$, total count rate $n = 30\ \text{ms}^{-1}$, histogram step $h = 1/41$, threshold value $N_T = 30$, and bin time $T = 1$ ms; (**b**) same as (**a**), except $h = 1/51$. Full line in (**a**) shows the Gaussian approximation, Eq. 16

The joint distribution can be expressed in terms of the total count rate, n, and the apparent FRET efficiency, ε, defined in Eq. 5a, 5b. Replacing the count rates $n_A = \varepsilon n$ and $n_D = (1 - \varepsilon)n$, the joint distribution in Eq. 10 can be rewritten as:

$$P(N_A, N_D) = \frac{(nT)^{N_A+N_D}}{(N_A + N_D)!} e^{-nT} \quad \frac{(N_A + N_D)!}{N_A! N_D!} \varepsilon^{N_A} (1 - \varepsilon)^{N_D} \tag{11}$$

Thus, the joint distribution is the product of a Poisson distribution of the total number of photons, $N_A + N_D$, and a binomial distribution, which is the probability of detecting N_A acceptor and N_D donor photons on condition that the total number of photons, $N_A + N_D$, is fixed.

The FRET efficiency histogram for Poisson statistics can be calculated from the joint distribution, Eqs. 10 or 11. In this way, we get the exact histograms that correspond to the limit of an infinite number of time bins. Examples of such histograms are shown in Fig. 3 (*see* **Note 4**). The FRET efficiency histogram of a single conformer is centered at the apparent FRET efficiency, ε, (*see* Eq. 13 below) and broadened by shot noise (i.e., due to the fluctuations in photon counts, which are random integers).

Now consider a molecule that diffuses through the laser spot. In this case, both acceptor, $n_A(t)$, and donor, $n_D(t)$, count rates depend on the location of the molecule in the confocal volume, and therefore fluctuate. The joint distribution of photons can be found only numerically by solving a reaction-diffusion equation [63] or by simulating photon counts modulated by translational diffusion. However, the problem can be simplified if we assume that the ratio of the count rates does not depend on the location of the molecule in the laser spot. In this case, the apparent FRET efficiency $\boldsymbol{\varepsilon} = n_A(t)/[n_A(t) + n_D(t)]$ does not change and the joint distribution can be written as:

$$P(N_A, N_D) = P(N_A + N_D) \quad \frac{(N_A + N_D)!}{N_A!N_D!} \boldsymbol{\mathcal{E}}^{N_A}(1 - \boldsymbol{\mathcal{E}})^{N_D} \tag{12}$$

Here $P(N_A + N_D)$ is the distribution of the sum of donor and acceptor photons:

$$P(N_A + N_D) \quad = \left\langle \frac{(\overline{n}T)^{N_A+N_D}}{(N_A + N_D)!} \mathrm{e}^{-\overline{n}T} \right\rangle$$

where $\overline{n} \equiv \int_0^T (n_A(t) + n_A(t))\, dt/T$ is the average count rate for a single trajectory of length T and $\langle \cdots \rangle$ means averaging over the trajectories of the diffusing molecule through the laser spot that result in detected photons. The distribution $P(N_A + N_D)$ depends on the parameters of the laser spot and on the diffusion coefficient of the molecule.

The distribution in Eq. 12 generalizes the result for immobilized molecules in Eq. 11. The Poisson distribution of the sum of donor and acceptor photon counts has been replaced by the non-Poissonian distribution for diffusing molecules, $P(N_A + N_D)$. Note that although the theory for calculating this distribution for binned photon sequences exists, it can be readily obtained directly from the experimental data [22, 23].

The mean and the standard deviation of the FRET efficiency can be found using the joint distribution in Eq. 12. The mean FRET efficiency is equal to the apparent efficiency and does not depend on any other parameter (i.e., the total count rate or the threshold value):

$$\langle E \rangle = \boldsymbol{\mathcal{E}} \tag{13}$$

The FRET efficiency variance σ_{sn}^2 (the subscript "sn" emphasizes that the variance is due to shot noise) is:

$$\sigma_{sn}^2 = \langle E^2 \rangle - \langle E \rangle^2 = \langle E \rangle (1 - \langle E \rangle) \left\langle (N_A + N_D)^{-1} \right\rangle \tag{14}$$

Here $\left\langle (N_A + N_D)^{-1} \right\rangle = \sum_{N=N_T}^{\infty} N^{-1} P(N) / \sum_{N=N_T}^{\infty} P(N)$ is the average of the reciprocal of the total number of photons, and N_T is the threshold value. The average $\langle (N_A + N_D)^{-1} \rangle$ can be readily calculated from the measured photon sequences used to construct FRET efficiency histograms. The influence of diffusion on the FRET efficiency variance is reflected only in the value of $\langle (N_A + N_D)^{-1} \rangle$.

Since the total number of photons in a bin is always larger than or equal to the threshold value, $N_A + N_D \geq N_T$, we can find an upper bound to the shot noise variance:

$$\sigma_{sn}^2 \leq \frac{\langle E\rangle(1-\langle E\rangle)}{N_T} \tag{15}$$

This upper bound provides a simple estimate for the width of the FRET efficiency histogram due to shot noise [17, 64]. By examining the variance of FRET efficiency distributions, one can establish the existence of multiple conformers. Specifically, if the width of the FRET efficiency histogram is larger than that due to shot noise, then multiple conformations with different efficiencies must be present.

Given the mean and variance, one can approximate the FRET efficiency histogram (*FEH*) by a Gaussian function:

$$FEH(E) \approx \left(2\pi\sigma_{sn}^2\right)^{-1/2} \exp\left(-\frac{(E-\mathcal{E})^2}{2\sigma_{sn}^2}\right) \tag{16}$$

where ${\sigma_{sn}}^2$ is given by Eq. 14. As can be seen in Fig. 3a, the Gaussian approximation describes the FRET efficiency histogram reasonably well.

3.1.2 Multiple Non-interconverting Conformational States

When there are several conformational states, acceptor, n_{Ai}, and donor, n_{Di}, count rates depend on a conformational state i (*see* **Note 5**). Conformational changes lead to additional fluctuations of the acceptor and donor count rates and cause broadening of the FRET efficiency histograms beyond that expected from shot noise. The type of broadening depends on the time scale of the conformational dynamics compared to the bin time.

When the ensemble of conformations does not interconvert during the bin time, the joint distribution for a single diffusing conformer, Eq. 12, can be readily generalized:

$$P(N_A, N_D) = P(N_A + N_D)\ \frac{(N_A+N_D)!}{N_A!N_D!}\sum_i \mathcal{E}_i^{N_A}(1-\mathcal{E}_i)^{N_D} p_{eq}(i) \tag{17}$$

where $\boldsymbol{\varepsilon}_i$ is the apparent efficiency when the molecule is in state i and $p_{eq}(i)$ is the equilibrium population of state i. The above joint distribution simplifies the analysis of FRET efficiency histograms of diffusing molecules because the distribution of the total number of photons does not need to be modeled but can be obtained directly from the experimental data [22, 23].

The above relation is based on the assumption that (1) the apparent FRET efficiency does not depend on the location in the laser spot R (and hence does not fluctuate when the molecule diffuses through the laser spot) and (2) the sum of the acceptor and donor count rates does not depend on the conformational state i (and hence the distribution of the sum of donor and acceptor photons is the same for all states) [17] (*see* **Note 6**):

$$\frac{n_{Ai}(R)}{n_{Ai}(R)+n_{Di}(R)}=\mathcal{E}_i \\ n_{Ai}(R)+n_{Di}(R)=n(R) \tag{18}$$

The FRET efficiency distribution is calculated from the joint distribution in Eq. (17) by summing over donor and acceptor counts. Alternatively, one can approximate the distribution by a weighted sum of Gaussians

$$FEH(E)\approx\sum_i\left(2\pi\sigma_i^2\right)^{-1/2}\exp\left(-\frac{(E-\mathcal{E}_i)^2}{2\sigma_i^2}\right)p_{eq}(i) \tag{19}$$

where $p_{\text{eq}}(i)$ is the equilibrium populations of state i. Each Gaussian results from the bins with photons emitted by the molecule in a single state. It has the mean FRET efficiency, ε_i, and the shot noise variance $\sigma_i^2=\mathcal{E}_i(1-\mathcal{E}_i)\left\langle(N_A+N_D)^{-1}\right\rangle$. This approximation is accurate for both immobilized and diffusing molecules and implies the same assumptions, Eq. 18, as the joint distribution in Eq. 17.

3.1.3 Multiple States with Conformational Dynamics

In this section, we consider dynamics that may occur on an arbitrary time scale, including those comparable to the observation time. For immobilized molecules, an exact analytical joint distribution exists for a system with two interconverting states. For diffusing molecules, the joint distribution invokes an approximation based on the factorization into diffusion-dependent and conformation-dependent factors.

When conformations interconvert during the observation time, the joint distribution cannot be factored as in Eq. 17. Diffusion through the laser spot and conformational dynamics cannot be treated separately. A rigorous theory for the joint distribution that treats conformational dynamics and diffusion through the laser spot has been developed [16]. It relates the joint distribution to the solution of a reaction-diffusion equation for the generating function. However, in general, even though the exact formalism has an elegant mathematical structure, it is not easy to use it to interpret experimental data. Even if one is willing to use numerical methods, the precise dependence of the count rate on the shape of the laser spot is required. To avoid the challenging rigorous treatment of diffusion, the joint distribution is approximated (*see* Eq. 20). The approximation generalizes the approach in the previous section to handle dynamics occurring during the observation time. Kalinin et al. [26] and Santoso et al. [61] presented a closely related but operationally different generalization, which is called dynamic PDA (photon distribution analysis), which invokes essentially the same approximation.

The joint distribution of photons in a bin of size T can be written as the product of the distribution of the sum of acceptor and donor photons, $P(N_A + N_D)$, and the dynamical generalization of the binomial distribution that involves averaging over fluctuations due to conformational changes [17]:

$$P(N_A, N_D) \approx P(N_A + N_D)\ \frac{(N_A + N_D)!}{N_A! N_D!} \left\langle \overline{\mathcal{E}}^{N_A} \left(1 - \overline{\mathcal{E}}\right)^{N_D} \right\rangle_c \tag{20}$$

Here $\overline{\mathcal{E}} \equiv \int_0^T \mathcal{E}(t)\, dt/T$ is the time average of the apparent FRET efficiency $\mathcal{E}(t) = n_A(t)/(n_A(t) + n_D(t))$ over the bin time. The efficiency, $\overline{\mathcal{E}}$, depends on the conformations that are explored during the bin time T. When the molecule interconverts between discrete states with the apparent FRET efficiencies $\boldsymbol{\varepsilon}_i$, then $\overline{\mathcal{E}} = \sum_i \mathcal{E}_i t_i / T$, where t_i is the time spent in state i during the bin time T. Since these times change from bin to bin, $\overline{\mathcal{E}}$ is a random quantity. It differs from the efficiency E defined in Eq. 9, which is the ratio of photon counts in a bin, because $\overline{\mathcal{E}}$ fluctuates only due to conformational changes, whereas E fluctuates due to both shot noise and conformational changes. In the limit of high photon intensity when shot noise is negligible, these two efficiencies coincide. $\langle \ldots \rangle_c$ is the average over conformational (state) trajectories. This average can be obtained analytically for two conformational states (see below). For a larger number of states, this can be done only numerically by simulating state trajectories [26, 61].

The above joint distribution allows one to circumvent analysis of diffusion through the spot by hiding everything that depends on the diffusion into the distribution of the sum of acceptor and donor counts, which can be determined experimentally. The approximation is more accurate when the molecule is quasi-immobilized during the observation time, i.e., the molecule only explores a region of the observation volume where the total count rate does not change significantly. The bins selected for analysis should have fairly uniform photon intensity with no significant gaps. To reduce the error introduced by the above approximation, Kalinin et al. proposed an empirical correction of the extracted interconversion rates [26]. The assumption of quasi-immobilization is not required in the likelihood-based analysis and recoloring considered below.

3.1.4 Two-State Dynamics

Consider two discrete states (e.g., an unfolded and a folded state of a protein), which are characterized by apparent FRET efficiencies $\boldsymbol{\varepsilon}_1$ and $\boldsymbol{\varepsilon}_2$. The states interconvert with the rates k_1 $(1 \to 2)$ and k_2 $(2 \to 1)$. For this model, the average in the joint distribution, Eq. 20, can be found analytically [15] using the distribution of the fraction of time spent in one of two interconverting states obtained by Berezhkovskii et al. [65]. For $\boldsymbol{\varepsilon}_2 > \boldsymbol{\varepsilon}_1$, we have [17]:

$$\begin{aligned}\left\langle \overline{\mathcal{E}}^{N_A} \left(1 - \overline{\mathcal{E}}\right)^{N_D} \right\rangle_c &= p_1 e^{-k_1 T} \mathcal{E}_1^{N_A} (1 - \mathcal{E}_1)^{N_D} + p_2 e^{-k_2 T} \mathcal{E}_2^{N_A} (1 - \mathcal{E}_2)^{N_D} \\ &+ 2kT p_1 p_2 \int_{\mathcal{E}_1}^{\mathcal{E}_2} x^{N_A} (1 - x)^{N_D} [I_0(y) + kT(1 - z) I_1(y)/y] e^{-kzT} dx/\delta\mathcal{E}\end{aligned} \tag{21}$$

where $k = k_1 + k_2$, $p_1 \equiv p_{eq}(1) = 1 - p_2 = k_2/k$, $y = 2kT\left[p_1 p_2(\mathcal{E}_2 - x)(x - \mathcal{E}_1)/(\delta\mathcal{E})^2\right]^{1/2}$, $z = p_1(x - \mathcal{E}_1)/\delta\mathcal{E} + p_2(\mathcal{E}_2 - x)/\delta\mathcal{E}$, $\delta\mathcal{E} = \mathcal{E}_2 - \mathcal{E}_1$, and $I_n(y)$ are modified Bessel functions of the first kind. This expression together with Eq. 20 gives the joint distribution of photons for all values of the interconversion rates. The first two terms in Eq. 21 result from the bins where the initial state does not change during the bin time. The last term (with the integral) results from the bins with one or more transitions between the two states during the bin time. A distribution similar to that in Eqs. 20 and 21 has been exploited in the dynamic PDA for diffusing molecules [26].

The shape of the histograms depends on whether the conformational relaxation time is shorter than, of the order of, or longer than the bin time. This is illustrated in Fig. 4, where the rate of the transitions between the states, k, is varied while keeping the bin size constant. If, instead, k is fixed and the bin size T is varied, the resulting histograms look similar. When conformational dynamics are slow compared to the bin time ($kT \ll 1$), the FRET efficiency distribution is a superposition of two peaks with widths determined by shot noise. When the conformational relaxation time is comparable to the bin time, $kT \sim 1$, transitions between the two states give rise to a plateau between the peaks. This is the case when the shape of the distribution is particularly sensitive to the value of the observation (bin) time. As kT increases, the distribution eventually becomes a Gaussian centered on the average FRET efficiency (the last row). The width of this Gaussian is determined by both shot noise and conformational dynamics.

3.1.5 Gaussian Approximation

A simpler approach to study FRET efficiency histograms is based on an approximation of the FRET efficiency distribution rather than the joint distribution of photon counts. The basic idea is to describe the contribution of the bins that contain no transitions between the states and those that involve transitions by Gaussians with the appropriate mean and variance. This approximation is the extension of the Gaussian approximation in Eqs. 16 and 19 to account for conformational dynamics during the observation time. Below we present the expressions for the FRET efficiency histograms of a molecule with two interconverting conformational states with transition rates k_1 ($1 \to 2$) and k_2 ($2 \to 1$) and with apparent FRET efficiencies $\boldsymbol{\varepsilon}_1$ and $\boldsymbol{\varepsilon}_2$ in each state.

Our approximation for two-state dynamics is the sum of three Gaussian distributions that combines the short- and long-time behavior [18]:

$$FEH(E) \approx \sum_{i=0}^{2} c_i \left(2\pi\sigma_i^2\right)^{-1/2} \exp\left(-\frac{(E - \mathcal{E}_i)^2}{2\sigma_i^2}\right) \tag{22}$$

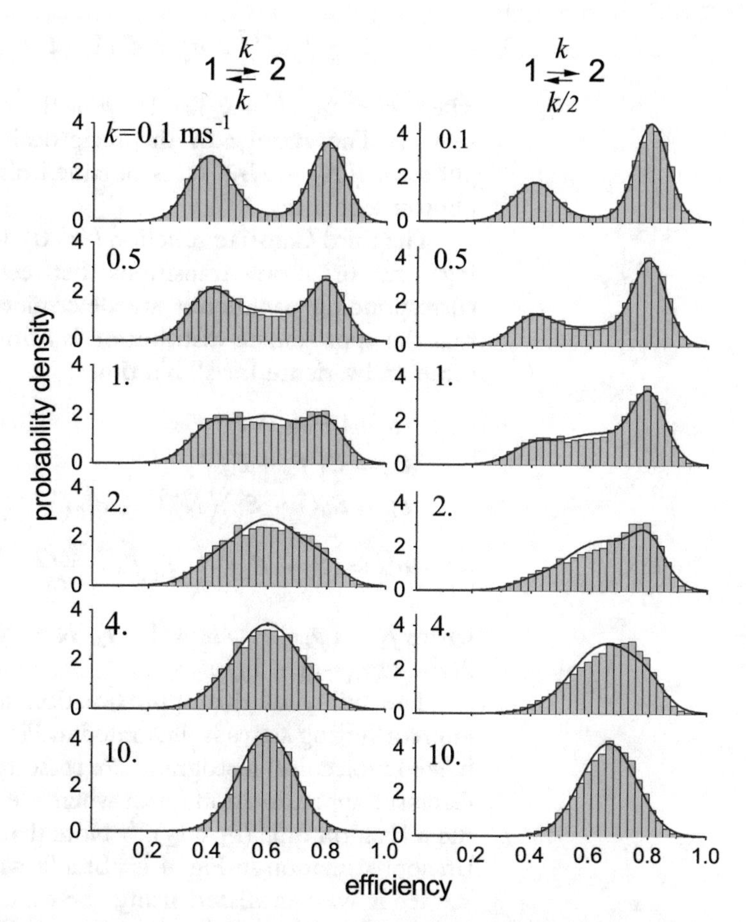

Fig. 4 FRET efficiency histograms for a molecule interconverting between two conformational states with low ($\varepsilon_1 = 0.4$) and high ($\varepsilon_2 = 0.8$) FRET efficiencies. The histograms are shown for different values of the transition rate given in the upper left corner of each histogram. In the first column, the equilibrium populations of the two states are the same. In the second column, the populations differ by a factor of 2. The histograms (bars) are obtained using Eqs. 20 and 21. They are compared with the Gaussian approximation (full curves, Eqs. 20–24). The distribution of the total number of photons is Poissonian with count rate $n = 50$ ms^{-1}. Other parameters: bin size $T = 1$ ms, threshold value $N_T = 30$, histogram step size $h = 1/41$. (Adapted from Ref. 18)

where ε_1 and ε_2 are the efficiencies in the two states. The remaining seven parameters, $c_0, c_1, c_2, \varepsilon_0, \sigma_0, \sigma_1$, and σ_2 are given below. They are explicit algebraic functions of the model parameters ε_1, ε_2, k_1, and k_2.

The Gaussians with $i = 1, 2$ are similar to those in Eq. 19 and describe the bins without transitions during the bin time. Coefficients c_i, $i = 1, 2$, are the probabilities to stay in state i during the entire bin time. The mean FRET efficiency obtained from such bins is exactly ε_1 and ε_2, and the variance results from shot noise. Therefore, c_1, c_2, σ_1^2, and σ_2^2 are given by ($i = 1, 2$).

$$c_i = p_i e^{-k_i T}, \quad \sigma_i^2 = \mathcal{E}_i(1 - \mathcal{E}_i)\left\langle (N_A + N_D)^{-1} \right\rangle \tag{23}$$

where $p_1 = k_2/(k_1 + k_2) = 1 - p_2$ is the equilibrium population of state 1. The average of the reciprocal of the total number of photons, $\langle (N_A + N_D)^{-1} \rangle$, is obtained directly from the measured photon sequences.

The third Gaussian function ($i = 0$) describes the bins containing one or more transitions between the two states. The corresponding parameters are determined from the requirement that the approximate distribution is normalized and has the exact mean and variance for all bin times:

$$\begin{aligned} c_0 &= 1 - c_1 - c_2 \\ \mathcal{E}_0 &= \mathcal{E}_1 f_1 + \mathcal{E}_2 f_2 \\ \sigma_0^2 &= \mathcal{E}_0(1 - \mathcal{E}_0)\langle N^{-1} \rangle + \sigma_{c0}^2(1 - \langle N^{-1} \rangle) \\ \sigma_{c0}^2 &= (\mathcal{E}_1 - \mathcal{E}_2)^2 \left(f_1 f_2 - \frac{p_1 p_2}{c_0} \phi((k_1 + k_2)T) \right) \end{aligned} , \tag{24}$$

where $f_1 = (p_1 - c_1)/c_0 = 1 - f_2$, $N = N_A + N_D$, and $\phi(\tau) = 1 - 2(\tau + \exp(-\tau) - 1)/\tau^2$.

The utility of this approximation for a molecule with two interconverting states is illustrated in Fig. 4. The exact (for immobilized molecules) histograms are reasonably well described by the Gaussian approximation, even when the bin time, T, is similar to the relaxation time $(k_1 + k_2)^{-1}$. Note that the profile by the Gaussian approximation in Fig. 4 is not a fit with adjustable parameters. Rather it was calculated using the same model parameters (i.e., FRET efficiencies of the two states and the rates of transitions) that were used to generate the exact histograms.

The Gaussian approximation in Eqs. 22–24 has been used to extract folding and unfolding rates and mean FRET efficiencies of the folded and unfolded subpopulations of a protein, α_3D [19]. Fitting FRET efficiency histograms to the approximation yields surprisingly accurate rates for immobilized molecules when the population of the folded or unfolded components is not small (*see* Fig. 5). Extracted parameters are most accurate when there are comparable contributions to the FRET efficiency histograms from all three Gaussian components. In free diffusion experiments, the rate coefficients obtained by the three-Gaussian method require correction and should be taken as a measure of the time scales and not as an exact parameter [19] (*see* **Note** 7).

The above approximation for two interconverting states can be extended to multiple states [18]. For example, the FRET efficiency histogram for a three-state molecule is a sum of seven Gaussians (six Gaussians for a linear three-state model 1↔2↔3). This approximation correctly reduces to the appropriate two-state approximation in all limiting cases when the three-state kinetic scheme reduces to a two-state one (e.g., transitions between any two states become sufficiently fast) (*see* **Note 8**).

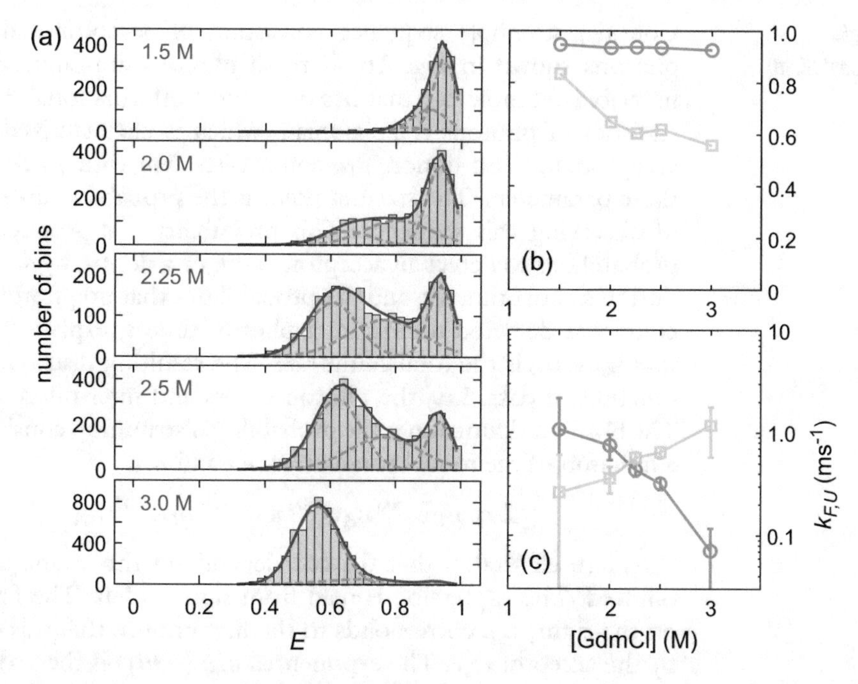

Fig. 5 Gaussian approximation to FRET efficiency histograms of Alexa 488/Alexa 594-labeled α_3D. (**a**) FRET efficiency histograms constructed by binning photons from immobilized molecules with bin time 1 ms at different concentrations of GdmCl. Dashed green curves are individual Gaussian components and solid red curves are their sum calculated for a two-state model from Eqs. 22–24. (**b**) Apparent FRET efficiencies and (**c**) folding (circle) and unfolding (square) rates obtained by fitting the Gaussian approximation to the histograms with 1 ms bin time. (Adapted from Ref. 19)

3.2 Likelihood-Based Analysis of Photon Sequences

Photon sequences can be analyzed without binning using the likelihood that a given set of model parameters can describe the observed sequence of photons. The likelihood function is maximized to obtain the parameters that best describe the data in the framework of the adopted model. The advantage of this method is that it uses maximum information from photon sequences and allows one to analyze diffusion experiments in a rigorous way. It is especially useful when the photon sequences do not have any apparent structure, due to high background noise, states with similar FRET efficiencies and/or fast conformational transition rates.

Below, we first explain the basic ideas by considering the simplest case of one conformational state and then consider how to deal with conformational dynamics and translational diffusion. The observed data is a sequence of photons with specified photon color and inter-photon times. We address the question of what is the appropriate likelihood function that describes the data and how this depends on the model parameters, i.e., the acceptor and donor count rates (or, the total count rate and the apparent FRET efficiency) and the parameters describing conformational dynamics.

3.2.1 Single Conformational State

Consider a simple sequence consisting of acceptor and donor photons shown in Fig. 1b. If these photons are emitted by an immobilized molecule that has only one conformational state, the statistics of photons is Poissonian, which is characterized by the acceptor, n_A, and donor, n_D, count rates. We wish to determine these parameters. The starting point is the probability distribution of observing this sequence. This probability is a product of the probabilities to detect an acceptor, $n_A dt$, or a donor, $n_D dt$, photon during a short time dt, and the probabilities that no photon of any color was detected during inter-photon time t, $\exp(-nt)$, where $n = n_A + n_D$ is the total count rate. The resulting distribution is a function of data, i.e., the photon colors and inter-photon times. The likelihood function is the probability distribution considered as a function of the model parameters, n_A and n_D:

$$L \propto n_D \mathrm{e}^{-nt_4} n_A \mathrm{e}^{-nt_3} n_A \mathrm{e}^{-nt_2} n_D \mathrm{e}^{-nt_1} n_A \tag{25}$$

Here all factors that do not depend on the count rates are omitted. The expression is read from right to left. The first term on the right, n_A, corresponds to the first photon that was emitted by the acceptor dye. The exponential exp.$(-nt_1)$ is the probability that no photons were detected during the time interval t_1, the second donor photon results in n_D, and so on. By varying n_A and n_D and maximizing the likelihood function, one can find the most likely count rates that are consistent with the observed sequence of photons.

Using the total count rate, $n = n_A + n_D$, and the apparent FRET efficiency, $\varepsilon = n_A/(n_A + n_D)$, the count rates can be presented as $n_A = \varepsilon n$ and $n_D = (1 - \varepsilon)n$. Then the above likelihood function can be written as a product of two factors

$$\begin{aligned} L &\propto n_A^3 n_D^2 \mathrm{e}^{-n(t_1+t_2+t_3+t_4)} \\ &= \left(n^5 \mathrm{e}^{-n(t_1+t_2+t_3+t_4)}\right) \left(\mathcal{E}^3 (1-\mathcal{E})^2\right) \end{aligned} \tag{26}$$

A particularly advantageous feature of this presentation is that the first factor involves only the total count rate, n, and the second one depends only on the apparent FRET efficiency, ε. Therefore, if one is interested only in the FRET efficiency, the first factor can be omitted when finding the maximum of the likelihood function.

Generalizing the above likelihood for a sequence containing N_A acceptor and N_D donor photons and keeping only the factor that depends on the apparent FRET efficiency, we have

$$L \propto \mathcal{E}^{N_A} (1 - \mathcal{E})^{N_D} \tag{27}$$

The maximum of the likelihood function (or, more conveniently, of the log-likelihood) can be found by setting the derivative with respect to ε to zero, $d \ln L / d\mathcal{E} = 0$. In simple cases such as above, this equation can be solved analytically. Using the likelihood in Eq. 27, we find the estimate for the apparent FRET efficiency ε:

$$\mathcal{E}_{\max} = \frac{N_A}{N_A + N_D}$$

The error of this estimate, $\sqrt{\sigma^2}$, can be found by estimating the curvature if the above likelihood function at the maximum, $\sigma^2 = 1/\mid d^2 \ln L/d\mathcal{E}^2 \mid$, which results in:

$$\sigma^2 = \frac{\mathcal{E}_{\max}(1 - \mathcal{E}_{\max})}{N_A + N_D}$$

Note that this result is similar to the variance in Eq. 14. In more complicated cases, the estimates for the model parameters cannot be found analytically and numerical methods of optimization should be invoked.

3.2.2 Multiple Conformations

Now we consider the likelihood function for an immobilized molecule that have several interconverting conformational states (*see* **Note 5**). When the molecule is in state i, the detected photons have Poisson statistics, which are characterized by the acceptor and donor count rates, n_{Ai} and n_{Di}. The transitions between the states are described by the rate matrix **K**. The matrix element K_{ij} is the rate constant of the transition from state j to state i (*see* **Note 9**). This process is also known as the Markov-modulated Poisson process in queuing theory [66].

The probability density of observing the sequence of photons in Fig. 1b, since it may result from any state trajectory, is

$$L \propto \sum_{i,j,k,l,m} n_{Dm} G_{ml}(t_4) n_{Al} G_{lk}(t_3) n_{Ak} G_{kj}(t_2) n_{Dj} G_{ji}(t_1) n_{Ai} p_{eq}(i) \tag{28}$$

where $G_{ji}(t)$ is the probability of finding the molecule in state j at time t given it was in state i initially and no photons were detected during this time interval, and $p_{\text{eq}}(i)$ is the equilibrium probability of state i. The summation is performed over all possible conformational states. The above expression (read from right to left) has the following interpretation, similar to that in Eq. 25. The first term on the right, $p_{eq}(i)$, arises because, initially, the system is in equilibrium. The next term corresponds to the first acceptor photon emitted by the molecule in state i, n_{Ai}; then the molecule evolves from state i to j without photons being detected during time t_1, $G_{ji}(t_1)$; a donor photon emitted from state j is detected at time t_1, n_{Dj}; and so on.

The probability $G_{ji}(t)$ describes transitions between the states, given that no photons are detected during time t. It can be shown [29] that $G_{ji}(t)$ is the ji element of the matrix exponential, $\exp((\mathbf{K} - \boldsymbol{\mathcal{N}})t)$, where $\boldsymbol{\mathcal{N}}$ is a diagonal matrix with the total count rates $n_i = n_{Ai} + n_{Di}$ on the diagonal. Introducing the diagonal matrices $\mathcal{N}_A$ and $\mathcal{N}_D$ with the elements n_{Ai} and n_{Di} on the diagonal, the likelihood in Eq. 28 can be written in matrix notation as follows:

$$L \propto \mathbf{1}^{\mathrm{T}}\mathcal{N}_D e^{(\mathbf{K}-\mathcal{N})t_4}\mathcal{N}_A e^{(\mathbf{K}-\mathcal{N})t_3}\mathcal{N}_A e^{(\mathbf{K}-\mathcal{N})t_2}\mathcal{N}_D e^{(\mathbf{K}-\mathcal{N})t_1}\mathcal{N}_A \mathbf{p}_{eq} \tag{29}$$

where $\mathbf{p}_{\mathrm{eq}}$ is the vector of equilibrium probabilities, T means transpose, and $\mathbf{1}^{\mathrm{T}}$ is a row vector with all its elements equal to 1.

Generalizing the above expression to an arbitrary number of photons N_{ph} in a sequence, we have [29]:

$$L \propto \mathbf{1}^{\mathrm{T}}\mathcal{N}(c_{N_{ph}})\ \Pi_{k=1}^{N_{ph}-1}\left(e^{(\mathbf{K}-\mathcal{N})t_k}\mathcal{N}(c_k)\right)\ \mathbf{p}_{eq} \tag{30}$$

where matrix $\mathcal{N}(c_i)$ depends on the color c_i of ith photon, i.e., $\mathcal{N}(\text{acceptor}) = \mathcal{N}_A$ and $\mathcal{N}(\text{donor}) = \mathcal{N}_D$, $\mathcal{N} = \mathcal{N}_A + \mathcal{N}_D$. When there are several photon sequences from different molecules, the likelihood is the product of the likelihoods in Eq. 30, corresponding to different sequences.

The likelihood function in Eq. 30 is exact when the photon statistics in each conformational state are Poissonian. The parameters of the model (i.e., the photon count rates n_{Ai} and n_{Di} and the transition rates K_{ij}) are found by optimizing numerically the logarithm of the likelihood function (the log-likelihood) (*see* **Note 10**).

3.2.3 Reduced Likelihood

When the total count rate $n = n_{Ai} + n_{Di}$ does not depend on conformational states i, the likelihood function can be factored similar to that in Eq. 26. Specifically, the likelihood function in Eq. 29 can be expressed as a product of two factors. One factor involves only the total count rate n and does not depend on the conformational parameters. Therefore, it can be ignored and only the other factor must be optimized with respect to the FRET efficiencies and transition rates:

$$L \propto \mathbf{1}^{\mathrm{T}}(\mathbf{I}-\mathbf{E})\ e^{\mathbf{K}t_4}\mathbf{E}\ e^{\mathbf{K}t_3}\mathbf{E}e^{\mathbf{K}t_2}(\mathbf{I}-\mathbf{E})\ e^{\mathbf{K}t_1}\mathbf{E}\ \mathbf{p}_{eq} \tag{31}$$

where $\mathbf{E}$ is the diagonal matrix with apparent FRET efficiencies $\boldsymbol{\varepsilon}_i$ on the diagonal and $\mathbf{I}$ is the unity matrix. The first term on the right, $\mathbf{p}_{eq}$, arises because initially the system is in equilibrium. The next term $\mathbf{E}$ corresponds to the first detected photon (the first acceptor photon in Fig. 1b). The evolution of the conformational states until the next photon is detected at time t_1 is described by the matrix exponential, exp.($\mathbf{K}t_1$). The next term $\mathbf{I} - \mathbf{E}$ corresponds to the donor photon, and so on. The final multiplication by $\mathbf{1}^{\mathrm{T}}$ sums over all conformational states.

Generalizing Eq. 31 to an arbitrary number of photons in a burst, the likelihood can be written as [29]:

$$L \propto \mathbf{1}^{\mathrm{T}} \mathbf{F}(c_{N_{ph}}) \, \Pi_{k=1}^{N_{ph}-1} \left(\mathrm{e}^{\mathbf{K} t_k} \mathbf{F}(c_k) \right) \, \mathbf{p}_{eq} \tag{32}$$

where $\mathbf{F}(c_k)$ is a matrix which depends on the color of kth photon c_k: $\mathbf{F}(acceptor) = \mathbf{E}$ and $\mathbf{F}(donor) = \mathbf{I} - \mathbf{E}$, t_k is the time between the $(k+1)$th and kth photon. The product is taken over all photons in the burst. Note that Eqs. 32 and 30 reduce the calculation of the likelihood to successive matrix-vector multiplication, which can be performed iteratively.

For the two-state model describing folded and unfolded states of a protein, the matrices in Eq. 32 are:

$$\mathbf{E} = \begin{pmatrix} \mathcal{E}_1 & 0 \\ 0 & \mathcal{E}_2 \end{pmatrix}, \quad \mathbf{K} = \begin{pmatrix} -k_1 & 0 \\ 0 & -k_2 \end{pmatrix}, \quad \mathbf{p}_{eq} = \begin{pmatrix} p_1 \\ p_2 \end{pmatrix} \tag{33}$$

where ε_1 and ε_2 are the FRET efficiencies in the folded and unfolded states, k_1 and k_2 are the unfolding and folding rates, and $p_1 = 1 - p_2 = k_2/(k_1 + k_2)$ is the equilibrium population in state 1. For more complex models, the matrices change to account for additional states. The above likelihood function can also be extended to the models with a continuous conformational coordinate (e.g., diffusion on a potential of mean force). The rate matrix is specified by a few parameters, which are found by optimizing the likelihood function [40].

The likelihood function for many bursts is a product of the likelihoods of individual bursts in Eq. 32. This product (or, equivalently, the sum of the log-likelihoods) is optimized numerically by varying the parameters, i.e., the apparent FRET efficiencies of the states, ε_i, and the transition rates K_{ij}. To get true FRET efficiency or the inter-dye distance, the obtained apparent FRET efficiency should be corrected for background noise, crosstalk, etc. (*see* **Note 11**).

The reduced likelihood function in Eq. 32 is particularly advantageous for the analysis of diffusing molecules since it does not involve photon count rates. The complete likelihood function in Eq. 30 for a molecule that diffuses through the confocal spot should be integrated over all possible paths of the diffusing molecule. The problem of the likelihood optimization becomes computationally challenging. However, the reduced likelihood function in Eq. 32 allows one to bypass modeling of translational diffusion. This likelihood function is valid when the conditions in Eq. 18 hold (*see* **Note 6**).

3.2.4 Non-interconverting Species

When conformational dynamics are on a time scale much longer than the bin or burst duration (static limit), the likelihood function in Eq. 32 simplifies to [17, 58]:

$$L \propto \sum_i \mathcal{E}_i^{N_A} (1 - \mathcal{E}_i)^{N_D} p_{eq}(i) \tag{34}$$

where N_A (N_D) is the number of acceptor (donor) photons in a burst. The summation is performed over all conformational states i. Note that the above likelihood function does not involve the interphoton times. The likelihood function for many bursts is a product of the likelihoods in Eq. 34. The likelihood function for static species in Eq. 34 was used to analyze photon sequences from polyproline20, which is a stiff rod on a time scale of burst duration [58].

3.2.5 Accuracy of the Parameter Estimation

The error of the extracted parameters x_i is estimated by evaluating the curvature of the likelihood function at the maximum [67]. Assuming that there are many data and the likelihood function near the maximum is roughly Gaussian, the log-likelihood is approximated as $-\log L \approx \frac{1}{2}\sum_{ij} H_{ij}\delta x_i \delta x_j + c$, where $\delta x_i = x_i - m_i$ is the difference of the parameter x_i and its value at the maximum m_i, $H_{ij} = -\partial_{x_i}\partial_{x_j} \log L$ is an element of the Hessian matrix, $\mathbf{H}$, and c is a constant. The Hessian matrix is an inverse of the covariance matrix, so the variance of x_i is $\langle \delta x_i^2 \rangle = \left[\mathbf{H}^{-1}\right]_{ii}$. The standard deviation of the extracted parameter x_i is $\left[\mathbf{H}^{-1}\right]_{ii}^{1/2}$. The elements H_{ij} involve partial derivatives of the log-likelihood at the maximum, which are evaluated numerically. When the parameters are independent, the calculation simplifies since only diagonal terms H_{ii} are involved (*see* **Note 12**).

The accuracy of the extracted parameters depends on how fast the transition rates are compared to the photon count rate [30, 49]. When the transitions between the states are slow and there are many photons detected while the molecule stays in one state, the key parameters are the number of transitions that occur during all photon sequences, N_{tr}, and the number of independent photon sequences, M. More specifically, consider the errors of the parameters for a two-state molecule with equally populated states $p_1 = p_2 = 1/2$, the relaxation rate k (the transition rates are $k/2$), and apparent FRET efficiencies ε_1 and ε_2. The variances of the extracted parameters in the slow transition limit are ($i = 1, 2$) [49]:

$$\frac{\langle \delta k^2 \rangle}{k^2} = \frac{1}{N_{tr}} = \frac{2}{kT}, \quad \frac{\langle \delta p_i^2 \rangle}{p_i^2} = \frac{1}{M + N_{tr}}, \quad \langle \delta \mathcal{E}_i^2 \rangle = 2\frac{\mathcal{E}_i(1-\mathcal{E}_i)}{N_{ph}} \tag{35}$$

where T is the duration of all photon sequences, so that $N_{tr} = kT/2$ is the mean number of transitions between the two states during T, and N_{ph} is the number of photons in all sequences.

When the transitions between the states are very fast and the states cannot be directly identified, the accuracy is determined by the small fraction of photons that are correlated with their neighbors. In this regime, the errors of the parameters are very sensitive to the separation of the FRET efficiencies in the states [30]. For example, for the two-state system, the variances of parameters in the fast transition limit are $\langle \delta k^2 \rangle \propto (\mathcal{E}_2 - \mathcal{E}_1)^{-4}$, $\langle \delta p_i^2 \rangle \propto (\mathcal{E}_2 - \mathcal{E}_1)^{-6}$, and $\langle \delta \mathcal{E}_i^2 \rangle \propto (\mathcal{E}_2 - \mathcal{E}_1)^{-4}$.

In the special case when the FRET efficiencies in the two-state system are 0 and 1, the variance of the relaxation rate and populations can be found analytically for arbitrary values of the relaxation rate k. Assuming equal populations in the two states ($p_1 = p_2 = 1/2$), we have [49]:

$$\begin{aligned} \frac{\langle \delta k^2 \rangle}{k^2} &= \left(N_{ph} \frac{n}{4k} \zeta\left(3, 1 + \frac{n}{2k}\right) \right)^{-1} \\ \frac{\langle \delta p_1^2 \rangle}{p_1^2} &= \left(N_{ph} \left(\frac{n}{k} \psi\left(1 + \frac{n}{2k}\right) - \frac{n}{k} \psi\left(\frac{n}{2k}\right) - 1 \right) + M \right)^{-1} \end{aligned} \tag{36}$$

where n is the photon count rate, $N_{ph} = nT$ is the total number of photons, $\zeta(z, a) \equiv \sum_{i=0}^{\infty} 1/(i + a)^z$ is the Hurwitz zeta function, and $\psi(z) = \Gamma'(z)/\Gamma(z)$ is the digamma function. These variances provide the estimate for the error for arbitrary k, including fast and slow transition limits. The variances in Eq. 36 reduce to those in Eq. 35 in the slow transition limit when $k \ll n$. It is interesting that the standard deviation of the relaxation rate has a "chevron" shape as a function of the transition rate in the log-log scale (*see* Fig. 6).

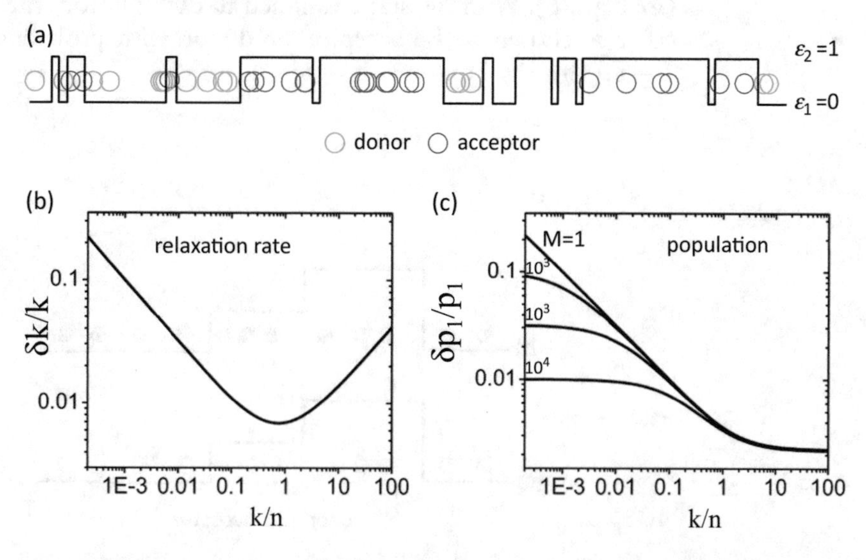

Fig. 6 Transitions between two states with FRET efficiencies 0 and 1. (**a**) Acceptor (red) and donor (green) photon and state (black) trajectories. The states are equally populated, total photon count rate $n = 100$ in arbitrary time units, two-state relaxation rate $k = 100$. (**b**) and (**c**) Relative standard deviation of the maximum likelihood estimate of the relaxation rate (**b**) and population (**c**) as a function of the ratio of the relaxation rate, k, and photon count rate, n (the square roots of the variances in Eq. 36). The number of photon sequences is $M = 1, 10^2, 10^3, 10^4$; the number of photons in all sequences is $N = 200{,}000$. (Adapted from Ref. 49)

3.3 Recoloring Photon Sequences

The model and model parameters can be validated by recoloring photon sequences [29]. In this method, the colors of experimental photon sequences are erased without changing the times between photons. Then the photons are recolored consistent with the model of conformational dynamics. In this way, we get a new photon sequence with the same times between photons as the measured ones, but with different colors. The recolored sequence of photons is then used to obtain a new FRET efficiency histogram or a new correlation function, which is compared with the observed one. The recoloring method is closely related to the likelihood function in Eq. 32 and is rigorous under the same conditions in Eq. 18.

The algorithm for recoloring is as follows. For each sequence of photons, the initial state is drawn from the equilibrium probability. If this state is i, the color of the first photon in the sequence is chosen to be acceptor or donor with probabilities ($\boldsymbol{\varepsilon}_i$, $1 - \boldsymbol{\varepsilon}_i$). The state j at the moment the second photon is detected is chosen with the probability $[\exp(\mathbf{K}t)]_{ji}$, where t is the time between the first and the second photon. This procedure is then repeated until the last photon of the sequence. The next sequence is recolored by the same way, starting with choosing the initial state.

An equivalent recoloring scheme is shown in Fig. 7. For each photon burst or fragment of a burst selected for the analysis (Fig. 7a), a conformational state trajectory of length equal to the burst duration is generated using the model parameters. The state trajectory and the colorless experimental burst are superimposed (*see* Fig. 7b). With the states assigned to each photon, the photon color is chosen to be acceptor or donor with probabilities ($\boldsymbol{\varepsilon}_i$, $1 - \boldsymbol{\varepsilon}_i$) (*see* Fig. 7c).

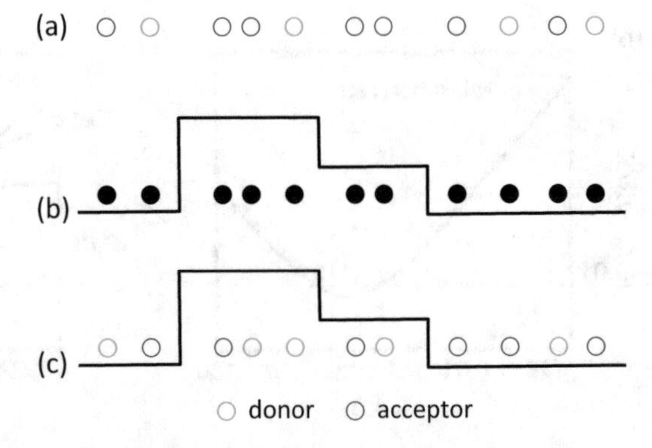

Fig. 7 The photon-by-photon recoloring algorithm for data analysis. (**a**) A sequence of donor (green) and acceptor (red) photons in a burst of duration T. (**b**) The superposition of the photon sequence, from which the colors have been erased, with a three-state conformational trajectory. (**c**) Photons are recolored using the appropriate apparent FRET efficiency as the probability of red (acceptor) color. (Adapted from Ref. 77)

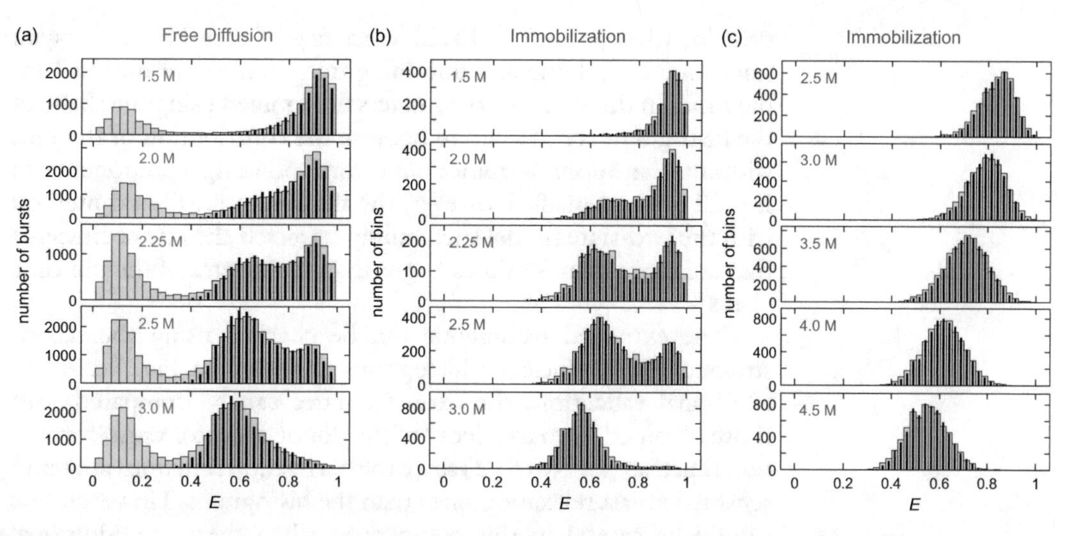

Fig. 8 Recolored FRET efficiency histograms of α_3D ((**a**) and (**b**)) and villin (**c**), at different concentrations of GdmCl in the free diffusion experiment (**a**) and immobilization experiment ((**b**) and (**c**)). Histograms with wide bars are experimental data, and those with narrow bars are calculated by recoloring the photons using a two-state model with parameters obtained by maximizing the likelihood function. (Adapted from Refs. 19, 68)

Application of the recoloring method to test a two-state model for protein folding kinetics is illustrated in Fig. 8. The histograms from both diffusing and immobilized molecules were compared to those obtained from the recolored photon sequences [19, 68]. The parameters (i.e., the FRET efficiencies of the folded and unfolded states and interconversion rates) were found by optimizing the likelihood function for the two-state model in Eqs. 32–33 (*see* **Note 13**). As can be seen from the comparison, the experimental and recolored histograms are in good agreement, demonstrating the adequacy of the two-state model.

Recoloring photon sequences can be also considered as an independent method to construct FRET efficiency histograms without calculating the joint distribution or the Gaussian approximation. Unlike these two methods, the recoloring method does not involve approximate factorization as in Eq. 20, so that it is valid also for diffusing molecules when conditions in Eq. 18 are satisfied. The best model parameters can be found by minimizing the deviation between the experimental and recolored FRET efficiency histogram [62].

3.4 Detection of Invisible States Using Likelihood Method and Recoloring

Since the likelihood analysis utilizes maximum information of single-molecule data by analyzing single photons, it is possible to probe dynamics that are so fast that cannot be visualized in the FRET efficiency histogram. Analysis of fast two-state folding trajectories is such an example [19, 68]. As shown in Fig. 8c, there is only one peak in the FRET efficiency histogram at all denaturant concentrations because of fast folding and unfolding rates of the protein, villin [68]. The FRET efficiency of each bin is averaged since multiple folding and unfolding transitions occur during 1 ms bin time. In this case, the parameters determined using the shape of the histograms are inaccurate because the contribution of the bins without transitions is minor and corresponding components in Eq. 22 are too small. However, the maximum likelihood method with the two-state model successfully extracted the rate coefficients that are more than 30 times larger than the inverse of the bin time (1 ms).

The extracted parameters can be checked using the reconstructed FRET efficiency histograms as shown in Fig. 8c. As an additional validation, the extracted rates can be compared with those obtained from the decay of the donor-acceptor cross-correlation function (*see* **Note 12**) since the correlation functions are more sensitive to fast relaxation rates than the histograms. However, one should be careful in this comparison when there are additional processes that change the correlation function (but do not contribute to the shape of the histogram). For example, fluorophores always blink and this affects the determination of the rates between states. This effect becomes larger as protein dynamics becomes faster and more comparable to the blinking kinetics [30, 49]. The advantage of using the maximum likelihood method is that the data can be analyzed using the model including fast photoblinking of dyes.

It should be noted that the donor-acceptor cross-correlation function is affected by both donor and acceptor blinking. However, in the maximum likelihood analysis using the reduced likelihood function without photon count rates (Eq. 32), only acceptor blinking should be explicitly included into the model [68]. Figure 9 compares the experimental donor-acceptor cross-correlation function with those obtained from the recolored photon sequences using a two-state model without blinking (Fig. 9b) and a four-

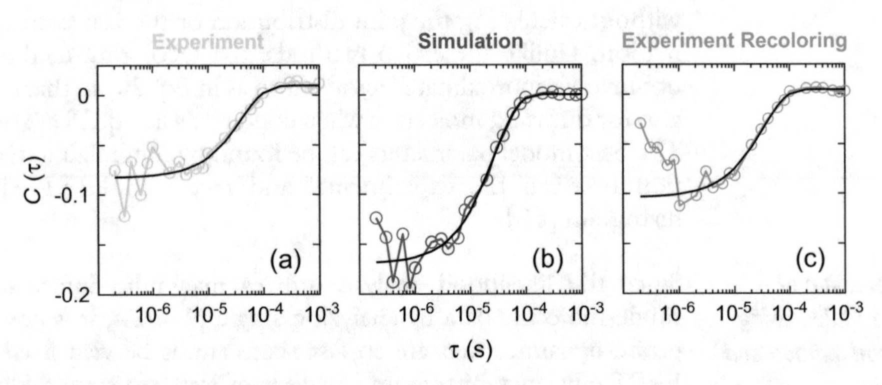

Fig. 9 Donor-acceptor cross-correlation functions calculated from (**a**) the experimental data in Fig. 8 (**c**) (3.5 M GdmCl), (**b**) simulated photon trajectories using a two-state kinetic model with the parameters obtained from the maximum likelihood analysis without blinking, and (**c**) recolored experimental trajectories using the model with two states for protein folding kinetics and acceptor blinking (four-state model). Black solid lines are exponential fitting curves. (Adapted from Ref. 68)

state model that includes acceptor dark and bright states (Fig. 9c) [68]. The parameters for the recoloring model were obtained by optimizing an appropriate likelihood function. In addition to the decay time, the amplitude of the correlation functions can be compared. The similar amplitudes of the experimental data in (a) and the recolored data in (c) indicate that the four-state model is a better model than the two-state model.

The maximum likelihood analysis of photon trajectories is also very useful for the detection of a transient intermediate state with a very low population or the transition path of macromolecular folding. Experimental detection of the transition path is very challenging because it is populated briefly only when transition occurs [69]. In other words, the fraction of photons emitted during the transition path is very small. One advantage of single-molecule spectroscopy is that the part of a trajectory near the transition can be analyzed selectively. In this way, the fraction of the photons emitted from the transition path can be increased. So far, average transition path times or its upper bounds have been measured for folding of several two-state proteins [70–72] and a DNA hairpin [73] by analyzing a large number of transitions using proper kinetic models. However, the fraction of photons during the transition path is still very low, and the conclusion sometimes relies on a likelihood value that is barely larger than the confidence level. In this situation, recoloring simulation of the photon trajectories is very useful. Photon trajectories can be recolored using the same model with different transition path times and analyzed again and compared with the result of the analysis of experimental data. This increases the reliability of the analysis result [70, 71].

Figure 10 shows the results of the transition path analysis of recolored photon sequences with different assumed transition path times as indicated in each panel for the folding data of a protein, α_3D, measured in a viscous solution (15 times higher viscosity than water) at low pH [70]. In this analysis, the average transition path time is measured from the maximum of the likelihood plot. In the analysis of the recolored data, when the assumed transition path time is 1 and 3 μs, there is no peak in the likelihood plot or the peak height is lower than the 95% confidence level (upper horizontal dashed line), indicating that the transition path time cannot be measured. At 5 μs, although there are cases that the likelihood at the maximum is higher than the confidence level, the transition path time cannot be determined with sufficiently high confidence. On the other hand, when the transition path time is 10 μs and longer, the peak height is much higher than the confidence level in all simulations, and therefore, the mean transition path time can be determined from the maximum of the peak. This analysis result of the recolored data is consistent with the experiment, in which the predicted transition path time was 6 μs, but it could not be measured because the maximum of the likelihood is lower than the confidence level [70].

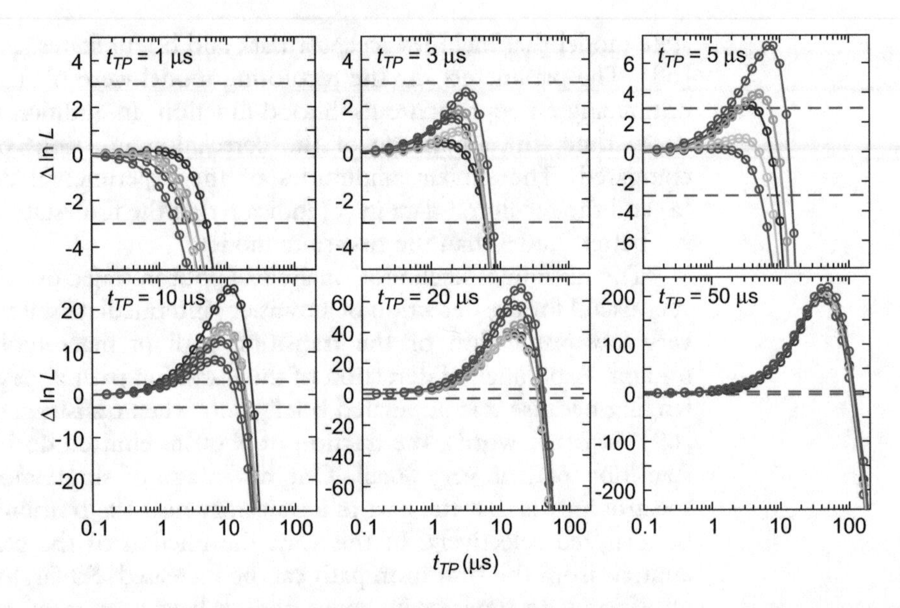

Fig. 10 Calculation of the difference of the log likelihood, $\Delta \ln L = \ln L(t_{TP}) - \ln L(0)$ for the recolored photon trajectories. $L(t_{TP})$ is the likelihood calculated for a model (three-state model) with a finite transition path time t_{TP}, and $L(0)$ is that for a model with an instantaneous transition (two-state model). If $\Delta \ln L$ is sufficiently large (larger than the upper confidence level, $\Delta L = 3$), the average transition path time can be determined from t_{TP} at the maximum. The lower confidence level at $\Delta L = -3$ is the 95% confidence for the determination of the upper bound of transition path time when there is no peak. Five set of photon trajectories for each assumed transition path time were obtained by recoloring the experimental photon trajectories with conserved photon intervals. (Adapted from Ref. 70)

3.5 Practical Issues in the Analysis of Free-Diffusion and Immobilization Data

The methods developed for the analysis of the FRET efficiency histograms and photon sequences described above can be applied to both free-diffusion experiments, in which molecules freely diffuse and emit a burst of photons during their occasional trip into the laser focus, and immobilization experiments, in which molecules are fixed on a surface. However, there are limits and different aspects in the analysis between the two experiments, which need to be carefully considered before drawing a conclusion from the analysis. In this last section, we briefly discuss these practical issues as listed below.

Free diffusion experiment has several advantages over the immobilization experiment such as no need for an additional immobilization tag in a molecule (e.g., histidine tag, biotin tag), relatively simple data collection and processing, and no potential artifacts caused by immobilization. However, relatively short burst duration (~ 1 ms or shorter in average) and small number of photons in a burst (typically less than 100) limit the detailed analysis.

In the free-diffusion experiment, it is more difficult to distinguish bursts of dye-labeled molecules from those of unknown impurities. In the immobilization experiment of the system with conformational dynamics, trajectories that do not exhibit apparent FRET efficiency transitions may result from impurities and can be excluded from the analysis. For the case of fast dynamics described in Subheading 3.4, in which apparent transitions are not observable in a trajectory, donor-acceptor cross-correlation can be calculated for each molecular trajectory and those that do not exhibit an apparent decay in the correlation function can be excluded. When a pulsed excitation is used, the average fluorescence lifetime can also be used to detect impurities.

Usually, the FRET efficiency histograms in the free-diffusion experiment are broader than that of the immobilization experiment (*see* Fig. 8a, b) because of the lower count rate, the contribution of impurity, and photophysics. Broadening by lower count rate can be accounted for in both histogram and photon trajectory analyses. However, the contributions of impurity or photophysics are very difficult to remove. The extra broadening by these contributions may result in the overestimation of the transition rates.

4 Notes

1. It should be noted that even for fast dynamics, the distribution of photons is not exactly Poissonian. When rare events are monitored, deviations from Poisson statistics can be detected. For example, photons separated by nanoseconds are correlated because of antibunching [74] and/or conformational dynamics on the nanosecond time scale [20]. To interpret such experiments, a more general theory should be used [20, 52, 75].
2. When the FRET efficiency is high (e.g., >0.8), donor lifetime can be shorter than the reorientational time of the donor, and the rotational averaging approximation does not hold. This effect reduces the FRET efficiency value and has been theoretically described [55, 76]. In addition, the point dipole approximation breaks at a short distance. Therefore, one needs to carefully consider these effects when translating the experimentally determined FRET efficiency into the distance.
3. FRET efficiency is often defined as $E = N_A/(N_A + \gamma N_D)$, i.e., corrected for the gamma factor, γ. Photon counts in a bin are also corrected for background noise, crosstalk, etc., before plotting FRET efficiency histograms. In this way, the mean FRET efficiency is approximately equal to the "true" FRET efficiency (*see* Eq. 8), but photon statistics are affected by all these corrections so that the methods discussed in this chapter are not rigorously applicable to the corrected data. In the alternative approach, raw data are used to construct FRET efficiency histograms with the FRET efficiency defined in

Eq. 9. The parameters extracted from the histogram or likelihood analyses are then corrected for the gamma factor, background noise, etc. Thus, the gamma-corrected FRET efficiency is more convenient to visualize data, for qualitative studies, or when only peak position is analyzed. The corrected data are also used in the two-dimensional FRET efficiency-lifetime histograms to visualize the relationship between donor lifetimes and FRET efficiencies [77–79]. Raw data and the FRET efficiency without gamma-factor correction are used for quantitative studies when the histogram width and shape are analyzed.

4. The histogram in Fig. 3b looks "noisy" even though it is exact. This is because the photon counts are integers and so FRET efficiencies can only be certain rational numbers. In FRET efficiency histograms obtained from real data or from simulations, there is an additional factor that leads to a jagged histogram, namely, the finite number of time bins. This "noise" (jagged appearance) in the exact histograms depends on the histogram step size h (cf. Figure 3a, b) and decreases when the number of photons in a bin is increased (by increasing the count rate or the bin size or the histogram step size).

5. We define a "conformation" as any state of the molecule that has a different apparent efficiency. This may be due to not only a different inter-dye distance but also "long-lived" photophysical states of the fluorophores [19, 69, 80], labeling permutations [2, 11], slow protein–dye interactions, and other effects. Consequently, we define a "conformation" in a broad sense as any state of the molecule that has a different apparent FRET efficiency.

6. The conditions in Eq. 18 are required for the histogram analyses as well as for the analysis using the reduced likelihood function in Eq. 32 and recoloring. The first condition in Eq. 18 is usually met when the laser intensity is low. The second condition implies that the gamma factor is equal to one in all conformational states (*see* Eq. 7). To understand how stringent the second condition of equal total count rates is, we simulated photon sequences for a two-state protein with different count rates in the folded and unfolded states and examined the accuracy of the parameters determined using the reduced likelihood function in Eq. 32, which is not exact anymore [30]. Nevertheless, the extracted parameters turned out to be very accurate even when the count rates in the folded and unfolded states differ considerably. The fraction of the folded state is the most sensitive parameter. Thus, the requirement of equal total count rates in different states can be relaxed, especially when the transitions between the states are slow compared to the photon count rates.

7. The three-Gaussian method requires the analysis of FRET efficiency histograms with a fixed bin time. Since the burst duration varies for freely diffusing molecules, it can be shorter than the bin time. In this case, the effective detection time is shorter than the bin time, which reduces the chance to observe transitions. Therefore, the interconverting rate is underestimated. The rate coefficients can be corrected by using the effective bin time, but the accurate determination of the effective bin time is difficult [19].
8. We have generalized the above formalism to more than three states [18]. After Ref. 18 was published, we found a few cases where the conformational variance of the zeroth Gaussian, σ_{c0}^2, is negative. The reason for this was that we evaluated certain conditional averages approximately. Fortunately, it appears that such cases can be handled by simply setting $c_0 = 0$ when $\sigma_{c0}^2 < 0$ and renormalizing the multi-Gaussian approximation for the FRET efficiency distribution so that it has unit area. The agreement between the calculated (with $c_0 = 0$) and simulated histograms is similar to that in the two-state model.
9. The off-diagonal elements K_{ij} of the rate matrix $\mathbf{K}$ are the transition rates from j to i, and the diagonal elements are $K_{ii} = -\Sigma_{j \neq i} K_{ji}$. The rate matrix satisfies $\Sigma_j K_{ji} = 0$, which in matrix notation is $\mathbf{1}^{\mathrm{T}}\mathbf{K} = 0$, where $\mathbf{1}$ is a column vector with all its elements equal to 1 and T denotes the transpose, so that $\mathbf{1}^{\mathrm{T}}$ is a row vector with all elements equal to one.
10. Numerical optimization of the likelihood function can be performed in various ways. A straightforward way to do that is to use standard functions of optimization (FindMaximum in Mathematica [29] or fminsearch in MatLab [19]). Alternatively, one could maximize it by adapting the expectation-maximization (EM) algorithm developed for Markov modulated Poisson processes [81]. One could also explore the landscape of the likelihood function using the Metropolis algorithm [58].
11. To obtain the true FRET efficiency, the extracted apparent FRET efficiency needs to be corrected for several factors as listed below [55, 79]. In the immobilization experiment, background photon count rates can be obtained from the part of trajectories after both donor and acceptor are photobleached. The crosstalk (i.e., the fraction of donor photons detected in the acceptor channel) is the same as the FRET efficiency of the donor-only segment of a trajectory, in which only the acceptor is photobleached. The gamma factor (i.e., the ratio of the quantum yields and detection efficiencies of the donor and acceptor) can be obtained by comparing the count rates before and after acceptor photobleaching. Finally, sometimes,

photoblinking of the acceptor, which is much faster than the bin time and therefore invisible, lowers the apparent FRET efficiency. This can be corrected by using the population of the acceptor bright state, which is found using the maximum likelihood method [79]. In the free diffusion experiment, background photon count rates can be obtained by constructing a histogram of photon counts in a bin for the entire data and fitting the low count rate part, which corresponds to background photons, to the Poissonian distribution. The donor leak can be obtained from the mean FRET efficiency of the peak corresponding to donor-only bursts. A predetermined gamma factor should be used because it cannot be easily determined directly from the data as in the immobilization experiment. Finally, correction for acceptor blinking is very difficult in the free diffusion experiment, but the chance of blinking may be lower than that in the immobilization experiment because the time of illumination would be shorter.

12. The parameter errors determined from the curvature of the likelihood function are meaningful only when the model is correct. There are several ways to test the validity of a model. First, one can compare different models using the values of the likelihood functions at the maximum. If the number of the parameters in different models is the same, the model with the highest likelihood value would be the best. However, if the number of model parameters is different, additional quantity needs to be compared such as Bayesian Information Criterion (BIC) because the likelihood will always be higher for the model with more model parameters. Practically, however, there is always a possibility that the data is contaminated by impurities, unknown photophysics, and protein–surface interactions for the case of immobilization, which may result in the preference for an unnecessarily complex (and possibly wrong) model with more parameters. In this case, cross validation of the likelihood analysis result using different methods can be useful. One of such methods is recoloring as described in Subheading 3.3. Recolored photon sequences are used to construct a FRET efficiency histogram or a donor-acceptor cross-correlation function. A good agreement between the reconstructed and experimental histograms (or correlation functions) indicates the reliability of the model.

13. The model was modified for the free diffusion data at 3M GdmCl to account for the contribution of a long-lived light-induced photophysical state (characterized by red shift in the donor spectrum [58]) that cannot be removed as in the immobilization experiments.

Acknowledgments

We thank Attila Szabo for many helpful discussions and comments. This work was supported by the Intramural Research Program of the National Institute of Diabetes and Digestive and Kidney Diseases (NIDDK), National Institutes of Health.

References

1. Michalet X, Weiss S, Jäger M (2006) Single-molecule fluorescence studies of protein folding and conformational dynamics. Chem Rev 106:1785–1813. https://doi.org/10.1021/cr0404343
2. Schuler B, Eaton WA (2008) Protein folding studied by single-molecule FRET. Curr Opin Struct Biol 18:16–26. https://doi.org/10.1016/j.sbi.2007.12.003
3. Schuler B, Hofmann H (2013) Single-molecule spectroscopy of protein folding dynamics-expanding scope and timescales. Curr Opin Struct Biol 23:36–47. https://doi.org/10.1016/j.sbi.2012.10.008
4. Dimura M, Peulen TO, Hanke CA et al (2016) Quantitative FRET studies and integrative modeling unravel the structure and dynamics of biomolecular systems. Curr Opin Struct Biol 40:163–185. https://doi.org/10.1016/j.sbi.2016.11.012
5. Wang Z, Campos LA, Muñoz V (2016) Single-molecule fluorescence studies of fast protein folding. In: Spies M, Chemla YR (eds) Single-molecule enzymology: fluorescence-based and high-throughput methods. Elsevier Academic Press Inc., San Diego, pp 417–459
6. Talaga DS, Lau WL, Roder H et al (2000) Dynamics and folding of single two-stranded coiled-coil peptides studied by fluorescent energy transfer confocal microscopy. Proc Natl Acad Sci U S A 97:13021–13026. https://doi.org/10.1073/pnas.97.24.13021
7. Rhoades E, Gussakovsky E, Haran G (2003) Watching proteins fold one molecule at a time. Proc Natl Acad Sci U S A 100:3197–3202. https://doi.org/10.1073/pnas.2628068100
8. Goldner LS, Jofre AM, Tang J (2010) Droplet confinement and fluorescence measurement of single molecules. Methods Enzymol 472:61–88. https://doi.org/10.1016/S0076-6879(10)72015-2
9. Dahan M, Deniz AA, Ha T et al (1999) Ratiometric measurement and identification of single diffusing molecules. Chem Phys 247:85–106. https://doi.org/10.1016/S0301-0104(99)00132-9
10. Schuler B, Lipman EA, Eaton WA (2002) Probing the free-energy surface for protein folding with single-molecule fluorescence spectroscopy. Nature 419:743–747. https://doi.org/10.1038/nature01060
11. Hoffmann A, Nettels D, Clark J et al (2011) Quantifying heterogeneity and conformational dynamics from single molecule FRET of diffusing molecules: recurrence analysis of single particles (RASP). Phys Chem Chem Phys 13:1857–1871. https://doi.org/10.1039/c0cp01911a
12. Fries JR, Brand L, Eggeling C et al (1998) Quantitative identification of different single molecules by selective time-resolved confocal fluorescence spectroscopy. J Phys Chem A 102:6601–6613. https://doi.org/10.1021/jp980965t
13. Zhang K, Yang H (2005) Photon-by-photon determination of emission bursts from diffusing single chromophores. J Phys Chem B 109:21930–21937. https://doi.org/10.1021/jp0546047
14. Hoffmann A, Kane A, Nettels D et al (2007) Mapping protein collapse with single-molecule fluorescence and kinetic synchrotron radiation circular dichroism spectroscopy. Proc Natl Acad Sci U S A 104:105–110. https://doi.org/10.1073/pnas.0604353104
15. Gopich IV, Szabo A (2003) Single-macromolecule fluorescence resonance energy transfer and free-energy profiles. J Phys Chem B 107:5058–5063. https://doi.org/10.1021/jp027481o
16. Gopich I, Szabo A (2005) Theory of photon statistics in single-molecule Förster resonance energy transfer. J Chem Phys 122:14707. https://doi.org/10.1063/1.1812746
17. Gopich IV, Szabo A (2007) Single-molecule FRET with diffusion and conformational dynamics. J Phys Chem B 111:12925–12932. https://doi.org/10.1021/jp075255e
18. Gopich IV, Szabo A (2010) FRET efficiency distributions of multistate single molecules. J Phys Chem B 114:15221–15226. https://doi.org/10.1021/jp105359z

19. Chung HS, Gopich IV, McHale K et al (2011) Extracting rate coefficients from single-molecule photon trajectories and FRET efficiency histograms for a fast-folding protein. J Phys Chem A 115:3642–3656. https://doi.org/10.1021/jp1009669
20. Nettels D, Gopich IV, Hoffmann A, Schuler B (2007) Ultrafast dynamics of protein collapse from single-molecule photon statistics. Proc Natl Acad Sci U S A 104:2655–2660. https://doi.org/10.1073/pnas.0611093104
21. Borgia A, Wensley BG, Soranno A et al (2012) Localizing internal friction along the reaction coordinate of protein folding by combining ensemble and single-molecule fluorescence spectroscopy. Nat Commun 3:1195. https://doi.org/10.1038/ncomms2204
22. Antonik M, Felekyan S, Gaiduk A, Seidel CAM (2006) Separating structural heterogeneities from stochastic variations in fluorescence resonance energy transfer distributions via photon distribution analysis. J Phys Chem B 110:6970–6978. https://doi.org/10.1021/jp057257+
23. Nir E, Michalet X, Hamadani KM et al (2006) Shot-noise limited single-molecule FRET histograms: comparison between theory and experiments. J Phys Chem B 110:22103–22124. https://doi.org/10.1021/jp063483n
24. Watkins LP, Chang H, Yang H (2006) Quantitative single-molecule conformational distributions: a case study with poly-(L-proline). J Phys Chem A 110:5191–5203. https://doi.org/10.1021/jp055886d
25. Hanson JA, Duderstadt K, Watkins LP et al (2007) Illuminating the mechanistic roles of enzyme conformational dynamics. Proc Natl Acad Sci U S A 104:18055–18060. https://doi.org/10.1073/pnas.0708600104
26. Kalinin S, Valeri A, Antonik M et al (2010) Detection of structural dynamics by FRET: a photon distribution and fluorescence lifetime analysis of systems with multiple states. J Phys Chem B 114:7983–7995. https://doi.org/10.1021/jp102156t
27. Santoso Y, Joyce CM, Potapova O et al (2010) Conformational transitions in DNA polymerase I revealed by single-molecule FRET. Proc Natl Acad Sci U S A 107:715–720. https://doi.org/10.1073/pnas.0910909107
28. Sisamakis E, Valeri A, Kalinin S et al (2010) Accurate single-molecule FRET studies using multiparameter fluorescence detection. Methods Enzymol 475:455–514. https://doi.org/10.1016/S0076-6879(10)75018-7
29. Gopich IV, Szabo A (2009) Decoding the pattern of photon colors in single-molecule FRET. J Phys Chem B 113:10965–10973. https://doi.org/10.1021/jp903671p
30. Chung HS, Gopich IV (2014) Fast single-molecule FRET spectroscopy: theory and experiment. Phys Chem Chem Phys 34:18644–18657. https://doi.org/10.1039/c4cp02489c
31. Chung HS (2017) Transition path times measured by single-molecule spectroscopy. J Mol Biol 430:409–423. https://doi.org/10.1016/J.JMB.2017.05.018
32. Chung HS, Eaton WA (2018) Protein folding transition path times from single molecule FRET. Curr Opin Struct Biol 48:30–39. https://doi.org/10.1016/J.SBI.2017.10.007
33. Andrec M, Levy RM, Talaga DS (2003) Direct determination of kinetic rates from single-molecule photon arrival trajectories using hidden Markov models. J Phys Chem A 107:7454–7464. https://doi.org/10.1021/jp035514+
34. Schröder GF, Grubmüller H (2003) Maximum likelihood trajectories from single molecule fluorescence resonance energy transfer experiments. J Chem Phys 119:9920–9924. https://doi.org/10.1063/1.1616511
35. Watkins LP, Yang H (2005) Detection of intensity change points in time-resolved single-molecule measurements. J Phys Chem B 109:617–628. https://doi.org/10.1021/jp0467548
36. Kou SC, Xie XS, Liu JS (2005) Bayesian analysis of single-molecule experimental data. J R Stat Soc Ser C Appl Stat 54:469–506. https://doi.org/10.1111/j.1467-9876.2005.00509.x
37. Jäger M, Kiel A, Herten D-P, Hamprecht FA (2009) Analysis of single-molecule fluorescence spectroscopic data with a Markov-modulated Poisson process. ChemPhysChem 10:2486–2495. https://doi.org/10.1002/cphc.200900331
38. Hoefling M, Lima N, Haenni D et al (2011) Structural heterogeneity and quantitative FRET efficiency distributions of polyprolines through a hybrid atomistic simulation and Monte Carlo approach. PLoS One 6:e19791. https://doi.org/10.1371/journal.pone.0019791
39. Haas KR, Yang H, Chu J-W (2013) Expectation-maximization of the potential of mean force and diffusion coefficient in Langevin dynamics from single molecule FRET data photon by photon. J Phys Chem B

117:15591–15605. https://doi.org/10.1021/jp405983d

40. Ramanathan R, Muñoz V (2015) A method for extracting the free energy surface and conformational dynamics of fast-folding proteins from single molecule photon trajectories. J Phys Chem B 119:7944–7956. https://doi.org/10.1021/acs.jpcb.5b03176
41. McKinney SA, Joo C, Ha T (2006) Analysis of single-molecule FRET trajectories using hidden Markov modeling. Biophys J 91:1941–1951. https://doi.org/10.1529/biophysj.106.082487
42. Lee T-H (2009) Extracting kinetics information from single-molecule fluorescence resonance energy transfer data using hidden Markov models. J Phys Chem B 113:11535–11542. https://doi.org/10.1021/jp903831z
43. Liu Y, Park J, Dahmen KA et al (2010) A comparative study of multivariate and univariate hidden Markov modelings in time-binned single-molecule FRET data analysis. J Phys Chem B 114:5386–5403. https://doi.org/10.1021/jp9057669
44. Pirchi M, Ziv G, Riven I et al (2011) Single-molecule fluorescence spectroscopy maps the folding landscape of a large protein. Nat Commun 2:493. https://doi.org/10.1038/ncomms1504
45. Keller BG, Kobitski A, Jäschke A et al (2014) Complex RNA folding kinetics revealed by single-molecule FRET and hidden Markov models. J Am Chem Soc 136:4534–4543. https://doi.org/10.1021/ja4098719
46. Schmid S, Götz M, Hugel T (2016) Single-molecule analysis beyond dwell times: demonstration and assessment in and out of equilibrium. Biophys J 111:1375–1384. https://doi.org/10.1016/j.bpj.2016.08.023
47. Gopich IV (2012) Likelihood functions for the analysis of single-molecule binned photon sequences. Chem Phys 396:53–60. https://doi.org/10.1016/j.chemphys.2011.06.006
48. Pirchi M, Tsukanov R, Khamis R et al (2016) Photon-by-photon hidden Markov model analysis for microsecond single-molecule FRET kinetics. J Phys Chem B 120:13065–13075. https://doi.org/10.1021/acs.jpcb.6b10726
49. Gopich IV (2015) Accuracy of maximum likelihood estimates of a two-state model in single-molecule FRET. J Chem Phys 142:34110. https://doi.org/10.1063/1.4904381
50. Lakowicz JR (2006) Principles of fluorescence spectroscopy, 3rd edn. Springer, New York
51. Gopich IV, Szabo A (2006) Theory of the statistics of kinetic transitions with application to single-molecule enzyme catalysis. J Chem Phys 124:154712. https://doi.org/10.1063/1.2180770
52. Gopich IV, Szabo A (2008) Theory of photon counting in single-molecule spectroscopy. In: Barkai E, Brown FLH, Orrit M, Yang H (eds) Theory and evaluation of single-molecule signals. World Scientific Publishing Co. Ltd., Singapore, pp 181–244
53. Haenni D, Zosel F, Reymond L et al (2013) Intramolecular distances and dynamics from the combined photon statistics of single-molecule FRET and photoinduced electron transfer. J Phys Chem B 117:13015–13028. https://doi.org/10.1021/jp402352s
54. Nettels D, Haenni D, Maillot S et al (2015) Excited state annihilation reduces power dependence of single-molecule FRET experiments. Phys Chem Chem Phys 17:32304–32315. https://doi.org/10.1039/C5CP05321H
55. Gopich IV, Szabo A (2012) Theory of single-molecule FRET efficiency histograms. Adv Chem Phys 146:245–297
56. Roy R, Hohng S, Ha T (2008) A practical guide to single-molecule FRET. Nat Methods 5:507–516. https://doi.org/10.1038/nmeth.1208
57. Schuler B, Lipman EA, Steinbach PJ et al (2005) Polyproline and the "spectroscopic ruler" revisited with single-molecule fluorescence. Proc Natl Acad Sci U S A 102:2754–2759. https://doi.org/10.1073/pnas.0408164102
58. Best RB, Merchant KA, Gopich IV et al (2007) Effect of flexibility and cis residues in single-molecule FRET studies of polyproline. Proc Natl Acad Sci U S A 104:18964–18969. https://doi.org/10.1073/pnas.0709567104
59. Camley BA, Brown FLH, Lipman EA (2009) Förster transfer outside the weak-excitation limit. J Chem Phys 131:104509. https://doi.org/10.1063/1.3230974
60. Makarov DE, Plaxco KW (2009) Measuring distances within unfolded biopolymers using fluorescence resonance energy transfer: the effect of polymer chain dynamics on the observed fluorescence resonance energy transfer efficiency. J Chem Phys 131:85105. https://doi.org/10.1063/1.3212602
61. Santoso Y, Torella JP, Kapanidis AN (2010) Characterizing single-molecule FRET dynamics with probability distribution analysis.

ChemPhysChem 11:2209–2219. https://doi.org/10.1002/cphc.201000129

62. Torella JP, Holden SJ, Santoso Y et al (2011) Identifying molecular dynamics in single-molecule fret experiments with burst variance analysis. Biophys J 100:1568–1577. https://doi.org/10.1016/j.bpj.2011.01.066
63. Gopich IV, Szabo A (2005) Photon counting histograms for diffusing fluorophores. J Phys Chem B 109:17683–17688. https://doi.org/10.1021/jp052345f
64. Deniz AA, Laurence TA, Dahan M et al (2001) Ratiometric single-molecule studies of freely diffusing biomolecules. Annu Rev Phys Chem 53:233–253. https://doi.org/10.1146/annurev.physchem.54.011002.103816
65. Berezhkovskii AM, Szabo A, Weiss GH (1999) Theory of single-molecule fluorescence spectroscopy of two-state systems. J Chem Phys 110:9145. https://doi.org/10.1063/1.478836
66. Fischer W, Meier-Hellstern K (1993) The Markov-modulated Poisson process (MMPP) cookbook. Perform Eval 18:149–171. https://doi.org/10.1016/0166-5316(93)90035-S
67. D'Agostini G (2003) Bayesian reasoning in data analysis - a critical introduction. World Scientific Publishing Co. Ltd., Singapore
68. Chung HS, Cellmer T, Louis JM, Eaton WA (2013) Measuring ultrafast protein folding rates from photon-by-photon analysis of single molecule fluorescence trajectories. Chem Phys 422:229–237. https://doi.org/10.1016/j.chemphys.2012.08.005
69. Chung HS, Louis JM, Eaton WA (2009) Experimental determination of upper bound for transition path times in protein folding from single-molecule photon-by-photon trajectories. Proc Natl Acad Sci U S A 106:11837–11844. https://doi.org/10.1073/pnas.0901178106
70. Chung HS, Piana-Agostinetti S, Shaw DE, Eaton WA (2015) Structural origin of slow diffusion in protein folding. Science 349:1504–1510. https://doi.org/10.1126/science.aab1369
71. Chung HS, McHale K, Louis JM, Eaton WA (2012) Single-molecule fluorescence experiments determine protein folding transition path times. Science 335:981–984. https://doi.org/10.1126/science.1215768
72. Chung HS, Eaton WA (2013) Single-molecule fluorescence probes dynamics of barrier crossing. Nature 502:685–688. https://doi.org/10.1038/nature12649
73. Truex K, Chung HS, Louis JM, Eaton WA (2015) Testing landscape theory for biomolecular processes with single molecule fluorescence spectroscopy. Phys Rev Lett 115:18101. https://doi.org/10.1103/PhysRevLett.115.018101
74. Basché T, Moerner WE, Orrit M, Talon H (1992) Photon antibunching in the fluorescence of a single dye molecule trapped in a solid. Phys Rev Lett 69:1516–1519. https://doi.org/10.1103/PhysRevLett.69.1516
75. Gopich IV, Nettels D, Schuler B, Szabo A (2009) Protein dynamics from single-molecule fluorescence intensity correlation functions. J Chem Phys 131:95102. https://doi.org/10.1063/1.3212597
76. Hummer G, Szabo A (2017) Dynamics of the orientational factor in fluorescence resonance energy transfer. J Phys Chem B 121:3331–3339. https://doi.org/10.1021/acs.jpcb.6b08345
77. Gopich IV, Szabo A (2012) Theory of the energy transfer efficiency and fluorescence lifetime distribution in single-molecule FRET. Proc Natl Acad Sci U S A 109:7747–7752. https://doi.org/10.1073/pnas.1205120109
78. Chung HS, Louis JM, Gopich IV (2016) Analysis of fluorescence lifetime and energy transfer efficiency in single-molecule photon trajectories of fast-folding proteins. J Phys Chem B 120:680–699. https://doi.org/10.1021/acs.jpcb.5b11351
79. Chung HS, Meng F, Kim J-Y et al (2017) Oligomerization of the tetramerization domain of p53 probed by two- and three-color single-molecule FRET. Proc Natl Acad Sci U S A 114:E6812–E6821
80. Kalinin S, Sisamakis E, Magennis SW et al (2010) On the origin of broadening of single-molecule FRET efficiency distributions beyond shot noise limits. J Phys Chem B 114:6197–6206. https://doi.org/10.1021/jp100025v
81. Ryden T (1996) An EM algorithm for estimation in Markov-modulated Poisson processes. Comput Stat Data An 21:431–447

Chapter 15

Mechanochemical Evolution of Disulfide Bonds in Proteins

Jörg Schönfelder, Alvaro Alonso-Caballero, and Raul Perez-Jimenez

Abstract

Disulfide bonds play a pivotal role in the mechanical stability of proteins. Numerous proteins that are known to be exposed to mechanical forces in vivo contain disulfide bonds. The presence of cryptic disulfide bonds in a protein structure may be related to its resistance to an applied mechanical force. Disulfide bonds in proteins tend to be highly conserved but their evolution might be directly related to the evolution of the protein mechanical stability. Hence, tracking the evolution of disulfide bonds in a protein can help to derive crucial stability/function correlations in proteins that are exposed to mechanical forces. Phylogenic analysis and ancestral sequence reconstruction (ASR) allow tracking the evolution of proteins from the past ancestors to our modern days and also establish correlations between proteins from different species. In addition, ASR can be combined with single-molecule force spectroscopy (smFS) to investigate the mechanical properties of proteins including the occurrence and function of disulfide bonds. Here we present a detailed protocol to study the mechanochemical evolution of proteins using a fragment of the giant muscle protein titin as example. The protocol can be easily adapted to AFS studies of any resurrected mechanical force bearing protein of interest.

Key words Mechanochemistry, Phylogenic analysis, Ancestral sequence reconstruction, Single-molecule force spectroscopy, Atomic force spectroscope, Mechanical stability, Disulfide bond

1 Introduction

Methods in phylogenic analysis and ASR have become powerful tools in biochemistry, as they offer the possibility to track the evolutionary history of protein genes in order to obtain information about their composition millions of years ago [1, 2]. In ASR, proteins sequences and genes from modern species are used to build a phylogenetic tree from which the sequence of the common ancestor of diverging species can be estimated. These ancestral proteins can be resurrected and experimentally studied in the laboratory as valuable tools to understand protein evolution and engineer optimized proteins. The information derived from ancestral proteins often refers to functional changes related to metabolic activity, physiological alterations, or even environmental adaptations to the harsh conditions of the planet billions of years ago.

Victor Muñoz (ed.), *Protein Folding: Methods and Protocols*, Methods in Molecular Biology, vol. 2376, https://doi.org/10.1007/978-1-0716-1716-8_15,

Several studies have shown that the application of ASR on a given protein sequence reveals changes in its physiological and metabolic features [3, 4] as well as details about the environmental conditions present at the time of the resurrected protein sequence [5, 6]. In the recent work of Manteca et al. [7], a correlation was found between the mechanochemical evolution of the muscle protein titin and the physiological features of extinct species. Whatever the goal of studying ancestral proteins is, the researcher first needs to build a robust phylogeny, and secondly, use available tools to estimate the amino acid sequence of the protein of interest within the phylogeny [1, 2].

In recent years, ASR has been successfully combined with single-molecule force spectroscopy to reveal mechanochemical aspects of proteins. These aspects include the occurrence and function of disulfide bonds. In order to track mechanochemical properties like the reduction of disulfide bonds of resurrected proteins, smFS experiments using an atomic force spectrometer (AFS) have been applied to measure the enzymatic disulfide bond reduction activity of ancestral thioredoxin [6] and the mechanical stability of ancestral titin [7]. Especially, the AFS has been proven to be the instrument of choice when measuring single chemical bond reactions [8].

In a classical AFS experiment, a polyprotein sample is adsorbed on a gold surface via disulfide bonds mounted on a piezoelectric actuator. The AFS cantilever is brought into contact with the adsorbed protein sample. In an event of unspecific adsorption of the polyprotein sample to the cantilever, the retraction of the piezoelectric actuator causes the cantilever to apply a mechanical force on the polyprotein sample [9]. The usage of a polyprotein sample, instead of a single domain, is preferred in smFS AFS experiments as it provides a unique mechanical fingerprint for studying mechanical properties of the protein [10]. The bending of the cantilever is being detected with the reflected laser beam on the backside of the cantilever, which changes the reflected signal on the photodiode. This signal is then transformed into force by calibrating the cantilever's spring constant k and the slope s of the bending of the cantilever while in contact with the gold substrate [9].

Different modes exist of applying a mechanical force to the polyprotein sample with the AFS, namely force-extension and force-clamp. In force-extension mode, the cantilever is being approached and retracted from the piezoelectric actuator with a constant velocity, which results in a typical sawtooth pattern in a force–distance curve. Therein each peak height corresponds to the unfolding force, F_u, of a single domain in the polyprotein construct. The inter-peak distance reflects the length of the fully stretched amino acid chain that is unfolding relative to that of its two ends in the 3D native structure and is called difference in contour length ΔL_c [11]. Hence, the measured ΔL_c value unmistakable identifies the fingerprint of the unfolded protein. ΔL_c between two unfolding peaks in the sawtooth pattern is calculated as the difference between the contour lengths of the two unfolding peaks. Contour lengths are estimated by fitting each unfolding peak with the so-called worm-like chain model (WLC), which has been found to be a good approximation of the polymeric properties of proteins [12] and DNA [13]. In the force-clamp mode, a feedback loop is used to hold the deflection of the cantilever fixed at a desired force setpoint. Depending on the mechanical stability of the polyprotein sample, and therefore the applied constant force, the protein domains contained in the polyprotein will unfold stochastically during the time of the applied force, which results in a typical staircase pattern in a distance vs. time trace [9]. The length between two steps, ΔL, corresponds again to the difference between the distance of the fully stretched domain and that of the two ends in the native structure of the domain before unfolding. Another important variable that is obtained from the length vs time trace is the dwell time Δt, which is related to the mechanical unfolding rate. Furthermore, when using force time segments with different forces, protein folding can be also induced and tracked on a single-molecule level [14].

When using the AFS technique with proteins that include a disulfide bond, the applied force will usually just unravel the polyprotein substrate until reaching the disulfide bond, because covalent bonds need much higher forces to be broken [15] than the forces that are typically applied with the AFS in smFS experiments. This event results in an extension length that is shorter than the ΔL of the protein. When the disulfide bond is exposed to a reducing agent, the disulfide bond reduction can be detected by the occurrence of an unfolding event that amounts exactly to the extension of the protein segment that was previously protected from unfolding by the disulfide bond. This will result in a second peak in the force extension, or in a second step in the force-clamp mode [8].

Here we describe a detailed protocol for the use of the AFS from Luigs and Neumann [9] to detect the mechanochemical evolution of the resurrected eight-domain containing fragment (I65–I72) of the giant muscle protein titin [7] from four extinct species (*see* Fig. 1). Furthermore, we compare the results with the titin counterpart fragment from five modern amniote species. The chapter starts with the materials and methods needed to do the phylogenetic analysis and ancestral sequence reconstruction. We then detail the protein cloning, expression, and purification steps required to obtain the titin fragment with oxidized disulfide bonds. Finally, we describe how to perform the AFS measurements and the data acquisition and analysis. Our versatile protocol can be applied to any protein sequence of interest for which a phylogenetic tree can be calculated.

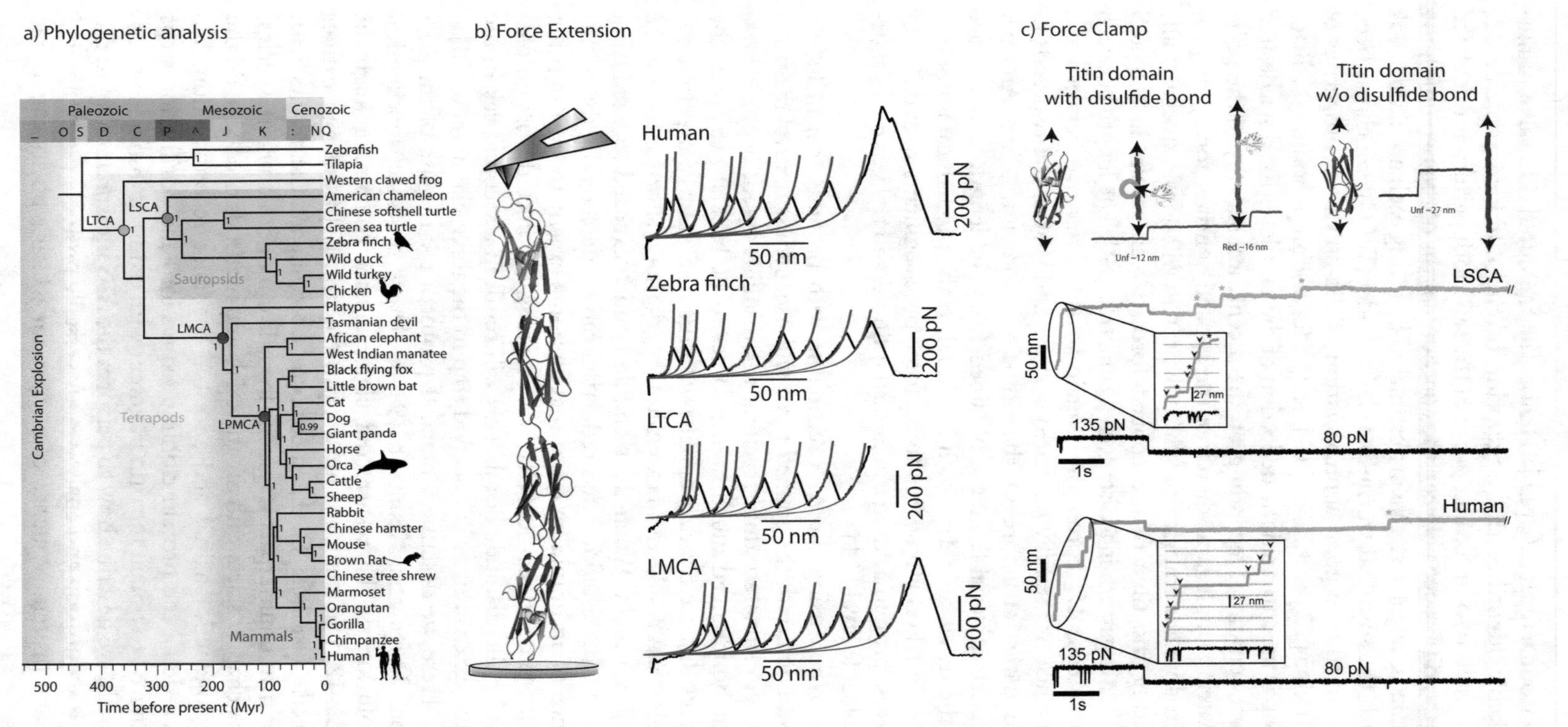

Fig. 1 (**a**) Reconstruction of ancestral titin fragments. Uncorrelated log-normal relaxed-clock chronogram of titin with geological time inferred with Bayesian inference. A total of 33 titin genes were used. The modern species studied are indicated by the animal outlines: zebra finch, chicken, orca, rat, and human. The internal nodes (circles) were selected for resurrection and laboratory testing and represent the last common ancestors of tetrapods (LTCA, 356 ± 11 Myr), sauropsids (LSCA, 278 ± 14 Myr), mammals (LMCA, 179 ± 38 Myr), and placental mammals (LPMCA, 105 ± 17 Myr). Posterior probabilities for branch support are shown in the nodes. Geological periods are shown in the upper bar. Outlines were retrieved from http:/www.phylopic.org. (**b**) Force-extension AFS measurements with titin at a speed of 400 nm/s. A schematic representation of a single-molecule experiment using the smFS (not to scale) is illustrated. Although the segment I65–I72 contains eight domains, for clarity only four are represented. Disulfide-bonded domains are depicted in red and gray with cysteines highlighted in yellow; non-disulfide–bonded domains are depicted in blue. The protein is mechanically stretched between a cantilever tip and a gold-coated surface. Four representative experimental force-extension traces of the polyprotein I65–I72 are shown. The unfolding of domains is monitored as a sawtooth

pattern of force versus extension peaks. The worm-like chain model was used to fit the data. Fully extended domains (blue lines) show extension of about 30 nm, whereas disulfide-bonded domains (red lines) show shorter contour lengths of 5–20 nm. (**c**) Force-clamp AFS measurements. A schematic representation illustrates the disulfide-bonded domains (red) revealing a two-step unfolding pattern. The first step (~12 nm) corresponds to the unfolding of the beta sheets that are not trapped in the disulfide bond while the second step (~15 nm) shows the unfolding of the rest of the protein after the reduction of the disulfide bond caused by thioredoxin. Not disulfide-bonded domains (blue) have a single-step (~27 nm) unfolding pattern that represents the stretching of the whole domain. Experimental force-clamp trace of LSCA and human titin are shown. We first apply a pulse of force of 135 pN during 2 s that triggers unfolding of non-disulfide–bonded domains (inset arrows) and disulfide-bonded domains up to their disulfide bond (asterisks in inset). The disulfide bonds can be reduced by Trx enzymes present at 10 μM concentration. The reduction events are monitored at a force of 80 pN (green line). (The panels (**a**–**c**) were adapted and reprinted with permission from Ref. 7)

2 Materials

2.1 Phylogenetic Analysis and ASR

1. Several software packages are required for ancestral sequence reconstruction.
2. MEGA 7 for alignment and database search (http://www.megasoftware.net/).
3. Uniprot database (http://www.uniprot.org/).
4. Model testing software ProTest (http://www.sigmazone.com/protest.htm).
5. BEAST software package for phylogenetic inference (https://www.beast2.org/).
6. FigTree software for tree representation (http://tree.bio.ed.ac.uk/software/figtree/).
7. PAML software for ancestral sequence reconstruction (http://abacus.gene.ucl.ac.uk/software/paml.html).
8. Multicore computer cluster to run the calculations in parallel.

2.2 Protein Cloning, Expression, Purification, and Oxidation

2.2.1 Protein Plasmid (See Note 1)

1. Genes encoding the titin protein fragments are synthesized and codon optimized for expression in *E. coli* competent cells (Life Technologies). The genes are flanked in 5′-end with a BamHI restriction site and in 3′-end with a KpnI restriction site to allow the directional cloning of the fragment into an expression plasmid. Just before the 3′-end of the gene, two codons encoding for two cysteine residues are placed. These cysteines placed in the C-terminal end of the protein allow immobilization of the proteins on the gold surfaces used in smFS experiments. Genes are provided inside commercial plasmids which additionally contain a gene for resistance to kanamycin (KanR), a plasmid origin of replication sequence (Col E1 origin), and two restriction sites for the enzyme SphI.
2. For gene expression, the plasmid vector pQE80L is used (Qiagen), which contains restriction enzyme sites for BamHI and KpnI. The genes cloned into this expression vector are under the control of the lac promoter. The promoter is under control of the lac repressor, which gene is also encoded in pQE80L. Besides the origin of replication (Col E1 origin) and a gene conferring antibiotic resistance (β-lactamase), the pQE80L plasmid has a 6× His-tag coding sequence which is transcribed just before the gene of interest.

2.2.2 Protein Cloning, Transformation, and Amplification

1. LB Broth: Casein peptone 10 g/L, yeast extract 5 g/L and NaCl 5 g/L (Fisher).
2. SOC medium (Invitrogen), agar (Fisher).

3. For the commercial plasmid, use 1% 100 mg/mL of kanamycin (Sigma-Aldrich).
4. For the pQE80L plasmid, use 0.1% 100 mg/mL carbenicillin (Fisher).
5. GeneJET Plasmid Miniprep and Gel Extraction Kit (Thermo Scientific).
6. FastDigest BamHI, KpnI, and SphI restriction enzymes (Thermo Scientific).
7. DNA gel electrophoresis (BioRad).
8. DNA agarose gel (Fisher).
9. 1× tris–acetate–EDTA buffer (Fisher).
10. Protein gel electrophoresis (BioRad).
11. SDS sample buffer solution: 4% w/v SDS, 20% glycerol, 10% 2-mercaptoethanol, 0.004% w/v bromophenol blue, and 0.125 M tris–HCl, pH 6.8.
12. 8% SDS-PAGE gels.
13. Bradford solution (Thermo Scientific).
14. T4-DNA Ligase (Invitrogen).
15. Incubator.

2.2.3 Protein Expression and Purification (See Note 2)

1. For the expression of the pQE80L-titin plasmid containing *E. coli* Origami B(DE3)pLysS cells (Merck Millipore), use LB broth–ampicillin–chloramphenicol: Casein peptone 10 g/L, yeast extract 5 g/L, NaCl 5 g/L, 0.1% 100 mg/mL carbenicillin, and 0.1% 50 mg/mL chloramphenicol (Fisher). Protein expression is induced with 1 mM of IPTG (Sigma-Aldrich).
2. Extraction buffer: 50 mM sodium phosphate pH 7.0 and 300 mM sodium chloride.
3. 1 mg/mL lysozyme (Thermo Scientific).
4. 1% Triton X-100 (Sigma-Aldrich).
5. 5 μg/mL DNAse I (Invitrogen).
6. 5 μg/mL RNAse A (Ambion).
7. 10 mM $MgCl_2$ (Sigma-Aldrich).
8. Protease inhibitor (Merck Millipore).
9. The mechanical lysis is performed in a French press machine (G. Heinemann HTU DIGI-F Press).
10. Syringe filters with 0.8, 0.45, and 0.22 μm pore size filters (Merck Millipore).
11. HisPur Cobalt resin (Thermo Scientific).
12. Pierce disposable 5 mL polypropylene column (Thermo Scientific).

13. Elution buffer: 50 mM sodium phosphate pH 7.0, 300 mM NaCl, and 150 mM imidazole.
14. Size-exclusion chromatography is performed in a Superdex 200HR column (GE Healthcare) in an ÄKTA pure fast protein liquid chromatography (FPLC) system (GE Healthcare).
15. Size-exclusion chromatography buffer: HEPES 10 mM pH 7.2, 150 mM NaCl, and 1 mM EDTA.
16. Disulfide bond formation is triggered with 0.5% of H_2O_2 (Sigma-Aldrich).

2.3 Preparation of AFS Gold Substrates

All the equipment for the AFS gold substrates preparation must be placed in a clean room (ISO 7). The final cutting of the substrate is carried out in the Material preparation lab.

1. Borosilicate glass waver (100 mm diameter, 1 mm thick) and gold wire (10 g, 1 mm diameter, 99.99% purity) (Pi-KEM).
2. Ultrasonic bath, isoropanol, acetone.
3. E-beam, thermal evaporator (Oerlikon UNIVX350/EPVD75 Kurt J. Lesker).
4. Protective organic resin (s1818).
5. Spin coater.
6. Automatic dicing saw (DISCO DAD 321).

2.4 SmFS-AFS Experiments

1. 10–20 μL of purified and oxidized I65–I72 titin sample at a concentration of 1–2 μM.
2. 100 μL of force-extension experimental buffer: 10 mM HEPES, pH 7.2, 150 mM NaCl, 1 mM EDTA, and 2 mM NADPH.
3. 100 μL of force-clamp experimental buffer: 10 mM HEPES, pH 7.2, 150 mM NaCl, 1 mM EDTA, 2 mM NADPH, 10 μM eukaryotic thioredoxin (Trx) enzyme, and 50 nM Trx reductase.
4. 0.22 μm syringe filter (Merck Millipore).
5. Small magnet and degassing station (TA Instruments).
6. AFS gold substrates, as described in Subheading 2.3.
7. Acetone, Milli-Q water, and dry nitrogen gas.
8. Vacuum grease (Dow Corning).
9. AFS cantilevers OBL-10 Biolever or MLCT (Bruker).
10. Ultrasharp tweezers for handling.
11. Table light microscope/magnifying glass.
12. ASF apparatus (Luigs and Neumann).
13. Vibration Isolation table.

2.5 Data Analysis of smFS-AFS Experiments

All collected smFS data were analyzed with an analytical procedure written in IgorPro 6.37 (Wavenetrics), which is also online available (AFM_Analysis_V2.40.ipf from Fernandez lab http://fernandezlab.biology.columbia.edu/downloads).

3 Methods

3.1 Phylogenetic Analysis and ASR

Ancestral reconstruction requires access to computer power proportionally to the size of the protein/gene to be reconstructed and the number of sequences to be handled. In general, a computer cluster with at least eight cores is required for most applications.

1. Download a set of sequences of the protein of interest from different taxa. Generally, over 30 sequences will be required from diverse taxonomic groups. BLAST is used to retrieve the sequences from databases such as Uniprot.
2. Align the sequences using ClustalW or MUSCLE software incorporated in the MEGA 7 package.
3. Test the alignment for the best evolution model to be used. The software ProTest can be used for model testing.
4. Construct a phylogeny by Bayesian inference using Markov Chain Monte Carlo. The BEAST software is used for Bayesian inference incorporating the BEAGLE library for parallel processing.
5. Start with the BEAST utility to create the XML file. Set the evolutionary model to be used and the number of desired gamma categories.
6. Set the clock model for evolutionary rate.
7. Set tree prior.
8. Starting with ten million generations is generally enough. The actual required number will depend on the protein length and number of sequences used in the analysis.
9. Set prior for model parameters and distribution. Priors can be set from paleontological information.
10. Set length chain. By default, ten million may be enough for an initial test. Generate XML file and run BEAST.
11. Verify tree convergence using TreeStat from BEAST software package to check the effective sample size (ESS).
12. Use Tree Annotator from the BEAST software package for maximum clade credibility removing at least 24% of initial trees as burn-in.
13. The tree can be represented using FigTree Software (*see* Fig. 1a).

14. Divergence times can be collected either from different sources using molecular clocks and paleontological records or could be computed using fossil record that can be input in BEAST as priors.
15. Infer ancestral sequences for each tree node using the PAML software incorporating the evolution model and number of gamma distribution.

3.2 Protein Cloning, Expression, Purification, and Oxidation

3.2.1 Cloning

1. Transform *E. coli* XL1-Blue competent cells with 50 ng of the commercial plasmid containing the titin gene following the manufacturer instructions. After transformation, grow the cells in SOC media for 1 h at 37 °C and 250 rpm of agitation. Then spread the cells on LB broth–Agar–kanamycin plates and incubate them overnight (o/n) at 37 °C.
2. Isolate single colonies from plates and grow them in 10 mL of LB broth + kanamycin o/n at 37 °C and 250 rpm.
3. Extract the amplified commercial plasmid following the GeneJET Plasmid Miniprep Kit manufacturer instructions.
4. Digest the commercial plasmid with BamHI-KpnI-SphI following the FastDigest protocol.
5. Separate the digestion reaction products on a 1% agarose gel electrophoresis in 1× tris–acetate–EDTA buffer.
6. Extract the titin gene from the agarose gel using the GeneJET Gel Extraction Kit following the protocol.
7. For the expression plasmid pQE80L, repeat the **steps 1–6**. Use as selection antibiotic carbenicillin (**steps 1** and **2**), only perform the BamHI-KpnI digestion (**step 4**), and extract the open plasmid band from the agarose gel (**step 6**).
8. Ligate the titin gene of interest and the expression plasmid pQE80L with T4-DNA Ligase following the manufacturer protocol. Use a 3:1 molar excess of gene respect to plasmid.
9. Amplify the ligation product pQE80L-titin following the **steps 1–3**, using carbenicillin as the selection antibiotic.

3.2.2 Screening Expression Test

1. Transform *E. coli* Origami B(DE3)pLysS cells with the amplified pQE80L-titin construct as explained in **step 1** from the previous cloning section using carbenicillin as antibiotic.
2. Isolate several colonies from the plate and grow them in 10 mL of LB Broth liquid medium with carbenicillin at 37 °C and agitation. When OD at 600 nm is 0.6, in half of the volume of each isolated colony add 1 mM IPTG for protein expression induction, leaving the other half untreated as a negative control. Protein expression is induced o/n with the same conditions.

3. From each colony isolated, use 1 mL of IPTG-treated and 1 mL of nontreated volumes for protein expression levels test.
4. Centrifuge 1 mL of each volume at 12,000 × *g* for 5 min at r.t.
5. Discard the supernatants and resuspend the cell pellets first in 10 μL of extraction buffer. Then, add 10 μL of SDS sample buffer and centrifuge for 30 min.
6. Boil the samples at 95 °C for 2 min.
7. Load the samples into 8% SDS-PAGE gels and perform the electrophoresis.
8. After electrophoresis, wash the gels with deionized water for 10 min and gentle shaking in an orbital shaker (10 rpm). Repeat three times.
9. Stain the gels with Bradford solution for 30 min and gentle shaking.
10. Rinse the gels in deionized water for 10 min and gentle shaking. Repeat three times.
11. Compare the bands from the noninduced and the induced samples from each colony. Select the colony showing the best overexpression levels at the expected molecular size, and showing the lowest degradation levels.
12. Take the noninduced sample from that colony and add fresh LB broth media with ampicillin.
13. Make 1 mL aliquots with 10% glycerol, freeze in liquid nitrogen, and store at −80 °C.

3.2.3 Large-Scale Expression

1. Inoculate 800 mL of LB broth–ampicillin–chloramphenicol with one aliquot. Incubate at 37 °C and 250 rpm.
2. When the OD at 600 nm is 0.6–0.8, add 1 mM IPTG and incubate o/n with the same conditions.

3.2.4 Purification (See Note 3)

1. Centrifuge the culture at 4000 × *g* for 20 min at 4 °C.
2. Discard the supernatant and resuspend the cell pellet in 16 mL of extraction buffer with 1:1000 protease inhibitor.
3. Lysate the cells with lysozyme for 30 min at 4 °C in an orbital shaker (5 rpm).
4. Add Triton X-100, DNAse I, RNAse I, and $MgCl_2$ and incubate for 10 min in ice.
5. Then start the mechanical lysis using the French press.
6. French press elements are made of steel, and they should be stored at 4 °C to minimize protein degradation during this process.
7. Insert the T-shape arm in the steel chamber.

8. Switch upside down the chamber and place it into assembling stand.
9. Fill the cell lysate into the chamber.
10. Close the cylinder with the closing smaller cylinder which contains two holes for connecting the valve and the output tube.
11. Screw the output tube.
12. Place a polystyrene bead on the valve screw. Screw the valve on its position.
13. Place the assembled chamber in the French press. Secure with clamps and screws its position.
14. Put a 50 mL tube at the opening end of the output tube to collect the lysate. Place the tube inside of a glass with ice to keep refrigerated the collected lysate.
15. Start cell breakage applying 18,000 psi of pressure.
16. Manually and slowly open the valve and regulate the flow for collecting around 0.5 mL/min of the lysate. Slow flows are better for cell breakage, improving the purification yield.
17. Centrifuge the collected lysate at 33,000 × *g* at 4 °C for 1 h 30 min.
18. Collect the supernatant and discard the cell pellet.
19. Filter the supernatant sequentially with filters of 0.8, 0.45, and 0.22 μm of pore size.
20. Incubate the supernatant with 3 mL of HisPur cobalt resin previously washed as specified by the manufacturer. Add 30 mL of extraction buffer with 1:1000 protease inhibitor and incubate at 4 °C for 1 h in an orbital shaker (5 rpm).
21. Centrifuge the mixture at 3000 × *g* at 4 °C for 10 min and discard the supernatant.
22. Resuspend the resin in 30 mL of extraction buffer with 1:1000 protease inhibitor.
23. Repeat three times **steps 9** and **10**.
24. Discard the supernatant and resuspend the 3 mL resin with 1 mL of extraction buffer.
25. Transfer the solution to a disposable 5 mL polypropylene column with a cap placed in its lower end. Let the resin to settle for 10 min.
26. Wash the resin from unspecific binding adding 30–60 mL extraction buffer through the upper open end of the column, and uncap the lower end. Before the column is emptied from extraction buffer, collect a few drops from the eluate and check with a spectrophotometer the absorbance at 280 nm. Do not

allow the resin to get dried (cap the bottom end of the column to stop the flow before losing the remaining extraction buffer). If the absorbance is below 0.02, proceed with the next step. If not, keep washing the resin with extraction buffer.

27. Let the rest of the extraction buffer flow through the column. After the elution of the last drops of extraction buffer, add 12 mL of elution buffer and collect 0.5 mL aliquots in individual tubes. When the elution buffer is added, the resin color changes from pale pink to dark pink.
28. Check the absorbance at 280 nm of the collected aliquots and select those ones with the highest absorbance.
29. Induce disulfide bond formation adding 0.5% of H_2O_2 and incubate o/n at r.t in those aliquots showing the highest absorbance.
30. Perform a size-exclusion chromatography in the FPLC with the selected aliquots. Ideally the chromatogram shows a high absorbance peak measured at 280 nm which corresponds to the polyprotein. Sometimes additional absorbance peaks are observed which usually corresponds to aggregated or degraded proteins. Collect all the FPLC fractions and check the presence of the protein of interest in the different peaks in an 8% SDS-PAGE.
31. Once the polyprotein has been identified in one of the absorbance peaks, use the collected fractions from the center of that peak for making aliquots and store them at −20 °C until use.

3.3 Preparation of AFS Gold Substrates

1. Clean the wafer in a glass beaker filled with acetone for 1 min in the ultrasonic bath, afterwards clean 1 min in isopropanol. In between, dry the wafer with compressed and filtered air.
2. Prepare the Oerlikon system for the gold deposition while weighting around 1 g of the gold wire and place it inside the holder. Place wafer on the corresponding plate for inserting it in the Oerlikon system.
3. After a vacuum of around 5–10 bar has been reached, start the deposition of Ti by E-beam evaporation at 10 kV and a deposition rate of 0.4 Å s^{-1} until a 7 nm layer is being reached.
4. Afterwards switch to the thermal evaporation of the gold sample at rate of 0.6 Å s^{-1} until a 40 nm layer is being reached.
5. Remove the deposited gold wafer from the Oerlikon system by opening the vacuum.
6. Place around 1 mL of the protective organic resist on the substrate using a spin coater (4000 rpm).
7. Afterwards use the dicing saw machine to cut the waver into 12 × 12 mm big single coverslips.

3.4 SmFS-AFS Experiments (See Notes 4 and 5)

1. Filter the experimental buffer for force extension or force clamp into a small Eppendorf tube.
2. Add a magnet to the Eppendorf tube and place it open in the degassing station.
3. Turn on the vacuum and the magnetic field and leave the buffer degassing for around 5–10 min.
4. In the meantime, immerse the prepared AFS gold substrate in acetone to remove the photo-protector resin from the surface.
5. Rinse the gold substrate with Milli-Q water and dry with filtered dry nitrogen gas.
6. Turn on the AFS controller using the switch at the front.
7. Turn on the vibration isolation table on the backside.
8. Turn on the PC.
9. Once in the Windows screen, open the IgorPro software and compile the data acquisition procedure.
10. Open the AFS closure head and turn the piezoelectric actuator block upwards using the screw.
11. Place the clean gold substrate on the piezoelectric actuator using a small amount of the vacuum grease. Press carefully on the substrate with the back of the tweezers to seal the grease.
12. Place a 20 μL drop of the titin protein sample in the middle of the gold substrate and spread it with the pipette tip without scratching the surface.
13. Leave it placed for around 10–20 min to allow the protein to adsorb, however do not let the solution dry out completely.
14. In the meantime, clean the fluid cell and the O-ring of the AFS while rinsing with 70% ethanol and Milli-Q water.
15. Insert the O-ring and cantilever using tweezers into the fluid cell, while lifting the clamp by pressing slightly the spring on the backside of the fluid cell.
16. Place the 100 μL of the experimental buffer on the cantilever. Take care that no bubbles remain between buffer and the fluid cell, cantilever, and O-ring.
17. Now place the filled fluid cell into the fluid cell holder of the AFS.
18. Move the camera into the place above the cantilever.
19. With the camera image, place the LASER spot onto the backside of the cantilever using the screw on the fluid cell or the LASER mounting screws. You should see now a round and bright reflection on the photodetector (PD) lid.
20. Wash the gold surface with 20 μL experimental buffer in order to remove unbound protein.

21. Turn the piezoelectric actuator block down and close carefully the fluid cell by approaching the gold substrate to the O-ring of the fluid cell using the control button in the IgorPro software.
22. Align the four quadrant PD so that the reflected laser beam is aimed at the center of the detector and the (A-B) display is close to zero while moving carefully the lever of the AFS mirror. Align also the (C-D) display to zero using the screw on the PD.
23. Increase the LASER intensity until a value of around 5–7 V is reached on the (A+B) display.
24. Now the thermal fluctuations (power) spectrum of the cantilever oscillations by Brownian motion has to be acquired and is calculated automatically using the equipartition theorem [9]. Select the main vibrational mode putting the first cursor at the beginning of the spectrum and the second after the main vibrational mode.
25. Calibrate the cantilever spring constant k. This is done by moving the cantilever into contact with the substrate and measuring the slope of the deflection vs distance curve using the two cursors. The calculated spring constant k needs to be in the range of 0.006 N/m for the OBL-10 Biolever cantilever B and 0.02 N/m for the MLCT cantilever C. Otherwise the cantilever might be defect and should be replaced.
26. Leave the instrument to equilibrate for 30–60 min in order to avoid thermal drift during the measurements.
27. Choose the smFS mode you want to measure (force extension or force clamp). In force-extension mode, you normally choose the pulling speed (nm/s) and the amplitude range (nm). In force-clamp mode, you choose the contact conditions between cantilever and gold substrate (contact force in pN and contact time in s), the PID feedback gain parameters to avoid a delay or overshoot, and the force sequence of the measurement, which contains the applied force and time values.
28. A stable set experiment can be let measuring continuously for up to 24 h.
29. When finishing the measurements, save the IgorPro file, retract the piezoelectric actuator, and take out the fluid cell from the fluid cell holder.
30. Remove the cantilever and clean the fluid cell with 70% ethanol and Milli-Q water.
31. Turn off the IgoPro software, the PC, the controller, and the vibration table.

3.5 Data Analysis of smFS-AFS Experiments

3.5.1 Force-Extension Data Analysis

1. All smFS force-extension data is analyzed by fitting each peak using the WLC model within the IgorPro analysis procedure file. However, traces that meet the following criteria need to be selected (*see* Fig. 1b).
2. The recordings should show a clean beginning, when the cantilever retracts the surface.
3. The recordings should show at least five from the eight possible domain unfolding peaks.
4. The recordings should show the detachment peak from the cantilever and the polyprotein sample at the end of the curve, which is usually higher than the individual unfolding peaks of the protein domains.
5. The fitting is manually adjusted by changing the persistence length ρ and total length L_c in the WLC model.
6. Then create a table and note the values for every force peak of F_u and ΔL_c.
7. The F_u data of the finished table can now be visualized in a cumulative histogram and divided into disulfide bond containing ($Force_{SS}$) and absent domains ($Force_{noSS}$).
8. Plot the average unfolding force with 95% confidence intervals of the ample mean versus the geological time of the resurrected protein and compare it to the corresponding value of the modern amniote species.
9. Plot the average unfolding force $Force_{SS}$ and $Force_{noSS}$ versus the body mass of the known species.
10. Fit the points to a power law curve. Now the values of body mass for the extinct species can be interpolated from the fitted curve.

3.5.2 Force-Clamp Data Analysis

1. All smFS force-clamp data is analyzed by measuring the distance ΔL between unfolding steps. Like when collecting force-extension traces, the selection needs to meet the following criteria (*see* Fig. 1c).
2. The recording should show at least five steps from the possible eight unfolding steps in the titin unfolding sequence of 135 pN (2 s).
3. After dropping the force to 80 pN to induce disulfide bond reduction, select traces that show the same amount of reduction steps as the number of identified shorter disulfide bond steps in the titin unfolding force sequence.
4. Then create a table with the measured step distances ΔL ordered by unfolding and reduction step sizes.
5. Do a histogram to visualize the data.

4 Notes

1. Because the digestion with BamHI-KpnI of the commercial plasmid yields two DNA fragments of similar size, digestion with a third enzyme (SphI) is needed in order to generate smaller DNA fragments of the plasmid. The three-digestion protocol facilitates separating the gene of interest from the plasmid sequences after agarose gel electrophoresis.
2. The purification of polyproteins such as titin can be approached with different methods, but the one described here yields the best results to obtain a highly pure protein and to minimize protein degradation during the process.
3. During purification, it is recommended to store all the buffers used along this process at 4 °C in order to minimize protein degradation.
4. The Biolever cantilever is extremely sensitive and when filling the fluid cell with the measuring buffer it often tilts. Sometimes this can cause a bad reflection spot of the laser when inserting the fluid cell into the AFS. In this case, it is highly recommended to change the cantilever.
5. When adjusting the PID feedback gain, simultaneously increase the proportional, integral, and differential gains until the Biolever starts to slightly vibrate. At this point, reduce the gains slowly until the vibration stops. Sometimes, the gains have to be readjusted during the measurement.

Acknowledgments

A.A.-C. is funded by the predoctoral program of the Basque Government. We acknowledge financial support from the Spanish Ministry of Economy, Industry and Competitiveness grant BIO2016-77390-R to R. P.-J. and Marie Curie Career Integration Grants (CIG) FP7-PEOPLE-2013-CIG from the European Commission to R. P.-J. This work was also supported by the Spanish Ministry of Economy, Industry and Competitiveness under the Maria de Maeztu Units of Excellence Program - MDM-2016-0618.

References

1. Hall BG (2006) Simple and accurate estimation of ancestral protein sequences. Proc Natl Acad Sci U S A 103(14):5431–5436
2. Merkl R, Sterner R (2016) Ancestral protein reconstruction: techniques and applications. Biol Chem 397(1):1–21
3. Kratzer JT, Lanaspa MA, Murphy MN, Cicerchi C, Graves CL et al (2014) Evolutionary history and metabolic insights of ancient mammalian uricases. Proc Natl Acad Sci U S A 111(10):3763–3768
4. Zakas PM, Brown HC, Knight K, Meeks SL, Spencer HT et al (2016) Enhancing the

pharmaceutical properties of protein drugs by ancestral sequence reconstruction. Nat Biotechnol 35(1):35–37

5. Gaucher EA, Govindarajan S, Ganesh OK (2008) Palaeotemperature trend for Precambrian life inferred from resurrected proteins. Nature 451(7179):704–707
6. Perez-Jimenez R, Inglés-Prieto A, Zhao Z-M, Sanchez-Romero I, Alegre-Cebollada J et al (2011) Single-molecule paleoenzymology probes the chemistry of resurrected enzymes. Nat Struct Mol Biol 18(5):592–596
7. Manteca A, Schonfelder J, Alonso-Caballero A, Fertin MJ, Barruetabena N et al (2017) Mechanochemical evolution of the giant muscle protein titin as inferred from resurrected proteins. Nat Struct Mol Biol 24:652–657
8. Liang J, Fernández JM (2009) Mechanochemistry: one bond at a time. ACS Nano 3 (7):1628–1645
9. Popa I, Kosuri P, Alegre-Cebollada J, Garcia-Manyes S, Fernandez JM (2013) Force dependency of biochemical reactions measured by single-molecule force-clamp spectroscopy. Nat Protoc 8(7):1261–1276
10. Hoffmann T, Dougan L (2012) Single molecule force spectroscopy using polyproteins. Chem Soc Rev 41(14):4781–4796
11. Carrion-Vazquez M, Oberhauser AF, Fisher TE, Marszalek PE, Li H et al (2000) Mechanical design of proteins studied by single-molecule force spectroscopy and protein engineering. Prog Biophys Mol Biol 74 (1–2):63–91
12. Rief M, Gautel M, Oesterhelt F, Fernandez JM, Gaub HE (1997) Reversible unfolding of individual titin immunoglobulin domains by AFM. Science 276(5315):1109–1112
13. Bustamante C, Marko JF, Siggia ED, Smith S (1994) Entropic elasticity of lambda-phage DNA. Science 265(5178):1599–1600
14. Garcia-Manyes S, Brujić J, Badilla CL, Fernández JM (2007) Force-clamp spectroscopy of single-protein monomers reveals the individual unfolding and folding pathways of I27 and ubiquitin. Biophys J 93(7):2436–2446
15. Grandbois M, Beyer M, Rief M, Clausen-Schaumann H, Gaub HE (1999) How strong is a covalent bond? Science 283 (5408):1727–1730

Part IV

Molecular Simulations

Chapter 16

Coarse-Grained Simulations of Protein Folding: Bridging Theory and Experiments

Vinícius G. Contessoto, Vinícius M. de Oliveira, and Vitor B. P. Leite

Abstract

Computational coarse-grained models play a fundamental role as a research tool in protein folding, and they are important in bridging theory and experiments. Folding mechanisms are generally discussed using the energy landscape framework, which is well mapped within a class of simplified structure-based models. In this chapter, simplified computer models are discussed with special focus on structure-based ones.

Key words Structure-based models, Molecular dynamics, Energy landscapes

1 Introduction

Proteins comprise the machinery that controls most of the functions in living organisms. Understanding the relationships between the structure, energetics, and functional dynamics of proteins is a major challenge in molecular biophysics. The fact that their activity depends on their three-dimensional structure and dynamics, and not simply on their amino-acid sequence, presents particular conceptual challenges. A central question, known as the protein folding problem, seeks to uncover the mechanism by which a functional conformation is reached.

Protein folding was the first area of molecular biophysics in which an approach based on principles of statistical mechanics led to a new qualitative and quantitative understanding of the problem [1]. The accepted formalism, known as the energy landscape, seeks a simplified view of the folding that reveals the general principles that govern this mechanism [2]. In general, potential energy surfaces are rough because of the many competing interactions between amino acids. In real proteins, the interactions between the amino acids in the native state follow the principle of minimal frustration, which guarantees the minimum global energy and thermodynamic stability of the native state [3–5]. These finely

Victor Muñoz (ed.), *Protein Folding: Methods and Protocols*, Methods in Molecular Biology, vol. 2376,
https://doi.org/10.1007/978-1-0716-1716-8_16, © Springer Science+Business Media, LLC, part of Springer Nature 2022

tuned interactions are not random but have been achieved through protein evolution. The potential energy surfaces in proteins have evolved towards a funnel, which has an energy gradient directed towards the region of the native state.

The energy landscape approach has contributed significantly to an understanding of complex molecular systems. The connection between landscape theory and real proteins was first established in the context of small fast folding proteins, which fold on millisecond time scales and have a single folding domain; i.e., they are two-state folders with a single, well-defined funnel. Nowadays, the theory has been extended to address all sorts of complex molecular systems [6], ranging from molecular machines [7], viruses [8], DNA and RNA dynamics [9, 10], assemblies [11], ribosome [12–14] to chromatin [15–21]. In addition to the theoretical literature describing this theory and its applications, a modern generation of powerful experiments, such as those discussed in this book (NMR dynamic spectroscopy, protein engineering, laser initiated folding, and ultrafast mixing) is providing the temporal and spatial detail needed to extend and elaborate upon it.

Computational studies have been essential in testing theories and corroborating experimental results, as well as in bridging the two approaches. The complexity of biomolecular problems led to the development of an arsenal of computational methods, each suited to a specific question. Starting with the early computational toy models, such as lattice ones, many fundamental problems were addressed [22–24], such as effective order parameters and reaction coordinates, transition temperatures, and stability criteria. These models were the foundations upon which a comprehensive theory was elaborated.

In this chapter, we focus on the practical details of how to apply the energy landscape approach to investigate protein folding using off-lattice protein coarse-grained models constructed with structure-based potentials. Models with different levels of structural detail have been explored with characteristics that are chosen in such a way as to facilitate comparison with experimental results obtained on particular protein systems. Here we address the practical aspects of these models, explaining how to use them, how to perform the simulations, and how to analyze their results.

2 Methods

2.1 Structure-Based Models (SBM): Potential and Cα Model

Structure-based models are grounded on two essential features. The first feature is the fact that the protein representation takes continuous values in three-dimensional space. Their second feature is a foundation on the principle of minimal frustration, which

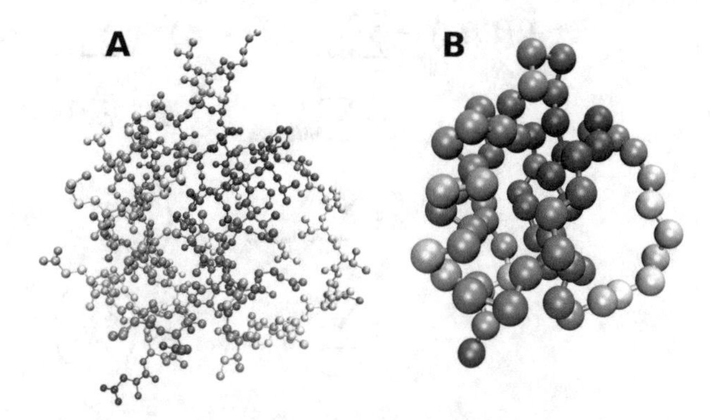

Fig. 1 Representation of protein components of the Serine Proteinase Inhibitor CI2 (PDB code: 2CI2) for different models. (**a**) All-atom model representation, each bead represents a heavy atom (no hydrogen); (**b**) C-alpha model, where every amino acid residue is presented as a sphere centered in alpha carbon

implies that the energy gap existing between the unfolded and folded configurations must be large relative to any energetic roughness arising from conflicting interactions or poor natural design. As a result, these models are constructed using forcefields designed to guarantee that the specific folded configuration corresponds to the global energy minimum, so that they drive the molecular simulation to fold into the correct native structure. These unfrustrated forcefields depict the limiting case of the principle of minimal frustration [25]. Such native structure-based models can be used to understand protein folding [2, 26–30], but also protein dimerization [31, 32], functional conformational changes and enzymatic reactions [12, 33–36], and other biomolecular processes.

The simplifications of SBM are related to the detail of representation of the protein components: all atoms, Cα + Cβ, or Cα [37–39]. The all atoms and Cα + Cβ approaches are used when information about side chains is important for the studied problem, e.g., packaging and specific interactions between amino acid residues. In the simplest case, the Cα model is utilized, which treats the residues as hard spheres centered on α-carbon, as shown in Fig. 1.

The target native protein structure is used to build an SBM potential, enabling the parameters of the potential expression to be obtained through the structural properties of the native conformation. Protein structures are resolved by experimental techniques such as X-ray crystallography and nuclear magnetic resonance (NMR) [40, 41]. The atom coordinates corresponding to the conformation of native protein are deposited in the protein data bank (PDB) [42].

$$V(\Gamma,\Gamma_o) = \sum_{bonds} \epsilon_r (r - r_o)^2 + \sum_{angles} \epsilon_\theta (\theta - \theta_o)^2 + \sum_{dihedrals} \epsilon_\phi \left\{ [1 - \cos(\phi - \phi_o)] + \frac{1}{2}[1 - \cos(3(\phi - \phi_o))] \right\} + \sum_{contacts} \epsilon_C \left[5\left(\frac{d_{ij}}{r_{ij}}\right)^{12} - 6\left(\frac{d_{ij}}{r_{ij}}\right)^{10} \right] + \sum_{non-contacts} \epsilon_{NC} \left(\frac{\sigma_{NC}}{r_{ij}}\right)^{12} \quad (1)$$

All ϵ parameters of Eq. (1) are defined in function of ϵ_C, which represents the interaction energy of van der Waals contacts (usually $\epsilon_C = 1.0$ kcal/mol [38]). ϵ_r, ϵ_θ, ϵ_ϕ, and ϵ_{NC} are usually equal to 100 ϵ_C, 20 ϵ_C, 40 ϵ_C, and 1 ϵ_C, respectively [2, 38]. In this equation, r_0 represents the native distance between two Cα atoms directly connected to each other, θ_0 is the angle between three consecutive Cα atoms in the native structure, and similarly, ϕ_0 is the dihedral angle between four Cα atoms. d_{ij} is the native distance between the pair of contacts i and j determined by the contact map of the native conformation. It is possible to generate a contact map using the web server SMOG—structure-based models for biomolecules (http://smog-server.org/) [43], which uses the shadow map algorithm to calculate the residue's contacts [44]. A fundamental variable in the description of any process is the order parameter or reaction coordinate. To indicate the folding stage of the protein during the simulation trajectory, one must use a structural parameter that represents a good reaction coordinate from a chemical–physical kinetics standpoint. The fraction of native contacts Q formed during the simulation is one of the most commonly used reaction coordinates in computational protein folding studies [2, 45, 46]. It is should be noted that Q is not directly measured experimentally, so the connection between simulation and experiment is not direct. There are other structural parameters that can be extracted from the protein folding simulations that may have a more direct relation with wet lab experiments, such as the radius of gyration [47, 48], the end-to-end distance [49, 50], and more recently, the calculation of FRET efficiencies [51]. The SMOG web server also uses the native structure of the protein to generate the input files required to perform SBM molecular dynamics in Gromacs and NAMD [52, 53].

2.2 Data Analysis and Connection to Experiments

In this section, we will discuss some features obtained from simulations using SBM which can be compared qualitatively and quantitatively with experiments. The analysis involves the data from thermodynamics and kinetic measurements.

2.2.1 Thermodynamics

Folding Temperature and Free Energy Profile

In simulations, deriving thermodynamic properties involves performing long simulation runs as needed to obtain sufficient statistics of the system. The data obtained from simulations are trajectories containing the ensemble of conformations of the protein in different snapshots. Each trajectory frame contains the position and the velocity of each atom of the protein in this particular simulation. Other parameters such as the forces acting within each atom, the total energy of the system or the energy contribution of a specific interaction may also be collected.

Usually, a set of simulations is carried out at different temperatures, which allows a better sampling of the protein conformations in the folded, the unfolded, and the folding transition states of the protein. The information obtained from the simulations helps to explore the energy landscape of the protein and to measure the relevant thermodynamic parameters that can be compared with experimental results, like the folding temperature (or midpoint denaturation temperature in the experimental jargon) and the heat capacity.

The folding free energy profile may be obtained via the useful methodology WHAM (weighted histogram analysis method), which helps to maximize the information provided by simulations [54, 55]. A Python script to calculate computational thermodynamic properties is available for download in the eSBMTools web link [56] and a Java version is available in the SMOG@ctbp [43].

ϕ-Value Analysis

ϕ-Value analysis was first developed as an experimental method by Fersht and coworkers for proteins that fold in a two-state manner [57, 58]. This method consists of measuring the contribution of each individual residue in the protein to the folding kinetics. Kinetic folding properties are measured for the wild-type protein and for a protein with a single mutation (substitution) in the residue of interest. The comparison between the WT and the mutated protein yields a value related to the free energy contribution of the replaced residue side chain to the folding transition state. Mathematically, the ϕ-value is defined as:

$$\phi = \frac{\Delta F_{WT}^{TS-U} - \Delta F_{M}^{TS-U}}{\Delta F_{WT}^{N-U} - \Delta F_{M}^{N-U}},$$

where ΔF_{WT}^{TS-U} is the free energy difference between the transition state and the unfolded state for the wild-type protein; ΔF_{M}^{TS-U} is the free energy difference for the mutated protein; and ΔF_{WT}^{N-U} and ΔF_{M}^{N-U} represent the free energy differences between the native and unfolded states for the wild-type and the mutant, respectively. The canonical ϕ-value range is between 0 and 1, where residues with a ϕ-value close to zero are those whose mutation does not cause significant variation in the free energy of the transition state relative to the unfolded state and thus result in an unvarying folding rate. On the other hand, residues with Φ-values near 1 are those important for the folding process, where their mutation has large effects on the folding transition state of the protein (and on the folding rate).

To calculate computational ϕ-values, one needs a simulation trajectory with a well-sampled folding transition state, which is usually best achieved at the folding temperature of the protein. To this end, expression 2 is used with a simplification that considers the free energy from the mutated protein to be equal to zero.

Folding Route

The folding route, $R(Q)$, is a parameter that makes it possible to specify the folding path of a protein. To calculate $R(Q)$, it is necessary to use trajectories where all the relevant folding stages are sufficiently sampled. The folding route is given by:

$$R(Q) = \sum_{i=1}^{M} \frac{\left\langle \left(\langle Q_i \rangle_Q - Q \right)^2 \right\rangle_Q}{MQ(1-Q)}$$

where M is the total number of native contacts; Q is the folding reaction coordinate (native contacts); and Q_i is equal to 1 when contact i is formed and equal to 0 when it is not formed. The average $\langle Q_i \rangle_Q$ is calculated using all configurations that have the same value of Q. $R(Q)$ is normalized between 1 and 0, where values close to zero mean that the folding path is not specific. In this case, the probability of a native contact being made is the same for all of the native contacts. When $R(Q)$ is near 1, the route is highly specific.

2.3 Kinetics

2.3.1 Folding Times and Diffusion

Kinetic measurements involve the calculation of the folding time and the intramolecular diffusion coefficient. The folding time is calculated from the Mean First Passage Time (MFPT) obtained from different runs. In each simulation, the protein starts in an open random configuration, and the simulation continues until the native state is reached. The results from the kinetic simulations lead to computational determination of the folding times that are directly comparable to the experimental folding rates determined from Chevron plots of the log of the relaxation rate versus denaturant concentration [59–63].

For the intramolecular diffusion coefficient D, the simulations need to be performed in such a way that ensures that the trajectory of correlated structures is measured within short time intervals. If D is assumed to be constant during the entire folding process, the equation which describes this property is related to the fluctuation of the reaction coordinate divided by the correlation time: $D = \frac{(\Delta Q)^2}{2\tau_C}$ [64, 65]. For more complex cases in which the intramolecular diffusion coefficient varies with the position along the reaction coordinate, there are alternative methods to calculate $D(Q)$ that are based on different premises and approaches [65–70], but they will not be discussed here.

2.3.2 Non-native Interactions and Frustration

The SBM is constructed in accordance with the principle of minimal frustration, i.e., the interactions are set in such a way that the native state corresponds to the global energy minimum of the system [38]. The contact map based on the 3D structure of the native protein contains all the interactions required to build the Hamiltonian equation (see above). This potential energy is oversimplified, but it has the advantage of ensuring that the protein folds to the correct native conformation. In this section, we briefly discuss additional non-native potentials, which can be added to the native-based SBM Hamiltonian. These additional terms introduce more complexity into the energy landscape, recreating different degrees of landscape roughness and representing protein folding mechanisms in a more realistic way.

Explicit electrostatic interactions between protein charged residues are considered a non-native interaction, as they are relatively nonspecific and act over long distances. Electrostatic terms can be added into the model to help in reproducing and understanding the effects of pH and salt concentration in the folding process. These added effective-electrostatic potentials make it possible to address the role of environmental parameters and charges in the stabilization/destabilization of the native state [36, 71–73]. The charge–charge interactions may be optimized by sequence engineering to increase the protein's thermostability [30, 35, 74–78]. Optimization of electrostatic interactions via modeling can be achieved using some tools available online [79, 80]. Hydrophobic-like non-native potentials can be also added to SBM and account for effects such as molten globule formation and nonspecific hydrophobic collapse. These type of more elaborate SBM potentials are generally called "flavored" potentials and are usually associated with phenomenological models of residue–residue interactions based on structural statistics, such as Miyazawa-Jernigan potential of mean-force [81–83]. Recent advances in the investigation of the role of non-native interactions using SBM have revealed the presence of cooperativity between electrostatic and hydrophobic interactions as the source of internal friction in alpha-spectrins domains [61].

As mentioned above, proteins are selected through natural evolution following the minimum frustration principle [4, 5]. However, even well-evolved proteins are likely to exhibit some degree of energetic frustration caused by non-native interactions, and thus non-native contributions to folding energy landscapes should not be neglected [84]. Frustration plays an important role in many different aspects of protein folding. Non-native interactions may lead the protein to traps or intermediate states [85]. On the other hand, the presence of frustration may be favorable to the protein folding dynamics, accelerating the process [86–92]. Moreover, a certain degree of local frustration on the protein surface has functional implications by providing binding sites for other proteins [93, 94]. In this regard, the frustratometer web server is a useful tool that allows to calculate the energetic frustration of a protein on the basis of its native structure [95, 96].

3 Conclusions

Computer simulations of coarse-grained models provide a powerful tool for investigating the conformational dynamics and thermodynamics of macromolecular systems in general and proteins in particular. This type of simulations provides an in silico platform that can be used to validate, test, and corroborate both theory and real (wet) lab experiments as well as to help interpreting the latter in mechanistic terms. There is no single protocol that fits all questions, and the approach requires a certain degree of model customization to best represent the available experimental data. SBM approaches are too simple to produce ab-initio predictions, but their simplicity and customizability make them ideal tools for the interpretation and analysis of experimental results. As word of caution, we should say that direct comparison between SBM simulations and experiments is, in general, not straightforward, and one should not expect to be able to compare exact values and units. However, the simulations should be able to capture trends, reproduce behaviors at least at a qualitative level, and facilitate the search for correlations that may provide remarkable insights about the underlying molecular mechanisms.

Acknowledgments

VGC was funded by Grant 2016/13998-8 and 2017/09662-7, FAPESP (São Paulo Research Foundation and Higher Education Personnel) and CAPES and (Higher Education Personnel Improvement Coordination) and also acknowledges NSF (National Science Foundation) Grants PHY-2019745 and CHE-1614101. VMO was supported by the CNPq (National Council for Scientific and Technological Development) Grant Process No. 141985/2013-5, and FAPESP 2018/11614-3. VBPL was supported by the CNPq and FAPESP Grant 2014/06862-7, 2016/19766-1, and 2019/22540-3.

References

1. Onuchic JN (2014) Physics transforming the life sciences. Phys Biol 11:053006. https://doi.org/10.1088/1478-3975/11/5/053006
2. Onuchic JN, Nymeyer H, García AE, Chahine J, Socci ND (2000) The energy landscape theory of protein folding: insights into folding mechanisms and scenarios. Adv Protein Chem 53:87–152
3. Goldstein RA, Luthey-Schulten ZA, Wolynes PG (1992) Protein tertiary structure recognition using optimized Hamiltonians with local interactions. Proc Natl Acad Sci 89:9029–9033. https://doi.org/10.1073/pnas.89.19.9029
4. Bryngelson JD, Wolynes PG (1987) Spin glasses and the statistical mechanics of protein folding. Proc Natl Acad Sci U S A 84:7524–7528. https://doi.org/10.1073/pnas.84.21.7524
5. Bryngelson JD, Wolynes PG (1989) Intermediates and barrier crossing in a random energy model (with applications to protein folding). J Phys Chem 93:6902–6915. https://doi.org/10.1021/j100356a007
6. Jin S, Contessoto VG, Chen M, Schafer NP, Lu W, Chen X, Bueno C, Hajitaheri A, Sirovetz BJ, Davtyan A, Papoian GA (2020) AWSEM-Suite: a protein structure prediction server based on template-guided, coevolutionary-enhanced optimized folding landscapes. Nucleic Acids Res 48(W1):W25–W30. https://doi.org/10.1093/nar/gkaa356
7. Wang Q, Diehl MR, Jana B, Cheung MS, Kolomeisky AB, Onuchic JN (2017) Molecular origin of the weak susceptibility of kinesin velocity to loads and its relation to the collective behavior of kinesins. Proc Natl Acad Sci U S A 114:201710328. https://doi.org/10.1073/pnas.1710328114
8. Staquicini DI, Tang FHF, Markosian C, Yao VJ, Staquicini FI, Dodero-Rojas E, Contessoto VG, Davis D, O'Brien P, Habib N, Smith TL, Bruiners N, Sidman RL, Gennaro ML, Lattime EC, Libutti SK, Whitford PC, Burley SK, Onuchic JN, Arap W (2021) Renata pasqualini design and proof of concept for targeted phage-based COVID-19 vaccination strategies with a streamlined cold-free supply chain. Proc Natl Acad Sci 118(30):e2105739118. https://doi.org/10.1073/pnas.2105739118
9. Roy S, Onuchic JN, Sanbonmatsu KY (2017) Cooperation between magnesium and metabolite controls collapse of the SAM-I riboswitch. Biophys J 113:348–359. https://doi.org/10.1016/j.bpj.2017.06.044
10. Echeverria I, Papoian G (2016) Perspectives on the coarse-grained models of DNA. In: Many-body effects and electrostatics in biomolecules. Pan Stanford, Singapore, pp 535–570
11. Zhao H, Winogradoff D, Bui M, Dalal Y, Papoian GA (2016) Promiscuous histone Mis-assembly is actively prevented by chaperones. J Am Chem Soc 138:13207–13218. https://doi.org/10.1021/jacs.6b05355
12. Nguyen K, Whitford PC (2017) Challenges in describing ribosome dynamics. Phys Biol 14:023001. https://doi.org/10.1088/1478-3975/aa626b
13. Whitford PC (2016) Quantifying the energy landscape of ribosome function. Biophys J 110:352a. https://doi.org/10.1016/j.bpj.2015.11.1898
14. Noel JK, Chahine J, Leite VBP, Whitford PC (2014) Capturing transition paths and transition states for conformational rearrangements in the ribosome. Biophys J 107:2881–2890. https://doi.org/10.1016/j.bpj.2014.10.022
15. Zhang B, Wolynes P (2015) Shape transitions and chiral symmetry breaking in the energy landscape of the mitotic chromosome. Phys Rev Lett 116(24):248101. https://doi.org/10.1101/031260
16. Di Pierro M, Zhang B, Aiden EL, Wolynes PG, Onuchic JN (2016) Transferable model for chromosome architecture. Proc Natil Acad Sci 113(43):12168–12173. https://doi.org/10.1073/pnas.1613607113
17. Pierro MD, Cheng RR, Aiden EL, Wolynes PG, Onuchic JN (2017) De novo prediction of human chromosome structures: epigenetic marking patterns encode genome architecture. Proc Natl Acad Sci U S A 114:12126–12131. https://doi.org/10.1073/pnas.1714980114
18. Cheng RR, Contessoto VG, Aiden EL, Wolynes PG, Di Pierro M, Onuchic JN (2020) Exploring chromosomal structural heterogeneity across multiple cell lines. Elife 9:e60312. https://doi.org/10.7554/eLife.60312
19. Contessoto VG, Cheng RR, Hajitaheri A, Dodero-Rojas, E, Mello MF, Lieberman-Aiden E, Wolynes PG, Di Pierro M, Onuchic

JN (2021) The Nucleome Data Bank: web-based resources to simulate and analyze the three-dimensional genome. Nucleic Acids Res 49(D1):D172–D182. https://doi.org/10.1093/nar/gkaa818

20. Junior ABO, Contessoto VG, Mello MF, Onuchic JN (2021) A scalable computational approach for simulating complexes of multiple chromosomes. J Mol Biol 433(6):166700. https://doi.org/10.1016/j.jmb.2020.10.034
21. Hoencamp C, Dudchenko O, Elbatsh AM, Brahmachari S, Raaijmakers JA, van Schaik T, Cacciatore ÁS, Contessoto VG, van Heesbeen RG, van den Broek B, Mhaskar AN (2021) 3D genomics across the tree of life reveals condensin II as a determinant of architecture type. Science 372(6545):984–989. https://doi.org/10.1126/science.abe2218
22. Chan HS, Dill KA (1996) Comparing folding codes for proteins and polymers. Proteins Struct Funct Genet 24:335–344. https://doi.org/10.1002/(sici)1097-0134(199603)24:3<335::aid-prot6>3.0.co;2-f
23. Shakhnovich E, Gutin A (1990) Enumeration of all compact conformations of copolymers with random sequence of links. J Chem Phys 93:5967–5971. https://doi.org/10.1063/1.459480
24. Socci ND, Onuchic JN (1994) Folding kinetics of proteinlike heteropolymers. J Chem Phys 101:1519–1528. https://doi.org/10.1063/1.467775
25. Whitford PC, Sanbonmatsu KY, Onuchic JN (2012) Biomolecular dynamics: orderdisorder transitions and energy landscapes. Rep Prog Phys 75:076601. https://doi.org/10.1088/0034-4885/75/7/076601
26. van der Spoel D, Seibert MM (2006) Protein folding kinetics and thermodynamics from atomistic simulations. Phys Rev Lett 96 (23):238102. https://doi.org/10.1103/physrevlett.96.238102
27. Wang J (2006) Diffusion and single molecule dynamics on biomolecular interface binding energy landscape. Chem Phys Lett 418:544–548. https://doi.org/10.1016/j.cplett.2005.11.016
28. Cieplak M, Hoang TX, Li MS (1999) Scaling of folding properties in simple models of proteins. Phys Rev Lett 83:1684–1687. https://doi.org/10.1103/physrevlett.83.1684
29. Leite VBP, Onuchic JN, Stell G, Wang J (2004) Probing the kinetics of single molecule protein folding. Biophys J 87:3633–3641. https://doi.org/10.1529/biophysj.104.046243
30. De Oliveira VM, Caetano DL, Da Silva FB, Mouro PR, de Oliveira Jr AB, De Carvalho SJ, Leite VB (2019) pH and charged mutations modulate cold shock protein folding and stability: A constant pH monte carlo study. J Chem Theory Comput 16(1):765–772
31. Moritsugu K, Kurkal-Siebert V, Smith JC (2009) REACH coarse-grained Normal mode analysis of protein dimer interaction dynamics. Biophys J 97:1158–1167. https://doi.org/10.1016/j.bpj.2009.05.015
32. Kaya H, Chan HS (2002) Towards a consistent modeling of protein thermodynamic and kinetic cooperativity: how applicable is the transition state picture to folding and unfolding? 1 1Edited by C. R Matthews. J Mol Biol 315:899–909. https://doi.org/10.1006/jmbi.2001.5266
33. Whitford PC, Miyashita O, Levy Y, Onuchic JN (2007) Conformational transitions of adenylate kinase: switching by cracking. J Mol Biol 366:1661–1671. https://doi.org/10.1016/j.jmb.2006.11.085
34. Contessoto VG, Ramos FC, de Melo RR, de Oliveira VM, Scarpassa JA, de Sousa AS, Zanphorlin LM, Slade GG, Leite VBP, Ruller R (2021) Electrostatic interaction optimization improves catalytic rates and thermotolerance on xylanases. Biophys J 120(11):2172–2180. https://doi.org/10.1016/j.bpj.2021.03.036
35. Ngo K, Bruno da Silva F, Leite VB, Contessoto VG, Onuchic JN (2021) Improving the thermostability of xylanase a from bacillus subtilis by combining bioinformatics and electrostatic interactions optimization. J Phys Chem B 125 (17):4359–4367. https://doi.org/10.1021/acs.jpcb.1c01253
36. Bruno da Silva F, Oliveira VM, Sanches MN, Contessoto VG, Leite VBP (2019) Rational design of chymotrypsin inhibitor 2 by optimizing non-native interactions. J Chem Inform Modeling 60(2):982–988. https://doi.org/10.1021/acs.jcim.9b00911
37. Clementi C, Garci'a AE, Onuchic JN (2003) Interplay among tertiary contacts secondary structure formation and side-chain packing in the protein folding mechanism: all-atom representation study of protein L. J Mol Biol 326:933–954. https://doi.org/10.1016/s0022-2836(02)01379-7
38. Clementi C, Nymeyer H, Onuchic JN (2000) Topological and energetic factors: what determines the structural details of the transition state ensemble and en-route intermediates for protein folding? An investigation for small globular proteins. J Mol Biol 298:937–953. https://doi.org/10.1006/jmbi.2000.3693
39. Whitford PC, Noel JK, Gosavi S, Schug A, Sanbonmatsu KY, Onuchic JN (2009) An all-atom structure-based potential for proteins:

bridging minimal models with all-atom empirical forcefields. Proteins 75:430–441. https://doi.org/10.1002/prot.22253

40. Zhao B, Yi G, Du F, Chuang Y-C, Vaughan RC, Sankaran B, Kao CC, Li P (2017) Structure and function of the Zika virus full-length NS5 protein. Nat Commun 8:14762. https://doi.org/10.1038/ncomms14762
41. Wishart D (2005) NMR spectroscopy and protein structure determination: applications to drug discovery and development. Curr Pharm Biotechnol 6:105–120. https://doi.org/10.2174/1389201053642367
42. Rose PW, Prlić A, Altunkaya A, Bi C, Bradley AR, Christie CH, Costanzo LD, Duarte JM, Dutta S, Feng Z, Green RK, Goodsell DS, Hudson B, Kalro T, Lowe R, Peisach E, Randle C, Rose AS, Shao C, Tao YP, Valasatava Y, Voigt M, Westbrook JD, Woo J, Yang H, Young JY, Zardecki C, Berman HM, Burley SK (2017) The RCSB protein data bank: integrative view of protein, gene and 3D structural information. Nucleic Acids Res 45:D271–D281
43. Noel JK, Levi M, Raghunathan M, Lammert H, Hayes RL, Onuchic JN, Whitford PC (2016) SMOG 2: a versatile software package for generating structure-based models. PLoS Comput Biol 12:e1004794. https://doi.org/10.1371/journal.pcbi.1004794
44. Noel JK, Whitford PC, Onuchic JN (2012) The shadow map: a general contact definition for capturing the dynamics of biomolecular folding and function. J Phys Chem B 116:8692–8702. https://doi.org/10.1021/jp300852d
45. Clementi C, Jennings PA, Onuchic JN (2001) Prediction of folding mechanism for circular-permuted proteins. J Mol Biol 311:879–890. https://doi.org/10.1006/jmbi.2001.4871
46. Cho SS, Levy Y, Wolynes PG (2006) P versus Q: structural reaction coordinates capture protein folding on smooth landscapes. Proc Natl Acad Sci 103:586–591. https://doi.org/10.1073/pnas.0509768103
47. Schaeffer RD, Fersht A, Daggett V (2008) Combining experiment and simulation in protein folding: closing the gap for small model systems. Curr Opin Struct Biol 18:4–9. https://doi.org/10.1016/j.sbi.2007.11.007
48. Kohn JE, Millett IS, Jacob J, Zagrovic B, Dillon TM, Cingel N, Dothager RS, Seifert S, Thiyagarajan P, Sosnick TR, Hasan MZ, Pande VS, Ruczinski I, Doniach S, Plaxco KW (2004) Random-coil behavior and the dimensions of chemically unfolded proteins. Proc Natl Acad Sci 101:12491–12496. https://doi.org/10.1073/pnas.0403643101
49. Moglich A, Joder K, Kiefhaber T (2006) End-to-end distance distributions and intrachain diffusion constants in unfolded polypeptide chains indicate intramolecular hydrogen bond formation. Proc Natl Acad Sci U S A 103:12394–12399. https://doi.org/10.1073/pnas.0604748103
50. Soranno A, Longhi R, Bellini T, Buscaglia M (2009) Kinetics of contact formation and end-to-end distance distributions of swollen disordered peptides. Biophys J 96:1515–1528. https://doi.org/10.1016/j.bpj.2008.11.014
51. Reinartz I, Sinner C, Schug A (2017) Simulation of FRET dyes allows direct comparison against experimental data. Biophys J 112:471a. https://doi.org/10.1016/j.bpj.2016.11.2528
52. Berendsen HJC, van der Spoel D, van Drunen R (1995) GROMACS: a message-passing parallel molecular dynamics implementation. Comput Phys Commun 91:43–56. https://doi.org/10.1016/0010-4655(95)00042-e
53. Phillips JC, Braun R, Wang W, Gumbart J, Tajkhorshid E, Villa E, Chipot C, Skeel RD, Kalé L, Schulten K (2005) Scalable molecular dynamics with NAMD. J Comput Chem 26:1781–1802. https://doi.org/10.1002/jcc.20289
54. Ferrenberg AM, Swendsen RH (1988) New Monte Carlo technique for studying phase transitions. Phys Rev Lett 61:2635–2638. https://doi.org/10.1103/physrevlett.61.2635
55. Kumar S, Rosenberg JM, Bouzida D, Swendsen RH, Kollman PA (1992) THE weighted histogram analysis method for free-energy calculations on biomolecules. I. The method. J Comput Chem 13:1011–1021. https://doi.org/10.1002/jcc.540130812
56. Lutz B, Sinner C, Heuermann G, Verma A, Schug A (2013) eSBMTools 1.0: enhanced native structure-based modeling tools. Bioinformatics 29:2795–2796. https://doi.org/10.1093/bioinformatics/btt478
57. Matouschek A, Kellis JT, Serrano L, Fersht AR (1989) Mapping the transition state and pathway of protein folding by protein engineering. Nature 340:122–126. https://doi.org/10.1038/340122a0
58. Fersht AR, Matouschek A, Serrano L (1992) The folding of an enzyme. J Mol Biol 224:771–782. https://doi.org/10.1016/0022-2836(92)90561-w
59. Wang J, Oliveira RJ, Chu X, Whitford PC, Chahine J, Han W, Wang E, Onuchic JN, Leite VBP (2012) Topography of funneled landscapes determines the thermodynamics

and kinetics of protein folding. Proc Natl Acad Sci U S A 109:15763–15768. https://doi.org/10.1073/pnas.1212842109

60. Chavez LL, Onuchic JN, Clementi C (2004) Quantifying the roughness on the free energy landscape: entropic bottlenecks and protein folding rates. J Am Chem Soc 126:8426–8432
61. Bruno da Silva F, Contessoto VG, De Oliveira VM, Clarke J, Leite VB (2018) Non-native cooperative interactions modulate protein folding rates. J Phys Chem B 122 (48):10817–10824. https://doi.org/10.1021/acs.jpcb.8b08990
62. Polotto F, Drigo Filho E, Chahine J, de Oliveira RJ (2018) Supersymmetric quantum mechanics method for the FokkerPlanck equation with applications to protein folding dynamics. Physica A 493:286–300. https://doi.org/10.1016/j.physa.2017.10.021
63. Xu W, Lai Z, Oliveira RJ, Leite VBP, Wang J (2012) Configuration-dependent diffusion dynamics of downhill and two-state protein folding. J Phys Chem B 116:5152–5159. https://doi.org/10.1021/jp212132v
64. Whitford PC, Blanchard SC, Cate JHD, Sanbonmatsu KY (2013) Connecting the kinetics and energy landscape of tRNA translocation on the ribosome. PLoS Comput Biol 9:e1003003. https://doi.org/10.1371/journal.pcbi.1003003
65. Chahine J, Oliveira RJ, Leite VBP, Wang J (2007) Configuration-dependent diffusion can shift the kinetic transition state and barrier height of protein folding. Proc Natl Acad Sci U S A 104:14646–14651. https://doi.org/10.1073/pnas.0606506104
66. Oliveira RJ, Whitford PC, Chahine J, Leite VBP, Wang J (2010) Coordinate and time-dependent diffusion dynamics in protein folding. Methods 52:91–98. https://doi.org/10.1016/j.ymeth.2010.04.016
67. Best RB, Hummer G (2009) Coordinate-dependent diffusion in protein folding. Proc Natl Acad Sci U S A 107:1088–1093. https://doi.org/10.1073/pnas.0910390107
68. Yang S, Onuchic JN, Levine H (2006) Effective stochastic dynamics on a protein folding energy landscape. J Chem Phys 125:054910. https://doi.org/10.1063/1.2229206
69. Contessoto VG, Oliveira ABD, Chahine J, Oliveira RJD, Pereira Leite VB (2018) Introdução ao problema de enovelamento de proteínas: uma abordagem utilizando modelos computacionais simplificados. Revista Brasileira de Ensino de Física, 40 https://doi.org/10.1590/1806-9126-RBEF-2018-0068
70. Freitas FC, Lima AN, Contessoto VDG, Whitford PC, Oliveira RJD (2019) Drift-diffusion (DrDiff) framework determines kinetics and thermodynamics of two-state folding trajectory and tunes diffusion models. J Chem Phys 151 (11):114106. https://doi.org/10.1063/1.5113499
71. Azia A, Levy Y (2009) Nonnative electrostatic interactions can modulate protein folding: molecular dynamics with a grain of salt. J Mol Biol 393:527–542. https://doi.org/10.1016/j.jmb.2009.08.010
72. Contessoto VG, de Oliveira VM, de Carvalho SJ, Oliveira LC, Leite VBP (2016) NTL9 folding at constant pH: the importance of electrostatic interaction and pH dependence. J Chem Theory Comput 12:3270–3277. https://doi.org/10.1021/acs.jctc.6b00399
73. de Oliveira VM, de Godoi Contessoto V, da Silva FB, Caetano DLZ, de Carvalho SJ, Leite VBP (2018) Effects of pH and salt concentration on stability of a protein G variant using coarse-grained models. Biophys J 114:65–75. https://doi.org/10.1016/j.bpj.2017.11.012
74. Ibarra-Molero B, Loladze VV, Makhatadze GI, Sanchez-Ruiz JM (1999) Thermal versus guanidine-induced unfolding of ubiquitin. An analysis in terms of the contributions from charge-charge interactions to protein stability. Biochemistry 38:8138–8149. https://doi.org/10.1021/bi9905819
75. Loladze VV, Ibarra-Molero B, Sanchez-Ruiz JM, Makhatadze GI (1999) Engineering a thermostable protein via optimization of charge-charge interactions on the protein surface. Biochemistry 38:16419–16423. https://doi.org/10.1021/bi992271w
76. Gribenko AV, Makhatadze GI (2007) Role of the ChargeCharge interactions in defining stability and Halophilicity of the CspB proteins. J Mol Biol 366:842–856. https://doi.org/10.1016/j.jmb.2006.11.061
77. Sanchez-Ruiz JM, Makhatadze GI (2001) To charge or not to charge? Trends Biotechnol 19:132–135. https://doi.org/10.1016/s0167-7799(00)01548-1
78. Gribenko AV, Patel MM, Liu J, McCallum SA, Wang C, Makhatadze GI (2009) Rational stabilization of enzymes by computational redesign of surface chargecharge interactions. Proc Natl Acad Sci U S A 106:2601–2606. https://doi.org/10.1073/pnas.0808220106
79. Gopi S, Devanshu D, Krishna P, Naganathan AN (2017) pStab: prediction of stable mutants unfolding curves, stability maps and protein electrostatic frustration. Bioinformatics 34 (5):875–877. https://doi.org/10.1093/bioinformatics/btx697

80. Contessoto VG, de Oliveira VM, Fernandes BR, Slade GG, VBP L (2017) TKSA-MC: A Web Server for rational mutation through the optimization of protein charge interactions. Proteins 86(11):1184–1188. https://doi.org/10.1101/221556
81. Cho SS, Levy Y, Wolynes PG (2008) Quantitative criteria for native energetic heterogeneity influences in the prediction of protein folding kinetics. Proc Natl Acad Sci U S A 106:434–439. https://doi.org/10.1073/pnas.0810218105
82. Chen T, Chan HS (2015) Native contact density and nonnative hydrophobic effects in the folding of bacterial immunity proteins. PLoS Comput Biol 11:e1004260. https://doi.org/10.1371/journal.pcbi.1004260
83. Chen T, Song J, Chan HS (2015) Theoretical perspectives on nonnative interactions and intrinsic disorder in protein folding and binding. Curr Opin Struct Biol 30:32–42. https://doi.org/10.1016/j.sbi.2014.12.002
84. Ferreiro DU, Komives EA, Wolynes PG (2018) Frustration function and folding. Curr Opin Struct Biol 48:68–73. https://doi.org/10.1016/j.sbi.2017.09.006
85. Sutto L, Latzer J, Hegler JA, Ferreiro DU, Wolynes PG (2007) Consequences of localized frustration for the folding mechanism of the IM7 protein. Proc Natl Acad Sci U S A 104:19825–19830. https://doi.org/10.1073/pnas.0709922104
86. Clementi C, Plotkin SS (2004) The effects of nonnative interactions on protein folding rates: theory and simulation. Protein Sci 13:1750–1766. https://doi.org/10.1110/ps.03580104
87. Plotkin SS (2001) Speeding protein folding beyond the Gö model: how a little frustration sometimes helps. Proteins Struct Funct Genet 45:337–345. https://doi.org/10.1002/prot.1154
88. Contessoto VG, Lima DT, Oliveira RJ, Bruni AT, Chahine J, Leite VBP (2013) Analyzing the effect of homogeneous frustration in protein folding. Proteins 81:1727–1737. https://doi.org/10.1002/prot.24309
89. Mouro PR, de Godoi CV, Chahine J, Junio de Oliveira R, Pereira Leite VB (2016) Quantifying nonnative interactions in the protein-folding free-energy landscape. Biophys J 111:287–293. https://doi.org/10.1016/j.bpj.2016.05.041
90. Tzul FO, Schweiker KL, Makhatadze GI (2015) Modulation of folding energy landscape by chargecharge interactions: linking experiments with computational modeling. Proc Natl Acad Sci U S A 112:E259–E266. https://doi.org/10.1073/pnas.1410424112
91. Oliveira RJ, Whitford PC, Chahine J, Wang J, Onuchic JN, Leite VBP (2010) The origin of nonmonotonic complex behavior and the effects of nonnative interactions on the diffusive properties of protein folding. Biophys J 99:600–608. https://doi.org/10.1016/j.bpj.2010.04.041
92. Oliveira LC, Silva RT, Leite VB, Chahine J (2006) Frustration and hydrophobicity interplay in protein folding and protein evolution. J Chem Phys 125:084904
93. Ferreiro DU, Hegler JA, Komives EA, Wolynes PG (2007) Localizing frustration in native proteins and protein assemblies. Proc Natl Acad Sci U S A 104:19819–19824. https://doi.org/10.1073/pnas.0709915104
94. Gosavi S, Chavez LL, Jennings PA, Onuchic JN (2006) Topological frustration and the folding of interleukin-1β. J Mol Biol 357:986–996. https://doi.org/10.1016/j.jmb.2005.11.074
95. Jenik M, Parra RG, Radusky LG, Turjanski A, Wolynes PG, Ferreiro DU (2012) Protein frustratometer: a tool to localize energetic frustration in protein molecules. Nucleic Acids Res 40:W348–W351. https://doi.org/10.1093/nar/gks447
96. Parra RG, Schafer NP, Radusky LG, Tsai M-Y, Guzovsky AB, Wolynes PG, Ferreiro DU (2016) Protein Frustratometer 2: a tool to localize energetic frustration in protein molecules now with electrostatics. Nucleic Acids Res 44:W356–W360. https://doi.org/10.1093/nar/gkw304

Chapter 17

Analysis of Molecular Dynamics Simulations of Protein Folding

Robert B. Best

Abstract

Unbiased molecular dynamics simulations of proteins can now capture spontaneous folding events. This provides a wealth of data reflecting information on folding mechanism, but raises the challenge of interpreting it in a meaningful way. Here, I describe how such simulations can be used to identify reactive states and reaction coordinates for describing folding, and how folding dynamics can be captured by projection onto those coordinates. Methods are described for quantifying the interactions important for defining the folding mechanism, and for comparison of simulations with experimental mechanistic probes, such as ϕ-values.

Key words Transition path, Reaction coordinate, Committor, Native contacts, Bayesian criterion, Markov state model

1 Introduction

Recent years have seen dramatic advances in folding simulations of small proteins using atomistic molecular dynamics simulations, to the point where it is now possible to perform simulations which are long enough to observe spontaneous folding events of small proteins with explicit representation of solvent molecules. This comes partly as the result of advances in computer hardware and accompanying algorithmic advances. Examples include the use of graphics processing units (GPUs) which have made microsecond simulations widely available, or the development of specialized hardware, such as the ANTON supercomputer which can perform simulations on a millisecond timescale [1]. It has also been made possible by improvements in the energy functions (or force fields) used in these simulations, which have removed some of the biases present in older force fields [2]. Earlier simulation work had, of necessity, relied on either much simplified models, in which assumptions were made about the important physical interactions (such as the assumption in Go models that only native contacts are favourable

Victor Muñoz (ed.), *Protein Folding: Methods and Protocols*, Methods in Molecular Biology, vol. 2376,
https://doi.org/10.1007/978-1-0716-1716-8_17,

[3, 4]), or enhanced sampling methods needed to be used, in which case the effect of these methods on the sampled mechanism must be considered. Thus, the availability of unbiased simulations, in which the mechanism is determined only by the physical forces captured by the molecular mechanics force field, is a major advance [5].

Although the advantage of molecular simulations is the atomistic detail in which folding mechanisms can be described, they also pose the challenge of interpreting the huge volume of data in a meaningful way. For example, it is possible to observe structure formation during a folding transition, but what are really the key events or interactions that determine whether the protein folds or not? It is also frequently the case that when different folding transitions are compared, they can be seen to be different just by using molecular graphics software. How can this transition path heterogeneity be interpreted? Lastly, because force field accuracy is still far from perfect, it is necessary to benchmark the results against experiment in some way. Observables which could reflect on the folding mechanism include the folding rates, as well as derived parameters such as ϕ-values, so methods are needed to compute these from simulations.

The methods for analyzing existing MD trajectories can be broadly divided into those which are based on reaction coordinates or progress variables, and those based on Markov state models. In the former, one tries to identify one or more key variables describing the state of the protein, which are functions of its current coordinates. The aim is to identify variables which accurately capture the folding dynamics of the protein. For example, given only the values of such reaction coordinate(s), one should be able to predict the most likely fate of the protein (will it fold, or unfold, first?). Projecting the folding trajectories onto such variables should also be captured by a simple dynamical model, such as one-dimensional diffusion. In this case, it should be possible to follow the progress of the protein by examining structures at different points along the reaction coordinate. In the case of Markov state models, coordinates are eschewed altogether, and the configuration space of the protein is partitioned into microscopic states. After a suitable definition of such states, one can compute the transition rates between them, and hence infer information about the most probable pathways for folding. The two approaches can yield similar information, so this chapter will focus on methods based on reaction coordinates. Markov state model methodology has been reviewed elsewhere [6].

In this chapter, I assume that the user already has a folding trajectory available, therefore I do not discuss methodology for running simulations. Rather, the focus is on interpreting them using methods based on reaction coordinates. Before describing these methods in detail, I will first outline the Bayesian procedure

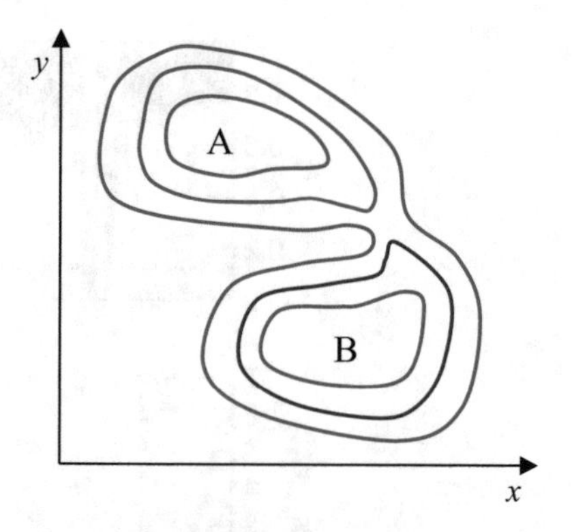

Fig. 1 Cartoon energy landscape. Illustration of good and bad reaction coordinates for a two-dimensional energy landscape. The stable states A and B, as well as the saddle connecting them, are overlapped on the coordinate "*x*," making it a poor coordinate to describe A–B transitions. Instead, on coordinate "*y*," A and B are well separated and the saddle between them can be separately resolved from the stable states, making it a good coordinate

for describing reaction coordinates which will be used extensively in what follows [7]. The method assumes we are trying to describe folding dynamics in terms of a reaction coordinate. In my discussion, I focus on a single coordinate, since that is often sufficient, but generalization to multiple coordinates is possible if enough data is available [8]. I also focus on two-state systems, with discussion of one-state or multistate systems saved for later. For our single coordinate, R, we would like to determine (1) whether this is a "good" coordinate for describing folding and (2) where are the most reactive states ("transition states") on this coordinate. What we mean by a good coordinate is that it should be able to separate these reactive events well from the equilibrium dynamics within the stable states. This is illustrated in Fig. 1 using a cartoon two-dimensional landscape. If we used the x coordinate to describe the dynamics, it would overlap the transition states with the stable states on the energy surface. That would mean that even if we knew the value of the coordinate corresponding to transition states, choosing structures with this value of the coordinate from trajectories would yield very few transition states. On the other hand, using the y coordinate, we find that the barrier top is well resolved from the stable states, and hence y is a useful folding coordinate. How can we determine if a given coordinate R is good or not, bearing in mind that we cannot visualize the highly multidimensional energy surface for protein folding as we can in Fig. 1? We do this by projecting the trajectory onto R, as shown in Fig. 2. What is obvious from observing the trajectory is that it spends most of its time in one of two stable states, in this case unfolded and folded, with the transitions between them being fleeting events of short

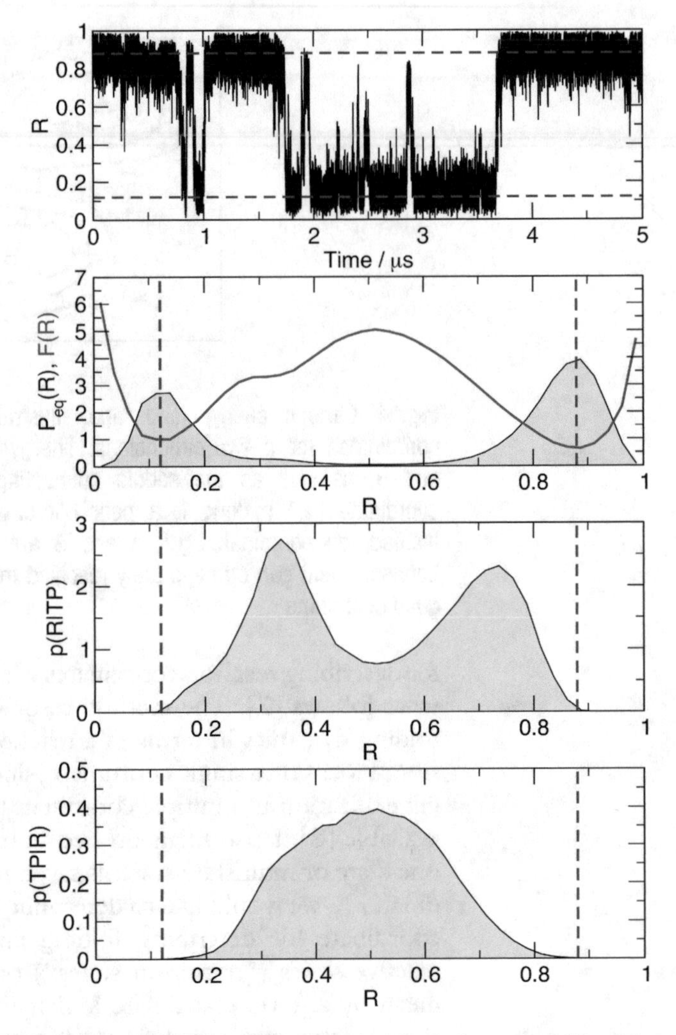

Fig. 2 Projection of a folding trajectory onto a reaction coordinate. (Top panel) Trajectory $R(t)$ obtained by projecting a protein folding trajectory onto a given reaction coordinate "R." The blue broken lines indicate the values of R used to define folding transitions and transition paths. (Second from top) Equilibrium probability density $P_{eq}(R)$ and corresponding free energy surface $F(R) = -\ln [P_{eq}(R)]$. (Third from top) Probability density of R on transition paths, $P(R \mid TP)$. (Bottom) Probability of being on a transition path given a value of R, $P(TP \vee R)$. This quantity, which identifies the most reactive states, has a theoretical maximum of 0.5 for diffusive dynamics, and so a peak of $P(TP \vee R)$ near this value suggests that the coordinate has gathered mostly reactive states together at the corresponding value of R

duration, which we term "transition paths" (TP). We assert that for a good coordinate, the conditional probability of being on a transition path given a particular value of that coordinate, $P(TP \mid R)$, can be used to assess coordinate quality. This quantity can be easily evaluated from an equilibrium trajectory by making histograms of the distribution of R at equilibrium (over the whole trajectory), $P_{eq}(R)$, and on transition paths, $P(R \mid TP)$, by applying Bayes' theorem,

$$P(TP|R) = \frac{P(R|TP)P(TP)}{P_{eq}(R)} \tag{1}$$

here, $P(TP)$ is the overall fraction of time spent on transition paths. A good coordinate should gather together the states which are likely to be on transition paths at a single value of R. The function $P(TP \mid R)$ will then have a maximum at this value of R, R_{max} (*see* Fig. 2). Thus, the best coordinates will be the highest value of $P(TP \mid R_{max})$. It can be shown that for diffusive trajectories, such as those for protein folding, the theoretical maximum of $P(TP \mid R_{max})$ is 0.5, allowing for direct quantification of reaction coordinates. Secondly, the reactive states are most likely to be found at R_{max}. A more formal version of the above discussion is given in Refs. 7, 9.

2 Materials

This chapter is written so as to not be specific to a particular computer code for analyzing trajectories, since a large number of such codes exist. What is needed is a way to quickly access the coordinates of each frame in the trajectory in such a way that intramolecular distances, contacts, and reaction coordinates can be calculated. One such tool which has been effectively used in the author's research group is the MDAnalysis package which provides an interface to MD trajectories in most common formats in the Python programming language [10]. However, most MD codes such as CHARMM [11], AMBER [12], and GROMACS [13] provide their own utilities, and users should consult the manuals for how to use these. In addition, the PLUMED plug-in code, although primarily intended for running simulations, can also be used to perform analysis and allows great flexibility in definitions of reaction coordinates [14].

3 Methods

3.1 Identifying or Optimizing Reaction Coordinates for Describing Folding

The first step to characterizing folding for two-state proteins is to identify a good reaction coordinate. The discussion below focuses on this situation, i.e., where the protein mainly populates two states, folded and unfolded, at equilibrium, and does not populate stable folding intermediates. At the end, we briefly discuss how to generalize for multistate folders, and how to deal with one-state (downhill) folders.

3.1.1 Identify Trial Coordinates for Testing

These could include coordinates such as contacts, key distances, or distance matrix-based coordinates. Experience has shown that for protein folding, coordinates based on the fraction of native contacts Q are often good [7, 15, 16], i.e., of the form:

$$Q = \sum\nolimits_{\mathrm{contacts}(i,j)} w_{ij} q(r_{ij}) / \sum\nolimits_{\mathrm{contacts}(i,j)} w_{ij} \tag{2}$$

In the above expression, the sums run over the pairs of atoms i,j in contact in the native state (*see* **Note 1**). The factor w_{ij} is a weight factor which can be optimized (see below) to improve coordinate quality. However, defining $w_{ij} \equiv 1$ is often sufficiently good for Q to be a useful reaction coordinate, with rare exceptions [16]. The function $q(r_{ij})$ is a smooth function which counts atoms which are close together, e.g., less than some cutoff distance r_0. A common example used is the Fermi function:

$$q(r) = \frac{1}{1 + e^{\beta(r - r_0)}} \tag{3}$$

In this function, r is the distance between the atoms, r_0 is the cutoff distance (*see* **Note 2**), and beta is a smoothing parameter, for which a value of 5 Å^{-1} is often good. Other functions, such as Gaussians, can also be used for $q(\mathrm{r})$, especially when applied to coarse-grained simulations.

Other coordinates which could be used are specific pair distances, the fraction of secondary structure content for different kinds of secondary structure, or a distance matrix RMSD from the native state, which is closely related to Q. Note that the conventional best fit RMSD to a specific structure is usually only a good coordinate very close to that structure and is not recommended as a folding coordinate.

3.1.2 Define Unfolded and Folded States in Terms of Given Reaction Coordinates

Next, project the trajectory onto the trial coordinate R, i.e., calculate $R(t)$ over the trajectory. If the coordinate does not clearly show folded and unfolded states at different values of the coordinate, the approach described here will not be successful. This may be simply because the chosen coordinate is too poor to even resolve the two states, so multiple choices should first be investigated. If these still fail to show evidence of two states, the protein may be a one-state folder; if there are clearly multiple stable states, it is probably a multistate folder. Here we focus on the two-state scenario. Computing an initial histogram of R will allow the regions of the coordinate which are most populated to be identified. This can be used to identify boundaries defining unfolded and folded states. A good choice is simply to take the peak of the population density in the unfolded and folded states as unfolded and folded boundaries, respectively (*see* **Note 3** regarding downhill folders or proteins with more than two stable states). While intuitively one would think using cutoffs in the low population region between the two states would be better, it is not, as outlined in the next step. Optimizing the choice of boundaries will be discussed in the next section.

3.1.3 Define Transition Paths Between Unfolded and Folded

Next, it is necessary to define reactive events, or transition paths, between the unfolded and folded states. Here, we define transition paths as the shortest segments of the trajectory crossing between the unfolded and folded states, as defined by the boundaries given in Subheading 3.1.2 above. With this definition, it is possible to separately analyze the properties of the transition paths. A good test for whether the boundaries chosen for unfolded and folded states are good is based on the estimated waiting times for folding and unfolding. If the boundaries are chosen too close to the barrier, the waiting times will be polluted by recrossing events in which a trajectory may briefly cross the barrier, but then come back again and return to the state from which it came, leading to an underestimation of unfolding and folding times. Thus, suitable boundaries can be identified by moving them systematically away from the barrier until the folding and unfolding times start to become independent of the definition of the boundaries.

3.1.4 Compute Histograms of Coordinate from Equilibrium Folding Trajectories

Once a suitable coordinate has been chosen, it is necessary to determine the quality of this coordinate for describing folding. We will do this using the Bayesian approach outlined in the introduction. In the case that a long trajectory containing multiple folding and unfolding events is available, this is quite straightforward. Divide the coordinate into bins spanning the full range of the coordinate sampled in the simulations (usually 0 to 1 for Q). A practical number of bins is often in the range 30–40. Using too few bins will blur the results too much; using too many will result in noisy histograms because of limited data points; experimenting to find the largest number that gives smooth histograms is recommended. Then, for your coordinate R, compute the histograms $P(R \mid \mathrm{TP})$ and $P_{\mathrm{eq}}(R)$ from the TP and from the whole trajectory, respectively.

3.1.5 Quantifying and Optimizing Reaction Coordinates

With $P(R \mid \mathrm{TP})$ and $P_{\mathrm{eq}}(R)$ available, Bayes theorem, Eq. (1), can be used to determine the function $P(\mathrm{TP} \mid R)$. The peak of this function $\max[P(\mathrm{TP} \mid R)]$ can be used to assess the quality of the coordinate, with the aim to be as close to the theoretical maximum of 0.5 as possible. As a practical guideline, the best coordinates for folding typically have $\max[P(\mathrm{TP} \mid R)] \approx 0.4$, while coordinates with $\max[P(\mathrm{TP} \mid R)] < 0.3$ are generally poor coordinates. Good coordinates can be found systematically by finding the one out of a trial set for which $\max[P(\mathrm{TP} \mid R)]$ is the largest. It is even possible to perform a variational optimization, for example, by varying the weights w_{ij} in Eq. (2). However, in this case, some care is needed to avoid overfitting; this situation is discussed in more detail in Ref. 7. An example of an all-atom trajectory in which nonuniform w_{ij} were necessary is given in Ref. 16.

3.2 Identifying Key Interactions Which Determine Folding Mechanism

An optimal reaction coordinate can help to identify key states in the folding pathway; however, one may also want to identify which are the key interactions that determine the folding mechanism. A Bayesian procedure similar to that for assessing reaction coordinates can also be applied to investigate the importance of individual contacts [15]. The procedure is outlined below.

3.2.1 Compute Contact Maps for Entire Trajectory

The contact map (including both native and non-native contacts) should be computed for all frames in the trajectory. In principle, this can be done for all pairs of contacts between heavy atoms; however, this could yield a rather large and hard to interpret contact map and will also suffer from sparsity of data (how many times a given atom–atom contact is actually formed early during folding events). Therefore, it is usually desirable to determine a residue-averaged contact map. A successful method of doing this is to define a residue–residue contact as occurring when any two heavy atoms, one from each residue (presumably those mediating the interactions between the residues), is less than some cutoff such as 4.5 Å. Then the contact map for an individual frame will be a binary matrix Θ indicating which residue pairs are in contact.

3.2.2 Compute Contact Map Averages

Using the transition path definitions from part Subheading 3.1 above, calculate the average $P(q_{ij}|\ TP) = \langle\theta_{ij}\rangle_{\mathrm{TP}}$, i.e., the average fraction of time that the residue pair *i,j* spends in contact on transition paths. Also compute the average $P(q_{ij})_{\mathrm{nn}} = \langle\theta_{ij}\rangle_{\mathrm{nn}}$, i.e., the fraction of time the residue pair *i,j* is in contact over all frames in the trajectory which are non-native (nn), i.e., those frames when the protein is unfolded or on transition paths. The reason for excluding the folded state is to remove any trivial dependence of the result on the protein stability. Lastly, also compute $p(\mathrm{TP})_{\mathrm{nn}}$, the fraction of time spent on transition paths when not in the native state.

3.2.3 Calculate Bayesian Criterion for Contacts

Analogous to the Bayesian criterion for reaction coordinates, a Bayesian criterion for contacts can be defined as:

$$P\left(TP|q_{ij}\right)_{\mathrm{nn}} = \frac{P\left(q_{ij}|TP\right)P(TP)_{\mathrm{nn}}}{P\left(q_{ij}\right)_{\mathrm{nn}}} \quad (4)$$

where the various quantities are defined in Subheading 3.2.2 above. This criterion tells us the probability of being on a transition path given that a contact between residues *i,j* is formed and that the protein is not already native. While there is no formal maximum (other than 1) for this criterion, unlike in the case of reaction coordinates, a higher value clearly means that a contact is more important in determining whether the protein folds. This analysis has generally shown native contacts to be most important; however, it will identify non-native contacts in certain cases, if they are significant [15].

3.2.4 Alternative Measure: Contact Lifetimes

A second method to determine the significance of contacts on transition paths is based on the contact lifetime, although it requires slightly more bookkeeping to calculate. The basis of the idea is that contacts which are important to the folding mechanism should have a longer lifetime on transition paths, t_{TP}, than in the unfolded state, t_U. Therefore, measuring the ratio t_{TP}/t_U should give an indication of which contacts are important in determining the folding mechanism.

3.2.5 Determining Contact Lifetimes

Contact lifetimes cannot be directly determined from the above contact map data $\theta_{ij}(t)$. This is because recrossings, in which contacts are broken for a very short time before being reformed, would reduce the effective contact lifetime. Therefore, for each residue pair i,j, the distance trajectory $r_{ij}(t)$ must be determined. Contact formation and breaking is determined from the distance trajectory in an analogous fashion to the determination of folding and unfolding events using a dual-cutoff scheme described above. Contact formation occurs when two heavy atoms approach within 3.5 Å of each other, while breaking occurs if they are separated by more than 8 Å. This allows a trajectory $\theta'_{ij}(t)$ to be determined, which is almost identical to $\theta_{ij}(t)$, but for the suppression of recrossing noise. The number of contact breaking events in the unfolded state, N_U, and on transition paths, N_{TP}, needs to be separately accumulated. Then, the average contact lifetime is $t_U = T_U \left\langle \theta'_{ij} \right\rangle_U / N_U$ in the unfolded state, where T_U is the total time spent in the unfolded state during the trajectory, $\left\langle \theta'_{ij} \right\rangle_U$ is the average fraction of time contact i,j is formed in the unfolded state, and N_U is the number of times contact i,j is broken in the unfolded state. An analogous formula is used to compute the lifetimes on transition paths.

3.3 Comparing with Experiment: Computing ϕ-Values

ϕ-value analysis is one of the few experimental methods for probing the folding mechanism of two-state proteins [17]. It is based on the measurement of folding rates and stabilities for wild-type and mutant proteins, where the ϕ-value for residue i is given by:

$$\phi_i = \frac{RT\ln[k_f^{mut}/k_f^{wt}]}{\Delta G_f^{mut} - \Delta G_f^{wt}} \tag{5}$$

where k_f^{wt} and k_f^{mut} are the folding rates of the wild-type and mutant protein, respectively, and ΔG_f^{wt} and ΔG_f^{mut} are the respective stabilities of the wild-type and mutant. There is not space to discuss the interpretation of ϕ_i here, except to say that for an idealized experiment, it is expected to have a value between 0 and 1, 0 if it is not formed in the folding transition state, and 1 if it is formed in the folding transition state (you can read more about ϕ-values in the context of experiments and prediction in Chapters 1 and 21).

3.3.1 Computing ϕ_i from Long MD Simulations from Putative Transition States

While ϕ_i is usually interpreted in terms of folding transition states, it is not possible to identify true transition states from existing long simulations. Doing so would involve computation of the committor, or p_{fold}, which would be much more computationally expensive than running the original folding simulation [18]. One approach may be to analyze the structures of the species extracted from a point on the reaction coordinate R where $P(\text{TP} \mid R)$ is maximal. If we call these structures "putative transition states" (PTS), then ϕ_i may be computed using the interpretation originally given by experimentalists, namely that ϕ_i is the fraction of native contacts formed by a residue in the transition state [19]. That is:

$$\phi_i^{\text{PTS}} = N_i^{-1} \sum_{j \in \text{native}} p(q_{ij} | \text{PTS}) \tag{6}$$

where the sum runs over the residues j contacting i in the native state, and $p(q_{ij} \mid \text{PTS})$ is the fraction of time a contacting pair i,j is formed in the structures comprising the PTS.

3.3.2 Computing ϕ_i from Long MD Simulations from Long Trajectories Using Transition Path Theory (TPT)

An approach to determining ϕ_i which is based less on ad hoc assumptions is one based on transition-path theory. While it also makes approximations, this approach has the merit of being derived directly from the experimental definition of ϕ_i. As will be seen below, it also does not require prior identification of a putative ensemble of "transition state" structures. We present two methods to estimate ϕ_i from transition paths. Each is derived (described in Ref. 20) from a slightly different expression for the folding rate, and the results are also slightly different. The first method, using transition-path theory to estimate the rate, yields an expression for ϕ_i of:

$$\phi_i^{\text{TPT}} = N_i^{-1} \sum_{j \in \text{native}} p(q_{ij} | \text{TP}) \tag{7}$$

This is very similar to the expression given in Subheading 3.3.1, but for $p(q_{ij} \mid \text{TP})$ being substituted for $p(q_{ij} \mid \text{PTS})$. The calculation of $p(q_{ij} \vee \text{TP})$ has already been described above, in Subheading 3.2.2.

3.3.3 Computing ϕ_i from Long MD Simulations from Long Trajectories Using Folding Flux (FF)

An alternative derivation of ϕ_i from folding flux [20] gives a very similar expression to that from transition-path theory:

$$\phi_i^{\text{FF}} = N_i^{-1} \sum_{j \in \text{native}} p'(q_{ij} | \text{TP}) \tag{8}$$

In this expression, the averaging over transition paths is performed slightly differently, with each reactive event contributing equally to $p'(q_{ij}|\text{TP})$ (while they contribute in proportion to transition-path length to $p(q_{ij} \mid \text{TP})$), i.e.,

$$p'(q_{ij}|\text{TP}) = N_{\text{TP}}^{-1} \sum_{k=1}^{N_{\text{TP}}} p(q_{ij}|\text{TP})_k \tag{9}$$

where $p(q_{ij} \mid \mathrm{TP})_k$ is the probability of contact q_{ij} being formed on transition path k. The advantage of this expression is that it allows for the effect of transition-path heterogeneity to be incorporated and the effect of reweighting different pathways on the observed ϕ_i to be assessed.

4 Notes

1. For all-atom simulations, defining the fraction of (native) contacts in terms of all heavy atom distances is much better than using a coarse-grained coordinate based on distances between alpha carbon atoms, for example. The reason is the degeneracy of all-atom configurations with respect to any particular coarse-grained representation, i.e., for a pair of alpha carbons a given distance apart, their side chains may be forming favourable contacts, or may be pointing in opposite directions.
2. The cutoff for counting native contacts is one free parameter in defining Q. Systematic testing of the quality of the reaction coordinate as a function of the cutoff found little difference for choices in the range 4.5–6.5 Å, for all-atom simulations. Commonly, a value of 4.5 Å is used, to reduce the number of contacts being computed. The optimal value for coarse-grained simulations will of course be larger and will depend on the model used.
3. The analysis described above applies most directly to two-state folding proteins. For proteins with more than two states (i.e., with stable folding intermediates), it can be generalized by considering separately the individual steps between each pair of states, and using the above methods to interpret the folding mechanism for each step. For one-state or downhill proteins, which lack a folding barrier, the stable states approximations used in many of the methods cannot easily be transferred. Nonetheless, the situation is in some sense easier because for true downhill folding proteins, the self-assembly mechanism can be inferred from the equilibrium distribution at different values of the reaction coordinate. Before concluding that a protein is downhill based on projection onto a given coordinate, some caution is needed, because projection onto a poor coordinate can give the impression that no barrier exists, when the coordinate is just unable to distinguish the stable states. However, if no barrier exists after projection onto Q, the protein likely is a downhill folder.

Acknowledgments

RB is supported by the Intramural Research Program of the National Institute of Diabetes and Digestive and Kidney Diseases of the National Institutes of Health. Victor Muñoz is thanked for his careful reading of the manuscript.

References

1. Klepeis JL, Lindorff-Larsen K, Dror RO, Shaw DE (2009) Long-timescale molecular dynamics simulations of protein structure and function. Curr Opin Struct Biol 19:1719–1722
2. Best RB (2012) Atomistic simulations of protein folding. Curr Opin Struct Biol 22:52–61
3. Ueda Y, Taketomi H, Go N (1975) Studies on protein folding, unfolding and fluctuations by computer simulation. I. The effects of specific amino acid sequence represented by specific inter-unit interactions. Int J Pept Res 7:445–459
4. Clementi C, Nymeyer H, Onuchic JN (2000) Topological and energetic factors: what determines the structural details of the transition state ensemble and "en-route" intermediates for protein folding? An investigation for small globular proteins. J Mol Biol 298:937–953
5. Best RB (2013) A "slow" protein folds quickly in the end. Proc Natl Acad Sci U S A 110:5744–5745
6. Chodera JD, Noé F (2014) Markov state models of biomolecular conformational dynamics. Curr Opin Struct Biol 25:135–144. https://doi.org/10.1016/J.SBI.2014.04.002
7. Best RB, Hummer G (2005) Reaction coordinates and rates from transition paths. Proc Natl Acad Sci U S A 102:6732–6737
8. Best RB, Chen Y-G, Hummer G (2005) Slow protein conformational dynamics from multiple experimental structures: the helix/sheet transition of Arc repressor. Structure 13:1755–1763
9. Hummer G (2004) From transition paths to transition states and rate coefficients. J Chem Phys 120:516–523
10. Gowers RJ, Linke M, Barnoud J, Reddy TJE, Melo MN, Seyler SL, Domański J, Dotson DL, Buchoux S, Kenney IM, Beckstein O (2016) MDAnalysis: A Python package for the rapid analysis of molecular dynamics simulations. In: Benthall S, Rostrup S (eds) Proc. 15th Python Sci. Conf., pp 98–105
11. Brooks BR, Brooks CL III, Mackerell AD Jr, Nilsson L, Petrella RJ, Roux B, Won Y, Archontis G, Bartels C, Boresch S, Caflisch A, Caves L, Cui Q, Dinner AR, Feig M, Fischer S, Gao J, Hodoscek M, Im W, Kuczera K, Lazaridis T, Ma J, Ovchinnikov V, Paci E, Pastor RW, Post CB, Pu JZ, Schaefer M, Tidor B, Venable RM, Woodcock HL, Wu X, Yang W, York DM, Karplus M (2009) CHARMM: the biomolecular simulation program. J Comp Chem 30:1545–1614
12. Salomon-Ferrer R, Case DA, Walker RC (2013) An overview of the Amber biomolecular simulation package. Wiley Interdiscip Rev Comput Mol Sci 3:198–210. https://doi.org/10.1002/wcms.1121
13. Hess B, Kutzner C, van der Spoel D, Lindahl E (2008) GROMACS 4: algorithms for highly efficient, load-balanced, and scalable molecular simulation. J Chem Theory Comput 4:435–447
14. Tribello GA, Bonomi M, Branduardi D, Camilloni C, Bussi G (2014) PLUMED 2: new feathers for an old bird. Comput Phys Commun 185:604–613. https://doi.org/10.1016/J.CPC.2013.09.018
15. Best RB, Hummer G, Eaton WA (2013) Native contacts determine protein folding mechanisms in atomistic simulations. Proc Natl Acad Sci U S A 110:17874–17879
16. Zheng W, Best RB (2015) Reduction of all-atom protein folding dynamics to one-dimensional diffusion. J Phys Chem B 119:15247–15255
17. Matouschek A, Kellis JT, Serrano L, Fersht AR (1989) Mapping the transition-state and pathway of protein folding by protein engineering. Nature 340:122–126
18. Du R, Pande VS, Grosberg AY, Tanaka T, Shakhnovich ES (1998) On the transition

coordinate for protein folding. J Chem Phys 108:334–350

19. Vendruscolo M, Paci E, Dobson CM, Karplus M (2001) Three key residues form a critical contact network in a protein folding transition state. Nature 409:641–645

20. Best RB, Hummer G (2016) Microscopic interpretation of folding φ-values using the transition path ensemble. Proc Natl Acad Sci U S A 113:3263–3268

Chapter 18

Atomistic Simulations of Thermal Unfolding

Angel E. Garcia

Abstract

This tutorial will provide a practical overview of the use of atomistic simulations to study thermal unfolding of biomolecules, in particular small proteins and RNA oligomers. The tutorial focuses on the use of atomistic, all atom simulations of biomolecules in explicit solvent, to study (reversible) thermal unfolding. The simulation methods described here have also been applied to study biomolecules using implicit solvent and coarse-grained models. We do not intend to provide an up-to-date review of the vast literature of biomolecular dynamics, enhanced sampling methods, force field developments, and applications of these methods. The purpose of this tutorial is to provide basic guidelines into the use of these methods to the starting scientist.

Key words Molecular simulations, Replica exchange molecular dynamics simulations, Hawley equation, Quantification of errors

1 Introduction

Molecular dynamics (MD) simulations have become a widely accepted and used technique to study biomolecular dynamics and thermodynamics. Today, it is common practice for experimental studies to also carry out MD simulations to test hypothesis and to complement the data analysis. The constant improvement of computers, the development of robust software, the refinement of phenomenological force fields, and the development of enhanced sampling methods have enabled the study of the unbiased folding of peptides, single-domain proteins, and small nucleic acids from the unfolded states.

The simulation of thermal unfolding of proteins is limited by the long timescale required to reach thermal equilibrium. The dynamics of biomolecules span a wide range of timescales ranging from femtoseconds (10^{-15} s) (intermolecular vibration including hydrogen atoms), to microseconds, milliseconds, (global

LA-UR-18-20660

Victor Muñoz (ed.), *Protein Folding: Methods and Protocols*, Methods in Molecular Biology, vol. 2376, https://doi.org/10.1007/978-1-0716-1716-8_18,

rearrangements like folding/unfolding and allosteric structural changes), and seconds (for multidomain proteins and molecular complexes). This broad timescale highlights the complexity of the highly dimensional energy landscape. Current simulations can commonly sample the microsecond timescale in a reasonable wall clock time. The development of specialized parallel computers, like Anton, has enabled the simulation of protein trajectories for milliseconds [1]. The simulation of a single (rare) event of protein folding from a random (unfolded) configuration may take 10^8 to 10^{12} integration steps (with integration steps $\Delta t \sim$ 1–4 fs). Therefore, depending on the computational resources and system size, this can take weeks, months, or years of wall clock time or computation time in a multiprocessor computer. The two most common methods for studying thermal unfolding of biomolecules are standard MD [2], and Replica Exchange MD (REMD) [3]. Currently, well-documented, highly parallelizable, optimized codes exist for performing MD and REMD [4–7]. REMD is best suited for calculating thermodynamics properties over a broad range of temperatures, such as melting profiles as a function of temperature. The REMD method has been widely used to study the free-energy landscape and folding/unfolding equilibrium of peptides, single-domain proteins, and RNA oligomers [8].

2 Performing Replica Exchange Molecular Dynamics Simulations

REMD provides an efficient way to sample the configurational space of complex systems [3]. In REMD, several copies (replicas) of identical systems (i.e., identical compositions) are simulated in parallel at different temperatures (or volumes). Exchange moves between replicas are attempted periodically and accepted with probability

$$P_{acc} = \min\left[1, \exp\left(-\Delta\right)\right]$$

with

$$\Delta = \exp\left[\left(\beta_i - \beta_j\right)\left(U\left(\vec{r}_J^{\,N}\right) - U\left(\vec{r}_i^{\,N}\right)\right]$$

here $U(r_j)$ is the potential energy of the system in the state j, and $\beta_j = 1/k_B T_j$, k_B is the Boltzmann constant, T_j is the temperature of the j *th* replica, and $\vec{r}_j^{\,N}$ is the set of coordinates for all atoms in the system j. When an exchange attempt is successful, the momenta of all atoms are scaled by the factor $(T_i/T_j)1/2$, such that the kinetic terms in the Boltzmann factor cancel out. [3] This algorithm generates correctly a Boltzmann weighted, canonical ensemble of configurations at each temperature. From these (properly sampled) ensembles we can calculate averages of desired quantities such as potential energy, fraction folded, and free energy landscape maps, for each temperature. For exchanges between replicas to occur, the energy distributions at the different temperatures need to overlap sufficiently. Typically, an exchange acceptance rate of 20% is desired. More important than the exchange rate is the transmission rate of replicas across the simulated temperatures. The use of REMD is limited to relatively small systems since the energy overlap decreases proportionally to $1/m^1/2$, with m being the mass of the system (or number of non-hydrogen atoms), thus requiring many more replicas to span the same temperature range. We have limited most of our studies to systems smaller than 25 K atoms, requiring fewer than 64 replicas. This limitation can be overcome by other methods that dynamically scale temperatures [9–11].

2.1 Practical Issues When Implementing REMD

2.1.1 Sampling from a Canonical Ensemble

The REMD algorithm requires that each replica simulated samples a Boltzmann-weighted, canonical, ensemble (constant temperature) ensemble [12]. Stochastic dynamics, Nose-Hoover [13], and Andersen thermostats satisfy this requirement. Berendsen [14] thermal coupling does not produce a canonical distribution and should be avoided for REMD production runs.

2.1.2 Determining the Temperature Distribution

The distribution of temperatures must have sufficient overlap to allow exchanges. An exponential distribution of temperatures, $T_{i+1} = T_i \times \varepsilon$, where ε is slightly larger than 1, is a quick way to determine the replica temperatures. ε can be determined by selecting two temperatures, $T_1 > T_0$, with sufficient overlap (i.e., a 2–5 Kelvin apart, depending on system size), such that $\varepsilon = T_1/T_0$, and then, for each replica N, $T_N = T_0\, \varepsilon^N$. This method works well and provides a reasonable temperature distribution, but can fail with implicit solvent and CG models in regions where the heat capacity peaks sharply. The exponential temperature distribution produces higher exchange rates at high temperatures and require a larger number of replicas to span the desired temperature range than other methods.

More efficient, approximate methods for determining the replica temperatures are based on estimates of the average energy at a given temperature, $<E(T)>$, and the width of the energy distribution, $\sigma(T)^2 = <E^2> - <E>^2$. Rathore et al. [15] found that the energy distribution overlap, A_{overlap}, at two temperatures is related to the exchange acceptance rate, P_{acc}, and suggested using a constant A_{overlap} to determine the temperature of the replicas. The A_{overlap} is given by:

$$A_{\text{overlap}} = erfc\left[\frac{\langle E_2\rangle - \langle E_1\rangle}{\sqrt[2]{8}\,\sigma}\right] = erfc\left[\frac{\Delta E}{\sqrt[2]{8}\,\sigma}\right],$$

where *erfc* is the complementary error function. Here $\langle E_1\rangle$ and $\langle E_2\rangle$ are the average energy at two neighboring temperatures T_1 and T_2, and σ is the standard deviation. Garcia et al. [16] obtained P_{acc}

analytically, with the assumption that energies at each temperature follows a Gaussian distribution, and found

$$P_{\mathrm{acc}}(\beta_1,\beta_2)=\frac{1}{2}\left[1+\mathrm{erf}\left[\frac{\langle E_2\rangle-\langle E_1\rangle}{\sqrt[2]{2(\sigma_1^2+\sigma_2^2)}}\right]\right]$$

$$+\frac{1}{2}\exp\left[\Delta\beta(\langle E_2\rangle-\langle E_1\rangle)+\frac{1}{2}(\Delta\beta)^2(\sigma_1^2+\sigma_2^2)\right]\mathrm{erfc}\left[\frac{\Delta\beta(\sigma_1^2+\sigma_2^2)+\langle E_2\rangle-\langle E_1\rangle}{\sqrt[2]{2(\sigma_1^2+\sigma_2^2)}}\right].$$

Here *erf* is the error function and $\Delta\beta = \beta_2 - \beta_1$. Other quantities are the same as defined before for A_{overlap}. To use the equations for A_{overlap} or P_{acc} described above, we need to get estimates for the average energies and their standard deviations over the temperature range of interest by performing multiple (~10–20 ns long) MD simulations of the system over a set of temperatures covering the desired range (i.e., 275–500 K), at large intervals (i.e., 25–50 K temperature steps). We will determine the average energies and standard deviations for each temperature, and then fit these to a cubic or quartic spline, which approximate $<E(T)>$ and $\sigma(T)$ for all temperatures of interest. These fitted functions are then used to iteratively determine pairs of temperatures that have a constant A_{overlap} or P_{acc}. These methods produce very similar temperatures for the replicas and give fairly constant A_{overlap} or P_{acc} values over these temperatures. For implicit solvent and CG models, the energy distributions are not necessarily Gaussian (i.e., they can be multimodal) and the methods described above fail. For these cases, the method of Trebts et al. [17] should be used.

2.1.3 Sampling Efficiency

The efficiency of REMD may be reduced at high temperatures where the system basin hoping rate may decrease with temperature (i.e., it exhibits non-Arrhenius behavior) [18]. As a corollary, it is not always better to simulate more replicas since sampling at higher temperatures may be slower above some temperature, which is system and force field dependent. In many instances, the efficiency of REMD is sacrificed by the need to calculate temperature or time-dependent properties of the system.

2.1.4 Exchange Attempt Frequency and Exchange Rates

Much debate has been generated on this subject. However, we recommend attempting exchanges often (1–5 ps) and getting average exchange times of ~10 ps. Longer exchange times provide short constant temperature trajectories that can be used to extract time-dependent properties of the system.

2.1.5 Selection of Force Fields

The energy parameters used to model biomolecular systems are under constant change [19–21]. As computational resources increase, longer, high-quality simulations have been able to find deficiencies in the force fields. In many instances, small changes in dihedral parameters are sufficient to improve the force fields. In other instances, the inclusion of polarizability or changes in Lennard Jones parameters have been suggested. For the user of MD simulation methods who wants to study a biomolecular system of interest and it is not interested in developing new methods, or force fields, I recommend the use of popular, widely used force fields validated against experimental data [22].

2.1.6 Reaching "Equilibrium"

The simulation results should be divided into equilibration and production segments. Averages are calculated over the production ensembles only. Equilibration ensembles can be used to monitor the dependence of the averages on simulation time. Once steady state for the desired quantity is reached, the production run is produced. Different quantities equilibrate at different rates. Average energies equilibrate fast (ns), standard deviations (second moment) equilibrate slower, and higher energy moments may not equilibrate even in microseconds. Free-energy profiles also equilibrate slowly, thus requiring extensive production runs. There will always be quantities that will not produce steady state averages in a simulation. Nevertheless, most quantities of interest will reach a steady state in reasonably long (microseconds) simulations. In general, time-dependent quantities are more difficult to obtain than equilibrium quantities.

3 Calculation of Thermodynamic Properties from REMD Simulations

The ensembles collected during REMD simulations can be used to calculate average properties of the system as a function of temperature. The thermodynamics of the system can be calculated under the assumption of two-state (folded/unfolded) thermodynamics. The definition of the folded state will depend on an order parameter that characterizes the structure of the system. These properties could be the helical or beta sheet content, radius of gyration, root mean square distance (*rmsd*) from the (known) folded state, etc. For REMD simulations, we monitored two quantities to assess the steady state equilibrium of the simulation, the number of replicas that have folded at least once during the simulation, and the average number of replicas that are folded at any time during the simulation. Figure 1 shows these two quantities calculated by Paschek et al. [23, 24] for the Trp-cage mini protein, using the Amberff94 force field [25]. Fig. 2 shows the distributions as a function of temperature of selected distances that can be used to characterize the folded and unfolded ensemble of Trp-cage. In these simulations, we found the *rmsd* to be a better descriptor of the transition than the other parameters. Nevertheless, *rmsd* is not always the best (or good) parameter to describe the folding transition.

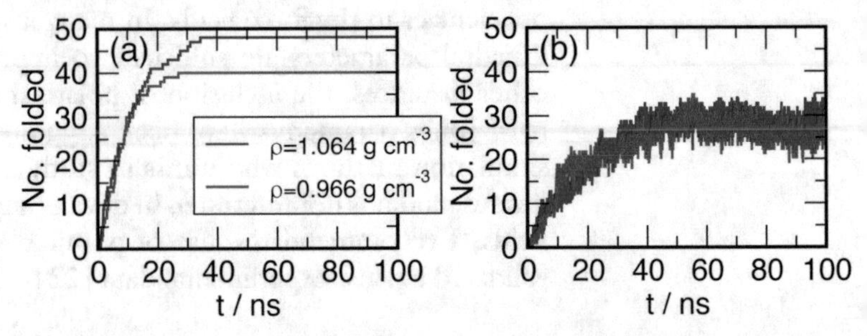

Fig. 1 *Reaching of steady state equilibrium in REMD simulations.* The calculations are for a fast folding, 20 amino-acid mini protein called Trp-cage. The simulations used the Amber ff94 [25] force field and extended for 100 ns/replica [23]. (**a**) Time history of the number of replicas that have folded (rmsd < 0.22 nm) at least once in the simulation. (**b**) Number of replicas sampling the folded state as a function of time. The total number of replicas sampling the 0.966 g/cm^3 and 1.064 g/cm^3 isochores is 40 and 48, respectively. After 40 ns all replicas reached the folded state at least once. REMD calculations using Amber ff99SB [29] took ~400 ns to reach steady state and required 1 μs per replica to describe the thermodynamics of the system [8]. Amber ff94 reaches equilibrium faster since it is biased for helix formation and the energy landscape is less rough than for ff99SB. (This figure is reproduced from Pascheck et al. [23], with permission. Copyright (2008) National Academy of Sciences, U.S.A.)

Under the two-state assumption, the change in Helmholtz free energy can be calculated from the population of folded and unfolded states as:

$$\Delta F_U(P,T) = F_{unfolded} - F_{folded} = -RT\ln\left[(1 - x_{folded})/x_{folded},\right.$$

where x_{folded} is the ensemble calculated average fraction of folded states at each temperature T. x_{folded} depends on the definition of the folded state in terms of the structure of the biomolecule. In constant volume REMD simulations, the pressure is not constant. We calculate Gibbs free energies by the transformation $G = F + PV$, where $P = \langle P \rangle$, such that: $\Delta G_U(P,T) = -RT\ln[(1 - x_{folded})/x_{folded} + V\Delta\langle P_U\rangle$ [24]

The Gibbs free energy is described by the Hawley equation [26]:

$$\begin{aligned}\Delta G_U(T,P) = {} & \Delta G_0 - \Delta S_0(T - T_0) + \Delta V_0(P - P_0) \\ & + \Delta\alpha_0(P - P_0)(T - T_0) \\ & - \Delta C_P\left[T\left(\ln\left(\frac{T}{T_0}\right) - 1\right) + T_0\right] \\ & + \frac{\Delta\beta_0}{2}(P - P_0)^2\end{aligned}$$

+ . . .here ΔG_0 is the change in free energy at the reference state (T_0, P_0), ΔS_0 is the change in entropy, $\Delta\alpha_0$ is the change in expansivity, ΔC_p is the change in heat capacity at constant pressure, and $\Delta\beta_0$ is

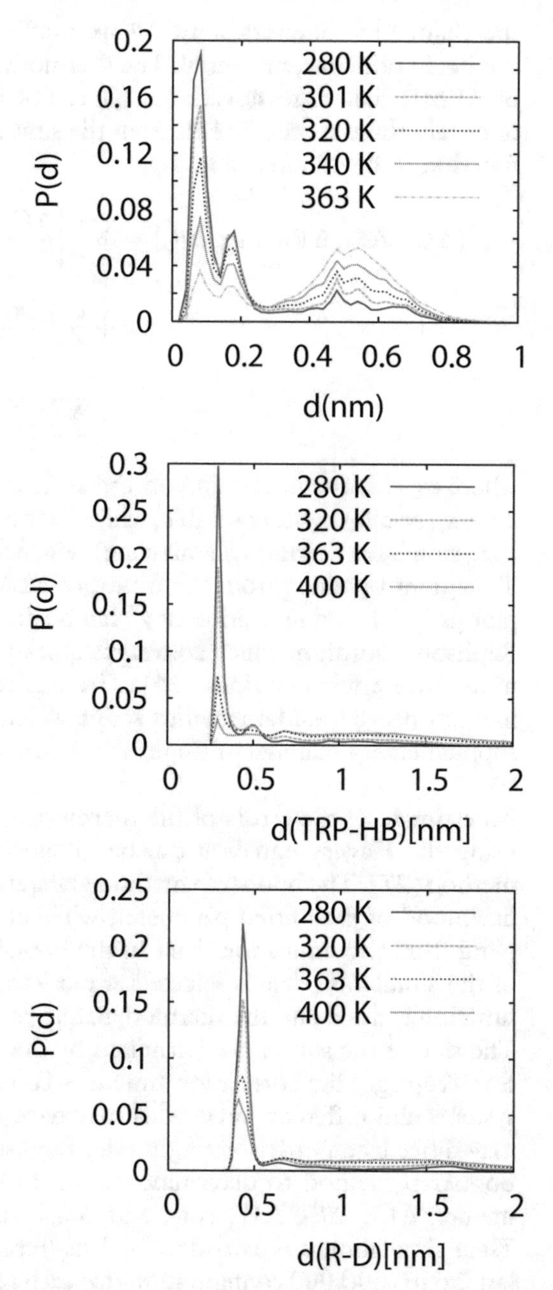

Fig. 2 *Distributions of properties characterizing the structure of the protein.* Temperature dependence of distributions of (**a**) rmsd, (**b**) Trp-Nε to Asp backbone carbonyl hydrogen bond, and (**c**) Arg-Asp ion pair distances obtained in REMD simulations of Trp-cage. All distances are in nanometers. (This figure is reproduced from Day et al. [8] with permission. Copyright (2010) John Wiley and Sons)

the change in compressibility. All six coefficients in this equation can be, in principle, measured. The thermodynamic states modeled by REMD (i.e., states at each (T_i,P_i)) can be fitted to this equation to obtain these six coefficients from the simulation ensembles. The function to be minimized is:

$$\chi^2(\Delta G_0, \Delta S_0, \Delta V_0, \Delta\alpha_0, \Delta\beta_0) = \sum_i \left[\frac{\Delta G_i - \Delta G(T_i, P_i)}{\sigma_{\Delta G_i}}\right]^2 + \sum_i \left[\frac{x_i - \mathrm{x}(T_i, P_i)}{\sigma_{x_i}}\right]^2 + \sum_i \left[\frac{\Delta E_i - \Delta E(T_i, P_i)}{\sigma_{\Delta E_i}}\right]^2,$$

where $\sigma_{\Delta \Upsilon_i}$ is the uncertainty in the average of a quantity Υ (i.e., ΔG_i, x_i, *or* ΔE_i) in the ensemble, while i is the corresponding replica (i.e., simulated thermodynamic state). Fig. 3d shows an elliptical P–T diagram obtained from the fitting of REMD data to a Hawley plot [23]. The minimization of χ^2 can be done using the Newton–Raphson algorithm, which converges quickly to the absolute minimum after a few iterations [24]. This algorithm has been implemented using a ©Mathematica script which was provided in the supplementary material of English and Garcia [24].

3.1 Data Analysis and Quantification of Errors

An estimate of the errors of the thermodynamic parameters fitted using the Hawley equation can be obtained using the bootstrap method [27]. The bootstrap method generates a distribution of the likelihood of each fitted parameter, while ensuring adequate sampling from the simulation data. In the bootstrap method, a subset of the simulation data is selected at random. For each subset, the function is fitted and the thermodynamic parameters are obtained. The size of the subset is determined by the data correlation time. For Trp-cage, the correlation time is ~100 ns (0.1 μs). Different systems and different force fields may have correlation times that may differ by an order of magnitude. English and Garcia used the bootstrap method to determine the distribution of the six parameters, ΔG_0, ΔS_0, ΔV_0, $\Delta\alpha_0$, and $\Delta\beta_0$,in the Hawley equation. Their simulation was extended for 1 μs/replica, and they used the last 0.6 μs (600,000 configurations for each replica) for production. With a correlation time of 0.1 μs, they produced multiple (300) random samples with 0.1 μs (60,000 configurations for each replica) worth of configurations and refitted the Hawley equation each time and used the resulting parameters to establish a distribution of the average thermodynamic parameters. The distributions for each for these parameters are shown in Fig. 4.

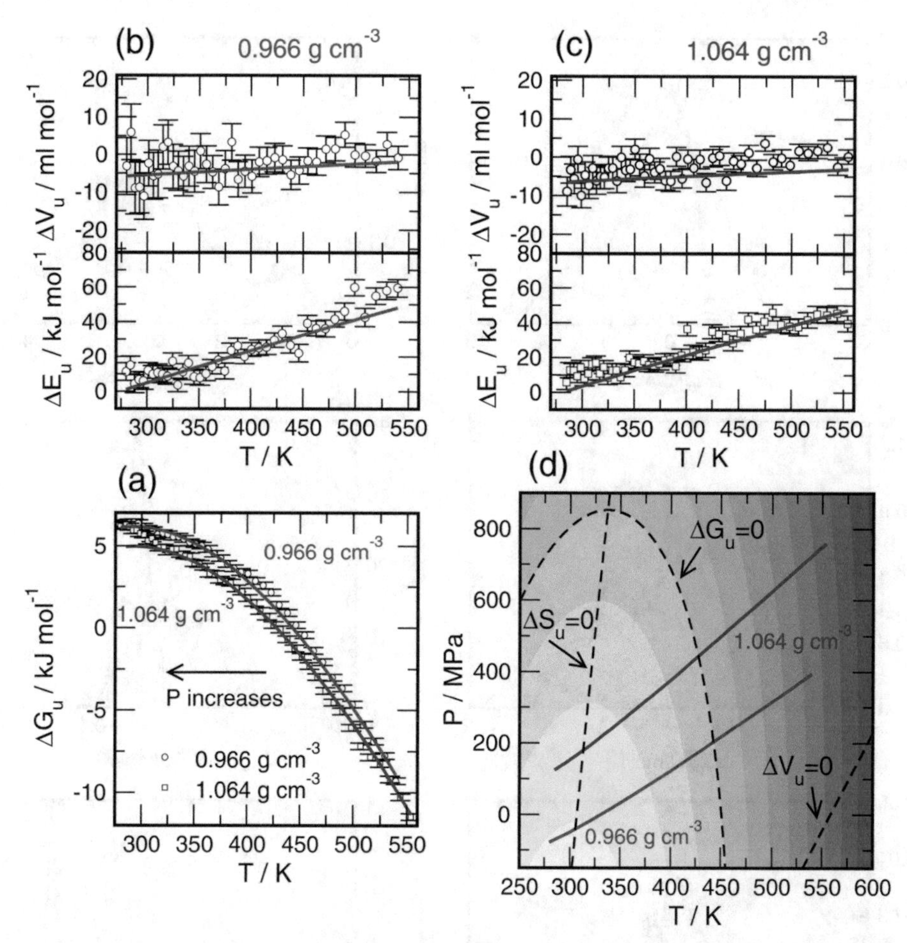

Fig. 3 Averages of the difference upon unfolding of the (**a**) free energy, (**b** and **c**) total energy, and specific volume, calculated from the steady state ensembles generated by REMD simulations for two isochores. Error bars were estimated using block averages over 10 ns blocks during the 60 ns production run. The average energy change, $\Delta E(T)$, and specific volume change, $\Delta V(T)$, can be fitted well to a straight line over the sampled temperature range, but show large deviations from the fitted line for each temperature. The slopes of these lines are the specific heat and compressibility changes. (**d**) Free energy surface $\Delta G_u(P,T)$ is obtained by fitting to a Hawley-type model the free energy and its derivatives calculated for the 2 isochores. The fit to the Hawley-type plot is a global fit to all folded fraction, and free energy, energy and volume differences. (This figure is reproduced from Pascheck et al. [23], with permission. Copyright (2008) National Academy of Sciences, U.S.A.)

3.2 Block Averages

The block average method is a simple method for obtaining estimates of the variation of the average from correlated data. In this method, a correlation time for the sampled data is calculated and independent data sets containing time blocks of twice this time are selected to get estimates of the average. The variation of the averages is then calculated from the block averages. A robust method for doing block averages has been described by Flyvbjerg and Petersen [28]. However, given that correlation times for

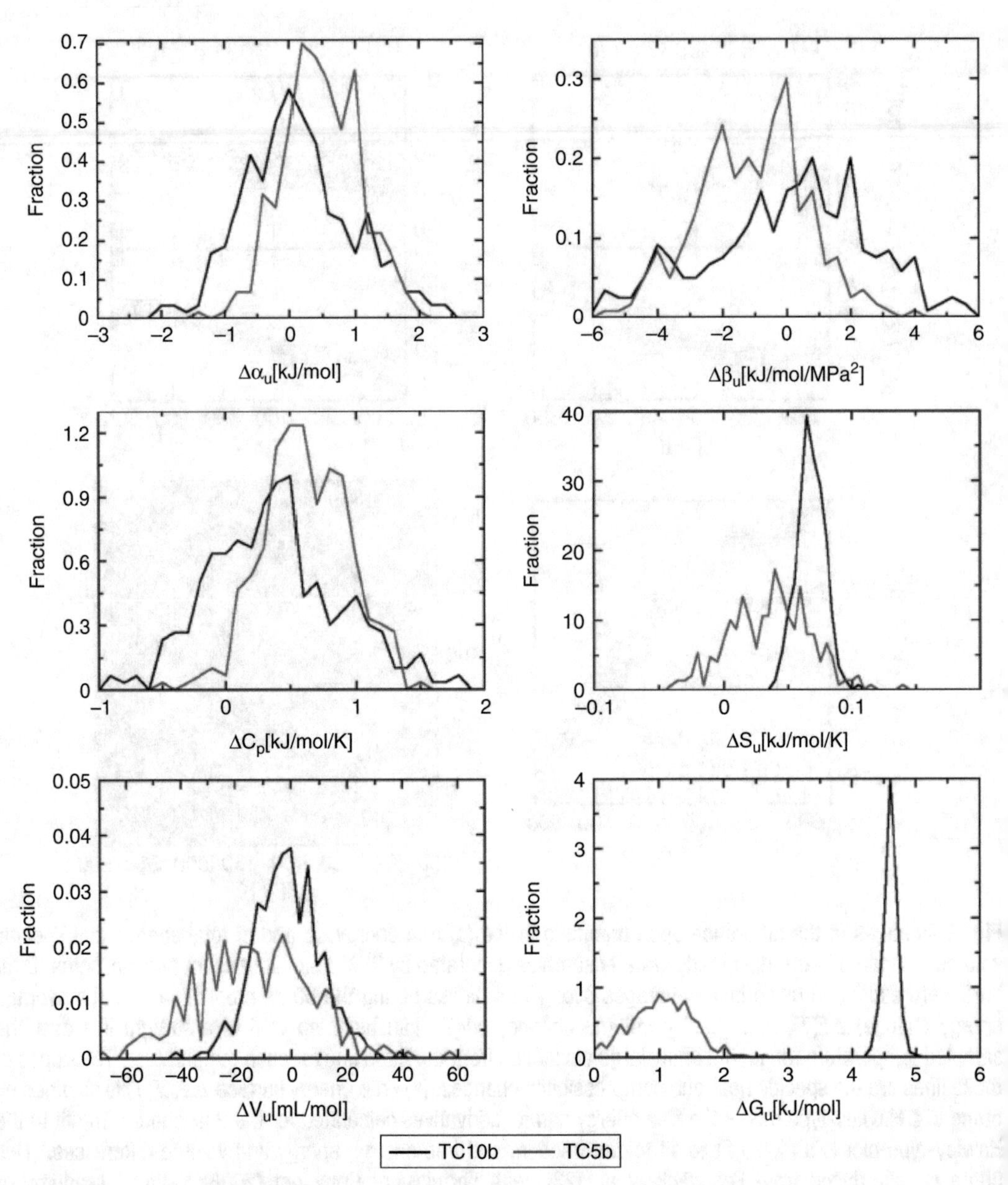

Fig. 4 Bootstrap analysis of the fitting of the Hawley equation to REMD data for Trp-cage. The plots show the distributions of the XXXX thermodynamic parameters for two variants of the Trp-cage (TC10b (black curves) and TC5b (red curves)). (Reproduced from Ref. [24] with permission from the PCCP Owner Societies)

structural properties of molecules are correlated in the 100 ns timescale, we can have 5–10 blocks, which sometimes is not sufficient to obtain good estimates of the average variation. The error bars shown in Fig. 4 were obtained using block averages.

References

1. Lindorff-Larsen K et al (2011) How fast-folding proteins fold. Science 334 (6055):517–520
2. Frenkel D, Smit B (2001) Understanding molecular simulation: from algorithms to applications. Academic Press, San Diego, Ca
3. Sugita Y, Okamoto Y (1999) Replica-exchange molecular dynamics method for protein folding. Chem Phys Lett 314:141–151
4. Phillips J et al (2005) Scalable molecular dynamics with NAMD. J Comput Chem 26 (16):1781–1802
5. Case D et al (2005) The Amber biomolecular simulation programs. J Comput Chem 26 (16):1668–1688
6. Brooks B et al (2009) CHARMM: the biomolecular simulation program. J Comput Chem 30(10):1545–1614
7. van der Spoel D, et al.2008 Gromacs User Manual version 3.3.
8. Day R, Paschek D, Garcia AE (2010) Microsecond simulations of the folding/unfolding thermodynamics of the Trp-cage miniprotein. Proteins 78(8):1889–1899
9. Rick S (2007) Replica exchange with dynamical scaling. J Chem Phys 126:054102
10. Zhang C, Ma J (2010) Enhanced sampling and applications in protein folding in explicit solvent. J Chem Phys 132:244101
11. Zang T et al (2014) Parallel continuous simulated tempering and its applications in large-scale molecular simulations. J Chem Phys 141:044113
12. Rosta E, Buchete N, Hummer G (2009) Thermostat artifacts in replica exchange molecular dynamics simulations. J Chem Theory Comput 5:1393–1399
13. Nosé S (1991) Constant temperature molecular dynamics methods. Prog Theor Phys Suppl 103:1–46
14. Berendsen HJC et al (1984) Molecular dynamics with coupling to an external bath. J Chem Phys 81:3684
15. Rathore N, Chopra M, de Pablo JJ (2005) Optimal allocation of replicas in parallel tempering simulations. J Chem Phys 122 (2):024111
16. García A, Herce H, Paschek D (2006) Simulations of temperature and pressure unfolding of peptides and proteins with replica exchange molecular dynamics. Annu Rep Comput Chem 2:83–95
17. Trebst S, Troyer M, Hansmann U (2006) Optimized parallel tempering simulations of proteins. J Chem Phys 124:174903
18. Rosta E, Hummer G (2009) Error and efficiency of replica exchange molecular dynamics simulations. J Chem Phys 131:165102
19. Huang J et al (2017) CHARMM36m: an improved force field for folded and intrinsically disordered proteins. Nat Methods 14:71–73
20. Maier J et al (2015) ff14SB: improving the accuracy of protein side chain and backbone parameters from ff99SB. J Chem Theor Comput 11(8):3696–3713
21. Lindorff-Larsen K et al (2010) Improved side-chain torsion potentials for the Amber ff99SB protein force field. Proteins 78(8):1950–1958
22. Lindorff-Larsen K et al (2012) Systematic validation of protein force fields against experimental data. PLoS One 8:e32131
23. Paschek D, Hempel S, Garcia AE (2008) Computing the stability diagram of the Trp-cage miniprotein. Proc Natl Acad Sci U S A 105 (46):17754–17759
24. English C, García AE (2014) Folding and unfolding thermodynamics of the TC10b Trp-cage miniprotein. Phys Chem Chem Phys 16:2748
25. Cornell WD et al (1995) A 2nd generation force-field for the simulation of proteins, nucleic-acids, and organic-molecules. J Am Chem Soc 117(19):5179–5197
26. Smeller L (2002) Pressure-temperature phase diagrams of biomolecules. Biochim Biophys Acta 1595:11–29
27. Efron B (1979) Bootstrap methods: another look at the jackknife. Annals Statistics 7 (1):1–26
28. Flyvbjerg H, Petersen HG (1989) Error estimates on averages of correlated data. J Chem Phys 91:461–466
29. Hornak V et al (2006) Comparison of multiple Amber force fields and development of improved protein backbone parameters. Proteins 65(3):712–725

Chapter 19

Molecular Simulations of Intrinsically Disordered Proteins and Their Binding Mechanisms

Xiakun Chu, Suhani Nagpal, and Victor Muñoz

Abstract

Intrinsically disordered proteins (IDPs) lack well-defined secondary or tertiary structures in solution but are found to be involved in a wide range of critical cellular processes that highlight their functional importance. IDPs usually undergo folding upon binding to their targets. Such binding coupled to folding behavior has widened our perspective on the protein structure–dynamics–function paradigm in molecular biology. However, characterizing the folding upon binding mechanism of IDPs experimentally remains quite challenging. Molecular simulations emerge as a potentially powerful tool that offers information complementary to experiments. Here we present a general computational framework for the molecular simulations of IDP folding upon binding processes that combines all-atom molecular dynamics (MD) and coarse-grained simulations. The classical all-atom molecular dynamics approach using GPU acceleration allows the researcher to explore the properties of the IDP conformational ensemble, whereas coarse-grained structure-based models implemented with parameters carefully calibrated to available experimental measurements can be used to simulate the entire folding upon binding process. We also discuss a set of tools for the analysis of MD trajectories and describe the details of the computational protocol to follow so that it can be adapted by the user to study any IDP in isolation and in complex with partners.

Key words Conformational disorder, Conformational dynamics, Folding coupled to binding, Induced-fit, Conformational selection, Structure-based model, Coarse-grained simulations, All-atom molecular dynamics

1 Introduction

Intrinsically disordered proteins (IDPs) are flexible polypeptide chains that do not form a stable unique three-dimensional structure in isolation, although they often present secondary and/or tertiary structural biases in solution. IDPs, which were conceptually introduced and experimentally observed over two decades ago [1–4], are particularly abundant in eukaryotic proteomes [5, 6]. IDPs widely participate in many pivotal physiological processes such as transcription, regulation, and signaling [7, 8], underlining their functional significance in biology [9, 10]. Investigations on IDPs have not only revolutionized our basic knowledge on the

Victor Muñoz (ed.), *Protein Folding: Methods and Protocols*, Methods in Molecular Biology, vol. 2376,
https://doi.org/10.1007/978-1-0716-1716-8_19,

fundamental protein folding process, but have also extended our general understanding of how proteins perform their functions. Conformational flexibility/disorder offers IDPs multiple functional advantages, such as the ability to bind promiscuously [11, 12], combine low affinity and high specificity during binding [13–15], achieve fast binding/unbinding rates to enable rapid functional on/off switching [16–20], and present tunable propensity as targets for posttranslational modifications [21, 22].

During binding, IDPs often undergo conformational transitions to folded forms aided by the interactions with their binding targets, which thermodynamically couple the binding event to the folding/ordering process of the IDP [23, 24]. Two extreme scenarios have been proposed to explain the folding upon binding mechanism of IDPs. One of them is the "induced-fit" mechanism whereby binding results in the formation of a fuzzy encounter complex in which the IDP becomes progressively structured as the complex is consolidated [25]. The other mechanism is known as "conformational selection," in which binding selects preexisting, but rare, conformations within the IDP ensemble that are binding-competent; folding occurs just prior to binding [26, 27]. These two extreme scenarios are seemingly unambiguous in general terms, but practically, IDPs tend to use both and also various intermediate levels, which makes their assignment challenging [28, 29]. Recent theoretical advances in this area indicate that the presence of strong long-range interactions at the binding interface [30], the fast conformational exchange within the IDP ensemble [31], and the presence of the IDP at high concentrations (relatively low affinity) [32] favor an induced-fit mechanism whereas the opposite is true for conformational selection. However, these patterns are very system-specific (depend on the IDP sequence and conformational ensemble as well as the structural properties of the complex with the partner), which highlights that having methods to characterize the process in mechanistic detail is of paramount importance.

IDPs in solution are not fully random statistical coils but do frequently exhibit a certain degree of transient secondary and/or tertiary structure in fast dynamic exchange [33–36]. Those preformed transient structural biases are supposed to strongly impact the binding process [37–39], but in which ways this happens is still unclear and somewhat controversial [17, 40, 41]. Current experimental approaches to study IDP folding upon binding structurally provide ensemble-averaged properties. Since these average properties arise from extremely large and diverse conformational pools, their mechanistic interpretation becomes difficult. Molecular dynamics (MD) simulations can track the details of the system evolving at the single-molecule level and with atomic and picosecond resolution, and hence are in principle an extremely powerful tool to investigate IDPs. However, the relevance of the results depends on the accuracy of the forcefield used in the MD simulation and the level of sampling. Current forcefields have been mostly parameterized and tested on well-folded proteins, which raises the issue of whether they are also adequate for the inherently flexible IDPs [42, 43]. Recent improvements on forcefields for IDPs have focused on tuning the balance between secondary and side chain structural propensities to match the experiments [44–46] and reshaping the protein–water interaction to avoid the excessive collapse that has been frequently observed in IDP simulations [43, 47].

The second limiting factor is the timescales that can be achieved with all-atom MD simulations using the computational resources that are currently available to most laboratories. Most potential users do not have access to the specialized supercomputer Anton [48, 49], which enormously extended the timescales accessible to all-atom MD thereby demonstrating successful folding of a group of fast-folding proteins, and which has also been applied to characterize the dynamics of an IDP-like acid-unfolded bovine acyl-coenzyme A binding protein (ACBP) [50]. An alternative is to develop new advanced sampling strategies, but such existing techniques either sacrifice the kinetic information of the system (replica-exchange molecular dynamics, REMD [51]) and/or work on some user-specified system-dependent reaction coordinates (umbrella sampling [52] and metadynamics [53]), which are difficult to implement to study the inherent conformational complexity of IDPs. Achieving the widespread availability of all-atom MD simulations of IDP folding upon binding is highly desirable, as they provide the level of structural insight needed to effectively dissect the molecular mechanisms. In this regard, some recent progress has come from the development of high-performance parallel algorithms to perform accelerated MD on Graphics Processing Units (GPUs) [54], which enable the easy integration of hundreds of tightly coupled processing units within a single GPU and result in large increases in the accessible timescales using relatively inexpensive computers.

An interesting alternative for achieving the longer timescales needed to resolve the entire folding upon binding process of IDPs is to use structure-based models (SBMs) in which a simplified and optimized ("unfrustrated") empirical potential is devised based on energy landscape theory and which only considers topology-based native interactions [55]. SBMs are naturally conducive to a coarse-graining strategy [56] that dramatically expands the accessible timescales of molecular simulations by simplifying the calculations and eliminating the higher frequency modes that define the ultra-short time integration step that is needed to perform all-atom MD. SBMs were initially designed for protein folding and have been later extended to binding with remarkable success [57–59]. In recent years, we have witnessed an increasing number of applications of SBMs for IDP research, which strongly suggests that energy landscape theory is also relevant to investigate IDP folding upon binding [60–64]. The key implication of such success is that the same type of native-like interactions is playing a major role in defining how IDPs fold upon binding.

In this chapter, we introduce standard protocols for performing hybrid all-atom MD and SBM simulations to investigate the conformational dynamics and binding mechanism of IDPs. The protocol involves (1) the use of GPU-accelerated all-atom explicit-solvent MD associated with the CHARMM22* forcefield to simulate the IDP conformational ensemble in isolation and (2) the use of coarse-grained SBM to simulate the entire IDP binding coupled to folding process. We also discuss some basic procedures to analyze and interpret all-atom MD simulations of flexible IDPs as well as the basic parameterization procedure of the SBM forcefield and how to perform a preliminary analysis of the binding mechanism from these simulations.

2 Materials

To implement the general simulation protocol, the user must install the MD engine GROMACS suite (versions later than 4.5 are recommended) (http://www.gromacs.org/) [65] and the free energy calculation plugin PLUMED (versions later than 2.0 are recommended) (http://www.plumed.org/) [66]. The structures of IDPs in isolation or in complex can be examined using the PyMOL software (http://pymol.org/) [67]. The initial preparation of SBM input files for GROMACS can be directly obtained from the SMOG webserver (http://smog-server.org/) [68]; further specific modifications of the topology files for the SBM will be required during the parameterization processes.

3 Methods

3.1 The Protein Model for Simulations

To illustrate the methods and overall protocol described here we selected one well-studied IDP system. This system consists of the phosphorylated kinase inducible domain (pKID) of cAMP response element-binding (CREB) protein, an IDP when isolated in solution [69], and its binding to the KIX domain of coactivator CREB-binding protein (CBP), a natively structured three-helix bundle protein (Fig. 1a) [70, 71]. Upon binding to KIX, pKID forms two perpendicularly arranged α-helices: α_A and α_B [70]. Posttranslational modification results on the phosphorylation of Serine-133, which is positioned at the beginning of the helix α_B in pKID. Experimentally, phosphorylation of S133 appears to increase the binding affinity for KIX, and hence it must be an important factor

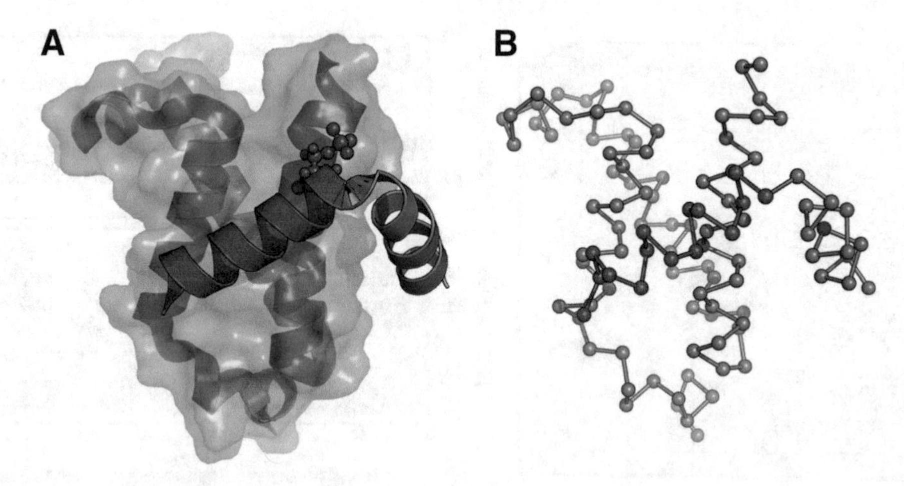

Fig. 1 Structural illustrations of the pKID-KIX complex. (**a**) Cartoon representation of the complex with a transparent surface model of KIX. The phosphorylated serine is shown with sticks-and-balls, and helices α_A and α_B are colored blue and red, respectively. (**b**) C_α-based representation used in the coarse-grained SBM simulations

for the binding mechanism [72]. The structure of the pKID-KIX complex has been determined by nuclear magnetic resonance (NMR) spectroscopy and is deposited in the Protein Data Bank with PDB-ID: 1KDX [70]. This complex coordinate file provides the initial structure for all-atom MD simulations and the structural input required for building the SBM.

3.2 All-Atom Molecular Dynamics Simulations

3.2.1 Input Protein Structure

In general, one can navigate through the PDB website to download the relevant atomic coordinate file to be used as input for the simulations. Here we use the PDB file: 1KDX. The file includes two chains: the KIX domain and pKID (Fig. 1a). The target for the all-atom MD simulations is pKID, and hence the KIX chain must be edited out of the file (*see* **Note 1**).

3.2.2 Generating the Topology File

The complete set up of all-atom explicit-solvent MD simulation can be performed using GROMACS, starting from the input structure. In GROMACS, the topology file includes all the interaction parameters for the protein of interest that are generated based on the forcefield to be used for the simulations. Subsequently, this file is updated to include other components (water, ions, etc.). The CHARMM22* forcefield files have to be provided to GROMACS. Further modifications for CHARMM22* forcefield to include the phosphorylation parameters at Serine-133 are required (*see* **Note 2**). As water model, a precise TIP3P model (inherently included in GROMACS) is chosen during the topology preparation process.

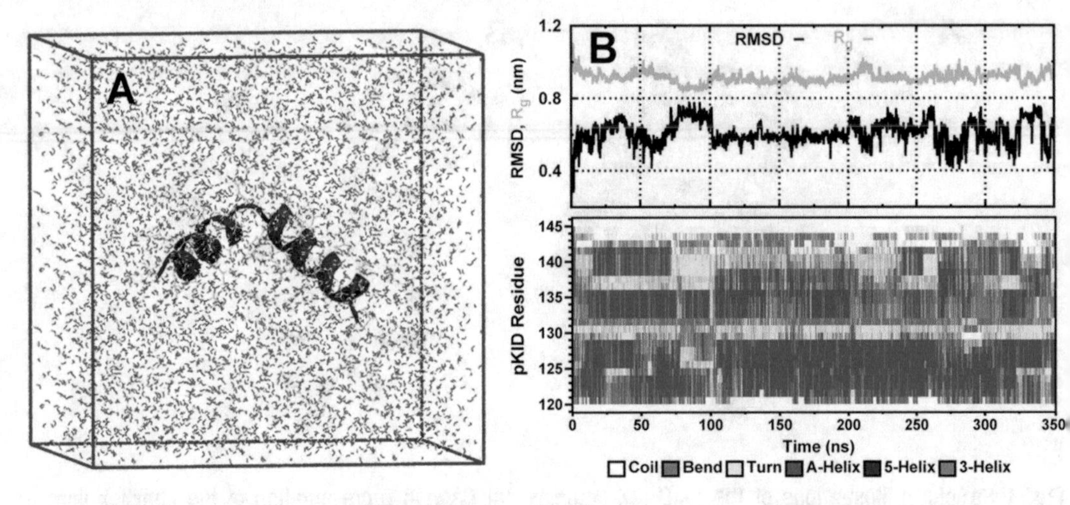

Fig. 2 All-atom molecular dynamics simulation. (**a**) Illustration of the simulation box including water molecules, ion, and centered pKID. (**b**) Projections of the trajectory on RMSD, R_g (top), and secondary structure (bottom)

3.2.3 System Preparation

For simplicity, a cubic box with periodic boundary conditions is used. pKID is then placed at the center and care must be taken to ensure a minimum of ~1 nm distance to all the sides of the box. Water molecules are subsequently added into the box in a number that is dependent on the defined volume around the protein box (e.g., ~5200 water molecules for the box used here). To neutralize the system, a single Cl^{-1} ion is added to the box replacing one water molecule (Fig. 2a). Energy minimization of the system is then carried out to relax abnormally large forces in the system using the steepest descent algorithm.

3.2.4 Simulation Protocols

The integration time step is set to 2 fs, and the LINCS algorithm is employed to constrain the geometry of covalent bonds with hydrogen atoms [73]. Temperature coupling is done by using velocity rescaling with a stochastic term and a coupling time of 0.1 ps [74]. Van der Waals interactions are truncated at 1.0 nm, and electrostatic interactions are treated using the Particle Mesh Ewald (PME) method with a grid size of 0.12 nm and a real-space cutoff of 1.0 nm [75]. In constant pressure simulations, the pressure is controlled by the Parrinello-Rahman method with a coupling time of 0.2 ps [76] (*see* **Note 3**).

3.2.5 Equilibration Simulation

Equilibration simulations are conducted in two steps to optimize the interactions between water, ion(s), and protein (pKID). To avoid unnecessary conformational distortions of the protein during equilibration, a position restraining force is applied on the heavy atoms of pKID. At first at constant temperature, a volume simulation (NVT ensemble) starting from the end configuration of the energy minimization step is performed for 1 ns. Then at constant temperature, a pressure simulation (NPT ensemble) starting from the end configuration of the NVT equilibration step is performed for 1 ns. Finally, the last configuration is collected as the starting input for the production MD simulation.

3.2.6 Production Molecular Dynamics Simulation

The production MD simulation of the pKID system is conducted at constant temperature and pressure. Ideally, the length of the production MD simulation should be longer than the intrinsic timescale of the protein's conformational dynamics (typically longer than microseconds), but often times it is further restricted by the available computational resources. GPU accelerated MD, which efficiently uses hundreds of computational cores within the GPU in parallel, is performed at the production step to achieve longer timescale simulations. Here, a pKID simulation of 350 ns was produced using a combination of GPU and CPU threads (*see* **Note 4**). The output trajectory is saved at a user-defined frequent time step (e.g., 100 ps) and used for preliminary analysis.

3.2.7 Trajectory Analysis

All-atom MD simulations in explicit solvent record the positions of every atom in the system evolving with time. However, it is practically inconvenient to analyze protein conformational dynamics directly using Cartesian coordinates. Instead, some key structural quantities most relevant to the system under study, such as the root-mean-square deviation (RMSD), radius of gyration (R_g), or the protein secondary structure, are introduced as order parameters to describe the dynamics of the system (here pKID, Fig. 2b) (*see* **Note 5**). Overall, pKID exhibits highly fluctuating conformational dynamics as judged from RMSD and R_g trajectories. Such enhanced dynamics are a hallmark of IDPs. The simulations show that the helix α_A is more stable than helix α_B, consistent with the NMR experimental findings [69].

3.3 Molecular Simulations with Structure-Based Models

3.3.1 A C_α Structure-Based Model

The Structural Representation

The binding simulations are conducted at a C_α coarse-grained level where each residue is represented by one single bead located at the C_α position and without any heterogeneity, regardless of geometry or weight of the residue. This model is identical to the original plain SBM (Fig. 1b) [56].

Potentials

The forcefield used in the simulations for the pKID-KIX system is made up by two intrachain (pKID and KIX) and one interchain (intermolecular interactions) SBM potentials, together with a general electrostatic potential:

$$V = V_{\mathrm{SBM}}{}^{\mathrm{KIX}} + V_{\mathrm{SBM}}{}^{\mathrm{pKID}} + V_{\mathrm{SBM}}{}^{\mathrm{pKID-KIX}} + V_{\mathrm{Ele}}$$

The SBM potential is native-centric and for each protein/structure can be written as [56]:

$$V_{SBM}^{(KIX,pKID,pKID-KIX)} = \sum_{bonds} K_r(r - r_0)^2 + \sum_{angles} K_\theta(\theta - \theta_0)^2 + \sum_{dihedrals} K_\phi^{(n)}[1 - cos\,(n \times (\phi - \phi_0))] + \sum_{i<j-3}^{native} \varepsilon_n \left[5\left(\frac{r_{0ij}}{r_{ij}}\right)^{12} - 6\left(\frac{r_{0ij}}{r_{ij}}\right)^{10}\right] + \sum_{i<j-3}^{non-native} \varepsilon_{nn}\left(\frac{\sigma_{ex}}{r_{ij}}\right)^{12}$$

where the first three terms describe local interactions, including bond stretching, angle bending, and dihedral rotation, whereas the last two terms describe nonlocal interactions that include native contact Lennard-Jones (LJ) interactions and non-native, repulsive volume interactions (*see* **Note 6**). Note that interchain SBM potential only contains nonlocal interactions. The variables with subscript "0" correspond to the specific values found in the PDB structure. By definition, the potential has its global minimum at the native structure. The native contact maps are built using the Contacts of Structural Units (CSU) software [77], which renders 159 contacts for KIX, 25 for pKID, and 50 interchain contacts between pKID and KIX. The used values for the relevant parameters are: $K_r = 10000.0$, $K_\theta = 20.0$, $K_\phi{}^{(1)} = 1.0$, $K_\phi{}^{(3)} = 0.5$, $\varepsilon_n = 1.0$, $\varepsilon_{nn} = 1.0$, and $\sigma_{ex} = 0.4$ nm. Length is expressed in nm, whereas energy, temperature, and time scale are expressed in reduced units (*see* **Note 7**). Ready-to-go SBM input files for GROMACS simulations are directly obtained by simply uploading the PDB file into the SMOG webserver (*see* **Note 1**). Further tweaks of the TOP file are necessary to calibrate the SBM parameters to match the experimental results.

3.3.2 Structure-Based Model Simulation Procedures

The SBM simulations can be performed using the GROMACS MD engine implemented with an MDP file that sets up the MD parameters [65]. A stochastic dynamics integrator is used at a time step of $\delta t = 0.0005$ (reduced units). The Langevin thermostat is applied with a friction coefficient of $\gamma = 1.0$. Nonbonded interactions including LJ and electrostatic interactions are cut off at a relatively long distance of 3 nm to better account for the severe coarse-graining of the model. A sample MDP file can be found in SMOG webserver. The calibration and production simulations are performed with the same MD parameters, and the length of the simulations will be set specifically for each simulation as needed.

3.3.3 Empirical Parameterization of the Structured-Based Model

Electrostatic Interactions

To include electrostatic interactions in the SBM, lysine and arginine residues are modeled with one positive charge, and glutamate and aspartate residues, as well as phosphorylated serine, are modeled with one negative charge. The rest of the residues are set to neutral charge. The electrostatic potential uses the Debye-Hückel model, which implicitly incorporates the effect of ionic strength:

$$V_{Ele} = K_{Coulomb} B(\kappa) \sum_{i<j-3} \frac{q_i q_j \exp\left(-\kappa r_{ij}\right)}{\varepsilon_r r_{ij}}$$

where q_i and q_j are respectively the charge of residues i and j, ε_r is the relative dielectric constant and is set to be 80, and $K_{Coulomb}$ is a constant equal to 138.94. $B(\kappa)$ is the ionic strength-dependent coefficient, approximately 1 at dilute solutions. κ^{-1} is the Debye screening length, which is directly affected by ionic strength. The relationship between κ and ionic strength I (in molar units) can be explicitly expressed as $\kappa \approx 3.2(I)^{1/2}$ nm^{-1}. In the present simulations, we use a moderate ionic strength of $I = 0.15$, so $\kappa \approx 1.24$ nm^{-1}. Therefore, the energetic contribution of the electrostatic potential for two oppositely charged residues located at the average distance for a native contact in this system (i.e., 0.76 nm) is about −0.89, which is close to the native contact LJ contribution (1.0). Native contacts that involve oppositely charged residues are then rescaled to ensure that the sum of electrostatic and LJ potentials is equal to 1.0. This rescaling aims to set the energetic contribution of each interacting pair to be equal at the native distance, thereby mimicking the original plain SBM, but also introducing separate LJ and electrostatic effects at distances shorter and longer than the average native contact distance, respectively [40, 78, 79]. The LJ and electrostatic interactions described here use user-defined potentials that are not included in GROMACS, so an extra parameter table file, which redefines the LJ and electrostatic interactions, should be supplied to GROMACS when launching the simulations (*see* **Note 8**).

Temperature

The coarse-grained representation and simplified physics of the SBM forcefield imply that a suitable calibration of the temperature scale is needed to compare with relevant experimental conditions. Such calibration is usually done using some thermodynamic property that is experimentally measurable as benchmark, such as the folding temperature or B-factor. Experimentally, the KIX domain has been proposed to fold cooperatively in a two-state manner [80], though the possible presence of a folding intermediate is under debate [81, 82]. The global midpoint denaturation temperature T_f^{Exp} of this protein is 65 ± 2 °C, as determined by tracking the thermal denaturation process using circular dichroism (CD) spectroscopy [83]. In SBM simulations of KIX alone, the observed unfolding temperature $T_f^{Sim} = 0.99$ (*see* **Note 9**).

Simulation temperatures in reduced units can be compared with the experimental scale using a simple linear extrapolation:

$$\frac{T^{Exp}}{T^{Sim}} = \frac{T_f^{Exp}}{T_f^{Sim}} = \frac{273 + 65}{0.99} = 341.41$$

Note that the relation is only approximate because solvent effects are nonlinear due to the presence of a significant change in denaturation heat capacity. However, this linear relationship is reasonably accurate for narrow temperature ranges, and it provides a simple way to correlate the experimental and SBM simulation in thermodynamic terms.

pKID Intrachain Interactions

NMR experiments indicate that the α_A segment of pKID has significant helical population (50% to 60%) whereas the α_B segment is mostly unstructured, exhibiting only 10% to 15% helix content at the relatively low temperature of 15 °C [69]. Since coarse-grained SBM treats all the protein residues as homogeneous (except for their charge), these two regions are likely to have similar stability in simulations that use the default parameters. One should then recalibrate the simulations to reproduce the existing experimental data. This means that the helical propensity must be mapped onto the simplistic C_α-only structural representation. An effective strategy for defining the amount of helical structure of any given conformation that is described in terms of a C_α trace has been proposed by de Sancho and Best, who used a Lifson-Roig description and determined that formation of helix structure requires that at least three contiguous C_α pseudo-dihedral angles fall between $-35°$ and $145°$ [61, 84]. Accordingly, the SBM potential for pKID can be separated into terms specific for α_A and α_B:

$$V_{\mathrm{SBM}}{}^{\mathrm{pKID}} = K_{\alpha\mathrm{A}} V_{\mathrm{SBM}}{}^{\mathrm{pKID}\alpha\mathrm{A}} + K_{\alpha\mathrm{B}} V_{\mathrm{SBM}}{}^{\mathrm{pKID}\alpha\mathrm{B}} + K_{\alpha\mathrm{A}-\alpha\mathrm{B}} V_{\mathrm{SBM}}{}^{\mathrm{pKID}\alpha\mathrm{A}-\mathrm{pKID}\alpha\mathrm{B}}$$

where $V_{SBM}{}^{pKID\alpha A}$ and $V_{SBM}{}^{pKID\alpha B}$ are intra-segment SBM potentials for the α_A and α_B segments and $V_{SBM}{}^{pKID\alpha A\text{-}pKID\alpha B}$ is the inter-segment SBM potential, which contains only nonlocal interactions. The pre-factors K for the different terms modulate their relative strengths. Since helical formation is mostly controlled by the strength of the intra-segment SBM (through $K_{\alpha A}$ and $K_{\alpha B}$), $K_{\alpha A\text{-}\alpha B}$ can be kept unchanged for simplicity. The optimal intra-chain scaling factors can be determined from a series of simulations implemented with different values of $K_{\alpha A}$ and $K_{\alpha B}$ at a constant temperature $T = 0.84$ that best corresponds to the reference NMR experimental temperature (15 °C) (Fig. 3a).

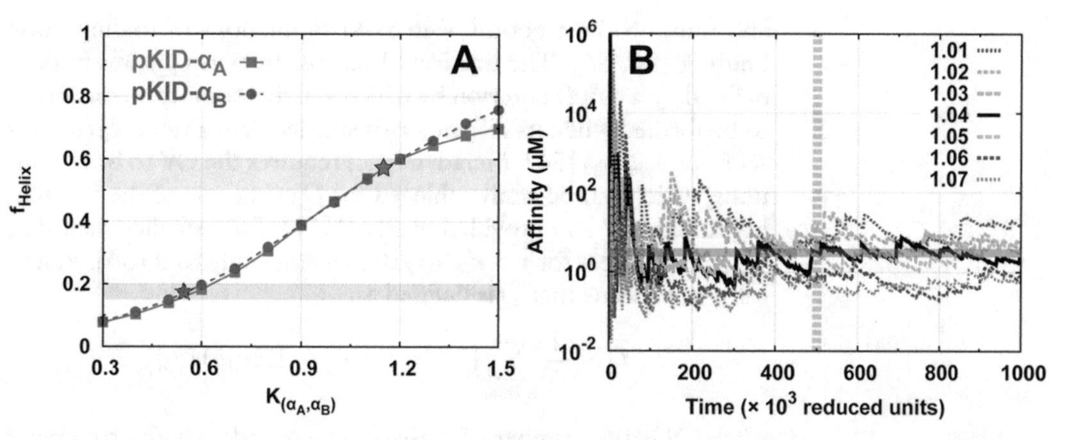

Fig. 3 Parameterization of the SBM. (**a**) Modulation of the interaction strength of helices α_A and α_B in pKID. The blue and red regions are the helical propensities measured in experiments for α_A and α_B, respectively. The final calibrated parameters $K_{\alpha A} = 1.15$ and $K_{\alpha B} = 0.55$ are shown as stars. (**b**) Modulation of the strength of native contacts ε_n in $V_{SBM}{}^{pKID\text{-}KIX}$ using metadynamics simulations. The metadynamics simulations converged after time 500×10^3. The gray region is the range in experimental binding affinity, and the final value obtained with $\varepsilon_n = 1.04$ is shown as a black thin line

pKID–KIX Interchain Interactions

The binding interaction should be calibrated to reproduce the experimentally determined affinity: $K_D = 3.10 \pm 0.60\mu\text{M}$ [72]. In simulations, the binding affinity can be calculated as:

$$K_D = \frac{[\text{pKID}][\text{KIX}]}{[\text{pKID} - \text{KIX}]} = \frac{1.66}{V_0} \frac{p_{ub}^2}{1 - p_{ub}}$$

where [pKID], [KIX], and [pKID–KIX] are the equilibrium concentrations of pKID, KIX, and binding complex pKID–KIX, respectively; V_0 is the volume that is effectively achieved by setting periodic boundary conditions for the simulation box in nm^3, and 1.66 is a normalization constant (in $\text{M}\cdot\text{nm}^{-3}$). Since there is no water in coarse-grained SBM simulations, the box is simply set to be a cube with an edge length of 10 nm, i.e., length much longer than the center of mass between the pKID and KIX at native complex (1.44 nm, *see* **Note 10**); p_{ub} is the probability of finding the two proteins unbound (calculated after the simulation has converged). Changes in binding affinity are obtained via modulation of the strength of the native contacts ε_n in $V_{\text{SBM}}{}^{\text{pKID-KIX}}$. To achieve fast convergence in the multiple runs performed to determine the optimal value of ε_n, we recommend the implementation of any suitable advanced sampling method. Here we use metadynamics. During metadynamics simulations, small Gaussian potentials are sequentially added along the specific reaction coordinates (collective variables, CV) [53]. For binding simulations, a simple intuitive way to choose suitable CV is to focus on those properties that best describe the binding process. The fraction of intermolecular native contacts (here referring to *Q(pKID–KIX)*) has been widely used and found to be a good CV in SBM simulations of folding upon binding [85, 86]. The fraction of native contacts (Q) was initially defined by a cutoff criterion by which a native contact is considered to be formed when its distance r_{ij} is smaller than a value d_O (usually set to be $1.2r_{Oij}$) [56]. Metadynamics requires the CV to be continuous, which can be easily achieved for Q using one of the "switching functions" $s(r)$ provided in PLUMED; for r smaller than d_O, $s(r) = 1$, whereas for $r > d_O$, $s(r)$ decays smoothly to 0 [66]. Practically, this means that Q is defined as:

$$Q = \frac{1}{N} \sum_{native} \{1 - \tanh[(r_{ij} - 1.2r_{Oij})/r_O]\}$$

where N is the number of native contacts and r_O is the parameter that controls the shape of the switching function (usually set to 0.01 nm). The equivalence between the continuous and discrete Q functions as descriptors of protein folding in SBM has been confirmed by Oliveira et al. [87]. The metadynamics simulations are performed in a standard way with different strengths ε_n for binding contacts (*see* **Note 11**). The value of ε_n that most closely reproduces the experimental binding affinity (within experimental error) after convergence is chosen for the production MD (Fig. 3b).

3.3.4 Production Coarse-Grained Simulation

Once the parameter calibration has been achieved, the resulting SBM model is used to perform tens to hundreds of independent molecular simulations, each one starting from a different bound/unbound configuration. Starting conformations can be directly extracted from the previous metadynamics simulations, or obtained from scratch by performing a short high-temperature simulation (*see* **Note 12**). These simulations are performed at a constant reduced temperature that closely represents the experimental temperature (25 °C), in this case $T = 0.87$. Each simulation should be sufficiently long as to sample a good number of binding and unbinding events so that there is thermodynamic consistency between simulations to calculate accurate free energy landscapes. The thermodynamics and kinetics of binding can be obtained directly from the constant temperature trajectories.

3.3.5 Analysis

The trajectories are usually projected onto a set of user-specified order parameters to track desired features of the system's time evolution. To describe the binding mechanism, one can use the fraction of intrachain native contacts within pKID (*Q(pKID)*) for folding, and the fraction of intermolecular native contacts between pKID and KIX (*Q(pKID-KIX)*) for binding (Fig. 3a). The free energy landscape is then calculated from the Boltzmann inversion of the probability distribution projected onto such folding and binding reaction coordinates (Fig. 3b). The extreme scenarios

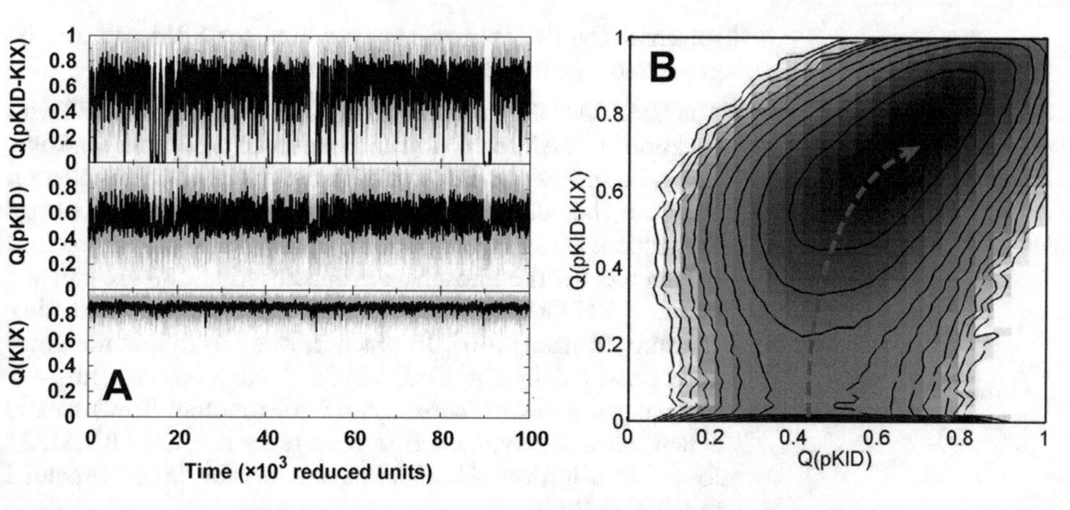

Fig. 4 Production SBM coarse-grained simulations. (**a**) One sample trajectory is shown as fraction of native contacts for interfacial binding *Q(pKID-KIX)* and intra-chain folding *Q(pKID), Q(KIX)*. (**b**) Two-dimensional free energy landscape projected along the binding *Q(pKID-KIX)* and folding *Q(pKID)* reactions

defined by the induced-fit and conformational selection mechanisms will go along the two rectangular edges of such two-dimensional free energy landscape (binding followed by folding or folding followed by binding). However, in most cases, the simulations produce results that lie in between these two extremes. The binding mechanism(s) of the complex of interest (here pKID binding to KIX) can be inferred from the analysis of the obtained energy landscape and a close structural analysis of all the binding transition paths; e.g., in a given path a partially folded unbound pKID binds to KIX without significantly changing its degree of folding upon formation of the highest energy region connecting the minima in the landscape (i.e., transition state region) (*Q (pKID–KIX)* ~ 0.1 from Fig. 4b), and then finally folds up while it remains bound to KIX.

4 Notes

1. There are 17 conformers deposited in the 1KDX PDB file because it is an NMR-determined structure. The first conformer is chosen to prepare the topology files for all-atom MD simulations and SBM simulations. For all-atom simulations, the initial structure does not influence the process of building the topology file. For SBM, using other alternative conformers from the PDB will not change the results since at the coarse-grained level they are virtually identical. In some cases, the PDB protein structures may have missing residues or side chain atoms. This problem can be addressed using the MODELER software [88] to add the missing parts.

Alternatively, the contact map required for SBM can also be generated by other advanced methods [89].

2. The CHARMM22* forcefield is a modified CHARMM22 with backbone CMAP correction and side chain parameter improvements [46]. It is not included with the GROMACS installation and can be downloaded from http://www.gromacs.org/Downloads/User_contributions/Force_fields. Forcefield parameters for the phosphate group at serine-133 are patched from CHARMM36 (http://mackerell.umaryland.edu/charmm_ff.shtml) [45]. In practice, one can follow the guidance provided by the GROMACS documentation (https://manual.gromacs.org/current/how-to/topology.html) to add a new phosphorylated serine residue in the CHARMM22* forcefield using the forcefield parameters for "SP2" provided in CHARMM36.
3. All of the GROMACS simulation parameters are included in an MDP file. A sample MDP file can be found in the GROMACS documentation. Specific CHARMM parameters can be found at https://manual.gromacs.org/current/reference-manual/functions/force-field.html#charmm.
4. Two Nvidia GeForce GTX 980 Ti GPU cards have been used for the GPU accelerated MD simulations discussed here. The performance of such minimalistic GPU workstation was of 175 ns/day of pKID simulation.
5. Calculations of RMSD, R_g, and secondary structure from trajectories can be directly carried out using the GROMACS in-home commands "rms," "gyrate," and "do_dssp," respectively. Use of the command "do_dssp" requires the addition of the independent dssp program, which can be directly downloaded from the DSSP website (https://swift.cmbi.umcn.nl/gv/dssp/) [90].
6. The LJ potential used for the C_α coarse-grained SBM has a 10–12 interaction form instead of the traditional 6–12 powers of the distance for the attractive and repulsive terms, respectively. The 6–12 LJ potential has a longer and wider interacting distance range than the 10–12 one does, which may result in unrealistically strong attractive native interactions in unfolded conformations. This problem is particularly severe for simplified C_α coarse-grained SBM in which the interacting centers are virtual and averaged over the entire residue, and thus the contacts are much less geometry-specific. The 10–12 type LJ potential is used in this case as a correction of this effect when using a C_α coarse-grained SBM.
7. In GROMACS, the default units for length, time, mass, and energy scale are nm, ps, amu, and kJ/mol. In general, SBMs are performed with reduced units and self-consistency in all scales.

Following the suggestions of the SMOG webserver, we still define the length scale in nm, whereas the mass, time, and energy scales are set to reduced units. Temperature is defined in energy units divided by the Boltzmann constant. Therefore, there is no direct connection between the scales in SBM and reality. However, one can estimate the relevant thermodynamics and kinetics through different strategies. For the thermodynamics, the easiest way to correlate simulation and experimental temperatures is described in Subheading 3.3.3. For kinetics, we recommend to use the methods proposed in the literature [91, 92]. For instance, the reduced time scale (τ) can be expressed as:

$$\tau = \left(\frac{m_0 a_0^2}{\varepsilon} \right)^{1/2}$$

where m_0 and a_0 are the mass of the beads and bond length, and ε is the reduced energy unit. In C_α coarse-grained SBM, the typical values of m_0 and a_0 are 1.8×10^{-22} g and 3.8×10^{-10} m and $\varepsilon = 1.0$ kJ/mol, and hence $\tau = 4.0$ ps.

8. An example of typical table file for SBM simulations can be found in the SMOG webserver, but the electrostatic potential must be adapted to the Debye-Hückel expression as explained in the text. One can follow the guide provided within the GROMACS documentation to do the modifications (https://manual.gromacs.org/documentation/current/reference-manual/special/tabulated-interaction-functions.html#user-specified-potential-functions).

9. The actual folding temperature of the KIX domain in the simulations is determined by performing REMD where a series of simulations (replicas) are run at exponentially distributed temperatures centered around $T = 1.0$ (usually proteins fold around $T = 1.0$ when using an SBM). The attempt to exchange between different neighboring replicas obeys Metropolis criterion and should occur frequently [51]. The data from the REMD simulations at all different temperatures are collected to calculate the heat capacity using the Weighted Histogram Analysis Method (WHAM) [93]. The WHAM program for analyzing the data can be downloaded from the SMOG webserver. The heat capacity curve of the KIX domain folding has one single peak that indicates the midpoint for its (un)folding transition.

10. There is an alternative intuitive approach to effectively limit the region of the space that the two chains need to explore, i.e., directly adding distance restraints between the center of mass of the interacting chains. Such restraints can be practically implemented with the PLUMED plugin by setting the

"UPPER_WALLS." The "wall" will kick in when the two chains are moving away from the user-defined distance, and therefore simulates an effective concentration of the two proteins. Both methods are equivalent when the simulations are performed with a sufficiently large volume so that boundary effects are minimized.

11. Here we performed standard well-tempered metadynamics simulations. The height of the elementary Gaussian hills was set to 0.1 and rescaled with a bias factor provided through the well-tempered strategy [94]. Their widths were set to 0.02 based on the fact that there are 50 binding contacts. The bias factor, which is the ratio between the temperature of the CV and the system temperature, was set to 10.0. The bias energies were stored at the user-defined grids along the CV to be able to make a fast evaluation of the large number of Gaussian hills that are added during a metadynamics simulation. The program "sum_hills" provided in PLUMED was used to do the free energy calculation.

12. To perform even faster simulations of folding upon binding, one can set the simulations so that the partner is kept more or less rigid throughout the simulation. One option is to add extra "native" forces in the partner to keep it completely rigid throughout the simulation, as it has been done before for KIX [17, 61, 95]. This is a very convenient and efficient approach that is a reasonable approximation when the IDP-binding target is a well folded and stable protein in the timescales relevant to the binding event. Previous theoretical work has also shown that the conformational dynamics of KIX plays little role on the thermodynamics and kinetics of binding to pKID [20, 96]. However, recent research suggests allosteric regulation in KIX may contribute to the macroscopic binding promiscuity by modulating the structural ensembles of KIX in the microscopic scale [97]. Therefore, we decided to perform the simulations with the KIX domain unrestrained and set the simulation temperature well below the KIX folding temperature to ensure it remains folded. The strategy used in this protocol is general and therefore can be adapted to any other IDP and its partners in straightforward ways.

Acknowledgments

This work was funded by Advanced Grant ERC-2012-ADG-323059 from the European Research Council to V. M. V. M. also acknowledges support from the Keck foundation, the CREST Center for Cellular and Biomolecular Machines (NSF-CREST-1547848) and the NSF (NSF-MCB-1616759).

References

1. Wright PE, Dyson HJ (1999) Intrinsically unstructured proteins: re-assessing the protein structure-function paradigm. J Mol Biol 293 (2):321–331
2. Dunker AK, Lawson JD, Brown CJ, Williams RM, Romero P, Oh JS, Oldfield CJ, Campen AM, Ratliff CM, Hipps KW, Ausio J, Nissen MS, Reeves R, Kang C, Kissinger CR, Bailey RW, Griswold MD, Chiu W, Garner EC, Obradovic Z (2001) Intrinsically disordered protein. J Mol Graph Model 19(1):26–59
3. Tompa P (2002) Intrinsically unstructured proteins. Trends Biochem Sci 27(10):527–533
4. Papoian GA (2008) Proteins with weakly funneled energy landscapes challenge the classical structure–function paradigm. Proc Natl Acad Sci U S A 105(38):14237–14238
5. Obradovic Z, Peng K, Vucetic S, Radivojac P, Brown CJ, Dunker AK (2003) Predicting intrinsic disorder from amino acid sequence. Proteins 53(S6):566–572
6. Ward JJ, Sodhi JS, McGuffin LJ, Buxton BF, Jones DT (2004) Prediction and functional analysis of native disorder in proteins from the three kingdoms of life. J Mol Biol 337 (3):635–645
7. Dunker AK, Brown CJ, Lawson JD, Iakoucheva LM, Obradovic Z (2002) Intrinsic disorder and protein function. Biochemistry 41 (21):6573–6582
8. Shammas SL (2017) Mechanistic roles of protein disorder within transcription. Curr Opin Struct Biol 42:155–161
9. Dyson HJ, Wright PE (2005) Intrinsically unstructured proteins and their functions. Nat Rev Mol Cell Biol 6(3):197–208
10. DeForte S, Uversky VN (2016) Intrinsically disordered proteins in PubMed: what can the tip of the iceberg tell us about what lies below? RSC Adv 6(14):11513–11521
11. Tokuriki N, Tawfik DS (2009) Protein dynamism and evolvability. Science 324 (5924):203–207
12. Zhou H-X (2012) Intrinsic disorder: signaling via highly specific but short-lived association. Trends Biochem Sci 37(2):43–48
13. Teilum K, Olsen JG, Kragelund BB (2009) Functional aspects of protein flexibility. Cell Mol Life Sci 66(14):2231
14. Dunker AK, Garner E, Guilliot S, Romero P, Albrecht K, Hart J, Obradovic Z, Kissinger C, Villafranca JE (1998) Protein disorder and the evolution of molecular recognition: theory, predictions and observations. Pac Symp Biocomput:473–484
15. Chu X, Wang J (2014) Specificity and affinity quantification of flexible recognition from underlying energy landscape topography. PLoS Comput Biol 10(8):e1003782
16. Shoemaker BA, Portman JJ, Wolynes PG (2000) Speeding molecular recognition by using the folding funnel: the fly-casting mechanism. Proc Natl Acad Sci U S A 97 (16):8868–8873
17. Huang Y, Liu Z (2009) Kinetic advantage of intrinsically disordered proteins in coupled folding–binding process: a critical assessment of the “fly-casting” mechanism. J Mol Biol 393 (5):1143–1159
18. Uversky VN, Oldfield CJ, Dunker AK (2005) Showing your ID: intrinsic disorder as an ID for recognition, regulation and cell signaling. J Mol Recognit 18(5):343–384
19. Pontius BW (1993) Close encounters: why unstructured, polymeric domains can increase rates of specific macromolecular association. Trends Biochem Sci 18(5):181–186
20. Umezawa K, Ohnuki J, Higo J, Takano M (2016) Intrinsic disorder accelerates dissociation rather than association. Proteins 84 (8):1124–1133
21. Xie H, Vucetic S, Iakoucheva LM, Oldfield CJ, Dunker AK, Obradovic Z, Uversky VN (2007) Functional anthology of intrinsic disorder. 3. Ligands, post-translational modifications, and diseases associated with intrinsically disordered proteins. J Proteome Res 6 (5):1917–1932
22. Csizmok V, Follis AV, Kriwacki RW, Forman-Kay JD (2016) Dynamic protein interaction networks and new structural paradigms in signaling. Chem Rev 116(11):6424–6462
23. Dyson HJ, Wright PE (2002) Coupling of folding and binding for unstructured proteins. Curr Opin Struct Biol 12(1):54–60
24. Wright PE, Dyson HJ (2009) Linking folding and binding. Curr Opin Struct Biol 19 (1):31–38
25. Koshland DE (1958) Application of a theory of enzyme specificity to protein synthesis. Proc Natl Acad Sci U S A 44(2):98–104
26. Tsai C-J, Ma B, Nussinov R (1999) Folding and binding cascades: shifts in energy landscapes. Proc Natl Acad Sci U S A 96 (18):9970–9972
27. Bosshard HR (2001) Molecular recognition by induced fit: how fit is the concept? Physiology 16(4):171–173
28. Csermely P, Palotai R, Nussinov R (2010) Induced fit, conformational selection and

independent dynamic segments: an extended view of binding events. Trends Biochem Sci 35(10):539–546
29. Kiefhaber T, Bachmann A, Jensen KS (2012) Dynamics and mechanisms of coupled protein folding and binding reactions. Curr Opin Struct Biol 22(1):21–29
30. Okazaki K-I, Takada S (2008) Dynamic energy landscape view of coupled binding and protein conformational change: induced-fit versus population-shift mechanisms. Proc Natl Acad Sci U S A 105(32):11182–11187
31. Zhou H-X (2010) From induced fit to conformational selection: a continuum of binding mechanism controlled by the timescale of conformational transitions. Biophys J 98(6): L15–L17
32. Hammes GG, Chang Y-C, Oas TG (2009) Conformational selection or induced fit: a flux description of reaction mechanism. Proc Natl Acad Sci U S A 106(33):13737–13741
33. Dyson HJ, Wright PE (1998) Equilibrium NMR studies of unfolded and partially folded proteins. Nat Struct Biol 5:499–503
34. Mukhopadhyay S, Krishnan R, Lemke EA, Lindquist S, Deniz AA (2007) A natively unfolded yeast prion monomer adopts an ensemble of collapsed and rapidly fluctuating structures. Proc Natl Acad Sci U S A 104 (8):2649–2654
35. Eliezer D (2009) Biophysical characterization of intrinsically disordered proteins. Curr Opin Struct Biol 19(1):23–30
36. Zhu F, Kapitan J, Tranter GE, Pudney PD, Isaacs NW, Hecht L, Barron LD (2008) Residual structure in disordered peptides and unfolded proteins from multivariate analysis and ab initio simulation of Raman optical activity data. Proteins 70(3):823–833
37. Fuxreiter M, Simon I, Friedrich P, Tompa P (2004) Preformed structural elements feature in partner recognition by intrinsically unstructured proteins. J Mol Biol 338(5):1015–1026
38. Knott M, Best RB (2012) A preformed binding interface in the unbound ensemble of an intrinsically disordered protein: evidence from molecular simulations. PLoS Comput Biol 8 (7):e1002605
39. Arai M, Sugase K, Dyson HJ, Wright PE (2015) Conformational propensities of intrinsically disordered proteins influence the mechanism of binding and folding. Proc Natl Acad Sci U S A 112(31):9614–9619
40. Chu X, Wang Y, Gan L, Bai Y, Han W, Wang E, Wang J (2012) Importance of electrostatic interactions in the association of intrinsically disordered histone chaperone Chz1 and histone H2A. Z-H2B. PLoS Comput Biol 8(7): e1002608
41. Iešmantavičius V, Dogan J, Jemth P, Teilum K, Kjaergaard M (2014) Helical propensity in an intrinsically disordered protein accelerates ligand binding. Angew Chem Int Ed 53 (6):1548–1551
42. Rauscher S, Gapsys V, Gajda MJ, Zweckstetter M, de Groot BL, Grubmüller H (2015) Structural ensembles of intrinsically disordered proteins depend strongly on force field: a comparison to experiment. J Chem Theory Comput 11(11):5513–5524
43. Best RB, Zheng W, Mittal J (2014) Balanced protein–water interactions improve properties of disordered proteins and non-specific protein association. J Chem Theory Comput 10 (11):5113–5124
44. Huang J, Rauscher S, Nawrocki G, Ran T, Feig M, de Groot BL, Grubmüller H, MacKerell A (2017) CHARMM36m: an improved force field for folded and intrinsically disordered proteins. Nat Methods 14(1):71–73
45. Best RB, Zhu X, Shim J, Lopes PE, Mittal J, Feig M, MacKerell AD Jr (2012) Optimization of the additive CHARMM all-atom protein force field targeting improved sampling of the backbone ϕ, ψ and side-chain $\chi 1$ and $\chi 2$ dihedral angles. J Chem Theory Comput 8 (9):3257–3273
46. Piana S, Lindorff-Larsen K, Shaw DE (2011) How robust are protein folding simulations with respect to force field parameterization? Biophys J 100(9):L47–L49
47. Piana S, Donchev AG, Robustelli P, Shaw DE (2015) Water dispersion interactions strongly influence simulated structural properties of disordered protein states. J Phys Chem B 119 (16):5113–5123
48. Shaw DE, Deneroff MM, Dror RO, Kuskin JS, Larson RH, Salmon JK, Young C, Batson B, Bowers KJ, Chao JC (2008) Anton, a special-purpose machine for molecular dynamics simulation. Commun ACM 51(7):91–97
49. Lindorff-Larsen K, Piana S, Dror RO, Shaw DE (2011) How fast-folding proteins fold. Science 334(6055):517–520
50. Lindorff-Larsen K, Trbovic N, Maragakis P, Piana S, Shaw DE (2012) Structure and dynamics of an unfolded protein examined by

molecular dynamics simulation. J Am Chem Soc 134(8):3787–3791

51. Sugita Y, Okamoto Y (1999) Replica-exchange molecular dynamics method for protein folding. Chem Phys Lett 314(1):141–151
52. Torrie GM, Valleau JP (1977) Nonphysical sampling distributions in Monte Carlo free-energy estimation: umbrella sampling. J Comput Phys 23(2):187–199
53. Laio A, Parrinello M (2002) Escaping free-energy minima. Proc Natl Acad Sci U S A 99 (20):12562–12566
54. Stanley N, Esteban-Martín S, De Fabritiis G (2014) Kinetic modulation of a disordered protein domain by phosphorylation. Nat Commun 5:ncomms6272
55. Bryngelson JD, Onuchic JN, Socci ND, Wolynes PG (1995) Funnels, pathways, and the energy landscape of protein folding: a synthesis. Proteins 21(3):167–195
56. Clementi C, Nymeyer H, Onuchic JN (2000) Topological and energetic factors: what determines the structural details of the transition state ensemble and "en-route" intermediates for protein folding? An investigation for small globular proteins. J Mol Biol 298(5):937–953
57. Levy Y, Cho SS, Onuchic JN, Wolynes PG (2005) A survey of flexible protein binding mechanisms and their transition states using native topology based energy landscapes. J Mol Biol 346(4):1121–1145
58. Levy Y, Onuchic JN, Wolynes PG (2007) Fly-casting in protein-DNA binding: frustration between protein folding and electrostatics facilitates target recognition. J Am Chem Soc 129(4):738–739
59. Levy Y, Wolynes PG, Onuchic JN (2004) Protein topology determines binding mechanism. Proc Natl Acad Sci U S A 101(2):511–516
60. Chu X, Gan L, Wang E, Wang J (2013) Quantifying the topography of the intrinsic energy landscape of flexible biomolecular recognition. Proc Natl Acad Sci U S A 110(26): E2342–E2351
61. De Sancho D, Best RB (2012) Modulation of an IDP binding mechanism and rates by helix propensity and non-native interactions: association of HIF1α with CBP. Mol BioSyst 8 (1):256–267
62. Ganguly D, Chen J (2011) Topology-based modeling of intrinsically disordered proteins: balancing intrinsic folding and intermolecular interactions. Proteins 79(4):1251–1266
63. Turjanski AG, Gutkind JS, Best RB, Hummer G (2008) Binding-induced folding of a natively unstructured transcription factor. PLoS Comput Biol 4(4):e1000060
64. Wang J, Wang Y, Chu X, Hagen SJ, Han W, Wang E (2011) Multi-scaled explorations of binding-induced folding of intrinsically disordered protein inhibitor IA3 to its target enzyme. PLoS Comput Biol 7(4):e1001118
65. Van Der Spoel D, Lindahl E, Hess B, Groenhof G, Mark AE, Berendsen HJ (2005) GROMACS: fast, flexible, and free. J Comput Chem 26(16):1701–1718
66. Tribello GA, Bonomi M, Branduardi D, Camilloni C, Bussi G (2014) PLUMED 2: new feathers for an old bird. Comput Phys Commun 185(2):604–613
67. Schrodinger, LLC (2015) The PyMOL Molecular Graphics System, Version 1.8
68. Noel JK, Whitford PC, Sanbonmatsu KY, Onuchic JN (2010) SMOG@ ctbp: simplified deployment of structure-based models in GROMACS. Nucleic Acids Res 38(suppl 2): W657–W661
69. Radhakrishnan I, Pérez-Alvarado GC, Dyson HJ, Wright PE (1998) Conformational preferences in the Ser133-phosphorylated and non-phosphorylated forms of the kinase inducible transactivation domain of CREB. FEBS Lett 430(3):317–322
70. Radhakrishnan I, Pérez-Alvarado GC, Parker D, Dyson HJ, Montminy MR, Wright PE (1997) Solution structure of the KIX domain of CBP bound to the transactivation domain of CREB: a model for activator: coactivator interactions. Cell 91(6):741–752
71. Zor T, De Guzman RN, Dyson HJ, Wright PE (2004) Solution structure of the KIX domain of CBP bound to the transactivation domain of c-Myb. J Mol Biol 337(3):521–534
72. Zor T, Mayr BM, Dyson HJ, Montminy MR, Wright PE (2002) Roles of phosphorylation and helix propensity in the binding of the KIX domain of CREB-binding protein by constitutive (c-Myb) and inducible (CREB) activators. J Biol Chem 277(44):42241–42248
73. Hess B, Bekker H, Berendsen HJ, Fraaije JG (1997) LINCS: a linear constraint solver for molecular simulations. J Comput Chem 18 (12):1463–1472
74. Bussi G, Donadio D, Parrinello M (2007) Canonical sampling through velocity rescaling. J Chem Phys 126(1):014101
75. Essmann U, Perera L, Berkowitz ML, Darden T, Lee H, Pedersen LG (1995) A smooth particle mesh Ewald method. J Chem Phys 103(19):8577–8593
76. Parrinello M, Rahman A (1981) Polymorphic transitions in single crystals: a new molecular dynamics method. J Appl Phys 52 (12):7182–7190

77. Sobolev V, Sorokine A, Prilusky J, Abola EE, Edelman M (1999) Automated analysis of interatomic contacts in proteins. Bioinformatics 15(4):327–332
78. Azia A, Levy Y (2009) Nonnative electrostatic interactions can modulate protein folding: molecular dynamics with a grain of salt. J Mol Biol 393(2):527–542
79. Givaty O, Levy Y (2009) Protein sliding along DNA: dynamics and structural characterization. J Mol Biol 385(4):1087–1097
80. Morrone A, Giri R, Brunori M, Gianni S (2012) Reassessing the folding of the KIX domain: evidence for a two-state mechanism. Protein Sci 21(11):1775–1779
81. Horng J-C, Tracz SM, Lumb KJ, Raleigh DP (2005) Slow folding of a three-helix protein via a compact intermediate. Biochemistry 44 (2):627–634
82. Tollinger M, Kloiber K, Ágoston B, Dorigoni C, Lichtenecker R, Schmid W, Konrat R (2006) An isolated helix persists in a sparsely populated form of KIX under native conditions. Biochemistry 45(29):8885–8893
83. Wei Y, Horng J-C, Vendel AC, Raleigh DP, Lumb KJ (2003) Contribution to stability and folding of a buried polar residue at the CARM1 methylation site of the KIX domain of CBP. Biochemistry 42(23):7044–7049
84. Lifson S, Roig A (1961) On the theory of helix—coil transition in polypeptides. J Chem Res 34(6):1963–1974
85. Socci N, Onuchic JN, Wolynes PG (1996) Diffusive dynamics of the reaction coordinate for protein folding funnels. J Chem Res 104 (15):5860–5868
86. Cho SS, Levy Y, Wolynes PG (2006) P versus Q: structural reaction coordinates capture protein folding on smooth landscapes. Proc Natl Acad Sci U S A 103(3):586–591
87. Oliveira RJ, Whitford PC, Chahine J, Wang J, Onuchic JN, Leite VB (2010) The origin of nonmonotonic complex behavior and the effects of nonnative interactions on the diffusive properties of protein folding. Biophys J 99(2):600–608
88. Šali A, Blundell TL (1993) Comparative protein modelling by satisfaction of spatial restraints. J Mol Biol 234(3):779–815
89. Chwastyk M, Bernaola AP, Cieplak M (2015) Statistical radii associated with amino acids to determine the contact map: fixing the structure of a type I cohesin domain in the clostridium thermocellum cellulosome. Phys Biol 12 (4):046002
90. Kabsch W, Sander C (1983) Dictionary of protein secondary structure: pattern recognition of hydrogen-bonded and geometrical features. Biopolymers 22(12):2577–2637
91. Veitshans T, Klimov D, Thirumalai D (1997) Protein folding kinetics: timescales, pathways and energy landscapes in terms of sequence-dependent properties. Fold Des 2(1):1–22
92. Kouza M, Li MS, O'Brien EP, Hu C-K, Thirumalai D (2006) Effect of finite size on cooperativity and rates of protein folding. J Phys Chem A 110(2):671–676
93. Kumar S, Rosenberg JM, Bouzida D, Swendsen RH, Kollman PA (1992) The weighted histogram analysis method for free-energy calculations on biomolecules. I. The method. J Comput Chem 13(8):1011–1021
94. Barducci A, Bussi G, Parrinello M (2008) Well-tempered metadynamics: a smoothly converging and tunable free-energy method. Phys Rev Lett 100(2):020603
95. Sharma R, De Sancho D, Muñoz V (2017) Interplay between the folding mechanism and binding modes in folding coupled to binding processes. Phys Chem Chem Phys 19(42):28512–28516
96. Huang Y, Liu Z (2010) Smoothing molecular interactions: the "kinetic buffer" effect of intrinsically disordered proteins. Proteins 78 (16):3251–3259
97. Law SM, Gagnon JK, Mapp AK, Brooks CL (2014) Prepaying the entropic cost for allosteric regulation in KIX. Proc Natl Acad Sci U S A 111(33):12067–12072

Part V

Prediction Methods

Chapter 20

Prediction of Folding and Unfolding Rates of Proteins with Simple Models

David De Sancho and Victor Muñoz

1 Introduction

The study of the kinetics of folding and unfolding in bulk has been central for our understanding of protein biophysics for the last 30 years [1]. Most typically in these experiments a stopped-flow apparatus is used to introduce a perturbation in a diluted protein sample at a certain concentration of denaturant (urea or guanidinium chloride) that is either increased or decreased. Using a spectroscopic technique (usually fluorescence or circular dichroism) that probes the change in the population of the folded and unfolded states, the kinetics of the relaxation to equilibrium can be measured. By doing this at a number of final denaturant concentrations one can obtain the characteristic "chevron" plot, i.e. the dependence of the observed relaxation times with the concentration of denaturant. In this way, the folding and unfolding rates of tens of proteins have been measured over the last decades, constituting an invaluable resource for extracting general principles regarding protein folding mechanisms.

If we restrict ourselves to small, single-domain, two-state folding proteins [2], the range of experimental folding rates spans over six orders of magnitude, with timescales ranging from microseconds (μs) to seconds. These extremely large differences within an otherwise rather homogeneous dataset have inspired considerable efforts from the theoretical front. One of the most successful approaches has been the derivation of metrics from the native structures of proteins, obtained from X-ray crystallography or nuclear magnetic resonance (NMR) and deposited in the RCSB Protein Data Bank (PDB) [3]. The most famous of these metrics is the relative contact order (*RCO*) proposed by Plaxco, Simons, and Baker [4], which over 20 years later still enjoys considerable

Victor Muñoz (ed.), *Protein Folding: Methods and Protocols*, Methods in Molecular Biology, vol. 2376,
https://doi.org/10.1007/978-1-0716-1716-8_20,

popularity. This metric is defined as the average sequence separation between interacting residues in the native structure of a protein and is normalized by sequence length. Several other instances of this type of empirical metric have also been successful in predicting the folding rates for slightly different datasets [5–7].

An alternative possibility is to resort to general principles from the energy landscape theory of protein folding [8], which exploits the use of projections of the many-dimensional energy landscapes of proteins onto relevant progress coordinates for this process [9]. According to this, folding dynamics can be understood as a diffusive barrier crossing process, and the kinetic rate constant for the molecular transition can be expressed using Kramers' high friction rate expression [10].

Here we summarize a specific instance of this second type of approach. Using size scaling principles that are well established for protein folding [7, 11–14], some fundamental definitions of thermodynamic contributions to the free energy [15] and a set of empirical parameters [16], the method requires only the sequence length and the secondary structure type for making estimates of both the folding and the unfolding rates of two-state, single-domain proteins [17, 18]. Since the method requires very limited information about the protein of interest, large-scale predictions can be run with high throughput. For this reason, our method has been useful in the context of large-scale analysis of protein folding in vivo [19].

2 Theory

2.1 Thermodynamics

In our model we define an order parameter for the folding process, the "nativeness" (n), as the fraction of Ramachandran angles in a native-like configuration. Using this progress variable, the free energy ($\Delta G(T, n)$) at a given temperature T is defined as

$$\Delta G(T,n) = \Delta H(T,n) - T\Delta S(T,n), \tag{1}$$

where $\Delta H(T, n)$ and $\Delta S(T, n)$ are the values of the enthalpy and entropy differences at the same temperature. The temperature dependence of both contributions to the free energy is described with reference to a reference temperature (T_o) using the unfolding heat capacity change ($\Delta C_p(n)$), using

$$\Delta H(T,n) = \Delta H(T_o,n) + \Delta C_p(n)(T - T_o) \tag{2}$$

and

$$\Delta S(T,n) = \Delta S(T_o,n) + \Delta C_p(n)\ln(T/T_o). \tag{3}$$

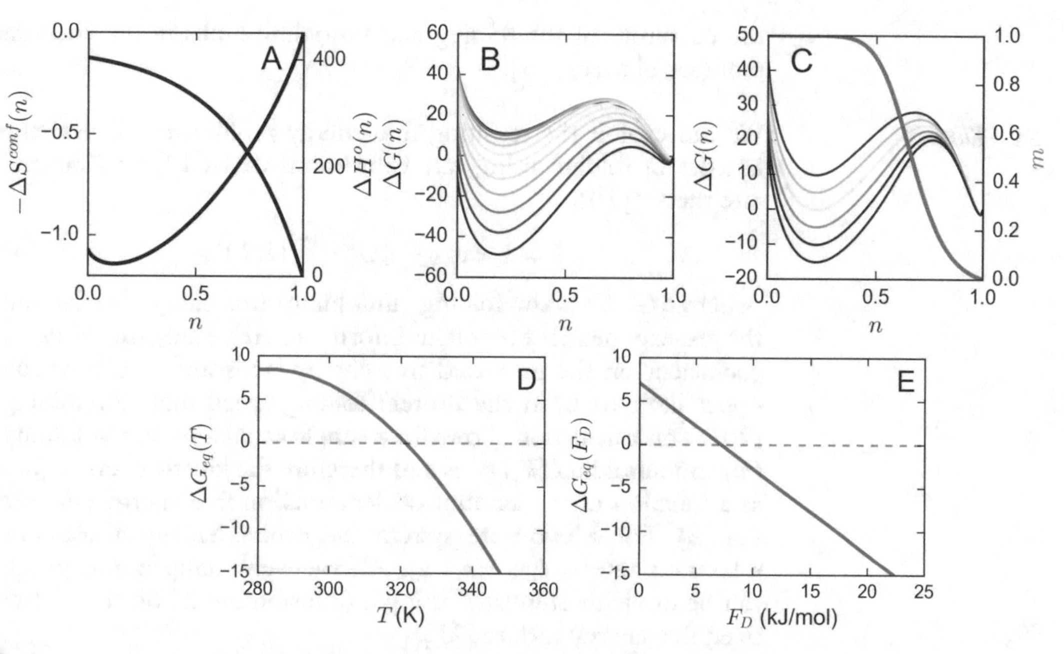

Fig. 1 (**a**) Enthalpic (blue) and entropic (red) contributions to the free energy as a function of the model order parameter (the nativeness, *n*). (**b**) Free energy profiles at multiple temperatures ranging from 285 K (blue) to 380 K (red). (**c**) Free energy of chemical denaturation (*m*, green) and free energy profiles at multiple concentrations of chemical denaturant. Units are in kJ/mol except for ΔS^{conf}, which is expressed in kJ/mol/K. (**d–e**) Temperature and chemical denaturant dependence of the protein stability

Here we assume for simplicity that $\Delta C_p(n)$ is independent of the temperature [16]. The dependence of the entropy with the nativeness is given by

$$\Delta S^{\text{conf}}(n) = N(-R[n\ln(n)+(1-n)\ln(1-n)] + n\Delta S_{\text{res}}^{n=1} + (1-n)\Delta S_{\text{res}}^{n=0}) \tag{4}$$

while the enthalpy and heat capacity are written down as exponential functions

$$\Delta H(T_{\text{o}}, n) = N\Delta H_{\text{res}}^{385\text{K}}[1 + (\exp(\kappa_{\Delta H} n) - 1)/(1 - \exp(\kappa_{\Delta H})] \tag{5}$$

$$\Delta C_p(n) = N\Delta C_{p,\text{res}}[1 + (\exp(\kappa_{\Delta C_p} n) - 1)/(1 - \exp(\kappa_{\Delta} C_p)] \tag{6}$$

(see Fig. 1a). Using these expressions we can recover a free energy profile like that shown in Fig. 1b. In addition to including the thermal effects on the free energy landscape we use an empirical expression for incorporating the chemical denaturation

$$\Delta G(F_D, n) = \Delta H(T, n) - T\Delta S(T, n) - mF_D, \tag{7}$$

where F_D is the denaturation free energy, $m = 1 - [(1 + C)(n^j/(n^j + C))]$, and C and j are adjustable parameters that determine the curvature of the folding and unfolding limbs in the chevron plot (see Fig. 1c) [15].

2.2 Kinetics

We can exploit the resulting free energy profiles to calculate the kinetics of folding using rate expressions derived from Kramers' rate theory [10],

$$k = k_o \exp\left(-\Delta G^{U(F)\ddagger}/(RT)\right), \tag{8}$$

where $\Delta G^{U(F)\ddagger}$ is the folding (unfolding) free energy barrier and the pre-exponential k_o contains information regarding the diffusion coefficient on the projected free energy landscape, which has an upper limit to k_o in the protein folding speed limit for folding [20]. This calculation allows for a simple estimation of the folding (k_f) and unfolding (k_u) rates and therefore the kinetic chevron plot as a function of the amount of denaturation free energy (F_D, see Fig. 2). For a two-state system the experimentally measurable relaxation rate is $k_{obs} = k_f + k_u$. Alternatively, temperature jumps can be explicitly simulated using a diffusion model on the discretized free energy surface [21].

3 Methods

3.1 Splitting Stabilization Energy in Local and Non-local Contributions

For the purpose of developing a method for the prediction of folding and unfolding rates of two-state proteins, the enthalpic contribution to the free energy (Eq. 5 above) was split into two different terms, corresponding to the local and non-local contributions to the stabilization energy [18]. This results in two equivalent expressions whose magnitude is determined by the values of ΔH_{loc} and ΔH_{nonloc}, respectively. Different degrees of local or non-local stabilization energy result in free energy profiles corresponding to different folding regimes. At room temperature we recover a downhill free energy profile when ΔH_{loc} becomes comparable to ΔH_{nonloc} and barrier limited when the non-local contribution dominates (see Fig. 3). The influence of these two net contributions to the stabilization energy has been considered before in the context of early on-lattice simulations of folding [22], the analysis of experiments [23], and Ising-like models [24].

3.2 Parametrization

Using a database of folding and unfolding rates for 52 single-domain proteins with no evidence of folding intermediates, and studied in similar experimental conditions [25], we calibrated the values of the model parameters $\kappa_{\Delta H,loc}$, $\kappa_{\Delta H,nonloc}$, $\Delta H_{loc,res}$, and $\Delta H_{nonloc,res}$ [18]. The values of these parameters are distinct for proteins of different structural classes, α, β, and $\alpha + \beta$. The rest of the thermodynamic parameters of the model, i.e. the entropy and heat capacity per residue, are derived from calorimetric data

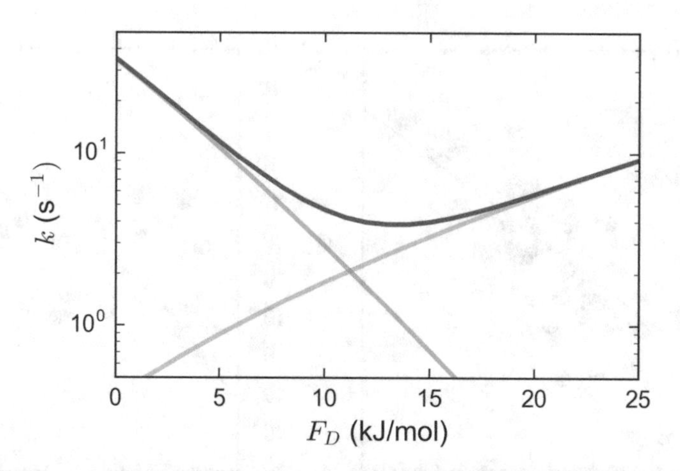

Fig. 2 Calculated chevron plot derived from the chemical denaturant dependent free energy profiles from Fig. 20.1c. Folding (k_f), unfolding (k_u), and observed (k_{obs}) rates are shown in blue, green, and red, respectively

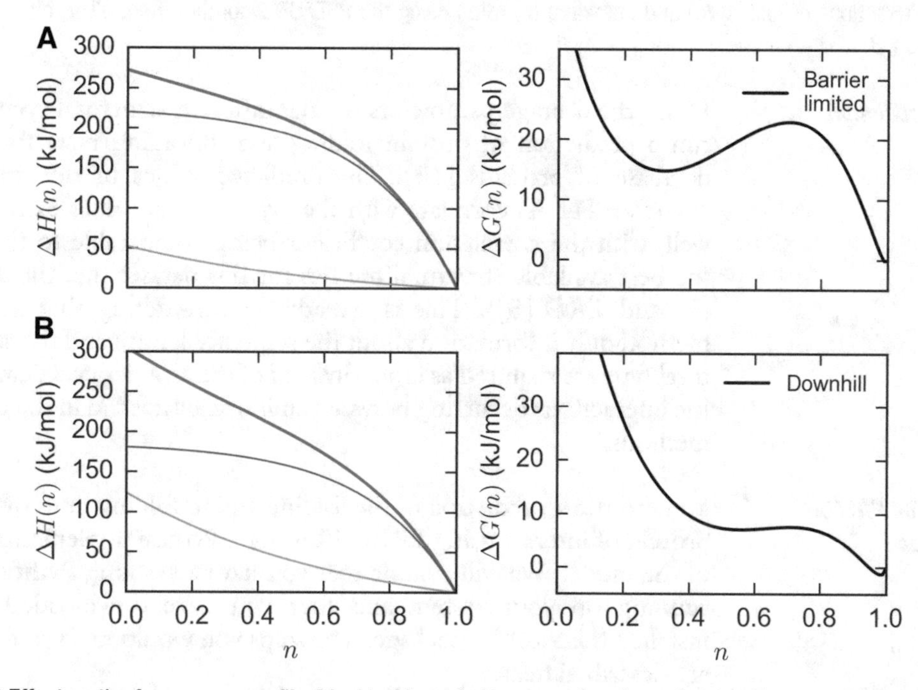

Fig. 3 Effect on the free energy profile (black) of local (ΔH_{loc}, green) and non-local (ΔH_{nonloc}, red) contributions to the total stabilization energy (blue). (**a**) When the non-local contribution dominates, we recover a two-state free energy profile (black). (**b**) When the two contributions are comparable, we recover a downhill free energy profile (right)

$\Delta S_{res}^{conf} = 16.5$ J/mol/K and $\Delta C_{p,res} = 58$ J/mol/K [16]. The curvature of the heat capacity is set to $\kappa_{\Delta C_p} = 4.3$ and the pre-exponential of the kinetic calculation is $k_o = 3.6 \times 10^6 / \text{Ns}^{-1}$ [17, 18].

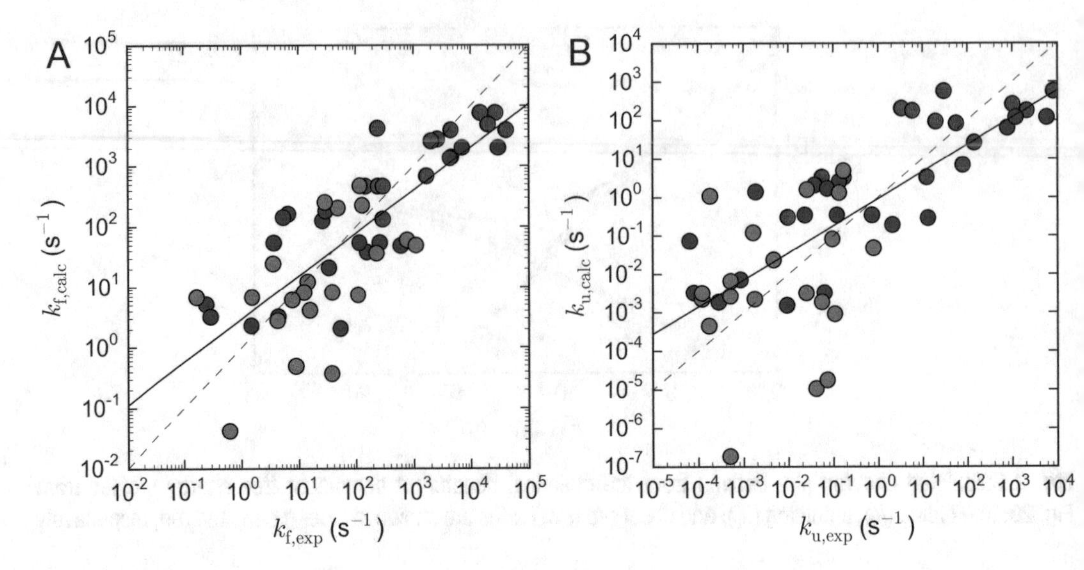

Fig. 4 Predictions of folding (**a**) and unfolding (**b**) rates using the PREFUR algorithm. Red: all α; blue: all β; green: $\alpha + \beta$

3.3 Prediction

Using the average parameters for the different structural types we run a prediction of protein folding and unfolding rates for our database of proteins [18]. The predicted values of the folding rates (see Fig. 4) correlate with the experimental values extremely well, with the correlation coefficient being comparable to that of the best available structural metrics for this dataset (i.e. the *ACO* [7] and *LRO* [5]). This is remarkable considering that in our method only information about the sequence length and the structural type are required as input, instead of the atomic-level detail on the interactions occurring between amino acids needed in the other methods.

3.4 The PREFUR Package

You can run a prediction of the folding and unfolding rates of your protein of interest using PREFUR using a Python implementation of the model. We will assume that you have a working Python 3.2 running on your system and that you have downloaded and installed the `PREFUR` package. The steps you would need to follow are described next:

1. **Protein selection:** First of all we will choose the protein of interest. In the Protein Data Bank (www.rcsb.org) we will be able to find which is the structural type for the corresponding protein under the "Annotations" header.
2. **Import the `kinetics` module** Although the `PREFUR` package has additional functionalities, all the bits that you need are conveniently incorporated in a single module. In a Python session, you can import it with the following statement.

```
from PREFUR import kinetics
```

3. **Run the rate prediction:** Using the number of amino acids of the protein (`nes`) and the structural type (`a`, `b` or `ab` for α, β or $\alpha + \beta$, respectively).

```
kf, ku, FES = kinetics.predict(nres=nres, \
                               struct='a')
```

This will return three Python objects; `kf` and `ku` are, respectively, the predicted folding and unfolding rates, and `FES` is the free energy surface object which you can be further explored.

4 Notes

1. The PREFUR method has been parametrized using a dataset of proteins with sizes ranging from 30 to 150 amino acids in length, having no evidence of stable intermediate states. Ideally, it should be used exclusively within this domain. The database is included as part of package.
2. The `PREFUR` package can be freely downloaded from GitHub[1] and is distributed under the GNU General Public License. It comes with simple installation instructions and worked out examples that demonstrate its capabilities.
3. In case the 3D structure of the protein has not been resolved experimentally, accurate predictions can be obtained from sequences alone using tools like PSIPRED [26].
4. After having run your calculation you may want to compare how the PREFUR prediction results with those of alternative descriptors. Some of these are accessible from different on-line servers, like the Contact Order Server[2] from Baker's laboratory or the PBDparam server [27].

Acknowledgements

DDS receives financial support from the Spanish Ministry of Science and Universities through the Office of Science Research (MINECO/FEDER) (Grants PGC2018-099321-B-I00 and RYC-2016-19590) and the Basque Government IT1254-19.

[1] https://doi.org/10.5281/zenodo.1038967.
[2] http://depts.washington.edu/bakerpg/contact_order.

References

1. Bachmann A, Kiefhaber T (2008) Kinetic mechanisms in protein folding. Wiley-VCH Verlag GmbH, Weinheim, pp 377–410. https://doi.org/10.1002/9783527619498.ch12a
2. Jackson SE (1998) Fold Des 3(4):R81. https://doi.org/10.1016/S1359-0278(98)00033-9
3. Berman HM, Westbrook J, Feng Z, Gilliland G, Bhat TN, Weissig H, Shindyalov IN, Bourne PE (2000) Nucleic Acids Res 28(1):235. https://doi.org/10.1093/nar/28.1.235
4. Plaxco KW, Simons KT, Baker D (1998) J Mol Biol 277(4):985. https://doi.org/10.1006/jmbi.1998.1645
5. Gromiha MM, Selvaraj S (2001) J Mol Biol 310(1):27. https://doi.org/10.1006/jmbi.2001.4775
6. Zhou H, Zhou Y (2002) Biophys J 82(1):458. https://doi.org/10.1016/S0006-3495(02)75410-6
7. Ivankov DN, Garbuzynskiy SO, Alm E, Plaxco KW, Baker D, Finkelstein AV (2003) Protein Sci 12(9):2057. https://doi.org/10.1110/ps.0302503
8. Bryngelson JD, Onuchic JN, Socci ND, Wolynes PG (1995) Proteins 21(3):167. https://doi.org/10.1002/prot.340210302
9. Socci ND, Onuchic JN, Wolynes PG (1996) J Chem Phys 104(15):5860. https://doi.org/10.1063/1.471317
10. Kramers HA (1940) Physica 7(4):284. https://doi.org/10.1016/S0031-8914(40)90098-2
11. Thirumalai D (1995) J Phys 5(11):1457. https://doi.org/10.1051/jp1:1995209
12. Gutin AM, Abkevich VI, Shakhnovich EI (1996) Phys Rev Lett 77(27):5433. https://doi.org/10.1103/PhysRevLett.77.5433
13. Koga N, Takada S (2001) J Mol Biol 313(1):171. https://doi.org/10.1006/jmbi.2001.5037
14. Naganathan AN, Muñoz V (2005) J Am Chem Soc 127(2):480. https://doi.org/10.1021/ja044449u
15. Naganathan A, Doshi U, Muñoz V (2007) J Am Chem Soc 129(17):5673. https://doi.org/10.1021/ja0689740
16. Robertson AD, Murphy KP (1997) Chem Rev 97(5):1251. https://doi.org/10.1021/cr960383c
17. De Sancho D, Doshi U, Muñoz V (2009) J Am Chem Soc 131(6):2074. https://doi.org/10.1021/ja808843h
18. De Sancho D, Muñoz V (2011) Phys Chem Chem Phys 13:17030. https://doi.org/10.1039/C1CP20402E
19. O'Brien EP, Ciryam P, Vendruscolo M, Dobson CM (2014) Acc Chem Res 47(5):1536. https://doi.org/10.1021/ar5000117
20. Kubelka J, Hofrichter J, Eaton WA (2004) Curr Opin Struct Biol 14(1):76. https://doi.org/10.1016/j.sbi.2004.01.013
21. Lapidus LJ, Steinbach PJ, Eaton WA, Szabo A, Hofrichter J (2002) J Phys Chem B 106(44):11628. https://doi.org/10.1021/jp020829v
22. Abkevich VI, Gutin AM, Shakhnovich EI (1995) J Mol Biol 252(4):460
23. Muñoz V, Serrano L (1996) Fold Des 1(4):R71
24. Muñoz V, Eaton WA (1999) Proc Natl Acad Sci U S A 96(20):11311. https://doi.org/10.1073/pnas.96.20.11311
25. Maxwell KL, Wildes D, Zarrine-Afsar A, De Los Rios MA, Brown AG, Friel CT, Hedberg L, Horng JC, Bona D, Miller EJ, Vallée-Bélisle A, Main ER, Bemporad F, Qiu L, Teilum K, Vu ND, Edwards AM, Ruczinski I, Poulsen FM, Kragelund BB, Michnick SW, Chiti F, Bai Y, Hagen SJ, Serrano L, Oliveberg M, Raleigh DP, Wittung-Stafshede P, Radford SE, Jackson SE, Sosnick TR, Marqusee S, Davidson AR, Plaxco KW (2005) Protein Sci 14(3):602. https://doi.org/10.1110/ps.041205405
26. McGuffin LJ, Bryson K, Jones DT (2000) Bioinformatics 16(4):404. https://doi.org/10.1093/bioinformatics/16.4.404
27. Nagarajan R, Archana A, Thangakani AM, Jemimah S, Velmurugan D, Gromiha MM (2016) Bioinf Biol Insights 10:73. https://dx.doi.org/10.4137%2FBBI.S38423

Chapter 21

Predicting and Simulating Mutational Effects on Protein Folding Kinetics

Athi N. Naganathan

Abstract

Mutational perturbations of protein structures, i.e., phi-value analysis, are commonly employed to probe the extent of involvement of a particular residue in the rate-determining step(s) of folding. This generally involves the measurement of folding thermodynamic parameters and kinetic rate constants for the wild-type and mutant proteins. While computational approaches have been reasonably successful in understanding and predicting the effect of mutations on folding thermodynamics, it has been challenging to explore the same on kinetics due to confounding structural, energetic, and dynamic factors. Accordingly, the frequent observation of fractional phi-values (mean of ~0.3) has resisted a precise and consistent interpretation. Here, we describe how to construct, parameterize, and employ a simple one-dimensional free energy surface model that is grounded in the basic tenets of the energy landscape theory to predict and simulate the effect of mutations on folding kinetics. As a proof of principle, we simulate one-dimensional free energy profiles of 806 mutations from 24 different proteins employing just the experimental destabilization as input, reproduce the relative unfolding activation free energies with a correlation of 0.91, and show that the mean phi-value of 0.3 essentially corresponds to the extent of stabilization energy gained at the barrier top while folding.

Key words Conformational entropy, Stabilization energy, Transition state ensemble, Microstates, Statistical mechanics, Diffusive kinetics

1 Introduction

The energy landscape theory of protein folding provides the analytical framework for understanding and interpreting folding mechanistic experiments [1, 2]. One of the main predictions of the theory is that the complex conformational landscape that determines folding can be simplified by projecting the conformations onto one or few order parameters that could in principle serve as folding reaction coordinate [3]. Results from lattice simulations and coarse-grained models support this prediction and highlight that folding kinetics can be simulated as a diffusive process on a

Victor Muñoz (ed.), *Protein Folding: Methods and Protocols*, Methods in Molecular Biology, vol. 2376,
https://doi.org/10.1007/978-1-0716-1716-8_21,

suitably chosen reaction coordinate [3, 4]. The free energy barrier (if any) in such a low-dimensional projection is primarily entropic in nature arising from an incomplete compensation between loss of entropy upon forming native interactions and the concomitant gain in stabilization free energy [1].

The successes of low dimensional approaches are evident in the ability of Ising-like statistical models to predict folding rates of proteins from one-dimensional (1D) representations [5, 6], and in understanding the folding mechanisms of α-helices [7] and numerous proteins at a quantitative level [8–12]. Such models, however, impose a mechanism that local structure forms first [13] and also require a protein structure as input. While coarse-grained and all-atom molecular simulations provide viable alternative avenues, they are time-intensive and are limited in their ability to quantitatively explain thermodynamics and kinetics of protein folding [14, 15]. In the absence of structural information or when the input experimental signal is complex, it is challenging to employ even such structure-based models.

An experimentalist, on the other hand, requires a simple model that is not only physically grounded but also flexible enough to reproduce and explain the observations from a variety of experimental probes without relying on any structural data or imposing mechanisms. In this chapter, we describe how to employ a minimalistic and physically realistic one-dimensional model of protein folding [16–18] that incorporates the basic tenets of the energy landscape theory—the energy/stability gap hypothesis, minimum frustration principle, entropic origins of folding free energy barriers—while also being consistent with empirical size-scaling laws of folding thermodynamic parameters [19]. We focus on simulating and predicting the effect of mutations on one-dimensional free energy profiles and the associated energetic consequences with implications in understanding the origin of fractional ϕ-values.

2 Materials

2.1 Reaction Coordinate

For any simple model that attempts to reduce the complex phase space accessible to protein conformations, the critical requirement is to define an order parameter that can potentially serve as a reaction coordinate (RC) to folding. In this regard, the simplest description of a protein chain involves the conformational status of the backbone dihedrals—the dihedral angles can be either native-like (i.e., sampling Ramachandran phi-psi angles [20] adjacent to that present in the folded state) or unfolded-like (sampling phi-psi angles that are non-native). Such a one-dimensional coordinate has been remarkably successfully in understanding helix-coil transitions [7, 21, 22]. Inspired by these successes, the free energy surface (FES) model employs a related order parameter termed nativeness

(n) defined as the average probability of finding any residue folded [16]. A value of *0* and *1* would therefore correspond to the fully unfolded and fully folded states, respectively. The average residue probability ranges from *0* to *1*, and fractional values represent a subset of folded residues without specifying the order in which they occur. The nativeness is therefore a continuous version of the "number of structured residues" frequently used in Ising-like protein folding models and also in the analytical model of Zwanzig [23].

2.2 Conformational Entropy and Enthalpy

The chain conformational entropy as a function of n and for a protein of length N is calculated as following:

$$\Delta S_{conf}(n) = N(-R[n\ln(n) + (1-n)\ln(1-n)] + (1-n)\Delta S_{conf,res}) \quad \text{for } 0 < n < 1 \tag{1}$$

$\Delta S_{conf,\ res}$ is the difference in entropy between the unfolded ($n = 0$) and folded ($n = 1$) state of a residue with the folded state as the reference. The first half of the expression above within the square brackets represents the entropy of mixing (which is symmetric around $n = 0.5$) that is then weighted by $\Delta S_{conf,\ res}$ (red in Fig. 1a). In fact, exact enumeration studies of the Wako-Sait-ô-Muñoz-Eaton (WSME) model [5, 24] reveal a near-identical dependence of the conformational entropy on the order parameter [6], attesting to the robustness of this simple functional.

In structure-based statistical mechanical models or coarse-grained models of folding, the folded state contacts (contact-map) are directly employed to identify interacting residues and their relative interaction energies [5, 24–26]. Since the FES model employs a functional-based approach eliminating the need for a structure, it is necessary to identify the dependence of enthalpy or stabilization free energy on n. One of the features of the energy landscape theory is the necessity to incorporate an "energy gap" between folded state and every other conformation to obtain a two-state–like behavior [1, 23]. This is required to stabilize the folded state, as the entropic stabilization is minimal. In view of this, the gain in stabilization energy or enthalpy arising from interactions between residues as the protein folds is modeled as a Markov chain (continuous blue and green curves in Fig. 1a). In other words, the probability of forming or breaking an interaction is assumed to be constant as a function of nativeness thus resulting in an exponential dependence:

$$\Delta H(n) = N\Delta H_{res}[1 + (\exp(\kappa_{\Delta H} n - 1)/(1 - \exp(\kappa_{\Delta H})))] \tag{2}$$

where ΔH_{res} is the mean-field stabilization energy per residue and $\kappa_{\Delta H}$ determines the shape or curvature of the function with smaller values resulting a near-linear dependence of the stabilization

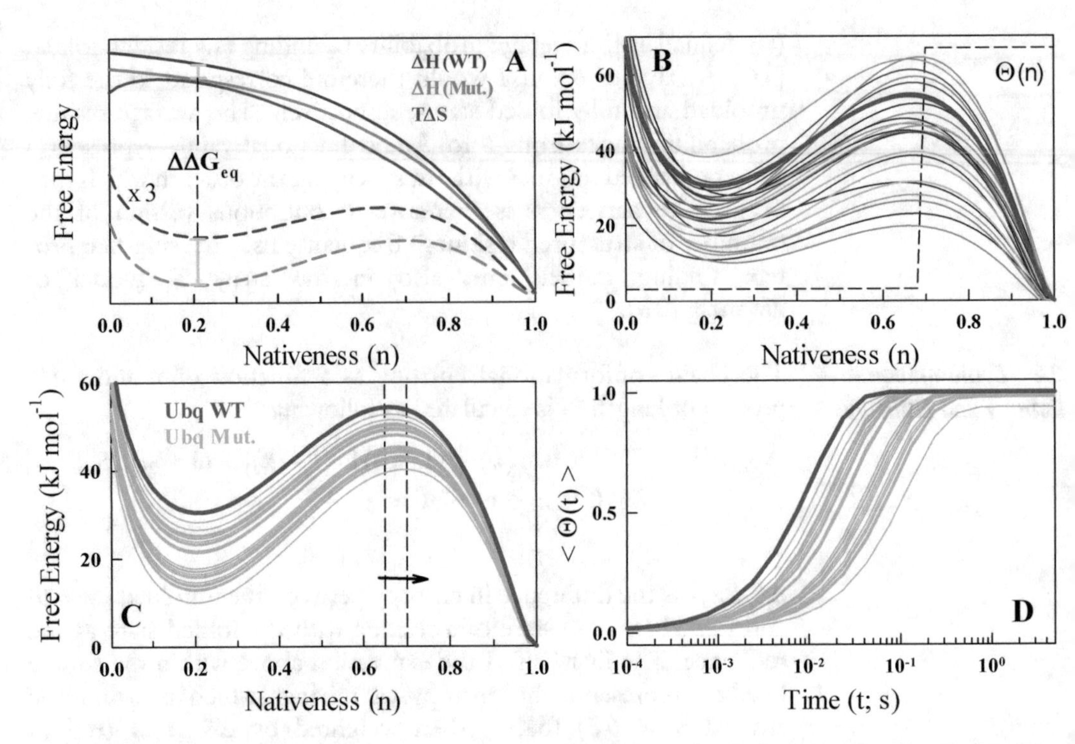

Fig. 1 Model construction and free energy profiles. (**a**) Illustrative entropic (red; Eq. 1) and enthalpic (blue; Eq. 2) functionals, together with the corresponding free energy profile (dashed blue; Eq. 3) as a function of the reaction coordinate, nativeness, for a WT protein. The mutant free energy profile (dashed green) is generated with the same entropic functional, but with an enthalpic dependence modified (continuous green) by the experimental destabilization ($\Delta\Delta G_{eq}$; cyan) following Eq. 9. Note that the free energy profiles (dashed lines) are scaled by a factor 3 to enable visualization. (**b**) The simulated free energy profiles of the 24 WT proteins display a range of stabilities and barrier heights despite minor modulations of the fundamental parameters (see text). An apparent signal dependence on the reaction coordinate, $\Theta(n)$, is shown as a dashed line. The free energy profile of WT ubiquitin (Ubq) is shown in thick magenta. (**c**) Simulated free energy profiles of WT and mutant ubiquitin proteins. The observed Hammond behavior upon mutational destabilization is also shown (arrow and vertical dashed lines). (**d**) Simulated single-exponential relaxation profiles (employing Eqs. 5–8) following the color code in panel **c**

enthalpy on the RC while larger relative values result in a sharper dependence [16]. Note that quadratic or higher order functionals on n can also be employed as long as there is sufficient difference between the enthalpy of the folded and unfolded states. The use of an exponential function simplifies the description to a single parameter that can be tuned to reproduce different folding scenarios (see below). The functional form is also consistent with the shape obtained upon projecting the energies of conformations generated in lattice and MD simulations onto a single-reaction coordinate [3, 27].

2.3 One-Dimensional Free Energy Profile

Since both enthalpy and entropy as a function of the order parameter are available, the free energy profile for folding is directly obtained from:

$$\Delta G(n) = \Delta H(n) - T\Delta S_{conf}(n) \tag{3}$$

where T is the temperature (dashed curves in Fig. 1a). The precise balance between the stabilization energy and conformational entropy determines the thermodynamic stability, ΔG_{eq}, while differences in their curvature determine the magnitude of the barrier height (β) [16]. Accordingly, the free energy barrier to folding emerges when the loss in conformational entropy is not compensated by energy (Fig. 1a), consistently with energy landscape descriptions of protein folding and inferences from empirical analysis of folding rates [28]. By tuning the magnitude of the curvature of the stabilization energy, it is therefore possible to simulate barriers ranging from zero (globally downhill or one-state) to $\gg RT$ (two-state) (*see* **Note 1**).

The thermodynamics of the model in this simple version is therefore determined merely by three parameters, $\Delta S_{conf,res}$, $\kappa_{\Delta H}$, and ΔH_{res} and incorporates the size-scaling effects intrinsic to protein folding thermodynamics [19] (*see* **Notes 2–4**). Importantly, it does not impose any mechanism unlike other statistical models and does not require any structural input.

2.4 Calculating Stabilities and Free Energy Barrier Heights

Once the free energy profiles are generated, there are two procedures to calculate the stability and barrier height from free energy profiles, one more rigorous than the other. In the first methodology, ΔG_{UF} is simply obtained as the difference in free energy between the folded and unfolded minima. Similarly, the folding and unfolding barrier heights (β_F and β_U) are calculated as the differences in free energies between the ground-state minima and the barrier top. However, this procedure is not applicable to downhill folders (small or zero barriers) or under conditions of low stability. Moreover, since the widths and curvatures are different between folded and unfolded wells, the procedure above will not give accurate stability values. A more rigorous approach is to set a dividing line at a value of the reaction coordinate that signals the barrier top in two-state systems (between a nativeness value of 0.6 and 0.8) and integrate the probabilities on either side of this line, the ratio of which is the stability when converted to energy units [16]. Extending this approach, the first step to calculate barrier heights is to define an area centered on the dividing line with a small width in nativeness units (W_{TS}). The integrated area on either side of this swath (in probability units) is referenced to W_{TS} to obtain β_F and β_U [16]. The magnitude of barriers estimated by this procedure converges with the first approach when $\beta > 3$ RT.

3 Methods

3.1 Simulating Diffusive Kinetics

For two-state–like systems, the relaxation rate constant (k) at a particular temperature T can be calculated by employing a simple rate equation with an appropriate pre-exponential factor:

$$k = k_0[\exp(-\beta_F/RT) + \exp(-\beta_U/RT)] \tag{4}$$

where k_0 is the pre-exponential factor and R is the universal gas constant. As with the calculation of barrier heights or stability, this procedure breaks down for marginal barrier or downhill systems because this representation does not capture barrierless diffusive dynamics. To this end, the relaxation rates are calculated by performing a diffusive calculation on the one-dimensional free energy surface employing the matrix method of diffusive kinetics [29]. The approach is presented below in detail:

1. To simulate the diffusive kinetics of a typical kinetic experiment, the probabilities as a function of nativeness are required for a starting condition (p_0; say high urea concentration where the unfolded well is highly populated) and an experimental condition (p_{eq}; low urea concentration at which the experimental rates are reported).
2. Construct an m x m rate matrix R where m is the dimension of the reaction coordinate to solve the rate equation:

$$\frac{\mathrm{d}p(t)}{\mathrm{d}t} = R \times p(t) \tag{5}$$

where $p(t)$ is the vector of probabilities of the different states along the reaction coordinate at time t. The elements of the rate matrix are obtained by numerically differentiating the one-dimensional Smoluchowski diffusion equation employing forward and backward difference operators resulting in diagonal and two off-diagonal relations as below [29]:

$$\begin{aligned} R_{i,i+1} &= \frac{1}{2}\left(D_{i+1} + D_i\frac{p_{eq,i}}{p_{eq,i+1}}\right) \\ R_{i+1,i} &= \frac{1}{2}\left(D_{i+1}\frac{p_{eq,i+1}}{p_{eq,i}} + D_i\right) \\ R_{i,i} &= -R_{i+1,i} - R_{i-1,i} \\ R_{1,1} &= -R_{2,1} \\ R_{m,m} &= -R_{m-1,m} \end{aligned} \tag{6}$$

For the sake of simplicity, the diffusion coefficients can be assumed to be independent of the reaction coordinate and can be set to $D_i = D = (k_0/N)/\Delta n^2$ where Δn is the nativeness interval and $1/k_0 = 440$ ns (*see* **Note 5**). The magnitude of the latter term sets the maximum rate possible in the absence of barrier and can be tuned if required, particularly if temperature-dependent data are employed.

3. Diagonalize $\boldsymbol{R}$ to obtain m eigenvalues λ and eigenvectors v (column vectors in the matrix $\mathbf{V}$). The eigenvector corresponding to the zero eigenvalue ($\lambda = 0$) will provide the final equilibrium probability distribution.
4. The rates are obtained from $k_i = -\lambda_i$. The $(m-1)$ eigenvectors (i.e., corresponding to nonzero eigenvalues) carry information on the changes in probability between the two states (and within them) in a two-state system and within the single well in a one-state system. Upon sorting, the first nonzero eigenvalue would correspond to the rate of equilibration between folded and unfolded wells with potentially a gap in the eigenvalue spectrum as expected for two-state systems (*see* **Note 6**).
5. Calculate the amplitude vector (a) and the effective time evolution of every species corresponding to every value of the coordinate n in the system (survival probability) following:

$$a = V^{-1} p_0$$

$$S(n,t) = a_n v_n \mathrm{e}^{-k_n t} \tag{7}$$

To simulate an experimental signal, $\langle \Theta(t) \rangle$, weigh the survival probability of each species with a coordinate dependent signal, $\Theta(n)$ (dashed line in Fig. 1b), or a signal switch located at a nativeness value corresponding to the barrier top:

$$\langle \Theta(t) \rangle = \sum_{n=0}^{n=m} \Theta(n) \tilde{n} S(n,t) \tag{8}$$

Fit $\langle \Theta(t) \rangle$ to a single- (Fig. 1d) or double-exponential function to obtain the predicted relaxation rate constant (k_{pred}). Generally, most proteins exhibit single-exponential relaxation phases (*see* **Note 6**).

3.2 Reproducing Wild-Type Stability, Relaxation Rate, and Simulating WT Free Energy Profiles

To predict the effect of mutations on protein folding kinetics, one must first reproduce the wild-type (WT) stability (ΔG_{UF}) and relaxation rate constant (k_{obs}). To do so, fix the entropic difference $\Delta S_{conf,\ res}$ to 16.5 J mol^{-1} K^{-1} [19] and float the two parameters, $\kappa_{\Delta H}$, and ΔH_{res}. Since the two parameters determine different physical features of the free energy profile, the fits will always be well determined. To reiterate, the ΔH functional whose shape is determined by $\kappa_{\Delta H}$ together with the entropic free energy functional ($T\Delta S_{conf}$) determines the barrier height while ΔH_{res} magnitude determines the stability. This procedure is sufficient to generate the free energy profile of the WT protein by combining Eqs. 1–3. As an example, this procedure was performed on a dataset of 24 different WT proteins whose thermodynamic stability and relaxation rates are available [30]. This results in diverse WT free energy profiles (Fig. 1b) but with $\kappa_{\Delta H}$ ranging merely from 2.68–3.52 while ΔH_{res} spans 5.33–5.77 kJ mol^{-1} per residue. The small spread is consistent with previous findings that size-scaling effects dominate the thermodynamics of folding with additional features (structure, sequence variation) contributing only a minor part [17].

3.3 Simulating Mutant Free Energy Profiles and Predicting Kinetics

Most mutational studies on single-domain proteins involve truncation of hydrophobic side chains to simplify the structural interpretation of the observed changes in free energy [31]. One simple strategy to introduce mutational effects in structure-based models is to destabilize the immediate neighborhood of contacts by a certain fraction [32, 33] or by perturbing the free energy profiles in a global manner [5, 9, 34]. In this regard, it is important to note that a truncation mutation not only destabilizes the immediate neighborhood but also percolates far and symmetrically around the mutation site (even up till 15–20 Å). This has been inferred from large-scale analysis of protein mutations and explicitly shown through network analysis of protein structures, long time-scale MD simulations, and reanalysis of NMR experiments [35, 36]. An empirical treatment of long-range mutational effects when introduced into a structure-based statistical mechanical model was able to reproduce the changes in stability of 375 mutations from 19 different proteins [35]. These observations therefore allow for a straightforward avenue to introduce destabilization effects of mutations into the FES model—modulate the WT's ΔH_{res} to reproduce the observed changes in experimental stability as was originally done [5]. The protocol to predict the mutant free energy profile and estimate the corresponding relaxation rate is discussed below:

1. Reproduce the WT stability and observed rate constant exactly by modulating $\kappa_{\Delta H}$, and ΔH_{res}.
2. Introduce mutational effect through Eq. 9 that essentially rescales the ΔH_{res} of the mutant based on the experimentally derived $\Delta\Delta G_{eq}$ (i.e., change in stability; Fig. 1a) without changing its curvature (i.e., $\kappa_{\Delta H}$ is fixed to WT values):

$$\Delta H_{res}^{mut} = \Delta H_{res} - (\Delta\Delta G_{eq} * \mathsf{x}/\mathsf{N}) \quad (9)$$

   and generate the free energy profiles of the mutants (Fig. 1c). A small rescaling factor x (one per protein) might need to be introduced to exactly reproduce the mutant stability as ΔH_{res} sets only the magnitude of the enthalpy functional (i.e., at nativeness $n = 0$), while stability is determined by the magnitude of the ΔH_{res} functional at the minimum in free energy on the unfolded well (typically found at $n \sim 0.2$–0.3).

3. Simulate the kinetic relaxation on the predicted mutant free energy profile employing the diffusive method discussed above. Fit the relaxation profiles (Fig. 1d) to single-exponential functions and estimate the mutant relaxation rates $\left(k_{pred}^{mut}\right)$ and hence the difference in activation free energies to folding ($\Delta\Delta G_{fol}$) and unfolding ($\Delta\Delta G_{unfol}$).

3.4 Φ-Values and the Folding Transition State (TS)

What can be inferred from the mutant free energy profiles? In this regard, it is informative to discuss the role of mutational studies in understanding folding mechanisms of single-domain proteins, i.e., ϕ-value analysis. In this methodology, mutations that typically correspond to side chain truncations are introduced in the protein of interest. The effect of such mutations on the stability and relaxation kinetics is studied relative to that of the wild-type and quantified in terms of the parameter ϕ, defined as the ratio of the changes in activation free energy during folding ($\Delta\Delta G_{fol}$) to changes in thermodynamic stability ($\Delta\Delta G_{eq}$) [31]. ϕ-values of *1* and *0* are generally seen as an evidence for a transition state in which the mutated residue is as structured as the folded and as unstructured as the unfolded state, respectively:

$$\phi = \frac{\Delta\Delta G_{fol}}{\Delta\Delta G_{eq}} = 1 - \phi_u = 1 - \frac{\Delta\Delta G_{unfol}}{\Delta\Delta G_{eq}} \quad (10)$$

Interestingly, a survey of 806 mutations from 24 different single-domain proteins that display linear chevron-plots led to a surprising observation: the overall effect of mutations can be simply estimated from the relation: $\Delta\Delta G_{unfol} = 0.76\Delta\Delta G_{eq} \pm 1.8$ kJ mol^{-1} or with a global ϕ-value of 0.24 ($r = 0.9$) from a Brønsted analysis under stabilizing conditions [30]. In other words, ϕ-values are mostly insensitive to the nature or location of the mutation, protein size, or structural class. Calculation of the same under iso-stability conditions increases the average ϕ-value to 0.36 ± 0.11, indicating Hammond behavior as the overall "structure" increases upon destabilization [30]. This remarkably universal behavior suggests that transition-state ensembles of two-state proteins are quite fluid-like with little tertiary structure (and hence the smaller ϕ-value). But why are the mutational effects uniform? Why do they lie between 0.24 and 0.36? And how can we reconcile this observation in terms of the energy landscape theory (see also [28, 34])?

3.5 Estimating Energetic Contributions to Φ-Values from Free Energy Profiles

Equation 10 highlights an important aspect that is generally underappreciated in ϕ-value analysis – the ratios of free energies are usually interpreted in structural terms. Given that protein folding involves the formation and breaking of numerous noncovalent interactions and hence large enthalpic-entropic compensations, any structural information is likely mixed with contributions from numerous factors including experimental noise [37, 38], mutation induced modulation of the folding diffusion coefficients [39], choice of reference conditions [30], Hammond behavior [30, 34, 40, 41], nontrivial changes in the end-states [42, 43], folding pathway heterogeneity contributing to an ensemble of transition state structures [32, 33, 44–46], and finally the nature and propagative effect of mutations [35]. Moreover, in many cases, the ϕ-values have been reported to be less than 0 and greater than 1, making it challenging to reconcile these observations in terms of a unique transition state that is structurally in-between that of a unfolded and folded state [47].

To understand the potential energetic contribution to ϕ-values, we predict the free energy profiles of 806 mutants from 24 different proteins following the protocol discussed in Subheadings 3.1–3.3 (illustrative examples for ubiquitin are shown in Fig. 1c and d) and estimate the relative activation free energies. Such a simple calculation alone reproduces the activation free energy due to unfolding ($\Delta\Delta G_{unfol}$) extremely well ($r \sim 0.91$, $p < 10^{-100}$) with a mean deviation of just ~1.5 kJ mol^{-1} (Fig. 2a). In other words, the wild-type free energy surface alone is sufficient to reproduce the mutational effects on average. The agreement between the

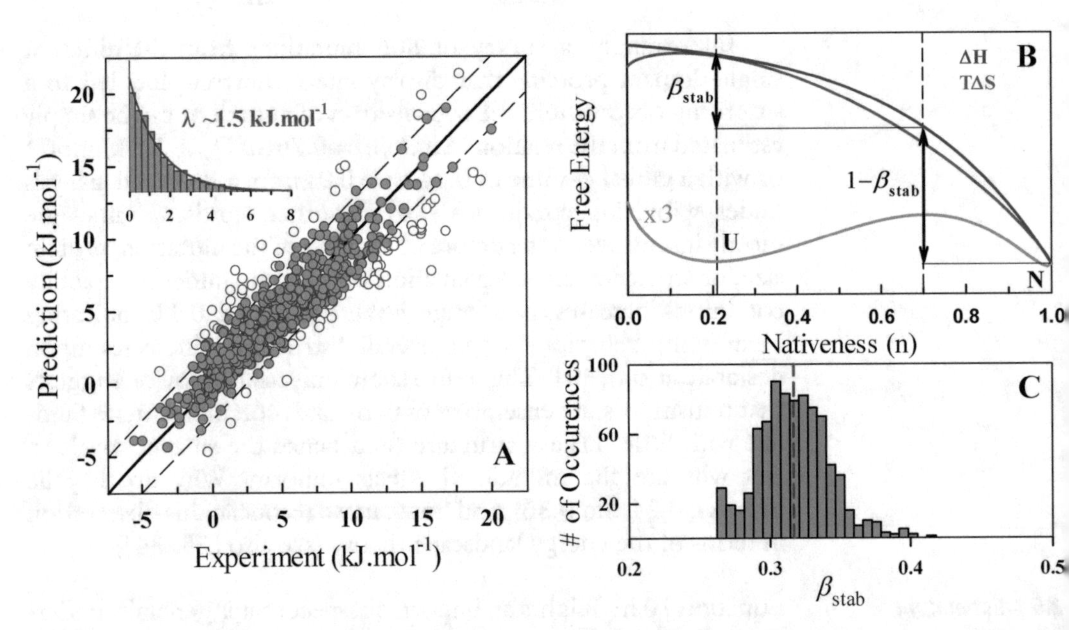

Fig. 2 Energetic contributions to ϕ-values. (**a**) Predicted unfolding activation free energies compared against the experimental values. The continuous black line represents the 1:1 correlation line. The open circles highlight ~10% of mutants (80 out of 806) whose prediction accuracy by the model falls beyond $\pm RT$ (dashed lines). Inset: Exponential-like absolute deviation between experimental and predicted unfolding activation free energies with an average deviation of 1.5 kJ mol^{-1}. (**b**) Estimating β_{stab} given the stabilization energy functional and the free energy profile following Eq. 11. (**c**) The distribution of β_{stab} values upon exactly reproducing the rates and stabilities for the 806 mutants

experimental and predicted values arises from a basic feature of the model defined as β_{stab}:

$$\beta_{stab} = \frac{\Delta H_U - \Delta H_{TS}}{\Delta H_U - \Delta H_N} \tag{11}$$

where the numerator is the degree of stabilization energy gained at the barrier top (note that ΔH_U represents the stabilization energy corresponding to the unfolded state minimum in the free energy profile and not at $n = 0$), and the denominator represents the total change (Fig. 2b) [34]. Thus, β_{stab} quantifies the fraction of stabilization energy gained at the barrier top during the folding process, while $1-\beta_{stab}$ accounts for the fraction lost during unfolding [28, 34]. β_{stab} therefore approximates the experimental global ϕ-value of a protein, i.e., obtained from a Brønsted analysis.

To further characterize the spread in this parameter, we exactly reproduce the mutant relaxation rates and stability employing the same procedure as above. This results in a β_{stab} of ~0.32 ± 0.03 for the 832 mutants and WT proteins, very similar to the mean global ϕ-value under native conditions (Fig. 2c). The minor spread in β_{stab} values is a manifestation of Hammond behavior—upon mutational destabilization, the mutant TS, or barrier top moves towards the native state and this results in a larger gain in the stabilization energy (Fig. 1c and also *see* Ref. 34). The folding transition state therefore acquires only a small and consistent amount of relative stabilization energy at the barrier top, in agreement with the results obtained from the empirical analysis of protein folding rates [28]. Since ϕ-values have also been shown to be insensitive to the nature of mutations, protein structure, or topology [30], the inference is that the tertiary structural information in ϕ-values is averaged out by variations in multiple energetic factors (potentially dominated by local interactions). The TS is therefore likely to be loosely packed and molten-globule–like thus consistent with the ensemble nature expected from the energy landscape theory.

4 Notes

1. If the absolute heat capacity of a protein is available, the FES model can be employed in conjunction with a multi-model Bayesian analysis to predict thermodynamic barrier heights [48]. In the version described above, only relative barrier heights can be extracted as this parameter is ultimately determined by the folding diffusion coefficient and $\kappa_{\Delta H}$.
2. The model is described here in its most simple version, but additional parameters can be introduced to describe solvation effects [16] and chemical denaturation [16] and also to differentiate between local and nonlocal stabilization energies

[18]. The latter approach has been employed to predict folding and unfolding rates of proteins with information on just the length of the protein and structural class [18].

3. It is possible to simultaneously quantify kinetic rate coefficients (as a function of temperature or denaturant) and amplitudes as a function of temperature with the FES model and hence obtain minimalistic free energy profiles and therefore barrier heights [49–51].
4. Three-state or higher order unfolding behavior, i.e., with one or more intermediates, cannot be reproduced with a single $\kappa_{\Delta H}$ value. It should, however, be possible to do so by employing a more complex dependence of enthalpy on the reaction coordinate.
5. It is straightforward to introduce a coordinate dependent diffusion coefficient into the model through an expansion of Eq. 6.
6. The current description does not allow for the presence of an intermediate or a local minimum (three-state behavior), but it can still produce bi-exponential kinetics under two scenarios: if the unfolded well moves with changing stability conditions or if the population on the barrier top is large enough to result in measurable amplitude changes, as in proteins with marginal thermodynamic barriers [52]. The latter case can be observed in the eigenvectors as a change in population across the barrier top and folded well (molecular phase) apart from the conventional exchange between folded and unfolded wells (activated phase). The signal switch functions along the coordinate can be tuned to reproduce the rate and amplitude of such phases [53].

References

1. Bryngelson JD, Onuchic JN, Socci ND, Wolynes PG (1995) Funnels, pathways, and the energy landscape of protein-folding - a synthesis. Proteins 21:167–195
2. Onuchic JN, LutheySchulten Z, Wolynes PG (1997) Theory of protein folding: the energy landscape perspective. Ann Rev Phys Chem 48:545–600
3. Socci ND, Onuchic JN, Wolynes PG (1996) Diffusive dynamics of the reaction coordinate for protein folding funnels. J Chem Phys 104:5860–5868
4. Cho SS, Levy Y, Wolynes PG (2006) P versus Q: structural reaction coordinates capture protein folding on smooth landscapes. Proc Natl Acad Sci U S A 103:586–591
5. Muñoz V, Eaton WA (1999) A simple model for calculating the kinetics of protein folding from three-dimensional structures. Proc Natl Acad Sci U S A 96:11311–11316
6. Henry ER, Eaton WA (2004) Combinatorial modeling of protein folding kinetics: free energy profiles and rates. Chem Phys 307:163–185
7. Doshi U, Muñoz V (2004) Kinetics of alpha-helix formation as diffusion on a one-dimensional free energy surface. Chem Phys 307:129–136

8. Kubelka J, Henry ER, Cellmer T, Hofrichter J, Eaton WA (2008) Chemical, physical, and theoretical kinetics of an ultrafast folding protein. Proc Natl Acad Sci U S A 105:18655–18662
9. Munshi S, Naganathan AN (2015) Imprints of function on the folding landscape: functional role for an intermediate in a conserved eukaryotic binding protein. Phys Chem Chem Phys 17:11042–11052
10. Sivanandan S, Naganathan AN (2013) A disorder-induced domino-like destabilization mechanism governs the folding and functional dynamics of the repeat protein IκBα. PLoS Comput Biol 9:e1003403
11. Naganathan AN, Sanchez-Ruiz JM, Munshi S, Suresh S (2015) Are protein folding intermediates the evolutionary consequence of functional constraints? J Phys Chem B 119:1323–1333
12. Narayan A, Campos LA, Bhatia S, Fushman D, Naganathan AN (2017) Graded structural polymorphism in a bacterial thermosensor protein. J Am Chem Soc 139:792–802
13. Muñoz V (2001) What can we learn about protein folding from Ising-like models? Curr Opin Struct Biol 11:212–216
14. Piana S, Klepeis JL, Shaw DE (2014) Assessing the accuracy of physical models used in protein-folding simulations: quantitative evidence from long molecular dynamics simulations. Curr Opin Struct Biol 24:98–105
15. Naganathan AN (2013) Coarse-grained models of protein folding as detailed tools to connect with experiments. WIREs Comput Mol Sci 3:504–514
16. Naganathan AN, Doshi U, Muñoz V (2007) Protein folding kinetics: barrier effects in chemical and thermal denaturation experiments. J Am Chem Soc 129:5673–5682
17. de Sancho D, Doshi U, Muñoz V (2009) Protein folding rates and stability: how much is there beyond size. J Am Chem Soc 131:2074–2075
18. De Sancho D, Muñoz V (2011) Integrated prediction of protein folding and unfolding rates from only size and structural class. Phys Chem Chem Phys 13:17030–17043
19. Robertson AD, Murphy KP (1997) Protein structure and the energetics of protein stability. Chem Rev 97:1251–1267
20. Ramachandran GN, Ramakrishnan C, Sasisekharan V (1963) Stereochemistry of polypeptide chain configurations. J Mol Biol 7:95–99
21. Naganathan AN, Doshi U, Fung A, Sadqi M, Muñoz V (2006) Dynamics, energetics, and structure in protein folding. Biochemistry 45:8466–8475
22. Doshi U, Muñoz V (2004) The principles of α-helix formation: explaining complex kinetics with nucleation-elongation theory. J Phys Chem B 108:8497–8506
23. Zwanzig R (1995) Simple model of protein folding kinetics. Proc Natl Acad Sci U S A 92:9801–9804
24. Wako H, Saito N (1978) Statistical mechanical theory of protein conformation .2. folding pathway for protein. J Phys Soc Jpn 44:1939–1945
25. Noel JK, Whitford PC, Sanbonmatsu KY, Onuchic JN (2010) SMOG@ctbp: simplified deployment of structure-based models in GROMACS. Nuc Acids Res 38:W657–W661
26. Naganathan AN (2012) Predictions from an Ising-like statistical mechanical model on the dynamic and thermodynamic effects of protein surface electrostatics. J Chem Theory Comput 8:4646–4656
27. Kim J, Keyes T (2007) Inherent structure analysis of protein folding. J Phys Chem B 111:2647–2657
28. Akmal A, Muñoz V (2004) The nature of the free energy barriers to two-state folding. Proteins 57:142–152
29. Lapidus LJ, Steinbach PJ, Eaton WA, Szabo A, Hofrichter J (2002) Effects of chain stiffness on the dynamics of loop formation in polypeptides. Appendix: testing a 1-dimensional diffusion model for peptide dynamics. J Phys Chem B 106:11628–11640
30. Naganathan AN, Muñoz V (2010) Insights into protein folding mechanisms from large scale analysis of mutational effects. Proc Natl Acad Sci U S A 107:8611–8616
31. Fersht AR, Matouschek A, Serrano L (1992) The folding of an enzyme .1. Theory of protein engineering analysis of stability and pathway of protein folding. J Mol Biol 224:771–782
32. Onuchic JN, Socci ND, LutheySchulten Z, Wolynes PG (1996) Protein folding funnels: the nature of the transition state ensemble. Fold Des 1:441–450
33. Naganathan AN, Orozco M (2011) The protein folding transition-state ensemble from a Gō-like model. Phys Chem Chem Phys 13:15166–15174
34. Muñoz V, Sadqi M, Naganathan AN, de Sancho D (2008) Exploiting the downhill folding regime via experiment. HFSP J 2:342–353
35. Rajasekaran N, Suresh S, Gopi S, Raman K, Naganathan AN (2017) A general mechanism for the propagation of mutational effects in proteins. Biochemistry 56:294–305
36. Rajasekaran N, Sekhar A, Naganathan AN (2017) A universal pattern in the percolation

and dissipation of protein structural perturbations. J Phys Chem Lett 8:4779–4784
37. Sanchez IE, Kiefhaber T (2003) Origin of unusual phi-values in protein folding: evidence against specific nucleation sites. J Mol Biol 334:1077–1085
38. De Los Rios MA, Muralidhara BK, Wildes D, Sosnick TR, Marqusee S, Wittung-Stafshede P, Plaxco KW, Ruczinski I (2006) On the precision of experimentally determined protein folding rates and phi-values. Protein Sci 15:553–563
39. Acharya S, Saha S, Ahmad B, Lapidus LJ (2015) Effects of mutations on the reconfiguration rate of α-Synuclein. J Phys Chem B 119:15443–15450
40. Ternstrom T, Mayor U, Akke M, Oliveberg M (1999) From snapshot to movie: phi analysis of protein folding transition states taken one step further. Proc Natl Acad Sci U S A 96:14854–14859
41. Pappenberger G, Saudan C, Becker M, Merbach AE, Kiefhaber T (2000) Denaturant-induced movement of the transition state of protein folding revealed by high-pressure stopped-flow measurements. Proc Natl Acad Sci U S A 97:17–22
42. Sanchez IE, Kiefhaber T (2003) Hammond behavior versus ground state effects in protein folding: evidence for narrow free energy barriers and residual structure in unfolded states. J Mol Biol 327:867–884
43. Cho JH, Raleigh DP (2006) Denatured state effects and the origin of nonclassical phi values in protein folding. J Am Chem Soc 128:16492–16493
44. Klimov DK, Thirumalai D (2001) Multiple protein folding nuclei and the transition state ensemble in two-state proteins. Proteins 43:465–475
45. Best RB, Hummer G (2016) Microscopic interpretation of folding phi-values using the transition path ensemble. Proc Natl Acad Sci U S A 113:3263–3268
46. Gopi S, Singh A, Suresh S, Paul S, Ranu S, Naganathan AN (2017) Toward a quantitative description of microscopic pathway heterogeneity in protein folding. Phys Chem Chem Phys 19:20891–20903
47. Raleigh DP, Plaxco KW (2005) The protein folding transition state: what are phi-values really telling us? Prot Pept Lett 12:117–122
48. Naganathan AN, Perez-Jimenez R, Muñoz V, Sanchez-Ruiz JM (2011) Estimation of protein folding free energy barriers from calorimetric data by multi-model Bayesian analysis. Phys Chem Chem Phys 13:17064–17076
49. Fung A, Li P, Godoy-Ruiz R, Sanchez-Ruiz JM, Muñoz V (2008) Expanding the realm of ultrafast protein folding: gpW, a midsize natural single-domain with alpha+beta topology that folds downhill. J Am Chem Soc 130:7489–7495
50. Li P, Oliva FY, Naganathan AN, Muñoz V (2009) Dynamics of one-state downhill protein folding. Proc Natl Acad Sci U S A 106:103–108
51. Naganathan AN, Muñoz V (2014) Thermodynamics of downhill folding: multi-probe analysis of PDD, a protein that folds over a marginal free energy barrier. J Phys Chem B 118:8982–8994
52. Yang WY, Gruebele M (2003) Folding at the speed limit. Nature 423:193–197
53. DeCamp SJ, Naganathan AN, Waldauer SA, Bakajin O, Lapidus LJ (2009) Direct observation of downhill folding of lambda-repressor in a microfluidic mixer. Biophys J 97:1772–1777

Chapter 22

Localization of Energetic Frustration in Proteins

A. Brenda Guzovsky, Nicholas P. Schafer, Peter G. Wolynes, and Diego U. Ferreiro

Abstract

We present a detailed heuristic method to quantify the degree of local energetic frustration manifested by protein molecules. Current applications are realized in computational experiments where a protein structure is visualized highlighting the energetic conflicts or the concordance of the local interactions in that structure. Minimally frustrated linkages highlight the stable folding core of the molecule. Sites of high local frustration, in contrast, often indicate functionally relevant regions such as binding, active, or allosteric sites.

Key words Protein folding, Protein function, Local frustration

1 Introduction

Biomolecules are made up of diverse parts, each falling in place, where small details of how they are put together are essential for action and life. Frustration occurs when a physical system is not able to simultaneously achieve the minimum energy for each and every part of it considered in isolation [1]. Frustration can arise from geometry or from competition between the interactions of the system's elements. The Energy Landscape Theory of protein folding applies this concept to protein molecules and provides powerful tools for understanding these evolved systems [2, 3]. Proteins are rare and wonderful polymers. They perform defined physical tasks that are seldom found by chance, such as accurate folding, specific binding, and powerful catalysis. These chemical activities are the outcome of protein natural history and must be, at least partly, encoded in their sequences. In order to fold robustly, proteins must satisfy a large number of local interactions simultaneously, an optimization task feasible when frustration is low [4, 5]. Other chemical activities that must be performed, related to biological function, impose further restrictions on the sequences, possibly conflicting with the necessity of self-assembly [6]. Therefore, looking for deviations from the expected structural stability upon folding can

Victor Muñoz (ed.), *Protein Folding: Methods and Protocols*, Methods in Molecular Biology, vol. 2376,
https://doi.org/10.1007/978-1-0716-1716-8_22,

give us hints that other teleonomic goals might be at play. Locating the frustration between conflicting goals in a recurrent system leads to fundamental insights about the chances and necessities that shape the encoding of biological information [7].

Since the global structure of a folded protein chain is the outcome of the cooperation among many local interacting units, it appears possible to decompose the global energy function into parts and quantify local conflicts as modulations of the energy change. The choice of the division into parts depends on the intention of the study, and it should be kept in mind that no subunit can be considered completely isolated. One could choose to quantify frustration of whole proteins, of domains or maybe even of "foldons", i.e., minimal folding units [8], or perhaps even go further down to contacts between amino-acid residues or atoms within residues. Having a reliable way of measuring the overall free energy of a protein structure at a given level of resolution allows one to explore how the free energy varies when the sequence or the structure of the protein changes. Years ago, a simple heuristic method to explore these relations was presented [9, 10]. To analyze the existence of energetic conflicts in a folded protein, the energy of structural or sequence decoys is determined and compared to that of the native state. A local frustration index is defined as the Z-score of the free energy of parts of the native structure with respect to the distribution of the energies of rearranged decoys. Interactions are divided into classes as being minimally frustrated, neutral, or highly frustrated according to this index. In this chapter, we will discuss the strategy to quantify local frustration, how to calculate a local frustration index, and how to analyze the results in relation to protein functions.

2 Materials

The basic elements needed for frustration analysis are a protein structure, an energy function (or, more precisely, a solvent-averaged free energy function) and a set of decoy protein structures. These elements should agree between each other in the level of description (or coarse-graining) of the system (Fig. 1). For example, if the energy function only takes into account the positions of the Cα and Cβ, a fully atomistic description of the protein structure is unnecessary. Alternatively, if a detailed quantum-mechanical energy function is desired, highly detailed structures of protein and solvent would be required. The choice of coarse-graining will depend on the specific questions being asked about particular systems. We will use as an example of the strategy the protein model implemented in the AWSEM system [11], although the basic strategy should be applicable at every level. Recently, an implementation of local frustration analysis at the all-atom level was presented [12].

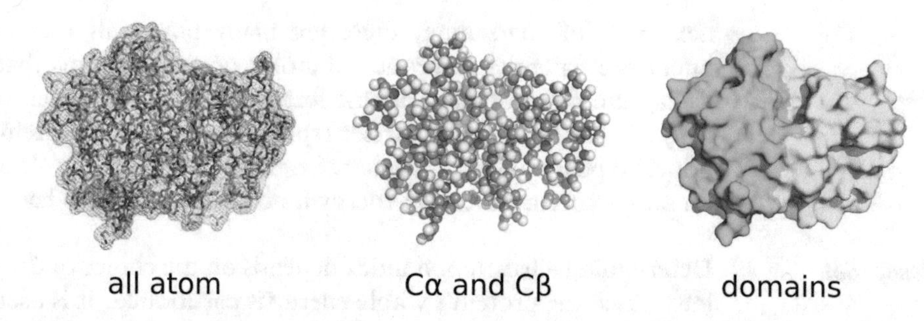

Fig. 1 Proteins can be described using different levels of detail, depending on the specific questions for the system. These are only three of the many possible representations for the structure of *B. Licheniformis* Beta-Lactamase (pdb 4BLM). An all-atom representation (*left*) colored according to atom type. A coarse-grained representation (*middle*) showing only the Alpha (dark gray) and Beta (light gray) Carbons. A domain-level representation (*right*) showing only the two subdomains of the protein without any atomic detail

2.1 Protein Structure

A physical description of the native reference structure of the protein is required. The RCSB Protein Data Bank [13] contains about 10^5 structures, where the interested researcher may find a satisfactory model. If there is no available experimental high-resolution structure, the analysis can alternatively be well performed with a good structural prediction, whose level of resolution will depend on the energy function to be used. We found that simple threading of a sequence in a known homologous structure is generally sufficient to evaluate the frustration of most proteins (*see* **Note 1**). In general, we recommend performing various "sanity checks" of the modeled structures [14], such as a thorough visual inspection and a structural comparison with the template and homologs [15].

2.2 Energy Function

There are several ways to quantify the "energy" of a particular protein structure, with varying degrees of detail and success in application, mostly depending on the question being asked. To quantify local frustration, an energy function is needed that can reliably assess how funnel-like a protein landscape is. Current approaches to analyze frustration from the ground-up, such as quantum-mechanical methods of energy evaluation, are still too computationally expensive for large macromolecules and, in any event, would require solvent averaging to be meaningful. Such a level of description is probably unnecessary for most applications related with the evaluation of local energetic frustration at the multiatom level. Evolution works on the residue level, not the atomic scale! So while there is the possibility to employ an all-atom energy function, like those routinely used to analyze the dynamics of folded proteins [16, 17], simulating completely atomistic models in their aqueous environment and with sufficient sampling to evaluate distinct conformational states still takes a lot of computing time and may not bring too many insights [18]. At the next level of coarsening, there are many potentials that assign interaction energies to groups of atoms or pseudoatoms that aim to capture most of the relevant features of the energy landscape [19, 20] (*see* **Note 2**). A coarser representation of the protein is in principle possible (pairs, triplets, or groups of residues), although no satisfactory transferable energy function is at hand today.

2.3 Decoy Set

Determining a frustration index depends on the choice of the parts into which the protein's whole energy is partitioned. It is useful to divide the energy up in a way that is at least roughly comparable to the energetic changes that occur in proteins in its natural or molecular history. Generally, one can examine the differences in energy upon changing the protein sequence, as relevant in evolution, or the energy changes in changing the structure, as relevant to dynamical motions. Here we present these two complementary ways for localizing frustration that differ in the way the set of decoys is constructed (Fig. 2):

Mutational frustration calculations are used to answer the question, "How favorable are the native residues relative to other residues that might have been found in that location?". For mutational frustration calculations, the decoy set is made by changing the identities of the interacting amino acids in every contact-pair or group of residues, fixing the other structural parameters at the reference value (*see* **Note 3**). Configurational frustration calculations are used to answer the question, "How favorable are the native interactions between two or more residues relative to other interactions these residues might have formed in other compact structures as the molecule folds?" For configurational frustration calculations, the decoy set is made such that the energy of the contact-pair or group of residues can be measured in globally different compact structures.

3 Methods

The general procedure for localizing energetic frustration consists of seven steps.

1. Get native (reference) structure.
2. Evaluate energy of reference.
3. Generate decoys.
4. Evaluate energies of decoys.
5. Assign frustration index.
6. Visualize and analyze results.
7. Interpret results in biological context.

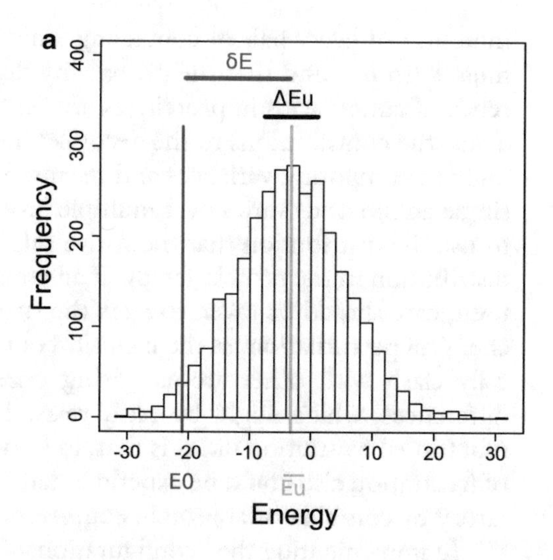

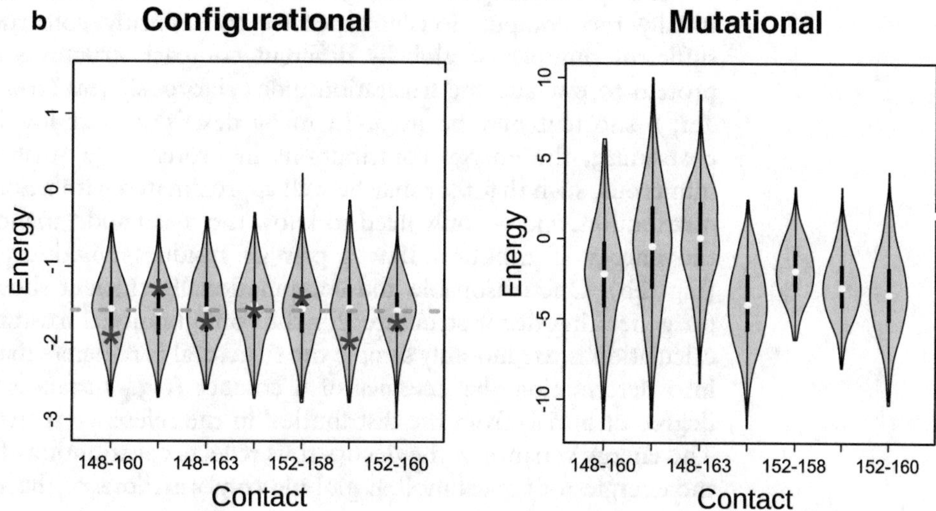

Fig. 2 (**a**) Energy distribution for the configurational decoys of an interaction. The energies have an approximately Gaussian distribution, with a mean $\bar{E}_u$ (orange) and a standard derivation ΔE_u (black). E_0 (blue) is the native energy of the protein. For the contact to contribute to robust folding, the native and mean decoy energy should be well separated. δE (gray) is the gap between these two energies. (**b**) Distribution of the decoy energies for a series of contacts. Both the configurational (*left*) and mutational (*right*) decoy energies are shown

Steps 1–3 are discussed in the previous section (*see* Subheading 2). **Steps 2–6** are implemented with the AWSEM energy function in a web server (frustratometer.tk) freely available for the community [10]. A stand-alone version is also available for download.

3.1 Energies of Decoys

For evaluating the energy of the decoys, care should be taken that the energy distribution of decoys reflects the natural distribution under scrutiny. In the "mutational frustration" scheme, exhaustive mutation of every pair of contacting amino acids is made, but we must keep in mind that the probability distribution of the occurrence of amino acids in proteins is not uniform [21] (*see* **Note 4**)! Thus, the contributions to the decoy set must be weighted accordingly in computing variances and means. Since the distribution of single amino acids varies over multiple protein families, we suggest to use the distribution that the native reference protein has or the distribution in the protein family. If an atomistic energy function is used, care should be taken to relax the structure upon making the chemical perturbation, as the inclusion of bulky residues may sterically clash with other atoms, giving unrealistically large energy differences, which might be easily relaxed. One advantage of the mutational frustration index is that, in principle, this local measure of frustration also could be experimentally determined in the laboratory by combinatorial protein engineering.

In implementing the "configurational frustration" scheme, it is usually too computationally expensive to explicitly construct a sufficient number of globally different compact structures of a protein to evaluate the frustration index rigorously (*see* **Note 5**). Yet, a shortcut may be used. In many descriptions at low-level coarsening, the energy contributions in proteins are small and numerous, such that they may be well approximated with a normal distribution, and we only need to know the mean and variance of the energy distribution that a pair of residues could explore (Fig. 2b). One reasonable and computationally efficient shortcut for generating decoy structures for the configurational frustration calculation is to randomly sample the structural parameters that go into determining the energies of a contact (e.g., distances and degree of burial) from the distribution in the reference structure. The energy variance in the decoys thus reflects contributions from the energies of typical molten globule conformations of the same polypeptide chain.

3.2 Frustration Index

A frustration index (Fi) is defined as the Z-score of the native reference energy to the decoys:

$$Fi = \frac{E_0 - \bar{E}_u}{\Delta E_u} \tag{1}$$

where E_0 is the native energy, $\bar{E}_u$ is the mean of the decoy energies, and ΔE_u is the square root of the variance of the decoy energies. In the case of the "mutational frustration" scheme, it is worth noting that the energy change upon pair-mutation not only comes directly from the particular interaction that is probed but it also changes through interactions of each residue with other residues not in the pair, as those contributions may also vary upon mutation. These changes are globally assigned to the pair-contact probed. In the "configurational frustration" scheme, the frustration index is assigned to each contact by quantifying its native energy relative to the structural decoys.

The frustration index takes on a continuous range of values. In order to decide whether the classified interactions favor robust folding of a domain or are in energetic conflict with folding, some cutoffs in the scale need to be made. We propose cutoffs that are not unique but that we have found to be useful for understanding patterns of localized frustration in many proteins. These cutoff values are physically motivated within the Energy Landscape Theory and are based on considering the expected ground state energy of a random heteropolymer. The basic parameters that go into determining these cutoffs are the ratio of the folding and glass transition temperatures of evolved proteins (Tf/Tg) and the configurational entropy change upon forming a pairwise interaction [22]. Several independent approximations to the Tf/Tg ratio indicate that the original estimate for the ratio of 1.6 is pretty good [23]. The entropy change for forming a single pairwise contact depends on the sequence separation between the residues and on the degree of consolidation of neighboring interactions. Ignoring these complexities, a crude upper limit for configurational entropy change is just the sum of the entropic cost for fixing individual parts, which was estimated to be ≈0.6 kb per residue in the case of amino acid-based coarsening [24]. Thus, to be minimally frustrated and fold at Tf, a given contact cannot be too unstable and must overcome:

$$E_0 > \bar{E}_u + \delta E \times \frac{Tf}{Tg} \times \sqrt{\frac{2S}{kb}x}\frac{1}{\sqrt{2}} \tag{2}$$

where δE is the energy gap between E_0 and Eg, the energy of the lowest energy misfolded state, and S is the configurational entropy change for forming a single pairwise contact. Note that this is not exactly the same δE as depicted in Fig. 2a, although the two are closely related. Satisfying this equation corresponds to having a frustration index larger than 0.78.

Conversely, a contact will be defined as highly frustrated, if E_0 is at the other end of the distribution with a local frustration index lower than -1; that is, unlike for a minimally frustrated pair, most other amino-acid pairs at that location would be more favorable for folding than the existing pair by more than one standard deviation of that distribution. Neutral interactions are defined to lie near the center of the distribution of possible energies in compact decoy states, between the highly and minimally frustrated cutoffs. Variations of these cutoffs can readily be applied and investigated, according to the variations given by the granularity and the energy function used.

3.3 Visualizing and Analyzing Results

An important aspect of frustration analysis is the visualization of the quantitative results. An intuitive representation highlighting these results can be given by coloring the pairwise contacts in the three-dimensional structure of the reference. Several computer visualization programs allow one to draw lines between atoms in a protein structure that can be used to represent "contacts" in 3D space. A psychologically (but unfortunately color-blind unfriendly) coloring scheme that became standard is to use green to represent favorable minimally frustrated contacts ("go" for folding) and red to highlight unfavorable highly frustrated contacts ("stop" for folding) (Fig. 3a) (*see* **Note 6**). The frustration patterns that emerge upon exploring the structures of particular systems often can spark the comprehension of the physiology of the protein under scrutiny, at least for experts on that system. An alternative representation of frustration patterns is provided by a "contact map" in which a matrix of interactions between every residue-pair is drawn and each contact (a point in the plot) colored according to the frustration index (Fig. 3b). Although natural proteins are clearly not one-dimensional objects [25], in order to compare the frustration results with other common sequence-based analysis tools, a 1D representation of frustration index along a primary structure may also be of use. One way of collapsing the local frustration information into one dimension is to count the number of contacts that fall into each frustration class around each amino acid (Fig. 3c) (*see* **Note 7**).

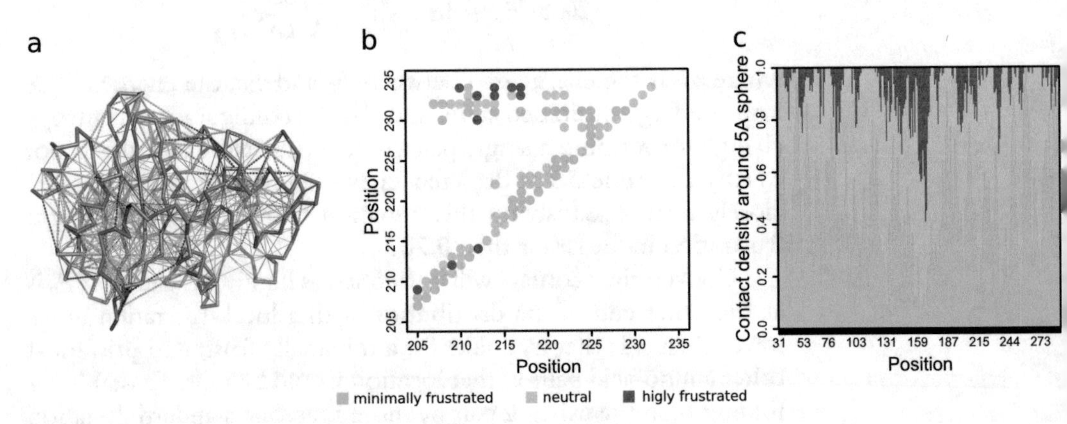

Fig. 3 Local frustration can be visualized in 3D, 2D, or 1D. In all cases, minimal frustration is represented in green, high frustration in red, and neutral frustration in gray. (**a**) Contacts are drawn on top of the 4BLM structure and colored according to their mutational frustration index. Neutral contacts are not shown in order to better enable the analysis of minimally and highly frustrated regions. (**b**) A portion of the contact map for 4BLM. Each dot represents a contact, which is colored according to its mutational frustration index. Contacts near the diagonal are short range, while those farther away from the diagonal make up elements of the secondary and tertiary structure. (**c**) The density of contacts of each frustration type, according to the mutational frustration index, around 5 Å of every residue. This is one way of representing the frustration index on a 1D sequence

Once the general aspects of the local frustration distributions are grasped, global statistical analysis of the interesting features has been performed. For example, minimally frustrated interactions tend to be found crosslinking the interior of domains, while highly frustrated interactions are found in patches at the surface of these domains [9]. A quantification of these observations can be made by calculating the pair-distribution functions of the contacts, either between each other or between these and other marked regions [26]. Analysis of the local frustration patterns over many members of protein families can identify the invariant aspects of the energy distributions and lead to the understanding of structure–function relationships [27] (*see* **Note 8**).

3.4 Interpreting Results

Interpreting the results in the biological context is the most challenging and interesting part of frustration analysis. Today, this is done by humans integrating knowledge from different sources about the particular system we are interested in. Anecdotes of local frustration distribution may be very useful to complement experimental and theoretical findings. We strongly recommend that the analysis is made having a clear hypothesis in mind with specific questions and tackling the analysis with statistics and appropriate controls [28].

The overall distribution of local frustration in proteins domains may be a useful guide for analyzing specific systems. In general, about 40% of the native interactions found in natural globular domains fall in the "minimally frustrated" class [7]. About half of the interactions can be labeled as neutral as they do not contribute distinctively to the total energy change, and around 10% of the interactions are found to be "highly frustrated." These are regions in which most local sequence or structural changes would lower the free energy of the system implying that these patterns have been held in place over evolutionary time as well as physiological time at the expense of other interactions, as they conflict with the robust folding of a domain. The adaptive value for a molecule to tolerate spatially localized frustration arises from the way such frustration sculpts protein dynamics for specific functions. In a monomeric protein, the alternate configurations caused by locally frustrating an otherwise largely unfrustrated structure provide specific control of the thermal motions, guiding them in useful directions. Alternatively, a site that is frustrated in a monomeric protein may become less frustrated in the larger assembly of this protein with partners, thus guiding specific association. For a detailed discussion of the basics of frustration biophysics, the reader is referred to ref. [7] and to a recent review of the outstanding applications [26].

4 Notes

1. Many algorithms are available to automatically generate protein models, with varying degrees of success [29]. We have found that the Modeller suite is reliable enough for most purposes. As in most protein model-building schemes, care should be taken in the choice of template structures, the detection of remote homologs, the crucial sequence-alignment result, and the completion of gaps, loops, and missing residues. None of these computational problems are today completely solved, so watch your tools!
2. To calculate energies, we use the AWSEM potential, which was effectively constructed to infer a transferable potential for protein folding [11, 19]. The forcefield treats the polymer as being composed of three atoms per amino acid including a side chain of a single Cβ whose interactions are amino-acid type-dependent [30]. The nonbonded interactions of the Cβ can fall in any of three "wells" of pairwise contacts—short range, long range, and water-mediated. The side chain degrees of freedom including the rotamer state have been effectively averaged over and are implicitly encoded in the Cβ interaction parameters.
3. A similar analysis can be carried out by mutating single residues. In this case, the set of decoys is constructed by shuffling the identity of the single amino acid i, keeping all other parameters and neighboring residues in the native state, and evaluating the total energy change upon mutation. We call the resulting ratio the "single-residue frustration index."
4. For the mutational scheme, one effective way to take into account the different amino-acid frequencies is to sample the space of mutations with a weighted probability for each amino acid. More detailed calculations could take into account the frequency of occurrence of pairs of amino acids or the frequencies at each position in the family, if known.
5. For the configurational decoys in the case of AWSEM coarsening, the residues i,j are not only changed in identity but also are displaced in location, randomizing the distances $r_{i,j}$ and densities ρ_i of the interacting amino acids according to the reference distribution.
6. To visualize the frustration upon the structure, we typically do not draw the neutral contacts, as the multitude of these cover over the more interesting minimally frustrated or highly frustrated contacts.
7. Care should be taken in analyzing this collapsed representation of the contacts, as their density is typically not uniform along the sequence and can lead to distortions in the interpretation of

the plots. It should be noted that the frustration index assigned to each contact is not additive with respect to the amino-acid pair, which precludes the averaging of the frustration indexes of the contacts that each residue contributes.

8. While analyzing the local frustration patterns in a protein family, care should be taken in the alignment of the sequences or structures used, the numbering schemes, indels, or gaps, and other bioinformatic details that often complicate and can even preclude the analysis. Serious bookkeeping may be tedious, but it is essential for making robust discoveries!

Acknowledgments

This work was supported by the Consejo de Investigaciones Científicas y Técnicas (CONICET); the Agencia Nacional de Promoción Científica y Tecnológica [PICT2016-1467 to D.U.F.] and Universidad de Buenos Aires (UBACYT 2018—20020170100540BA). Additional support was provided by D. R. Bullard-Welch Chair at Rice University [Grant C-0016 to P.G.W.]. We thank R. Gonzalo Parra, Maria Freiberger, Joe Hegler and Nacho Sanchez for their contribution and suggestions to this work.

References

1. Vannimenus J, Toulouse G (1977) Theory of the frustration effect. II. Ising spins on a square lattice. J Phys C: Solid State Phys 10: L537–L542
2. Wolynes PG (2015) Evolution, energy landscapes and the paradoxes of protein folding. Biochimie 119:218–230
3. Wei G, Xi W, Nussinov R, Ma B (2016) Protein ensembles: how does nature harness thermodynamic fluctuations for life? The diverse functional roles of conformational ensembles in the cell. Chem Rev 116:6516–6551
4. Bryngelson JD, Wolynes PG (1987) Spin glasses and the statistical mechanics of protein folding. Proc Natl Acad Sci U S A 84:7524–7528
5. Tzul FO, Vasilchuk D, Makhatadze GI (2017) Evidence for the principle of minimal frustration in the evolution of protein folding landscapes. Proc Natl Acad Sci U S A 114: E1627–E1632
6. Lubchenko V (2008) Competing interactions create functionality through frustration. Proc Natl Acad Sci U S A 105:10635–10636
7. Ferreiro DU, Komives EA, Wolynes PG (2014) Frustration in biomolecules. Q Rev Biophys 47:285–363
8. Panchenko AR, Luthey-Schulten Z, Wolynes PG (1996) Foldons, protein structural modules, and exons. Proc Natl Acad Sci U S A 93:2008–2013
9. Ferreiro DU, Hegler JA, Komives EA, Wolynes PG (2007) Localizing frustration in native proteins and protein assemblies. Proc Natl Acad Sci U S A 104:19819–19824
10. Parra RG, Schafer NP, Radusky LG, Tsai M-Y, Brenda Guzovsky A, Wolynes PG, Ferreiro DU (2016) Protein Frustratometer 2: a tool to localize energetic frustration in protein molecules, now with electrostatics. Nucleic Acids Res 44:W356–W360
11. Davtyan A, Zheng W, Schafer N, Wolynes P, Papoian G (2012) AWSEM-MD: coarse-grained protein structure prediction using

physical potentials and Bioinformatically based local structure biasing. Biophys J 102:619a

12. Chen M, Chen X, Schafer NP, Clementi C, Komives EA, Ferreiro DU, Wolynes PG (2020) Surveying biomolecular frustration at atomic resolution. Nat Commun 11(1):5944
13. Berman HM (2000) The Protein Data Bank. Nucleic Acids Res 28:235–242
14. Bank RPD RCSB PDB. https://www.rcsb.org/pdb/static.do?p=software/software_links/analysis_and_verification.html. Accessed 16 Dec 2017
15. Sippl MJ, Wiederstein M (2008) A note on difficult structure alignment problems. Bioinformatics 24:426–427
16. Kurplus M, McCammon JA (1983) Dynamics of proteins: elements and function. Annu Rev Biochem 52:263–300
17. Lindorff-Larsen K, Maragakis P, Piana S, Eastwood MP, Dror RO, Shaw DE (2012) Systematic validation of protein force fields against experimental data. PLoS One 7:e32131
18. Pan AC, Weinreich TM, Piana S, Shaw DE (2016) Demonstrating an order-of-magnitude sampling enhancement in molecular dynamics simulations of complex protein systems. J Chem Theory Comput 12:1360–1367
19. Schafer NP, Kim BL, Zheng W, Wolynes PG (2014) Learning to fold proteins using energy landscape theory. Isr J Chem 54:1311–1337
20. Capelli R, Paissoni C, Sormanni P, Tiana G (2014) Iterative derivation of effective potentials to sample the conformational space of proteins at atomistic scale. J Chem Phys 140:195101
21. Krick T, Verstraete N, Alonso LG, Shub DA, Ferreiro DU, Shub M, Sánchez IE (2014) Amino acid metabolism conflicts with protein diversity. Mol Biol Evol 31:2905–2912
22. Plotkin SS, Wang J, Wolynes PG (1996) Correlated energy landscape model for finite, random heteropolymers. Phys Rev E Stat Phys Plasmas Fluids Relat Interdiscip Topics 53:6271–6296
23. Onuchic JN, Wolynes PG, Luthey-Schulten Z, Socci ND (1995) Toward an outline of the topography of a realistic protein-folding funnel. Proc Natl Acad Sci U S A 92:3626–3630
24. Luthey-Schulten Z, Ramirez BE, Wolynes PG (1995) Helix-coil, liquid crystal, and spin glass transitions of a collapsed Heteropolymer. J Phys Chem 99:2177–2185
25. Chowdary PD, Gruebele M (2009) Molecules: what kind of a bag of atoms? J Phys Chem A 113:13139–13143
26. Ferreiro DU, Komives EA, Wolynes PG (2017) Frustration, function and folding. Curr Opin Struct Biol 48:68–73
27. Parra RG, Gonzalo Parra R, Espada R, Verstraete N, Ferreiro DU (2015) Structural and energetic characterization of the Ankyrin repeat protein family. PLoS Comput Biol 11:e1004659
28. Brenner S (2010) Sequences and consequences. Philos Trans R Soc Lond Ser B Biol Sci 365:207–212
29. Schwede T (2013) Protein modelling: what happened to the "protein structure gap"? Structure 21:1531–1540
30. Papoian GA, Ulander J, Wolynes PG (2003) Role of water mediated interactions in protein–protein recognition landscapes. J Am Chem Soc 125:9170–9178

Chapter 23

Modeling the Structure, Dynamics, and Transformations of Proteins with the UNRES Force Field

Adam K. Sieradzan, Cezary Czaplewski, Paweł Krupa, Magdalena A. Mozolewska, Agnieszka S. Karczyńska, Agnieszka G. Lipska, Emilia A. Lubecka, Ewa Gołaś, Tomasz Wirecki, Mariusz Makowski, Stanisław Ołdziej, and Adam Liwo

Abstract

The physics-based united-residue (UNRES) model of proteins (www.unres.pl) has been designed to carry out large-scale simulations of protein folding. The force field has been derived and parameterized based on the principles of statistical-mechanics, which makes it independent of structural databases and applicable to treat nonstandard situations such as, proteins that contain D-amino-acid residues. Powered by Langevin dynamics and its replica-exchange extensions, UNRES has found a variety of applications, including ab initio and database-assisted protein-structure prediction, simulating protein-folding pathways, exploring protein free-energy landscapes, and solving biological problems. This chapter provides a summary of UNRES and a guide for potential users regarding the application of the UNRES package in a variety of research tasks.

Key words Coarse graining, Molecular dynamics simulations, Protein folding, Protein–structure prediction, Protein dynamics

1 Introduction

Coarse-grained protein models, in which groups of atoms are combined into single interaction sites, have since long been developed and employed in the studies of protein structure, dynamics, and biochemical processes involving proteins, as well as in the prediction of protein structures [1]. Owing to the reduction of the polypeptide-chain representation, the timescale of simulations with coarse-grained models is 3–4 orders of magnitude greater compared to that of all-atom simulations, although this comes at the inevitable expense of accuracy. Compared to the design of all-atom force fields, the design of a coarse-grained force field poses a far greater problem because the interaction sites are more

Victor Muñoz (ed.), *Protein Folding: Methods and Protocols*, Methods in Molecular Biology, vol. 2376,
https://doi.org/10.1007/978-1-0716-1716-8_23,

complex. Statistical (database-derived) potentials are often used with success, while simple elastic-network potentials are very successful in studying protein fluctuations. The derivation of potentials that connect the coarse-grained representation to the all-atom one (the physics-based potential) is most difficult, because of the necessity of finding such a mapping, doing expensive all-atom calculations, and finding structural and thermodynamic data to parameterize the potentials. One such example is the well-known and widely used MARTINI force field [2], which uses functional expressions imported from all-atom force fields that are parameterized to reproduce the distribution functions and other quantities obtained from all-atom simulations and the experimental thermodynamic data of small molecules. MARTINI has been very successful in the simulations of lipid membranes, vesicles, etc.; however, to use it in the simulations of proteins, restraints have to be imposed to keep a structure native-like. For more information about the existing coarse-grained force fields for proteins, see the recent excellent review by Kmiecik et al. [1].

The united-residue (UNRES) model of proteins and force field developed in our laboratory [3] has probably the most rigorous connection to the physics of protein interactions. UNRES is part of the Unified Coarse-Grained Model of biological macromolecules which is being developed in our laboratory and covers proteins, nucleic acids, polysaccharides, and lipid membranes [4]. UNRES is based on the expansion of the potential of mean force (PMF) of polypeptide chains in water into Kubo cluster-cumulant functions [5], which are further identified with particular coarse-grained energy terms [6, 7]. Because of the close connection to the physics of given interactions, UNRES has been successful in a wide range of applications, including ab initio, database- and experimental-data-assisted prediction of protein structures [8–13], simulating protein folding pathways, kinetics, and free-energy landscapes [14–17], including the formation and breaking of disulfide bonds [18], and studying biological processes [19–22]. This chapter provides a guide that gives a succinct description of the UNRES model, the availability and how to install the UNRES software, as well as to some applications of UNRES in solving concrete problems. For more details on the theory of UNRES, the reader is referred to our recent work [7], while a more comprehensive description of UNRES is contained in an earlier book chapter [3] and review article [4].

2 Methods

2.1 UNRES Model

1. A polypeptide chain is reduced to α-carbon (C^{α}) trace with united side chains (SC) attached to the C^{α}s and united peptide groups (p) positioned in the middle between the two consecutive C^{α} atoms [3, 4] (Fig. 1).

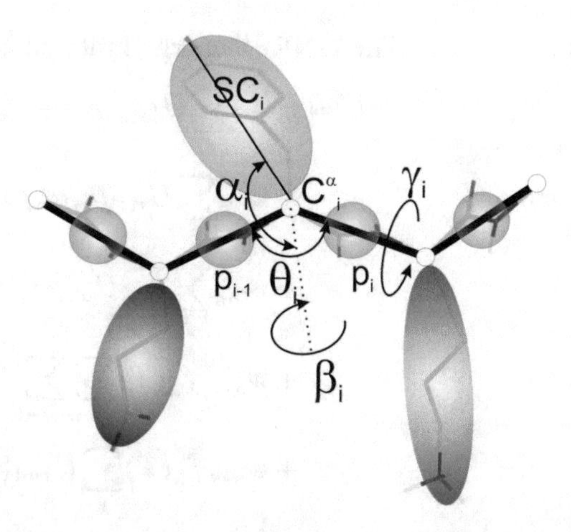

Fig. 1 UNRES model of polypeptide chains. The interaction sites are united peptide groups located between consecutive α-carbon atoms (light-blue spheres) and united side chains attached to the α-carbon atoms (spheroids with different colors and dimensions). Backbone geometry of the simplified polypeptide chain is defined by the $C^\alpha \cdots C^\alpha \cdots C^\alpha$ virtual-bond-angles θ (θ_i has the vertex at $C^\alpha{}_i$) and the $C^\alpha \cdots C^\alpha \cdots C^\alpha \cdots C^\alpha$ virtual-bond-dihedral angles γ (γ_i has the axis passing through $C^\alpha{}_i$ and $C^\alpha{}_{i+1}$). The local geometry of the *i*th side-chain center is defined by the zenith angle α_i (the angle between the bisector of the respective angle θ_i and the $C^\alpha{}_i \cdots SC_i$ vector) and the azimuth angle β_i (the angle of counter-clockwise rotation of the $C^\alpha{}_i \cdots SC_i$ vector about the bisector from the $C^\alpha{}_{i-1} \cdots C^\alpha{}_i \cdots C^\alpha{}_{i+1}$ plane, starting from $C^\alpha{}_{i-1}$. For illustration, the bonds of the all-atom chains, except for those to the hydrogen atoms connected with the carbon atoms, are superimposed on the coarse-grained picture. (Reproduced with permission from Zaborowski et al., J. Chem. Inf. Model. **55**, 2050 (2015). Copyright 2015 American Chemical Society)

2. The SCs and ps are the sole interaction sites, and the C^α atoms function is to define the backbone geometry.
3. The variables are the $C^\alpha \cdots C^\alpha$ and $C^\alpha \cdots SC$ virtual-bond vectors; these variables are used both in energy minimization and molecular dynamics.
4. For glycine, the SC center sits on the corresponding C^α atom. A "glycine" residue is placed as the first/last residue of the chain if the first/last peptide group is present (e.g., when the chain is terminally blocked with the *N*-acetyl group or *N*-methyl group), otherwise the first/last residue is a dummy residue with a dummy peptide group. The reason for this is that each side chain in UNRES has to have a preceding and succeeding residue to define its geometry.
5. Multichain systems are treated with UNRES [23]. Initially, restraints were imposed on the distances between the chains to keep a system together [23], and recently, periodic-boundary conditions and cutoff on nonbonded interactions have been introduced [24].

2.2 UNRES Force Field

The UNRES energy function is expressed by Eq. 1 [3, 4, 6, 7].

$$
\begin{aligned}
U = {} & w_{bond}\sum_i U_{bond}(d_i) + w_{bond}\sum_i U_b(d_i,\theta_i,\gamma_{i-1},\gamma_i) \\
& + w_{rot}\sum_i U_{rot}(\theta_i,\alpha_i,\beta) + w_{SCSC}\sum_j\sum_{i<j} U_{SC_iSC_j} \\
& + w_{SCp}\sum_j\sum_{i<j} U_{SC_iSC_j} + w_{pp}^{el} f_2(T)\sum_j\sum_{i<j-1} U_{p_ip_j}^{el} \\
& + w_{pp}^{vdW} f_2(T)\sum_j\sum_{i<j-1} U_{p_ip_j}^{vdW} + w_{tor} f_2(T)\sum_i U_{tor}(\gamma_i) \\
& + w_{tord} f_3(T)\sum_i U_{tord}(\gamma_i,\gamma_{i+1}) \\
& + w_{torSC} f_2(T)\sum_i U_{torSC;\,i,i+1} + \sum_{m=2}^{N_{corr}} w_{corr}^{(m)} f_m(T) U_{corr}^{(m)} \\
& + \sum_{\substack{disulfide \\ bonds}} U_{SS_i}
\end{aligned} \tag{1}
$$

1. The first row of Eq. 1 groups the "virtual-bond stretch" (U_{bond}) and (backbone) "virtual-bond-angle" (U_b) terms, while the U_{rot} terms account for the totality of interactions within the *i*th side chain and its immediate backbone neighborhood. In contrast to all-atom force fields, the expressions for U_{bond}, U_b, and U_{rot} are not simple harmonic potentials and possess, in general, multiple minima. The expressions for these terms were parameterized based on quantum mechanical calculations of model systems [25, 26].
2. The second row contains pairwise terms corresponding to the interactions between UNRES sites. It should be noted that the U_{SCiSCj} potentials contain contributions from interactions with water in an implicit manner [27, 28]. The $U_{pp}{}^{(el)}$ terms represent the energy of the interactions of peptide-group dipoles averaged over the rotation of the peptide groups about the $C^{\alpha}...C^{\alpha}$ virtual-bond axes, and the parameters of $U_{pp}{}^{(el)}$ and $U_{pp}{}^{(vdW)}$ were obtained by fitting the analytical expressions to the PMF of two interacting peptide groups [29]. The U_{SCp} terms are simple excluded-volume potentials to prevent the collapse of the peptide groups on the side chains [28, 29].
3. The terms U_{tor}, U_{tord}, and U_{torSC} are backbone virtual-bond torsional and double-torsional potentials, and the torsional potentials involving the virtual $C^{\alpha}\cdots SC$ bonds, respectively; these potentials were parameterized from quantum mechanical calculations [30, 31].

4. The $U_{corr}^{(m)}$ terms are the m-th order correlation (multibody) terms that account for the coupling between the backbone-local and backbone-electrostatic interactions; these terms are essential to reproduce the regular secondary structures of proteins. The analytical expressions were obtained [6, 7] from the Kubo's generalized cumulant expansion of the Kubo's cluster-cumulant functions [5] representing these terms. Their parameters were determined by fitting the analytical expressions to the PMFs of model system calculated with ab initio quantum mechanics [32].
5. The factors f are temperature-dependent multipliers that account for the fact that UNRES is an analytical approximation of the potential of mean force [33].
6. The U_{SS} terms account for the formation and breaking of the disulfide bonds between cysteine residues. The respective potential is bimodal, with one minimum at the SS-bond and another minimum corresponding to a pair of nonbonded cysteines. It was parameterized from quantum mechanical calculations and thermodynamic data of disulfide-bond formation [34].
7. The multipliers $w_{bond} - w_{SS}$ are the energy-term weights, which were determined by forcefield calibration [33]. At present, four versions of UNRES are in use: one calibrated with the 1GAB α-helical proteins (the force field is referred to as FFGAB) [33], which is a good solution to simulate proteins with helical structure; one general purpose force field, referred to as FF2 [31], with energy term weights determined by extensive search of parameter space to reproduce the structures of both α- and β-proteins; and two recent variants: one with parameters determined by maximum-likelihood optimization with the use of seven proteins of different structural classes, which is referred to as OPT-WTFSA-2 [35] and another one, NEWCT-9P [36], with energy terms based on the recently developed scale-consistent theory of forcefield derivation [7], and optimized with nine proteins.
8. UNRES handles both L- and D-amino-acid residues [30]. A number of nonstandard residues have also been included.
9. Recently, UNRES was extended to differentiate the strength of the electrostatic (U_{pp}) and correlation terms depending on shielding of the peptide groups from the solvent [37]. Along with the recent scale-consistent variant of the force field [7, 35], this modification resulted in the improvement of the quality of the secondary structure representation.
10. A continuous nanotube models has been developed to carry out simulations of peptides and proteins interacting with nanotubes with UNRES [38].

11. The lipid phase has also been recently added to UNRES to handle membrane proteins [39].
12. A version to handle phosphorylated amino-acid residue was also developed [40].

2.3 Types of Calculations

1. *Single energy minimization*. This calculation is suitable as a first step in order to relax the input conformation. Calculations of this type can be carried out in both unrestrained mode and with added restraints. The Secant Unconstrained Minimization Solver (SUMSL) quasi-Newton local minimizer [41] has been implemented for this purpose. Minimization can be carried out in the Cartesian coordinates or in the virtual-bond-angles and virtual-bond-dihedral angles (*see* Eq. 1); in the second case, all virtual-bond lengths are assigned standard values.
2. *Canonical molecular dynamics (constant temperature mode)*. The Berendsen, Nose-Hoover, Nose-Poincaré, and Langevin thermostats have been implemented [42, 43]; use of the Langevin thermostat is recommended. This option is suitable to study folding/unfolding pathways, folding/unfolding kinetics, and conformational changes.
3. *Global search of the energy minima with the conformational space annealing (CSA) method*. This option is suitable for estimating the native structure of a given protein. CSA [44] uses local minimization as an elementary step.
4. *Replica exchange (REMD) and multiplexed replica exchange (MREMD) molecular dynamics*. In these calculations, a number of MD trajectories are run in parallel at different temperatures (one per temperature for REMD [45] and several at one temperature for MREMD [46]). Both REMD and MREMD were implemented in UNRES [47], and this option is recommended when studying protein-folding thermodynamics, computing ensemble averages at multiple temperatures, and free-energy landscapes, as well as for the prediction of protein structures; this type of calculations is described in more detail in Subheading 3. The use of the weighted-histogram analysis method (WHAM) [48] is necessary to interpret the results.
5. *Calculations with restraints*. The abovementioned types of calculations can be run in unrestrained mode or with added restraints. The following restraints can be imposed:
   (a) Restraints on backbone virtual-bond-angles and virtual-bond-dihedral angles in a form of quartic functions. The centers of restraints can be taken from the results of secondary structure prediction with PSIPRED [49].
   (b) Restraints on the $C^{\alpha} \cdots C^{\alpha}$ or SC-SC distances. These are quartic restraints or Lorentz-like restraints [50, 51] which are bounded and, therefore, can handle contradictory

restraints. Recently [12], this feature has been extended to include the restraints from nuclear magnetic resonance (NMR) measurements.

(c) Distance- and angle restraints generated from multiple templates (server models). These restraints have a form of log-Gaussian or Lorentzian functions [52, 53].

(d) Distance distribution obtained from small X-ray scattering (SAXS) experiments. A maximum-likelihood penalty term has been introduced to enforce that the calculated distance distribution matches the experimental one [11].

(e) Restraints from chemical crosslink-mass spectroscopy (XL-MS) experiments [54].

(f) Steered molecular dynamics (with restraints varying during the course of the simulations) has also been implemented [50].

2.4 Analysis of UNRES Results

1. The output of MD and (M)REMD runs is, by default, the Cartesian coordinates of the coarse-grained sites in the compressed format. To convert a trajectory to PDB format, the xdrf2pdb program from the UNRES package should be used.
2. To construct conformational ensembles from the results of (M)REMD simulations, the WHAM [33, 47] calculation should be run. The WHAM has been implemented in the raw data mode so the probability of each conformation subjected to analysis can be computed at a given temperature. WHAM reads the MREMD output coordinates in the compressed format; therefore, no data format conversion is needed. WHAM recalculates the UNRES energies of the input conformations and checks if they are the same as those read from the MREMD file, issuing a warning if differences are found. The most common origin of the differences is from the use of different variants of the UNRES force field in MREMD simulations and the WHAM calculation.
3. After processing with WHAM, the ensemble at a given temperature can be clustered into families of conformations using the Hierarchical Clustering (HC) Fortran subroutine written by F. Murtagh [55]. Several hierarchical-clustering algorithms are available, with Ward's minimum-variance method being the default option. Clustering is temperature-dependent because the probabilities of given conformations depend on temperature. The population and the average root-mean-square deviation (rmsd) from the experimental structure, if the latter is available, are calculated for each family. The representative conformation of each family is determined to be the conformation from the family which is closest to the average conformation [10, 33].

4. All-atom conformations can be obtained from the UNRES reduced conformation by using available conversion software. Use of PULCHRA [56] and SCWRL [57] is recommended. To relax the converted structure, running restrained all-atom MD is recommended.

2.5 Programming Language

The components of the UNRES package have been originally written in FORTRAN 77; recently, a FORTRAN 90 version has been released [58]. To store coordinates from MD/MREMD, this new version implements the XDRF library written in ANSI C by F. van Hoesel (https://github.com/Pappulab/xdrf).

2.6 Parallelization

1. MPI libraries (mpi-forum.org) have been used to parallelize the code.
2. The UNRES program has two levels of parallelization: coarse-grained and fine-grained [59].
3. Coarse-grained parallelization: conformations or trajectories are distributed to the tasks/group of tasks.
   (a) CSA: a task/group of tasks handles energy minimization of a conformation.
   (b) MD, REMD, and MREMD: a task/group of tasks handles one trajectory. For MD, the tasks are run independently while, for (M)REMD, information is exchanged between trajectories every predefined number of steps.
   (c) Fine-grain parallelization: a group of tasks handles energy and force evaluation for a given conformation. Should be used for proteins larger than 50 amino-acid residues. For the Cray system, the scalability plot is presented in Fig. 2, and the recommended numbers of fine-grain tasks are summarized in Table 1.
   (d) WHAM and CLUSTER have also been parallelized (one parallelization level only).

2.7 Availability

1. The UNRES package, including the source code, is available for download at www.unres.pl.
2. The site contains the downloadable software, package description, installation guide, and input description.
3. Both serial and parallel versions of UNRES can be installed. MPI is required to install the parallel version. The (M)REMD and CSA calculations can only be run using parallel UNRES.
4. To run specific calculations, follow the detailed instructions in the input data manual. In this mode, input files need to be prepared by the user and all components of the package (UNRES, but, for MREMD, WHAM, and CLUSTER, as well as conversion to all-atom chains), are run by the user.

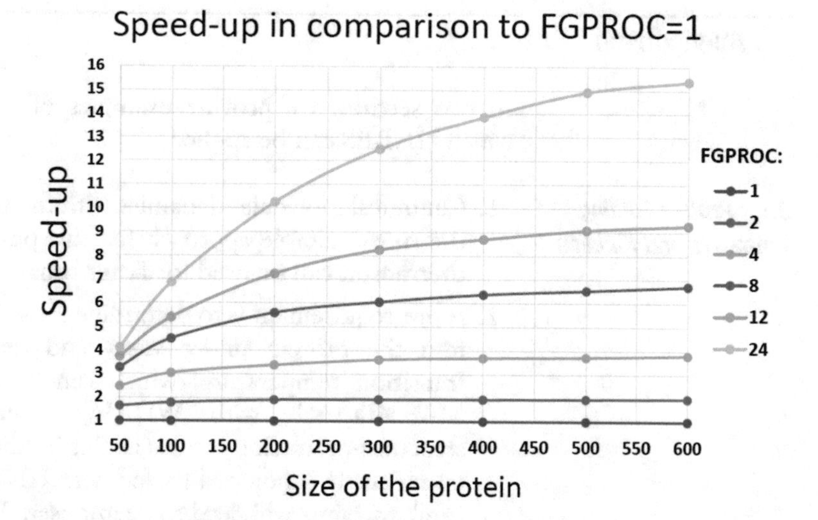

Fig. 2 Plot illustrating the speed-up of the parallel runs of canonical MD simulations with UNRES as a function of protein size (the number of amino-acid residues), for different numbers of cores per trajectory (fine-grain processors) on the Cray XC40 system

Table 1
Recommended values of cores to use for various numbers of amino-acid residues in the simulated systems (Cray XC40 system)

Residue #	Core #
<50	1
51–100	2
101–200	4
201–300	8
301–500	12
>500	24

5. An UNRES server has been created recently [60] with the UNRES-dock extension [61] and is available from www.unres.pl or directly at unres-server.chem.ug.edu.pl. Most types of possible UNRES calculations can be run on this server, including peptide–protein and protein–protein docking.
6. The user selects calculation type, inputs the amino-acid sequence, and selects the PDB file with the starting experimental structure, if applicable. Selecting MREMD calculations invokes automatically WHAM and CLUSTER after the MREMD is carried out.
7. The results are displayed graphically, and the resulting structures/trajectories can also be downloaded from the server.

3 Applications

In this section, we provide examples of particular problems to which UNRES can be applied.

3.1 Protein Folding Pathways and Kinetics

1. Canonical molecular dynamics with the Langevin thermostat is the most suitable approach for this purpose. The Berendsen thermostat can be used for faster runs.
2. A pre-requirement is to determine if the UNRES forcefield can fold the protein under study and determine the folding-transition temperature, which can be done by running an MREMD job with WHAM/cluster postprocessing in structure-prediction mode (*see* Subheading 3.2). If the protein under study is not well folded with UNRES, additional structural restraints will need to be imposed. However, in such case, the user needs to be cautioned that the emerging folding pathways can be distorted.
3. The low computational cost of UNRES calculations compared to those of all-atom simulations permits to run tens or hundreds of trajectories simultaneously, which will allow to draw reliable statistics. This can be optimally done by running a parallel job.
4. The obtained trajectories can be used to construct plots of the fraction of folded/intermediate conformations vs. time, from which the kinetics can be derived, or to construct free-energy landscapes. Some examples are: a study of the kinetics of folding of the mutants of FBP 28 WW domain [16], of the dynamics of disulfide bonds in the unfolding of ribonuclease A [18], and an evaluation of the effect of hydrodynamic interactions on the kinetics of folding of the N-terminal section of the B-domain of staphylococcal protein A and the FBP-28 WW domain [17]. Another application was the simulation of the folding pathway of the N-terminal portion of the B-domain of staphylococcal protein A in order to describe protein folding in terms of the formation of dark solitons [62].
5. Free-energy landscapes can also be produced from the trajectories. UNRES outputs summary files containing energies, rmsd from the experimental structure, and radius of gyration, from which the rmsd-radius of gyration histograms (*see* Ref. 16 as an example) can be constructed directly; other properties need to be calculated directly from the coordinate files. An example of the latter are the free-energy landscapes of protein A and FBP-28 WW domain constructed from UNRES in the principal components of the essential motions of these proteins extracted from UNRES trajectories [15].

3.2 Determination of Most Likely Conformations (Protein–Structure Prediction)

1. Prepare the data for MREMD simulations. When using the UNRES server, select "MREMD" as simulation type.
   (a) To speed-up the simulations, secondary structure information (e.g., from PSIPRED [49]) can be incorporated. Secondary structure information is especially useful for proteins containing a large degree of β-structure.
   (b) Knowledge-based models for a target protein, e.g., those from servers, distance restraints, e.g., from correlated mutations, and a SAXS distance distribution can be incorporated to restraint the simulations, if available. This is especially important for large proteins.
   (c) For oligomers, it is strongly recommended that the calculations are started from a predefined initial structure.
2. Run MREMD simulations.
   (a) Simulations have to be long enough to obtain convergence, which can be measured by changes of heat capacity plots every few million steps. Usually, 50–100 million steps are enough to obtain converged simulations for a medium-size system.
   (b) For larger proteins, run calculations in the fine-grain mode.
3. Run WHAM from the last segment of the simulations (typically the last 500–1000 snapshots per trajectory). Determine the position of the major heat-capacity peak.
   (a) Do cluster analysis at a temperature ~ 10 °C below the major heat capacity peak (usually 260–310 K) to produce five clusters.
4. Reconstruct all-atom structures from the UNRES coarse-grained representative models (models closest to the average cluster structures) using the PULCHRA [56] and SCWRL [57] programs.
5. Check the obtained structure for the presence of clashes visually and with the MolProbity score (http://molprobity.biochem.duke.edu/) and if necessary, run a short all-atom trajectory for refinement.
6. An example of the prediction with UNRES in the ab initio mode, with the CASP12 target T0663 [8], is shown in Fig. 3. The "Model 1" prediction of this target protein, which was performed by the Cornell-Gdansk group using UNRES, has the correct domain-packing topology, even though its overall resolution is coarse. Conversely, the best model as far as the local details are concerned (Model 4 from the LEEcon group) has an incorrect domain packing, with the first strand of the C-terminal domain oriented parallel to the last strand of the N-terminal domain.

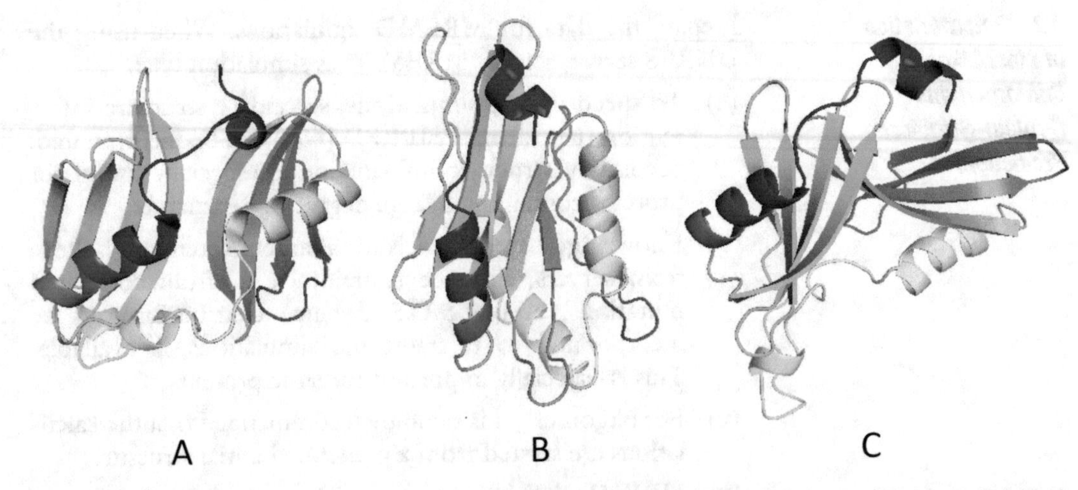

Fig. 3 (**a**) The experimental 4EXR structure of the CASP10 target T0663. (**b**) Model 1 of the Cornell-Gdansk group. (**c**) Model 4 of the LEEcon group. The chains are colored from blue to red from the N- to the C-terminus. The values of Global Distance Test Total Score (GDT_TS) are 23.19, 31.98, and 42.80 of the whole protein and its domains D1 and D2, respectively, for model 1 of the Cornell-Gdansk group (using UNRES) and 42.93, 60.76, and 89.02 of the whole protein and its domains D1 and D2, respectively for model 4 from the LEEcon group (the model that scored the highest overall GDT_TS). It should be noted that, even though the individual domains of model 4 from the LEEcon group closely match their experimental counterparts, the packing of the domain packing is incorrect. In contrast, the domain packing of Cornell-Gdansk model 1 is correct. (The picture has been drawn based on the data from Ref. 8)

3.3 Examples of Applications to Biological Problems

In this section, three examples of using UNRES to study elements of important biological processes are provided. In the first and second one, UNRES alone was used, while the third one is an example of how to use UNRES in combination with available bioinformatics tools.

1. *Amyloid formation from the $A\beta_{1\rightarrow 40}$ peptide.* The Aβ peptide is the main precursor of the formation of amyloid deposits in the nerve tissue, leading to the onset of human diseases such as Alzheimer and Parkinson diseases, type II diabetes, and spongiform encephalopathies [63]. Using the canonical and replica-exchange molecular dynamics with UNRES, we investigated the mechanism of the addition of another chain to the preformed stacks of $A\beta_{1\rightarrow 40.}$ The monomeric peptide forms a partially α-helical structure. It was found that the unstructured C-terminal part of the monomer attaches to the stack first to form a β-sheet, while the N-terminal helical section unfolds later to join the stack [19].
2. *Simulations of the opening of the yeast Hsp70 chaperone.* The Hsp70 molecular chaperones consist of a nucleotide-binding (NBD) (ATPase) and a substrate-binding (SBD) domain, the latter being subdivided into the α-(SBD-α) and (SBD-β) β-subdomains [64]. The ATPase domain is split into the NBD-I and NBD-II subdomains with a nucleotide (ATP or

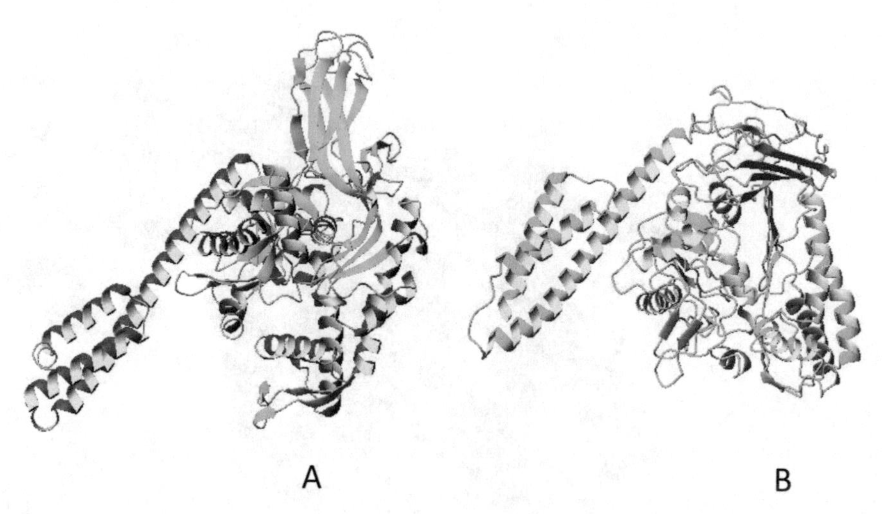

Fig. 4 Comparison of the experimental 4B9Q (**a**) [19] and predicted structure (**b**) [65] of the open form of yeast Hsp70 chaperone

ADP) bound to their interface. In the substrate-bound (closed) conformation (in which ADP is bound to the NDB), SBD-α is bound to SBD-β, holding tightly the substrate in between, while in the open conformation (in which ATP is bound to the NBD), SBD-α is bound to NBD-I and separated from SBD-β, thus releasing the substrate. Using canonical MD simulations with UNRES, we studied the conversion of the closed to the open conformation [21]. We started from the 2KHO structure of yeast Hsp70 chaperone in the closed conformation. Distance restraints were imposed on the NBD, SBD-α, and SBD-β; no restraints were imposed on the distances between these units. No experimental structure of Hsp70 in the open conformation was available at that time. We found that the system spontaneously converted into open conformations, part of which contained SBD-α bound to NBD-I, as in the experimental structure which was determined by the Mayer group [65] after our work had been published (Fig. 4).

3. *Interactions of the Isu-1 sulfur-binding protein with the Hsp40 co-chaperone.* Iron-sulfur proteins are key components of the respiratory chains in Eukaryotes and the transfer of iron-sulfur clusters to them, in which molecular chaperones are involved, is a key element of their biogenesis [66]. We studied [22] the interactions of the Isu1 iron-sulfur-binding protein to the Hsp40 co-chaperone (Jac1). Because no Isu1 structure was available, we obtained a structural model by homology modeling using the I-TASSER [67] and YASARA software [68]. We used the 3UO3 PDB structure of Jac1. Subsequently, we used ZDOCK [69] and YASARA [68] to obtain the initial structures of the Isu1-Hsp40 complex. Three plausible initial models

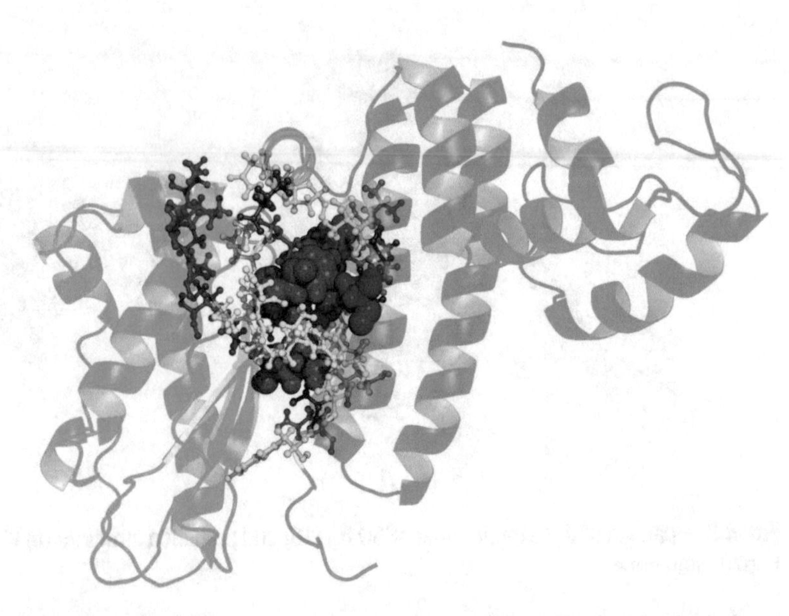

Fig. 5 A representative structure of the Isu1-Jac1 complex obtained by molecular docking followed by UNRES/MD simulations [22]. The bulk of the structure is shown in cartoon representation (blue: Jac1, red: Isu1), while the residues of the binding interface are shown in atomic-detailed representation. The Jac1 residues found to be important for binding both in this work and in earlier experimental studies (L_{105}, L_{109}, and Y_{163})[3] are colored in dark-blue and shown in space-filling representation; the Jac1 residues found important for binding in this study and suggested by experiment to be important in binding (L_{104}, K_{107}, D_{110}, D_{113}, E_{114}, and Q_{117}) are colored in blue and shown in ball-and-stick representation, residues that are predicted to be important for binding according to simulations (but not verified by experiment) are colored cyan and shown in ball-and stick representation (residues N_{95}, T_{98}, P_{102}, H_{112}, V_{159}, L_{167}, A_{170}, W_{174}), while the residues of Jac1 found by simulations to make less tight contacts with Isu1 but still be possibly important for binding (residues E_{91}, V_{108}, S_{116}, and E_{160}) are colored dark-gray and shown in ball-and-stick representation. The same hierarchy of representation and colors red (L_{63}, V_{72}, and F_{94}), yellow (V_{64}, A_{66}, D_{71}, M_{73}, R_{74}, K_{92}, T_{93}, C_{96}), and magenta (G_{65}, G_{70}, G_{95}, V_{135}, K_{136}, H_{138}, C_{139}, L_{142}) are used for the residues of Isu1 found to be important for binding by both simulation and experiment, by simulation and, to a lesser extent, by experiment, and by simulation only, respectively. (Reproduced with permission from Mozolewska et al., Proteins: Struct. Funct. Bioinf. 83:1414–1426 (2015). Copyright 2015 John Wiley & Sons)

were then selected, each of which was subjected to canonical molecular dynamics with UNRES, in which 16 independent trajectories were run (48 trajectories total). Although the three models were distinct (with the Isu1 protein attached to a different side of Jac1), they all converged to one model, which is shown in Fig. 5. The interface of the two proteins shown in the figure contains all residues found to be important in binding by mutational studies; moreover, more important residues, not yet tested experimentally, were detected [22].

Acknowledgments

This work was supported by grants UMO-2017/25/B/ST4/01026 (to AL), UMO-2017/27/B/ST4/00926 (to AKS), UMO-2015/17/N/ST4/03935 (to MAM), UMO-2015/17/N/ST4/03937 (to PK), UMO-2013/10/E/ST4/00755 (to MM), and UMO-2017/26/M/ST4/00044 (to CC) from the National Science Center of Poland (Narodowe Centrum Nauki). Calculations were carried out using the computational resources provided by (a) the supercomputer resources at the Informatics Center of the Metropolitan Academic Network (CI TASK) in Gdansk, (b) the supercomputer resources at the Interdisciplinary Center of Mathematical and Computer Modeling (ICM), University of Warsaw within grant GA76-11, (c) the Polish Grid Infrastructure (PL-GRID) at the the Academic Computer Centre Cyfronet AGH in Krakow under grants asunres18 and unres19, and (d) our 692-processor Beowulf cluster at the Faculty of Chemistry, University of Gdansk.

References

1. Kmiecik S, Gront D, Kolinski M, Wieteska L, Dawid AE, Kolinski A (2016) Coarse-grained protein models and their applications. Chem Rev 116:7898–7936
2. Monticelli L, Kandasamy SK, Periole X, Larson RG, Tieleman DP, Marrink S-J (2008) The MARTINI coarse-grained force field: extension to proteins. J Chem Theory Comput 4:819–834
3. Liwo A, Czaplewski C, Oldziej S, Rojas AV, Kazmierkiewicz R, Makowski M, Murarka RK, Scheraga HA (2008) Simulation of protein structure and dynamics with the coarse-grained UNRES force field. In: Voth G (ed) Coarse-graining of condensed phase and biomolecular systems. Taylor & Francis, Oxfordshire, pp 107–122
4. Liwo A, Baranowski M, Czaplewski C, Golas E, He Y, Jagiela D, Krupa P, Maciejczyk M, Makowski M, Mozolewska MA, Niadzvedtski A, Oldziej S, Scheraga HA, Sieradzan AK, Slusarz R, Wirecki T, Yin Y, Zaborowski B (2014) A unified coarse-grained model of biological macromolecules based on mean-field multipole.Multipole interactions. J Mol Model 20:2306
5. Kubo R (1962) Generalized cumulant expansion method. J Phys Soc Jpn 17:1100–1120
6. Liwo A, Czaplewski C, Pillardy J, Scheraga HA (2001) Cumulant-based expressions for the multibody terms for the correlation between local and electrostatic interactions in the united-residue force field. J Chem Phys 115:2323–2347
7. Sieradzan AK, Makowski M, Augustynowicz A, Liwo A (2017) A general method for the derivation of the functional forms of the effective energy terms in coarse-grained energy functions of polymers. I. Backbone potentials of coarse-grained polypeptide chains. J Chem Phys 146:124106
8. He Y, Mozolewska M, Krupa P, Sieradzan AK, Wirecki TK, Liwo A, Kachlishvili K, Rackovsky S, Jagiela D, Slusarz R, Czaplewski CR, Oldziej S, Scheraga HA (2013) Lessons from application of the UNRES force field to predictions of structures of CASP10 targets. Proc Natl Acad Sci U S A 110:14936–14941
9. Khoury GA, Liwo A, Khatib F, Zhou H, Chopra G, Bacardit J, Bortot LO, Faccioli RA, Deng X, He Y, Krupa P, Li J, Mozolewska MA, Sieradzan AK, Smadbeck J, Wirecki T, Cooper S, Flatten J, Xu F, Baker D, Cheng J, Delbem ACB, Floudas CA, Keasar C, Levitt M, Popovic Z, Scheraga HA, Skolnick J, Crivelli SN, Players F (2014) WeFold: a coopetition for

protein structure prediction. Proteins 82:1850–1868

10. Krupa P, Mozolewska MA, Wiśniewska M, Yin Y, He Y, Sieradzan AK, Ganzynkowicz R, Lipska AG, Karczynska A, Slusarz M, Slusarz R, Gieldon A, Czaplewski C, Jagiela D, Zaborowski B, Scheraga HA, Liwo A (2016) Performance of protein-structure predictions with the physics-based UNRES force field in CASP11. Bioinformatics 32:3270–3278
11. Karczynska AS, Mozolewska MA, Krupa P, Gieldon A, Liwo A, Czaplewski C (2018) Prediction of protein structure with the coarse-grained UNRES force field assisted by small X-ray scattering data and knowledge-based information. Proteins 86(S1):228–239
12. Lubecka EA, Karczynska AS, Lipska AG, Sieradzan AK, Zieba K, Sikorska C, Uciechowska U, Samsonov SA, Krupa P, Mozolewska MA, Golon L, Gieldon A, Czaplewski C, Slusarz R, Slusarz M, Crivelli SN, Liwo A (2019) Evaluation of the scale-consistent UNRES force field in template-free prediction of protein structures in the CASP13 experiment. J Mol Graph Model 92:154–166
13. Karczynska A, Zieba K, Uciechowska U, Mozolewska MA, Krupa P, Lubecka EA, Lipska AG, Sikorska C, Samsonov SA, Sieradzan AK, Gieldon A, Liwo A, Slusarz R, Slusarz M, Lee J, Joo K, Czaplewski C (2020) Improved consensus-fragment selection in template-assisted prediction of protein structures with the UNRES force field in CASP13. J Chem Inf Model 60:1844–1864
14. Liwo A, Khalili M, Scheraga HA (2005) Ab initio simulations of protein-folding pathways by molecular dynamics with the united-residue model of polypeptide chains. Proc Natl Acad Sci U S A 102:2362–2367
15. Maisuradze GG, Senet P, Czaplewski C, Liwo A, Scheraga HA (2010) Investigation of protein folding by coarse-grained molecular dynamics with the UNRES force field. J Phys Chem A 114:4471–4485
16. Zhou R, Maisuradze GG, Sunol D, Todorovski T, Macias MJ, Xiao Y, Scheraga HA, Czaplewski C, Liwo A (2014) Folding kinetics of WW domains with the united residue force field for bridging microscopic motions and experimental measurements. Proc Natl Acad Sci U S A 2014 (111):18243–18248
17. Lipska AG, Seidman SR, Sieradzan AK, Gieldon A, Liwo A, Scheraga HA (2016) Molecular dynamics of protein a and a WW domain with a united-residue model including hydrodynamic interaction. J Chem Phys 144:184110
18. Krupa P, Sieradzan AK, Mozolewska MA, Li H, Liwo A, Scheraga HA (2017) Dynamics of disulfide-bond disruption and formation in the thermal unfolding of ribonuclease A. J Chem Theory Comput 13:5721–5730
19. Rojas A, Liwo A, Browne D, Scheraga HA (2010) Mechanism of fiber assembly; treatment of A β-peptide aggregation with a coarse-grained united-residue force field. J Mol Biol 404:537–552
20. He Y, Liwo A, Weinstein H, Scheraga HA (2011) PDZ binding to the BAR domain of PICK1 is elucidated by coarse-grained molecular dynamics. J Mol Biol 405:298–314
21. Golas E, Maisuradze GG, Senet P, Oldziej S, Czaplewski C, Scheraga HA, Liwo A (2012) Simulation of the opening and closing of Hsp70 chaperones by coarse-grained molecular dynamics. J Chem Theory Comput 8:1750–1764
22. Mozolewska M, Krupa P, Scheraga HA, Liwo A (2015) Molecular modeling of the binding modes of the iron-sulfur protein to the Jac1 co-chaperone from Saccharomyces cerevisiae by all-atom and coarse-grained approaches. Proteins 83:1414–1426
23. Rojas AV, Liwo A, Scheraga HA (2007) Molecular dynamics with the united-residue force field: ab initio folding simulations of multichain proteins. J Phys Chem B 111:293–309
24. Sieradzan AK (2015) Introduction of periodic boundary conditions into UNRES force field. J Comput Chem 36:940–946
25. Kozlowska U, Maisuradze GG, Liwo A, Scheraga HA (2010) Determination of side-chain-rotamer and side-chain and backbone virtual-bond-stretching potentials of mean force from AM1 energy surfaces of terminally-blocked amino-acid residues, for coarse-grained simulations of protein structure and folding. 2. Results, comparison with statistical potentials, and implementation in the UNRES force field. J Comput Chem 31:1154–1167
26. Sieradzan AK, Niadzvedtski A, Scheraga HA, Liwo A (2014) Revised backbone-virtual-bond-angle potentials to treat the L- and D-amino-acid residues in the coarse-grained united residue (UNRES) force field. J Chem Theory Comput 10:2194–2203
27. Liwo A, Oldziej S, Pincus MR, Wawak RJ, Rackovsky S, Scheraga HA (1997) A united-residue force field for off-lattice protein-structure simulations. I: Functional forms and parameters of long-range side-chain interaction potentials from protein crystal data. I: Functional forms and parameters of long-range side-chain interaction potentials from protein crystal data. J Comput Chem 18:849–873

28. Makowski M, Liwo A, Scheraga HA (2017) Simple physics-based analytical formulas for the potentials of mean force of the interaction of amino acid side chains in water. VII. Charged-hydrophobic/polar and polar-hydrophobic/polar side chains. J Phys Chem B 121:379–390
29. Liwo A, Pincus MR, Wawak RJ, Rackovsky S, Scheraga HA (1993) Prediction of protein conformation on the basis of a search for compact structures; test on avian pancreatic polypeptide. Prot Sci 2:1715–1731
30. Sieradzan AK, Hansmann UHE, Scheraga HA, Liwo A (2012) Extension of UNRES force field to treat polypeptide chains with D-amino-acid residues. J Chem Theory Comput 8:4746–4757
31. Sieradzan AK, Krupa P, Scheraga HA, Liwo A, Czaplewski C (2015) Physics-based potentials for the coupling between backbone- and side-chain-local conformational states in the united residue (UNRES) force field for protein simulations. J Chem Theory Comput 11:817–831
32. Liwo A, Oldziej S, Czaplewski C, Kozlowska U, Scheraga HA (2004) Parameterization of backbone-electrostatic and multibody contributions to the UNRES force field for protein-structure prediction from ab initio energy surfaces of model systems. J Phys Chem B 108:9421–9438
33. Liwo A, Khalili M, Czaplewski C, Kalinowski S, Oldziej S, Wachucik K, Scheraga HA (2007) Modification and optimization of the united-residue (UNRES) potential energy function for canonical simulations. I. Temperature dependence of the effective energy function and tests of the optimization method with single training proteins. J Phys Chem B 111:260–285
34. Chinchio M, Czaplewski C, Liwo A, Oldziej S, Scheraga HA (2007) Dynamic formation and breaking of disulfide bonds in molecular dynamics simulations with the UNRES force field. J Chem Theory Comput 3:1236–1248
35. Krupa P, Halabis A, Zmudzinska W, Oldziej S, Scheraga HA, Liwo A (2017) Maximum likelihood calibration of the UNRES force field for simulation of protein structure and dynamics. J Chem Inf Model 57:2364–2377
36. Liwo A, Sieradzan AK, Lipska AG, Czaplewski C, Joung I, Zmudzinska W, Halabis A, Oldziej S (2019) A general method for the derivation of the functional forms of the effective energy terms in coarse-grained energy functions of polymers. III. Determination of scale-consistent backbone-local and correlation potentials in the UNRES force field and force-field calibration and validation. J Chem Phys 150:155104
37. Sieradzan AK, Lipska AG, Lubecka EA (2017) Shielding effect in protein folding. J Mol Graph Model 79:118–132
38. Sieradzan AK, Mozolewska M (2018) Extension of coarse-grained UNRES force field to treat carbon nanotubes. J Mol Model 24:121
39. Zieba K, Slusarz M, Slusarz R, Liwo A, Czaplewski C, Sieradzan AK (2019) Extension of the UNRES coarse-grained force field to membrane proteins in the lipid bilayer. J Phys Chem B 123:7829–7839
40. Sieradzan AK, Bogunia M, Mech P, Ganzynkowicz R, Gieldon A, Liwo A, Makowski M (2019) Introduction of phosphorylated residues into the UNRES coarse-grained model: toward modeling of signaling processes. J Phys Chem B 23:5721–5729
41. Gay DM (1983) Algorithm 611. Subroutines for unconstrained minimization using a model/trust-region approach. ACM Trans Math Software 9:503–524
42. Khalili M, Liwo A, Jagielska A, Scheraga HA (2005) Molecular dynamics with the united-residue model of polypeptide chains. II. Langevin and Berendsen-bath dynamics and tests on model α-helical systems. J Phys Chem B 109:13798–13810
43. Kleinerman DS, Czaplewski C, Liwo A, Scheraga HA (2008) Implementations of Nose-Hoover and Nose-Poincare termostats in mesoscopic dynamic simulations with the united-residue model of a polypeptide chain. J Chem Phys 128:245103
44. Lee J, Scheraga HA, Rackovsky S (1998) Conformational analysis of the 20-residue membrane-bound portion of melittin by conformational space annealing. Biopolymers 46:103–116
45. Hansmann UHE, Okamoto Y (1994) Comparative study of multicanonical algorithm and multicanonical replica exchange method for simulating systems with rough energy landscape. Physica A 212:415–437
46. Rhee YM, Pande VS (2003) Multiplexed replica exchange molecular dynamics method for protein folding simulations. Biophys J 84:775–786
47. Czaplewski C, Kalinowski S, Liwo A, Scheraga HA (2009) Application of multiplexed replica exchange molecular dynamics to the UNRES force field: tests with α and α+β proteins. J Chem Theor Comput 5:627–640
48. Kumar S, Bouzida D, Swendsen RH, Kollman PA, Rosenberg JM (1992) The weighted histogram analysis method for free energy calculations on biomolecules. I. The method. J Comput Chem 13:1011–1021

49. McGuffin L, Bryson K, Jones D (2000) The PSIPRED protein structure prediction server. Bioinformatics 16:404–405
50. Sieradzan AK, Jakubowski R (2017) Introduction of steered molecular dynamics into UNRES coarse-grained simulations package. J Comput Chem 38:553–562
51. Lubecka EA, Liwo A (2019) Introduction of a bounded penalty function in contact-assisted simulations of protein structures to omit false restraints. J Comput Chem 40:2164–2178
52. Mozolewska MA, Krupa P, Zaborowski B, Liwo A, Lee J, Joo K, Czaplewski C (2016) Use of restraints from consensus fragments of multiple server models to enhance protein-structure prediction capability of the UNRES force field. J Chem Inf Model 56:2263–2279
53. Karczynska AS, Czaplewski C, Krupa P, Mozolewska MA, Joo K, Lee J, Liwo A (2017) Ergodicity and model quality in template-restrained canonical and temperature/Hamiltonian replica exchange coarse-grained molecular dynamics simulations of proteins. J Comput Chem 38:2730–2746
54. Fajardo JE, Shrestha R, Gil N, Belsom A, Crivelli SN, Czaplewski C, Fidelis C, Grudinin S, Karasikov M, Karczynska AS, Kryshtafovych A, Leitner A, Liwo A, Lubecka EA, Monastyrskyy B, Pages G, Rappsilber J, Sieradzan AK, Sikorska C, Trabjerg E, Fiser A (2019) Assessment of chemical-crosslink-assisted protein structure modeling in CASP13. Proteins 87:1283–1297
55. Murtagh F (1985) Multidimensional clustering algorithms. Springer Verlag, Vienna
56. Rotkiewicz P, Skolnick J (2008) Fast procedure for reconstruction of full-atom protein models from reduced representations. J Comput Chem 29:1460–1465
57. Wang Q, Canutescu AA, Dunbrack RL (2008) SCWRL and MolIDE: computer programs for side-chain conformation prediction and homology modeling. Nat Protoc 3:1832–1847
58. Lubecka EA, Liwo A (2016) New UNRES force field package with FORTRAN 90. TASK Quart 20:399–407
59. Liwo A, Oldziej S, Czaplewski C, Kleinerman DS, Blood P, Scheraga HA (2010) Implementation of molecular dynamics and its extensions with the coarse-grained UNRES force field on massively parallel systems; towards millisecond-scale simulations of protein structure, dynamics, and thermodynamics. J ChemTheory Comput 6:890–909
60. Czaplewski C, Karczynska A, Sieradzan AK, Liwo A (2018) UNRES server for physics-based coarse-grained simulations and predictions of protein structure, dynamics, and thermodynaics. Nucleic Acids Res 46:W304–W309
61. Krupa P, Karczynska AS, Mozolewska MA, Liwo A, Czaplewski C (2020) UNRES-dock protein-protein and peptide-protein docking by coarse-grained replica-exchange MD simulations. Bioinformatics 2020:1–3. https://doi.org/10.1093/bioinformatics/btaa897
62. Krokhotin A, Liwo A, Maisuradze GG, Niemi AJ, Scheraga HA (2014) Kinks, loops, and protein folding, with protein a as an example. J Chem Phys 140:025101
63. Masters CL, Simms G, Weinman NA, Multhaup G, McDonald B, Beyreuther K (1985) Amyloid plaque core protein in Alzheimer disease and down syndrome. Proc Natl Acad Sci U S A 82:4245–4249
64. Kampinga HH, Craig EA (2010) The HSP70 chaperone machinery: J proteins as drivers of functional specificity. Nat Rev Mol Cell Biol 11:579–592
65. Kityk R, Kopp J, Sinning I, Mayer MP (2012) Structure and dynamics of the ATP-bound open conformation of Hsp70 chaperones. Mol Cell 48:863–874
66. Lill R, Muhlenhoff U (2008) Maturation of iron-sulfur proteins in eukaryotes: mechanisms, connected processes, and diseases. Annu Rev Biochem 77:669–700
67. Zhang Y (2008) ITASSER server for protein 3D structure prediction. BMC Bioinformatics 9:40
68. Venselaar H, Joosten R, Vroling B, Baakaman C, Hekkelman M, Krieger E, Vriend G (2010) Homology modelling and spectroscopy, a never-ending love story. Eur Biophys J 39:551–563
69. Pierce B, Hourai Y, Weng Z (2011) Accelerating protein docking in ZDOCK using an advanced 3D convolution library. PLoS One 6:e24657

Index

Victor Muñoz (ed.), *Protein Folding: Methods and Protocols*, Methods in Molecular Biology, vol. 2376, https://doi.org/10.1007/978-1-0716-1716-8,

Printed in the United States
by Baker & Taylor Publisher Services